International Symposium on

Nonlinear Differential Equations and Nonlinear Mechanics

International Symposium on

Nonlinear Differential Equations and Nonlinear Mechanics

Edited by

Joseph P. LaSalle and Solomon Lefschetz

Research Institute of Advanced Studies
Baltimore, Maryland

Research Institute of Advanced Studies
Baltimore, Maryland
and
National University of Mexico

ACADEMIC PRESS New York and London 1963

ACADEMIC PRESS INC.
111 FIFTH AVENUE
NEW YORK, NEW YORK 10003

United Kingdom Edition
Published by
ACADEMIC PRESS INC. (LONDON) LTD.
BERKELEY SQUARE HOUSE, LONDON, W. 1

Library of Congress Catalog Card Number 63-11696

First Printing, 1963
Second Printing, 1965

PRINTED IN THE UNITED STATES OF AMERICA

Contributors

H. A. ANTOSIEWICZ, Department of Mathematics, University of Southern California, Los Angeles, California. *Page 259.*

JOSEPH AUSLANDER, RIAS, Baltimore, Maryland. *Page 454.*

ROBERT W. BASS, Aeronca Manufacturing Corporation, Baltimore, Maryland. *Page 91.*

RICHARD BELLMAN, The RAND Corporation, Santa Monica, California. *Page 121.*

V. E. BENEŠ, Bell Telephone Laboratories, Incorporated, Murray Hill, New Jersey. *Page 384.*

A. BLAQUIÈRE, Institut National des Sciences et Techniques Nucléaires Saclay, Paris, France. *Page 403 and 413.*

PAUL J. BLATZ, Guggenheim Aeronautical Laboratory, California Institute of Technology, Pasadena, California. *Page 193.*

FRED BRAUER, Mathematics Department, University of Wisconsin, Madison, Wisconsin. *Page 435.*

LAMBERTO CESARI, Mathematics Department, University of Michigan, Ann Arbor, Michigan. *Page 33.*

J. D. COLE, Firestone Flight Sciences Laboratory, California Institute of Technology, Pasadena, California. *Page 113.*

COURTNEY COLEMAN, Mathematics Department, Harvey Mudd College, Claremont, California. *Page 445.*

C. C. CONLEY, Institute of Mathematical Sciences, New York University, New York, New York. *Page 86.*

KENNETH L. COOKE, Mathematics Department, Pomona College, Claremont, California. *Page 155.*

F. G. DRESSEL, Department of Mathematics, Duke University, Durham, North Carolina. *Page 218.*

RODNEY D. DRIVER, Mathematics Research Center, United States Army, University of Wisconsin, Madison Wisconsin. *Page 474.*

B. A. FLEISHMAN, Mathematics Department, Rensselaer Polytechnic Institute, Troy, New York. *Page 211.*

L. A. GARDNER, JR., Lincoln Laboratory, Massachusetts Institute of Technology, Lexington, Massachusetts. *Page 241.*

J. J. GERGEN, Department of Mathematics, Duke University, Durham, North Carolina. *Page 218.*

P. GRIVET, Department of Mathematics, University of Paris, Paris, France, *Page 403.*

HUBERT HALKIN, RIAS, Baltimore, Maryland. *Page 284.*

W. A. HARRIS, JR., Mathematics Department, University of Minnesota, Minneapolis, Minnesota. *Page 489.*

C. A. HARVEY, MPG Research Department, Minneapolis-Honeywell Regulator Company, Minneapolis, Minnesota. *Page 393.*

ROBERT HERMANN, Lincoln Laboratory, Massachusetts Institute of Technology, Lexington, Massachusetts, and University of California, Berkeley, California. *Page 325.*

G. STEPHEN JONES, RIAS, Baltimore, Maryland. *Page 105.*

ROBERT KALABA, The RAND Corporation, Santa Monica, California. *Page 135.*

J. KEVORKIAN, Firestone Flight Sciences Laboratory, California Institute of Technology, Pasadena, California. *Page 113.*

RICHARD E. KOPP, Grumman Aircraft Engineering Corporation, Bethpage, New York. *Page 395.*

M. Z. V. KRZYWOBLOCKI, Michigan State University, East Lansing, Michigan. *Page 231.*

DANIEL C. LEWIS, Department of Mathematics, Johns Hopkins University, Baltimore, Maryland. *Pages 91 and 99.*

LAWRENCE MARKUS, Mathematics Department, University of Minnesota, Minneapolis, Minnesota. *Page 22.*

Y. A. MITROPOL'SKIY, Institut Matematiki, Kiev, USSR. *Page 1.*

RICHARD M. MONRONEY, Massachusetts Institute of Technology, Cambridge, Massachusetts. *Page 224.*

JÜRGEN MOSER, Institute of Mathematical Sciences, New York University, New York, New York. *Pages 71 and 147.*

LUCIEN NEUSTADT, The Aerospace Corporation, El Segundo, California. *Page 267.*

B. PAIEWONSKY, Aeronautical Research Associates of Princeton, Inc., Princeton, New Jersey. *Page 333.*

M. M. PEIXOTO, Instituto de Matemática Pura e Aplicada, Rio de Janeiro, Brazil. *Page 23.*

E. R. RANG, Minneapolis-Honeywell Regulator Company, Minneapolis, Minnesota. *Page 377.*

GEORGES REEB, Institut Fourier, University of Grenoble, Grenoble, France. *Page 16.*

BRUCE L. REINHART, RIAS and University of Maryland, Baltimore, Maryland. *Page 442.*

R. REISSIG, Institut für Angewandte Mathematik und Mechanik, DAW und Humboldt-Universität zu Berlin, Germany. *Page 24.*

J. B. ROSEN, Shell Development Company, Emeryville, California. *Page 366.*

EMILIO ROXIN, RIAS, Baltimore, Maryland. *Page 303.*

PETER SEIBERT, RIAS, Baltimore, Maryland. *Page 454 and 463.*

P. R. Sethna, Department of Aeronautics and Engineering Mechanics, University of Minnesota, Minneapolis, Minnesota. *Page 58.*

Yasutaka Sibuya, Ochanomizu University, Tokyo, Japan. *Page 485.*

George P. Szegö, Control and Information Systems Laboratory, Purdue University, Lafayette, Indiana. *Page 421.*

Minoru Urabe, Mathematics Department, Hiroshima University, Hiroshima, Japan. *Page 184.*

R. A. Willoughby, International Business Machines Corporation, Thomas J. Watson Research Center, Yorktown Heights, New York. *Page 431.*

G. Milton Wing, Sandia Corporation, Albuquerque, New Mexico. *Page 172.*

Taro Yoshizawa, Mathematics Department, Nihon University, Tokyo, Japan. *Page 80.*

Preface

This volume is the proceedings of a Symposium which brought together a group of pure and applied mathematicians interested in nonlinear differential equations, nonlinear mechanics, control theory, and other related topics. The contents shows that the range covered by the Symposium was very large indeed.

Invited addresses, contained in the volume, were given by L. Markus, G. Reeb, R. Reissig, and Yu A. Mitropolskii. A major part of the meeting centered around the following very active round table discussions:

"Oscillations and Asymptotic Behavior."
Chairman: Jack K. Hale, RIAS.

"Applications of Differential Equations."
Chairman: Richard Bellman, the RAND Corporation.

"Control and Stability."
Chairman: R. E. Kalman, RIAS.

"Qualitative Theory."
Chairman: Shlomo Sternberg, Harvard University.

In each case the chairman was the organizer of his round table discussion. Each day concluded with a relatively short session of contributed papers.

The Symposium, officially known as the 1961 International Symposium on Nonlinear Differential Equations and Nonlinear Mechanics, was held 31 July through 4 August at the United States Air Force Academy, Colorado Springs, Colorado, and was sponsored by the Air Force Office of Scientific Research and RIAS (Research Institute for Advanced Study.)

The undersigned wish to express their great appreciation for the arduous work done by the organizers of the round table discussions and likewise the Air Force Office of Scientific Research for its constant backing towards the success of the meeting. We must not fail also to thank the Air Force Academy for their generous hospitality and all of the services which they so graciously provided.

S. Lefschetz.
J. P. LaSalle.

November, 1962

Contents

The Method of Integral Manifolds in the Theory of Nonlinear Oscillations

Y. A. MITROPOL'SKIY
Institut Matematiki, Kiev, USSR

In my report I shall give a short presentation of basic results obtained recently (in the Soviet Union and in the U.S.A.) concerning the further development, generalization, and application to new classes of problems of the method of integral manifolds. The fundamental concepts were formulated in 1945 by Bogoliubov [*2*] in whose monograph a particular problem was studied concerning the properties of solutions of equations in standard form in the infinite time interval. However, the ideas and methods for proving a series of theorems are very flexible and effective and may be applied to the study of quite a wide class of differential equations with a small parameter encountered in the study of oscillatory processes.

At the present time the ideas of the theory of integral manifolds have undergone considerable development and generalization and already are used fundamentally in the study of complex phenomena which are observed in the most varied dynamic systems described by differential equations with a small parameter.

I shall dwell briefly on the general idea of this method.

In the first place we shall give an analytical definition of an integral manifold for the system of differential equations

$$\frac{dx}{dt} = X(t, x, \epsilon) \tag{1}$$

where x, X are n-vectors of an n-dimensional euclidean space E_n, t is the time, ϵ is a small positive parameter, and the right-hand sides of the system satisfy certain quite general conditions for $-\infty < t < \infty$ and for x in an open set $U_n \subset E_n$.

Let each t from the interval $(-\infty, \infty)$ correspond to a certain set S_t of points x, which can be analytically represented by a parametric equation of the form

$$x = f(t, c_1, c_2, \ldots, c_s), \tag{2}$$

where $f(t, c_1, c_2, \ldots, c_s)$ satisfies a Lipschitz condition with respect to $c_1, c_2, \ldots, c_s$ in the entire domain of variability of $c_1, c_2, \ldots, c_s$.

We shall say then, that S_t is an s-dimensional ($s \leq n$) integral manifold for equation (1) if, for any solution $x = x(t)$ of this equation, there ensues from the validity of the relationship

$$x(t) \in S_t \tag{3}$$

at any moment of time $t = t_0$, its validity for any arbitrary real t.

Geometrically an integral manifold is a hypersurface having the property that if any value of the solution of the system of equations is located on the integral manifold (on the hypersurface), then the complete solution will be located on the integral manifold (on the hypersurface).

Originally in Bogoliubov's monograph [*2*], the idea of the theory of integral manifolds was applied to the following particular problem. The particular case of system (1) is investigated when the right-hand side of the system is proportional to a small parameter and the so-called equations are in standard form

$$\frac{dx}{dt} = \epsilon X(t, x). \tag{4}$$

Assuming that for any $x \in U_n$ the limit

$$\lim_{T\to\infty} \frac{1}{T} \int_t^{t+T} X(t, x)\, dt = X_0(x) \tag{5}$$

exists uniformly with respect to t in the interval $(-\infty, \infty)$, the averaged equation

$$\frac{dx}{dt} = \epsilon X_0(x) \tag{6}$$

is also investigated simultaneously with system (4). For system (4), a sufficiently complex problem is solved: determination of a series of correspondences between the properties of exact solutions (solutions of system (4)) and approximate solutions (solutions of the averaged system (6)), which depend on their behavior on an infinite interval.

Thus, for example, in many important cases the averaged equations (6) admit invariant manifolds of the torus type and the following problem is of interest: Will the integral manifolds for the exact equations (4) lie in a sufficiently small neighborhood of these manifolds, and to what degree will these manifolds be stable?

Here particularly it is necessary to note that, independently of the problem mentioned above concerning the correspondence between exact and approximate solutions of system (4) (between solutions of system (4) and solutions of system (6)) which is the basic problem of the averaging method (and in general of asymptotic methods), the construction of even a local theory of integral manifolds for system (4) is of great interest in itself in connection with the fact that qualitative investigation of the solutions of system (4) are considerably simplified when these solutions are located on a manifold of smaller dimensionality than that of initial phase space.

A special peculiarity of the ideas developed in the method of integral manifolds is a certain new approach in the qualitative theory of differential equations. We are studying here two systems of differential equations—exact equations (4) and approximate equations (6); the difference between their right-hand sides is asymptotically small.

As is known, individual solutions, as a rule, are very sensitive to small variations of the right-hand sides of equations.

In our theory we deal not with individual solutions, but with integral manifolds (not with curves, but with hypersurfaces). It appears that an integral manifold is a more stable formation with respect to small variations of the right-hand sides of equations as compared with individual solutions. In many cases we can prove theorems of the following type: if the approximate equations (6) have a certain integral manifold, then the exact equations (4) also admit an integral manifold located in an asymptotically narrow neighborhood of the integral manifold of the approximate system. At the same time, similar theorems for individual solutions can be obtained only for sufficiently rigid conditions laid upon the right-hand sides of equations (4) and (6).

Here it is appropriate to stress again that the existence of integral manifolds for exact equations (4) is also of great importance in the study of their individual solutions insofar as we can now concentrate on the study of solutions located on an integral manifold—on a hypersurface—instead of studying the entire phase space.

This relates particularly to the case of stable integral manifolds when there exists a contraction of the entire phase space toward the integral manifold.

After these preliminary remarks and the cursory characterization of basic peculiarities of the method of integral manifolds which we have illustrated as applied to equations of type (4), we go on to the study of the most characteristic types of differential equation systems with small parameter, for which the application of the ideas of the method of integral manifolds is quite effective and to the study of the characteristics of certain recently obtained results in the further development and generalization of this method.

In addition to the above-mentioned equations in standard form, more complex systems with a small parameter may be successfully studied by using the method of integral manifolds.

Thus we shall apply this method to systems which are close to autonomous of the type

$$\frac{dx}{dt} = X(x) + \epsilon X^*(t, x, \epsilon) \tag{7}$$

and to systems which are close to nonautonomous of the type

$$\frac{dx}{dt} = X(t, x) + \epsilon X^*(t, x, \epsilon) \tag{8}$$

where ϵ is a small positive parameter and x, X, X^* are n-dimensional vectors of the euclidean space E_n.

To such systems are reduced, on a series of assumptions, numerous problems of oscillation theory: the problem of oscillations which are not close to harmonic, oscillations in a system with many degrees of freedom, relaxational oscillations, and others.

Systems of equations of the type:

$$\begin{aligned} \frac{dx_s}{dt} &= X_s(\alpha, x_1, \dots, x_n, \epsilon), \\ \frac{d\alpha}{dt} &= \lambda\omega(x_1, \dots, x_n, t, \epsilon) + A(\alpha, x_1, \dots, x_n, t, \epsilon), \end{aligned} \tag{9}$$

can also be studied where λ is a large parameter and X_s, A are periodic (or almost periodic) functions of t. Systems of this type are encountered in the study of gyroscopic systems, the theory of accelerations, etc.

Using the method of integral manifolds, systems of a more general type may also be investigated:

$$\begin{aligned} \frac{dx}{dt} &= X(\theta, x) + \epsilon X^*(\theta, x, \epsilon), \\ \frac{d\theta}{dt} &= \Theta(\theta, t, x, \epsilon). \end{aligned} \tag{10}$$

Usually, for simplified application of the method of integral manifolds to the study of the types of equations presented, it is expedient to transform them into new variables, reducing them to a form which would exhibit specific features of the equations and make it possible to evaluate higher order infinitesimal terms when passing on to the approximate equations.

These transformations usually consist in reducing our systems of equations to cylindrical or polar coordinates, and formulas for substituting variables depend fundamentally on the form of the particular solutions of the corresponding nonperturbed equations, that is, of the systems of equations where $\epsilon = 0$.

After the appropriate transformations, our systems of equations in the general case are reduced to systems of the type:

$$\begin{aligned}
\frac{d\varphi}{dt} &= \alpha_1(t, \varphi, \psi, a, b) + S_1(t, \varphi, \psi, a, b, \epsilon), \\
\frac{d\psi}{dt} &= \alpha_2(t, \varphi, \psi, a, b) + S_2(t, \varphi, \psi, a, b, \epsilon), \\
\frac{da}{dt} &= \epsilon A_1(t, \varphi, \psi, a, b) + A_2(t, \varphi, \psi, a, b, \epsilon), \\
\frac{db}{dt} &= B_1(t, \varphi, \psi)\, b + B_2(t, \varphi, \psi, a, b, \epsilon),
\end{aligned} \tag{11}$$

where $0 \leq \epsilon \leq \epsilon_0$ and φ, ψ, a, b are vectors of the corresponding dimensions: k, l, m, n.

It is to be noted that in many works it is usual to investigate directly systems of type (11), and for such systems integral manifolds are investigated. However, reduction of equations to form (11) is not by any means a trivial problem.

For the types of differential equations presented above, numerous theorems have been proved on the existence and properties of integral manifolds, on the behavior of solutions located on the manifold, and also on the behavior of solutions in the neighborhood of integral manifolds. I shall dwell briefly on some typical theorems.

We shall study the system of equations in standard form

$$\frac{dx}{dt} = \epsilon X(t, x) \tag{12}$$

and the corresponding first approximation system of equations

$$\frac{dx}{dt} = \epsilon X_0(x) \tag{13}$$

where

$$X_0(x) = \lim_{T\to\infty} \frac{1}{T} \int_0^T X(t, x)\, dt. \tag{14}$$

Then the following well-known theorem by N. N. Bogolyubov is valid.

Theorem. Let equations (13) have the periodic solution $x = \xi(\omega t)$, $\xi(s) = \xi(s + 2\pi)$ and let the real parts of characteristic exponents of the linear variational equations for the solution $\xi(\omega t)$ be different from zero. We also assume that the functions $X(t, x)$ satisfy certain sufficiently general conditions in a ρ neighborhood U_ρ of the periodical solution $\xi(\omega t)$, and are almost periodic in t uniformly with respect to x.

Under these assumptions, it is possible to find positive numbers ϵ_0, σ_0 $(\sigma_0 < \rho)$ such that for every ϵ $(0 \leq \epsilon \leq \epsilon_0)$ the following statements are valid:

(1) Equations (12) have a unique integral manifold S_t located for all t in U_{σ_0} which can be represented in parametric form

$$x = f(t, \theta, \epsilon) \tag{15}$$

where f is defined for all t, θ, periodic with respect to θ with period 2π, almost periodic with respect to t, uniformly relative to θ with the frequency basis of the function $X(t, x)$; in addition, the following inequality is valid:

$$\| f(t, \theta, \epsilon) - \xi(\theta) \| \leq \delta(\epsilon), \tag{16}$$

where $\delta(\epsilon) \to 0$ as $\epsilon \to 0$ and $f(t, \theta, \epsilon)$ has uniformly continuous derivatives with respect to θ up to the second order inclusive.

(2) There exists a function $F(t, \theta, \epsilon)$ which is defined for all real t and θ, periodic relative to θ with period 2π, and almost periodic relative to t uniformly with respect to θ, which has continuous derivatives up to the second order inclusive, such that equations (12) on the manifold S_t is equivalent to the equation

$$\frac{d\theta}{dt} = \epsilon F(t, \theta, \epsilon). \tag{17}$$

(3) If the cylinder of periodic solutions $\xi(\omega t + \varphi)$ (in the space ξ, t), where φ is an arbitrary constant, of equations (13) is stable, nonstable, or conditionally stable with respect to an s dimensional point manifold, then the integral manifold S of equations (12) is stable, nonstable, or conditionally stable with respect to an s dimensional point manifold.

Recently a number of mathematicians, both in the Soviet Union and in the U.S.A., have proved a series of theorems of this type for the most varied forms of systems of differential equations, and with the aid of these theorems, the behavior and character of the solutions located on the manifold and also in its vicinity have been studied.

One of the essential stages in proving this kind of theorem is the reduction of systems of equations (12), (7), or (8) to form (11), which is convenient for the further study and proof of auxiliary lemmas.

In this reduction a fundamental difficulty arises (particularly in reduction of the equations to the standard form (12)) in the selection of such a substitution of variables that in the new [transformed] equations, the terms which depend upon the time t will be contained in terms of a higher order infinitesimal. After this the family of solutions of our system is sought in such a way that it will belong to a specially selected class of functions. For this the last equation (or the last two equations) of our system is investigated in which an arbitrary function from a given class is substituted for φ and ψ; the solution of a certain functional equation is then sought by the use of the principle of a fixed point. Further, for the manifold obtained, periodicity or almost periodicity with respect to t and stability properties are established from the fact that this manifold satisfies the equation.

Here it should be noted that it is an interesting and useful detail that (because the transformation obtained from the system of differential equations in the process of proving the existence and uniqueness of the integral manifold is a contraction [which transfers every term $C(D, \Delta)$ of the class of functions $F(t, \theta, \epsilon)$ into a term $C(D, \Delta)$]) the integral manifold can be calculated by means of successive approximations.

Of essential importance in the method of proving all the previous theorems is the fact that this method makes it possible to prove the existence of the integral manifold without any special assumptions concerning the dependence on time of the right-hand sides of the equations. This makes it possible to apply the method to the study of properties of integral manifolds for systems of differential equations of the type

$$\frac{dx}{dt} = X(t, x) + \epsilon X^*(t, x, \epsilon), \tag{18}$$

in which the right-hand sides depend arbitrarily on t. Assuming that systems of type (18) satisfy a series of conditions, theorems are obtained which fundamentally facilitate the study of complex phenomena observed in nonstationary oscillatory processes.

Interesting results are obtained in the study of an integral manifold for systems of equations of the type

$$\frac{dx}{dt} = X(x) + \epsilon X^*(\nu t, x, \epsilon) \tag{19}$$

which describe a weak periodic perturbation of relaxation oscillations. Here the following theorem is valid [9]:

Theorem. For a system of differential equations (19), let the following conditions be satisfied:

(a) the equations of "nonperturbed" motion

$$\frac{dx}{dt} = X(x) \tag{20}$$

have the periodic solution

$$x = x^0(\psi), \qquad x^0(\psi + 2\pi) = x^0(\psi) \tag{21}$$

which depends on one arbitrary constant φ ($\psi = \omega t + \varphi$);

(b) the real parts of all $n - 1$ characteristic exponents for the equations of variations,

$$\frac{d\delta x}{dt} = X'_x(x^0)\, \delta x \tag{22}$$

which correspond to the periodic solution (21), are different from zero;

(c) it is possible to find a ρ—neighborhood, D_ρ, of the orbit of the periodic solution (21) and an ϵ_0 such that in the domain $-\infty < t < \infty$, $x \in D_\rho$, $0 \leq \epsilon \leq \epsilon_0$, the functions

$$X(x) + \epsilon X^*(\nu t, x, \epsilon) \tag{23}$$

are infinitely differentiable with respect to x and ϵ and have in the same domain bounded and uniformly continuous derivatives;

(d) The functions $X^*(\nu t, x, \epsilon)$ are either almost periodic with respect to νt or has an average with respect to t.

Then there exist positive numbers ϵ' and σ ($\sigma < \zeta$, $\epsilon' < \epsilon_0$) such that for any positive $\epsilon < \epsilon'$ the following assertions are valid:

(1) Equation (19) has a unique integral manifold in the domain

$$-\infty < t < \infty, \qquad x \in D_\sigma, \qquad 0 < \epsilon < \epsilon'. \tag{24}$$

which can be represented in parametric form

$$x = x^0(\psi) + \tfrac{1}{2}\{A(\psi)\, h(\nu t, \psi, \epsilon) + \bar{A}(\psi)\, \bar{h}(\nu t, \psi, \epsilon)\} = \Phi(\nu t, \psi, \epsilon) \tag{25}$$

where $\Phi(\nu t, \psi, \epsilon)$ is defined in the domain $-\infty < t < \infty$, $\psi \in \Omega$, $0 < \epsilon < \epsilon'$ and has the period 2π with respect to ψ.

(2) For solutions located on the manifold (25), system (19) is equivalent to the equation

$$\frac{d\psi}{dt} = \omega + \epsilon F(\nu t, \psi, \epsilon) \tag{26}$$

in which the function $F(\nu t, \psi, \epsilon)$ is defined in the domain $-\infty < t < \infty$, $\psi \in \Omega$, $0 < \epsilon < \epsilon'$, is infinitely differentiable with respect to ψ and periodic with respect to ψ with period 2π.

(3) The manifold (25) has the same stability and instability or conditional stability properties as the cylinder of periodic solutions (21). This theorem permits a more detailed analysis of integral curves in the case when $X^*(\nu t, x, \epsilon)$ is a periodic function with respect to νt with period 2π. In this case the following theorem is valid:

Theorem. If in addition to the conditions of the preceding theorem, $X^*(\nu t, x, \epsilon)$ is a periodic function with respect to νt with period 2π, then it is possible to find positive numbers ϵ' and σ $(\epsilon' < \epsilon_0, \sigma < \rho)$, such that for any positive $\epsilon < \epsilon'$, the system has an integral manifold S_t (in the form of a torus) which satisfies the conditions of the preceding theorem.

Solutions of system (7) which are located on this integral manifold can be represented in the form

$$x(t) = \Phi(\alpha_e t, \alpha_\rho t + \psi_0), \qquad \psi_0 = \text{const.} \tag{27}$$

and are characterized by a certain number $\bar{\nu}$, that is, they are quasi-periodic or periodic depending on whether the number $\bar{\nu}$ is irrational or rational.

In the latter case, any non-periodic solution located on this integral manifold converges to one of the periodic solutions.

Finally, if $n-1$ of the characteristic exponents of (22) have negative real parts, then any solution of system (19) passing at $t = t_0$ through any point of the domain U_σ approaches as $t \to \infty$ one of the above-shown stationary solutions located on the manifold S_t.

In connection with this, the convergence will be sufficiently fast if the absolute value of the negative real parts of the $n-1$ characteristic exponents is sufficiently large.

More complex is the study of integral manifolds for system (19) in the case when the equations of "nonperturbed" motion (20) have a static solution or a family of periodic solutions which depends on more than one arbitrary variable (for example on two).

In such cases we are confronted with a series of difficulties which do not make it possible to study the properties of integral manifolds in the entire phase space and compel us to study only local manifolds. The extensive work by Lykova [6], which we shall not discuss here, is devoted to these questions.

We turn now to the interesting results obtained recently by Hale and

Stokes [4] on the behavior of solutions in the neighborhood of integral manifolds.

Let a system of autonomous differential equations

$$\frac{dx}{dt} = X(x), \tag{28}$$

be given where x, X, are n-dimensional vectors of euclidean space E_n.

We assume that $x = x^0(t)$ is a periodic solution of system (28). Then, as is known, this periodic solution is asymptotically orbitally stable (that is, in the x-space the closed curve C_0, determined by the equation $x = x^0(t)$, is asymptotically stable) if $n - 1$ of the characteristic exponents of the system of equations in variations

$$\frac{dy}{dt} = X'_x(x^0(t))y \tag{29}$$

corresponding to the solution $x = x^0(t)$, have negative real parts.

If $x^0(t)$ is a solution of equation (28), then $x^0(t + \varphi)$, where φ is an arbitrary constant, is also a solution, and therefore the function $x^0(t + \varphi)$ defines a cylinder in (x, t)-space.

If the closed curve in x-space is stable, then it is obvious that the cylinder C in (x, t) space is also stable. Under the same assumptions about the characteristic exponents of the system of equations in variations (29), as is known, there exists not only cylindrical asymptotic stability, but also so-called asymptotic orbital stability with the asymptotic phase consisting of the fact that for any solution $x(t)$ of equation (28) with x_0 sufficiently close to C, there exists such a φ_0 that $x(t) - x^0(t + \varphi_0) \to 0$ as $t \to \infty$.

J. K. Hale and A. Stokes studied the case when system (28) has a $(k + 1)$ parametric family of periodic solutions $x = x^0[\omega(a)\, t + \varphi, a]$, where $a = (a_1, a_2, ..., a_k)$ and φ is a real constant. In the (x, t)-space such a family of solutions defines a k-parameter family of cylinders C_a or a $(k + 1)$-dimensional manifold $\mathfrak{M}$ of solutions. For a fixed value of a, the equation in variations (29) can be formed and in the general case will have $(k + 1)$ characteristic exponents equal to zero.

Assuming that $n - (k + 1)$ characteristic exponents of the system (29) have negative real parts (for every value of the vector a), the authors show that under sufficiently general conditions upon the manifold $\mathfrak{M}$, for any solution $x(t)$ of equation (28) with $x(0)$ sufficiently close to $\mathfrak{M}$, there exist φ_0, a_0 such that $x(t) - x^0[\omega(a_0)\, t + \varphi_0, a_0] \to 0$ when $t \to \infty$; that is, $x(t)$ approaches the cylinder C_{a_0} with asymptotic amplitude and asymptotic phase.

The following interesting problem is further studied: we assume that system (28) has an asymptotic orbitally stable periodic solution $x^0(t)$ and the function $X^*(t, x)$, which describes a perturbation acting on system (28), converges uniformly to zero at $t \to \infty$ for all $t \geq T$ and $x \in U$, where U is a bounded set in x-space. In this case will the solution of the perturbed system, $x^*(t)$ approach the cylinder defined by function $x^0(t + \varphi)$ when $x^0(t)$ will not in general satisfy the perturbed equation?

In connection with this question, the following theorem is presented.

Theorem. We assume that the system of equations

$$\frac{dx}{dt} = X(x) \tag{30}$$

where x, X are real, n-dimensional vectors, $X \in C^2$ has the $(k + 1)$-parameter family of real periodic solutions:

$$\begin{gathered} x = x^0[\omega(a)t + \varphi, a], \\ x^0(s + 2\pi, a) = x^0(s, a), \end{gathered} \tag{31}$$

where $a = (a_1, ..., a_k)$, φ are real constants, and

$$x^0(s, a) \in C^3(s), \qquad x^0(s, a) \in C^2(a), \qquad \omega(a) > 0, \qquad \omega(a) \in C^2, \qquad a \in U,$$

where U is either an open set in E_k or a unique point; in the latter case $k = 0$.

If $n - (k + 1)$ characteristic exponents $\lambda_1(a), \lambda_2(a), ..., \lambda_{n-(k+1)}(a)$ of the system of linear equations in variations which corresponds to the periodic solution (31) have negative real parts for all $a \in U$ and

$$\text{rank} \left\| \frac{\partial x^0(s, a)}{\partial s}, \frac{\partial x^0(s, a)}{\partial a} \right\| = k + 1 \tag{32}$$

for all $a \in U$, $s \in R$, then the manifold $\mathfrak{M}$, determined by expressions (31), is asymptotically stable with asymptotic phase and amplitude realtive to the perturbed system $dx/dt = X(x) + X^*(t, x)$ provided that $X^*(t, x)$ has a finite double integral with respect to t for all $t \geq T$, $x \in U$.

The method of integral manifolds can be successfully applied to the study of complex oscillatory phenomena in relaxational systems.

As is known, in many cases, relaxational oscillations in systems with many degrees of freedom are described by systems of differential equations with a small parameter in certain derivatives. In connection with this fact, we shall present certain results of Zadiraka [*12*] related to the study of the system of equations

$$\frac{dx}{dt} = f\left(t, x, z, \frac{t}{\epsilon}\right), \qquad \epsilon \frac{dz}{dt} = F(t, x, z), \tag{33}$$

where x, f are n-dimensional vectors, and z and F are m-dimensional vectors in euclidean space E_{n+m}.

Assuming the existence of the

$$\lim_{T\to\infty} \frac{1}{T} \int_0^T f(t, x, z, \nu)\, d\nu = f_0(t, x, z) \tag{34}$$

uniformly with respect to $x, z \in E_{n+m}$, simultaneously with system (33), there is studied the degenerate system

$$\frac{dx}{dt} = f_0(t, x, z), \qquad z = \varphi(t, x) \tag{35}$$

where $z = \varphi(t, x)$ is determined as a root of the equation

$$F(t, x, z) = 0. \tag{36}$$

It is assumed that in the domain

$$-\infty < t < \infty, \qquad x \in U_n, \qquad \| z - \varphi(t, x) \| < \rho, \qquad 0 < \epsilon < \epsilon^* \tag{37}$$

in addition to conditions (34), conditions of continuity and boundedness of the right-hand sides of equations (33) are also satisfied, as well as the condition that the roots of the characteristic equation,

$$\text{Det}\, \| I_m p - A(t, x) \| = 0, \tag{38}$$

where $A(t, x) = F_z'(t, x, 0)$ is a square matrix of order m satisfy the condition

$$\text{Re}\, \{\rho_i(t, x)\} \leq -\alpha < 0. \tag{39}$$

Under these conditions a theorem is proved which establishes the existence and a series of properties of the integral manifold of system (33) which is close to the integral manifold

$$z = \varphi(t, x).$$

Theorem. Under the above-mentioned conditions, it is possible to find a positive ϵ_0 such that, for all positive $\epsilon < \epsilon_0$, system (33) has a unique local integral manifold of the form

$$z(t, x) = \varphi(t, x) + \psi(t, x, \epsilon), \tag{40}$$

where $\psi(t, x, \epsilon)$ is definite in the domain $-\infty < t < \infty$, $x \in U_n$ and satisfies the inequalities

$$| \psi(t, x, \epsilon) | \leq D(\epsilon), \qquad | \psi(t, x', \epsilon) - \psi(t, x'', \epsilon) | \leq \Delta(\epsilon) | x' - x'' | \tag{41}$$

where $D(\epsilon) \to 0$, $\Delta(\epsilon) \to 0$ for $\epsilon \to 0$, and $\varphi(t, x)$ is an isolated root of equation (36). Moreover, the manifold (40) has bounded and uniformly continuous derivatives with respect to x, the order of which is determined from the order of smoothness of the vectors f and F.

This theorem makes it possible to analyze the structure of the solutions of system (33). Thus, for example, if the vector functions $f(t, x, z, t/\epsilon)$, $F(t, x, z)$ are periodic with respect to t with period 2π, and system (35) has an isolated periodic solution $x = x(t)$, $z = z(t)$ with period 2π, then using the preceding theorem, no difficulties arise in proving [*12*] the existence of the unique periodic solution $x = x(t, \epsilon)$, $z = z(t, \epsilon)$ of system (33) with period 2π which, as $\epsilon \to 0$, converges to the periodic solution of system (35) uniformly with respect to t.

Autonomous relaxational systems may be studied analogously.

Due to the shortage of time in my report I have dwelt briefly on certain theorems and ideas of the theory of integral manifolds in conformity with the study of a series of classes of differential equations containing a small parameter, and I have had no opportunity to touch upon numerous interesting and important results recently obtained in our country and the U.S.A.

Here are pertinent results, for example, obtained in the past two or three years by Hale and Stokes [*4*], Markus [*8*], Diliberto [*3*], Koosis [*5*], Reinhart [*11*], and others, which concern problems of the existence of integral manifolds for special types of differential equations, study of the behavior of solutions located on manifolds and in the vicinity of manifolds, application of the method of integral manifolds to stability problems, etc.

The work of Lykova [*7*] is devoted to the study of local integral manifolds, that of K. V. Zadiraka to manifolds in relaxational systems; also, interesting results were obtained by Pontryagin and Mishchenko [*10*], by Anosov [*1*], and others, in the study of complex phenomena described by differential equations containing a small parameter attached to the derivative.

Finally, we shall dwell briefly on certain aspects of the further development of the theory of integral manifolds and its application to the study of numerous actual problems of nonlinear mechanics and related problems of differential equations containing a small parameter which are encountered in these problems.

An interesting trend is the extension of the integral manifolds methods to the study of infinite systems of differential equations. This is a pressing problem in connection with the study of oscillations in systems with distributed parameters (vibration of beams, shafts, turbine rotors, etc.) described by partial differential equations, and particularly in connection with the application of asymptotic methods to the study of single-frequency oscillations in electrodynamic problems, etc.

Certain essential difficulties, however, will apparently be encountered, for example, in regard to the reducibility of infinite systems with periodic coefficients and other problems; therefore, it is to be expected that effective results can be obtained only for particular cases when the oscillatory systems with distributed parameters studied are restricted by certain sufficiently rigid conditions.

A prospective trend is the extension of the integral manifolds method to systems of differential equations with delayed argument. Here in all probability, it will be possible to generalize most of the above-mentioned theorems on the existence and properties of integral manifolds for quite a wide class of differential equations with delayed argument, at least in cases of small delay, slowly varying delay, and others. Results which can be obtained here will unquestionably be of interest for the study of oscillatory phenomena and stability problems in complex control systems, servo-systems, etc.

Pressing but quite complicated is the problem of the development and application of the theory of integral manifolds to the study of systems of differential equations with a small parameter attached to the highest derivatives which describe relaxational oscillatory processes. In the case when for a degenerate oscillatory system it is comparatively simple to find an analytical expression for the limiting cycle, then, as is evident from the series of theorems presented above, it is comparatively easy to apply the method and to obtain a series of interesting results. For the case when the limiting cycle of a degenerate system is not sufficiently smooth, a series of fundamental difficulties arise. At the same time, the solution of this problem even for a series of particular cases is of great interest in connection with numerous problems in physics and engineering.

Also of interest are problems of the topological study of the structures of integral manifolds for various forms of differential equations which describe quite complex oscillatory phenomena connected with impact, instantaneous change of velocity, etc.

REFERENCES

[*1*] Anosov, D. V., Averaging in systems of ordinary differential equations with rapidly oscillating solutions. *Izvest. Akad. Nauk S.S.S.R., Ser. Mat.* **24**, 721-742 (1960).

[*2*] Bogoliubov, N. N., "O nekotorykh statisticheskikh Metodakh v matematicheskoy Fizike" (On Certain Statistical Methods in Mathematical Physics). *Akad. Nauk Ukr. S.S.R.*, Kiev, 1945.

[*3*] Diliberto, S. P., Perturbation Theory of Periodic Surfaces I, II, III, IV. Mimeographed O.N.R. reports, Berkeley, California, 1956-1957.

[*4*] Hale, J. K., and A. P. Stokes, Behavior of solutions near integral manifolds. *Arch. rational Mech. Anal.* **6**, 133-170 (1961).

[*5*] Koosis, P., One-dimensional repeating curves in the nondegenerate case. *In* "Contributions to the Theory of Nonlinear Oscillations" (S. Lefschetz, ed.), Vol. 3, pp. 277-285. Princeton Univ. Press, Princeton, New Jersey, 1956.

[*6*] Lykova, O. B., On the behavior of solutions of a system of differential equations in the neighborhood of the isolated static solution. *Ukr. Mat. Zhur.* **9**, No. 3 (1957).

[*7*] Lykova, O. B., On the behavior of solutions of a system of differential equations in the neighborhood of closed orbits. *Ukr. Mat. Zhur.* **9**, No. 4 (1957).

[*8*] Markus, M. D., An invariant surface theorem for a nondegenerate system. *In* "Contributions to the Theory of Nonlinear Oscillations" (S. Lefschetz, ed.), Vol. 3, pp. 243-256. Princeton Univ. Press, Princeton, New Jersey, 1956.

[*9*] Mitropol'skiy, Yu. A., On certain differential equations of the theory of relaxational oscillations. *Ukr. Mat. Zhur.* **9**, No. 3, 296-309 (1957).

[*10*] Pontryagin, L. S., and Ye. F. Mishchenko, Derivation of certain asymptotic evaluations for solutions of differential equations with a small parameter attached to derivatives. *Izvest. Akad. Nauk S.S.S.R. Ser. Mat.* **23**, 643-660 (1959).

[*11*] Reinhart, B. L., Periodic orbits on two manifolds. *Proc. intern. Symposium on Ordinary Differential Equations, Mexico City, 1959* in press.

[*12*] Zadiraka, K. V., On periodic solutions of nonlinear differential equations with small parameters attached to derivatives. *Dopovidi Akad. Nauk Ukr. S.S.R.* No. 2 (1958).

Sur un Théorème de Seifert sur les Trajectoires Fermées de Certains Champs de Vecteurs

GEORGES REEB
Institut Fourier, University of Grenoble, Grenoble, France and RIAS, Baltimore, Maryland

Introduction

La situation suivante se présente dans de nombreuses questions de Mécanique Analytique classique [*1*]:

Sur une variété compacte $\mathcal{V}_n$ (que l'on supposera orientée, munie d'une structure de variété différentiable de classe C_∞ et d'une structure d'espace de Riemann) est défini un champ de vecteurs E, de classe C_∞, sans singularités, dont les trajectoires sont homéomorphes au cercle T et constituent les fibres d'une structure fibrée de $\mathcal{V}_n$. La base de cette structure fibrée est alors une variété $\mathcal{V}_{n-1}$ de classe C_∞, orientée et compacte. On se propose l'étude des trajectoires d'un champ $E + \Delta E$ obtenu par perturbation de E; en fait ici on s'intéressera uniquement à l'étude de (certaines) trajectoires fermées. La petitesse de la perturbation sera mesurée, selon l'usage, par l'une des normes habituelles: les deux normes les plus intéressantes sont la norme habituelle (ou C_0-norme) notée $\| \ \|_0$ et la norme qui porte sur les composantes et leurs dérivées, appelée la C_1-norme et notée $\| \ \|_1$. Il convient d'attirer tout de suite l'attention sur un point fondamental qui a souvent été mal compris:

La locution "ΔE est une petite perturbation" signifie $\| \Delta E \|_i < \epsilon$ ($i = 0$ ou 1) où $\epsilon > 0$ et où ϵ ne dépend que du couple $(\mathcal{V}_n, E)$. Dans la méthode classique du "petit paramètre" la notion d'une petite perturbation est tout à fait différente: une perturbation de la forme $\lambda \Delta E$ (λ réel) est dite petite si $|\lambda| < \epsilon$, $\epsilon > 0$, où ϵ dépend du triplet $(\mathcal{V}_n, E, \Delta E)$.

En fait nous nous intéressons ici à une classe particulière de trajectoires fermées: la classe des trajectoires simplement fermées ou, si l'on préfère, la classe des trajectoires qui se ferment après un tour. Cette notion est précisée par la construction naturelle mais utile que voici: on définit dans $\mathcal{V}_n$ la famille de plaques $Q(x)$. La plaque $Q(x)$ est l'ensemble des points qu'il est possible de joindre au point x par un arc de géodésique orthogonal en x à $E(x)$, de longueur inférieure à η, $\eta > 0$, pour un η fixe convenable. Ces plaques sont homéomorphes à des cellules à $n - 1$ dimensions. Si la perturbation ΔE est petite, la trajectoire du champ perturbé $E + \Delta E$ issue de x suit approximativement la fibre Tx contenant x et recoupe $Q(x)$ une première fois en x'. On désigne par $V(x)$ le vecteur tangent en x à l'arc de géodésique orientée joignant x à x' et de longueur égale à la longueur de cet arc de géodésique. Si η est choisi convenablement $V(x)$ est bien défini et le champ de vecteurs $V(x)$ est continu. *La trajectoire issue de x est dite simplement fermée si* $V(x) = 0$. Les vecteurs $V(x)$ et $E(x)$ sont orthogonaux.

Il est quelquefois intéressant de considérer le cas où $\mathcal{V}_n$ est une variété à bord et où la perturbation ΔE vérifie l'hypothèse de "relaxation" suggérée par la mécanique classique:

H: *le champ* $V(x)$ *est dirigé, en tout point du bord de* $\mathcal{V}_n$, *vers l'intérieur de* $\mathcal{V}_n$.

On trouvera des exemples en [*1*].

Ici nous nous proposons de démontrer sous diverses hypothèses l'existence de trajectoires simplement fermées du champ perturbé. Le théorème le plus fin est certainement le théorème de Seifert [*2*] (cf. 1). La démonstration de ce théorème est délicate. Nous donnerons une démonstration plus directe de ce théorème, mais en substituant la C_1-norme à la C_0-norme. Notre démonstration offre l'avantage de pouvoir être transposée à certains cas où $\mathcal{V}_n$ est de dimension supérieure.

1. Examen de Quelques Cas Particuliers

Dans certains cas, l'existence d'au moins une trajectoire simplement fermée pour une petite perturbation (dans la C_0-norme) est évidente.

Il convient de signaler tout d'abord le cas où la structure fibrée associée à $(\mathcal{V}_n, E)$ est triviale et où la caractéristique d'Euler Poincaré $\chi(\mathcal{V}_{n-1})$ de l'espace de base est non nulle. L'existence de trajectoires simplement fermées résulte alors du théorème de points fixes de Lefschetz; d'ailleurs ce théorème montre de plus que la somme des index des trajectoires simplement fermées (supposées isolées) est égale à $\chi(\mathcal{V}_{n-1})$. Malheureusement le cas envisagé ne se présente guère dans les problèmes suggérés par la mécanique des systèmes conservatifs (l'invariant intégral d'Elie Cartan permet de le montrer).

Un deuxième cas intéressant correspond aux variétés $\mathcal{V}_n$ n'admettant pas deux champs de vecteurs linéairement indépendants: le champ V admet dans ce cas des points singuliers et par conséquent le champ perturbé admet au moins une trajectoire simplement fermée. Un exemple intéressant correspond au cas où $\mathcal{V}_n$ est la sphère S_{4p+1} (p entier, $p > 0$) et où E admet comme lignes intégrales les fibres de la fibration classique de Hopf. De même si $\mathcal{V}_n = S_{4p+1} \times [0, 1]$, si E désigne le champ associé à la fibration de Hopf du facteur S_{4p+1} et si ΔE vérifie l'hypothèse de relaxation H énoncée dans l'introduction, on peut affirmer que le champ perturbé admet au moins une trajectoire simplement fermée. On remarque cependant que dans le cas envisagé nous n'avons aucun résultat sur les index des trajectoires simplement fermées.

Le cas des variétés à trois dimensions échappe aux remarques précédentes. Mais le *théorème de Seifert* affirme précisément:

Si $n = 3$ et si la base $\mathcal{V}_2$ n'est pas le tore à deux dimensions, alors le champ perturbé admet au moins une trajectoire simplement fermée.

En fait le théorème de Seifert comporte un complément affirmant que la somme des index des trajectoires simplement fermées (supposées isolées) est égale à la caractéristique d'Euler Poincaré $\chi(\mathcal{V}_2)$.

2. Examen du Cas Où $n = 3$

Nous supposons à partir de maintenant $n = 3$; nous supposons également que le champ perturbé étudié n'admette pas de trajectoire simplement fermée (donc $V(x) \neq 0$ pour tout point x de $\mathcal{V}_3$); notre but est évidemment d'aboutir sous certaines hypothèses à une contradiction, établissant ainsi l'existence d'au moins une trajectoire fermée.

A tout point x de $\mathcal{V}_3$ associons l'entier $I(x)$ ainsi défini: soit p la projection canonique de $\mathcal{V}_3$ sur la base $\mathcal{V}_2$; l'entier $I(x)$ est la variation angulaire, divisée par 2π, du vecteur $p(V(y))$ lorsque y parcourt la fibre $T(x)$, contenant x, dans le sens direct. L'entier $I(x)$ est indépendant de x, on le désignera simplement par I, ou éventuellement par $I(\Delta E)$ pour rappeler que I dépend de la perturbation ΔE.

On peut supposer la base $\mathcal{V}_2$ triangulée; on distingue un simplexe à deux dimensions e_2 de cette triangulation et le complexe K obtenu en privant $\mathcal{V}_2$ des points internes de e_2. Le simplexe e_2 est muni de l'orientation induite par l'orientation de $\mathcal{V}_2$ et ∂e_2 désigne le bord de e_2 muni de l'orientation cohérente. La théorie de la construction d'une section de l'espace fibré $\mathcal{V}_3$ montre:

(i) qu'il existe des sections f_1 et f_2 au-dessus de e_2 et K. (Donc f_1 et f_2 sont des applications continues de e_2 et K dans $\mathcal{V}_3$ telles que pf_1 et pf_2 soient respectivement les applications identiques de e_2 et K).

(ii) que $(f_1, \partial e_2)$, et $(f_2, \partial e_2)$ définissent de façon naturelle deux éléments α_1 et α_2 du groupe de Poincaré $\pi(T)$ de la fibre. La différence $i = \alpha_1 - \alpha_2$ est donc un entier rationnel; cet entier i ne dépend pas du choix particuler des sections f_1 et f_2, cet entier caractérise le *premier obstacle* de la structure fibrée.

Les champs de vecteurs V_1 et V_2 sur e_2 et K qui associent à $x \in e_2$

le vecteur $p(V(f_1(x))$ et qui associent à $y \in K$ le vecteur $p(V(f_2(y))$ ne se raccordent pas le long de ∂e_2; la somme S des index des singularités des champs V_1 et V_2 est nulle d'après l'hypothèse $V(x) \neq 0$. D'autre part il est clair, d'après la formule d'Euler Poincaré pour les singularités d'un champ de vecteurs que:

$$S = \chi(\mathcal{V}_2) + iI$$

donc:

$$0 = \chi(\mathcal{V}_2) + iI. \tag{1}$$

La démonstration (de la première partie) *du théorème de Seifert résulterait donc de l'affirmation: $I = 0$.* En fait l'affirmation $I = 0$ constitue le *lemme fondamental* de l'article de Seifert. C'est également ce lemme fondamental que nous étudions dans la suite et que nous démontrons dans le cas d'une petite perturbation par rapport à la C_1-norme.

Signalons une conséquence directe de la formule (1). Le cas $i = 0$ correspond à une structure fibrée triviale de $\mathcal{V}_3$ et par conséquent nous laissons ce cas de côté. La formule (1) montre alors que l'absence de trajectoires simplement fermées implique: i est un diviseur de $\chi(\mathcal{V}_2)$. (On rappelle que i ne dépend que de la structure fibrée de $\mathcal{V}_3$). *Si $\mathcal{V}_2$ est la sphère à deux dimensions S_2 et si $i \neq 1$ et $i \neq 2$* (les cas $i = 1$ et $i = 2$ correspondent respectivement à $\mathcal{V}_3 = S_3$ et $\mathcal{V}_3 = P_3(R)$) *alors le champ perturbé $E + \Delta E$, où ΔE est petit dans la C_1-norme, admet au moins une trajectoire simplement fermée.*

On observera que ce dernier résultat est sans objet puisque le théorème de Seifert le contient comme cas particulier; mais ce résultat est obtenu par des moyens élémentaires; les raisonnements utilisés se transposent sans difficulté au cas où $\mathcal{V}_n = \mathcal{V}_3 \times [0, 1]$ et où le champ perturbateur ΔE vérifie l'hypothèse de relaxation énoncée dans l'introduction. Il n'en est pas de même des techniques de Seifert.

3. Etude du Lemme Fondamental ($I = 0$)

Dans le plan $x0y$ les inégalités

$$|y| \leq 0,9 \qquad |y| \leq 1 \qquad \text{et} \qquad |y| \leq 1,1$$

définissent trois bandes emboîtées D' D D''. Nous étudions un homéomorphisme φ de D dans D'' vérifiant, pour tout point m, les propriétés suivantes:

(i) $\varphi(m + e) = \varphi(m) + e$ où e est le vecteur $(1, 0)$.

(ii) $\| \varphi(m) - m \| < 0,1$.

(iii) $\varphi(m) \neq m$.

Cet homéomorphisme φ est compatible avec la relation d'équivalence ρ engendrée par le groupe des translations de vecteurs ne (n entier); l'homéomorphisme φ vérifie l'inclusion: $\psi(D) \supset D'$. Par passage au quotient selon la relation d'équivalence ρ, l'homéomorphisme φ définit un homéomorphisme d'une couronne circulaire dans une couronne circulaire un peu plus grande.

La construction classique suivante associe à φ un entier rationnel $I(\varphi)$: Faisons décrire au point m le segment 1 l'origine (0, 0) et d'extrémité (1, 0) et soit $U(m)$ le vecteur $\varphi(m) - m$. Au segment 1 correspond dans l'espace quotient un chemin fermé qui représente un générateur du groupe de Poincaré. L'entier $I(\varphi)$ est la variation angulaire, divisée par 2π, du vecteur $U(m)$ lorsque m décrit le segment 1.

Considérons maintenant un deuxième exemplaire des bandes emboîtées D' D D'' et un homéomorphismé ψ de D dans D'' vérifiant les conditions (i) (ii) et (iii) imposées à φ; soit $I(\psi)$ l'entier associé à ψ.

Supposons de plus qu'il existe un homéomorphisme Φ du rectangle fondamental $0 \leq x \leq 1$ de D dans D'' vérifiant les propiétés suivantes:

$$(\alpha) \quad \| \Phi(m) - m \| \leq 0,1$$

$$(\beta) \quad \psi(m) = \Phi^{-1} \varphi \Phi(m) \qquad (\text{modulo } e = (0, 1)).$$

En d'autres termes Φ transporte l'homéomorphisme φ sur l'homéomorphisme ψ. Nous affirmons que *le lemme fondamental du paragraphe précédent est une conséquence de l'égalité:*

$$I(\varphi) = I(\psi). \tag{1}$$

Cette implication sera établie au paragraphe 4.

Il est probable que l'égalité (1) soit susceptible d'une démonstration géométrique simple, mais je ne suis pas capable d'indiquer une telle démonstration. Par contre l'égalité (1) est évidente si, en plus des conditions α et β, on impose à Φ la condition γ, qui exprime la petitesse des dérivées partielles de Φ:

$$(\gamma) \; \| \Phi \|_1 < \epsilon' \text{ pour un } \epsilon'(\epsilon' > 0) \text{ convenable.}$$

4. L'Égalité $I(\varphi) = I(\psi)$ Implique $I = 0$

Pour établir le lemme: $I = 0$ il suffit de se restreindre au cas où $\mathcal{V}_3$ est la variété (à bord) $\mathcal{V}_3 = B_2 \times T$ où T est le cercle trigonométrique et B_2 le disque plan défini par $x^2 + y^2 \leq 2$; le champ E a pour composantes (0, 0, 1). Mieux encore, il suffit de considérer le cas où $\mathcal{V}_3 = \Gamma_2 \times T$; ici Γ_2 est la couronne circulaire définie par l'inégalité

$1 \leq x^2 + y^2 \leq 2$. Le champ V ne sera évidemment défini que pour les points d'un sous-espace $\Gamma_2' \times T$ de $\Gamma_2 \times T$, où Γ_2' est la couronne définie par les inégalités $1{,}1 \leq x^2 + y^2 \leq 1{,}9$. Dans l'espace $\Gamma_2 \times T$ nous construisons deux sections h_1 et h_2:

(i) h_1 est la section droite $h_1 : (x, y) \to (x, y, 0)$

(ii) h_2 décrit un tour autour de la fibre T, elle est définie par: $h_2 : (x, y) \to (x, y, \theta(x, y))$ où $\theta(x, y)$ est l'argument de (x, y).

Les deux sections h_1 et h_2 peuvent être envisagées comme des applications du recouvrement universel D (*cf.* 3) de Γ_2 dans $\Gamma_2 \times T$; on peut même supposer que D' est appliqué par h_1 et h_2 au-dessus de Γ_2. Nous définissons maintenant deux homéomorphismes φ et ψ dans Γ_2, ou — ce qui revient au même — dans D, par la construction suivante:

au point m de Γ_2' faisons correspondre le point m_i de la section h_i ($i = 1$ ou 2) qui recouvre m; la trajectoire du champ perturbé $E + \Delta E$ issue de m_i recoupe la section h_i une première fois en m_i', le point m_i' se projette en $p(m_i') = m_i''$. Le point m_i'' est précisément $\varphi(m)$ si $i = 1$ et $\psi(m)$ si $i = 2$. Il est clair que φ et ψ vérifient les propriétés (i) (ii) et (iii) requises en 3; la propriété (iii) résulte de l'hypothèse: $V(x) \neq 0$ pour tout x.

Il reste à voir que φ et ψ sont bien reliés par un homéomorphisme Φ vérifiant les propriétés α) et β) de 3. A cet effet il convient d'étudier le rectangle fondamental Q de D. Si m_0 est le point de Q de coordonnées $(0, \frac{1}{2})$ on désignera par $m_0' = \Phi(m_0)$ la projection $p(\mu')$ du premier point de rencontre μ' de $h_2(Q)$ avec la trajectoire de $E + \Delta E$ issue de $\mu = h_1(m_0)$. La correspondance $m \to \mu'$ peut être étendue de façon continue à Q de telle sorte que μ et μ' appartiennent à la même trajectoire; cette correspondance définit l'homéomorphisme Φ qui remplit les conditions requises.

Concluons en remarquant que si la perturbation ΔE est faible, dans la C_1-norme, alors Φ vérifie la condition γ du paragraphe 3; ce qui établit la version du théorème de Seifert obtenue en remplaçant la C_0-norme par la C_1-norme.

BIBLIOGRAPHIE

[*1*] Reeb, G., Sur certaines propriétés topologiques des trajectoires des systèmes dynamiques. *Mém. Acad. roy. belg., Cl. sci.* **27**, No. 9 (1952).

[*2*] Seifert, H., Closed integral curves in 3-space and isotopic two-dimensional deformations *Proc. AMS* **1**, 287-302 (1950).

Generic Properties of Differential Equations

LAWRENCE MARKUS
Mathematics Department
University of Minnesota, Minneapolis, Minnesota

Let M be a compact differential manifold and let B be some complete metric space of vector fields (first order, autonomous, differential equations) on M. We say that a property P of vector fields in B is generic in case the subset of B consisting of vector fields which fail to have P is of first category.

Thus a consequence of the important results of J. Moser, presented in these Proceedings, page 71, is that if M is a 3-dimensional Riemannian manifold and if B consists of all C^∞ volume preserving differential equations, then the property of topological transitivity is not generic in B.

Theorem. Let M be a compact differential manifold, and let B be the Banach space of all C^1 first-order differential equations, with the C^1 metric, on M. Then the property of having only elementary critical points and elementary periodic solutions is generic in B.

We note that a critical point of a differential system $\mathscr{S}$ in B is elementary in case its eigenvalues all have non-zero real parts. Thus an elementary critical point is isolated. Also a non-constant periodic solution S of a differential system $\mathscr{S}$ in B is elementary in case the non-trivial characteristic multipliers for S have moduli different from unity. Thus an elementary periodic solution S is isolated in that there is a tubular neighborhood of S which contains no other periodic solution of $\mathscr{S}$.

Therefore, a generic differential system in B has only a finite number of critical points and only a countable number of periodic solutions.

The proof of the theorem follows from certain local approximation theorems [*1*, *2*] and usual category arguments. Details will be published later.

REFERENCES

[*1*] Markus, L., The behavior of the solutions of a differential system near a periodic solution. *Ann. Math. (Princeton)* **72**, 245-266 (1960).

[*2*] Markus, L., Structurally stable differential systems. *Ann. Math. (Princeton)* **73**, 1-19 (1961).

The Density Theorem for Structurally Stable Systems on a Two Dimensional Manifold

M. M. PEIXOTO
Instituto de Matemática Pura e Aplicada
Rio de Janeiro, Brazil

The structurally stable vector fields on a differentiable two manifold M^2 are the ones which exhibit the simplest features and can be classified into equivalence classes modulo homeomorphisms of M^2 onto itself mapping trajectories onto trajectories (Smale). It becomes important to know how large is the set Σ of all structurally stable vector fields, considered as a subset of the set $\mathscr{B}$ of all vector fields defined on M^2, with the C^1-topology.

Theorem. Σ is open and dense in $\mathscr{B}$. This fact was proved previously by the author [*1*] in the case where M^2 is the sphere S^2. It shows that for dimension two the structurally stable vector fields are generic. The proof depends on several approximation lemmas, the crucial fact being that whenever we have a minimal set μ which is not a singular point or a closed orbit then by means of a C^1-small perturbation we can get out of it either a closed orbit or else a new connection between saddle points.

REFERENCE

[*1*] Peixoto, M. M., On Structural Stability. *Ann. Math. (Princeton)* **69**, (1959).

Beurteilung Nichtlinearer Systeme auf der Grundlage der Strukturstabilität

R. REISSIG
Institut für Angewandte Mathematik und Mechanik, DAW und Humboldt-Universität zu Berlin, Germany

Das Ziel des vorliegenden Berichtes ist es, die Bedeutung des Begriffes "*Strukturstabilität*"[1] bei der Untersuchung des asymptotischen Verhaltens von nichtlinearen dynamischen Systemen, die durch gewöhnliche Differentialgleichungen beschrieben werden, herauszustellen.

Dieser Begriff tritt auf, wenn man die Beschränktheit der Lösungen und die Stabilität einer Lösung erforscht, und er spielt sogar bei der Abschätzung des zeitlichen Verlaufes von Übergangslösungen eine Rolle.

Sei

$$x' = F(x, t)$$

die Vektor-Differentialgleichung, die das betrachtete dynamische System darstellt. Dann ist die rechte Seite $F(x, t)$ im allgemeinen nicht exakt bestimmt; vielmehr sind durch die physikalischen Bedingungen nur einige Merkmale dieser Funktion festgelegt. Das heißt: Wir haben nicht eine einzige Differentialgleichung zu behandeln, sondern eine gewisse Klasse von Gleichungen; das Problem besteht nun darin, Bedingungen aufzustellen, unter denen in der Klasse ein gemeinsames asymptotisches Verhalten der Lösungen, z. B. Beschränktheit, vorliegt. Die Klasse muß hinsichtlich der betrachteten Lösungseigenschaft *strukturstabil* sein.

Wenn man sich für die Stabilität einer Lösung $x(t)$ interessiert, dann ist es üblich, die Nachbarlösungen (die sogenannten gestörten Bewegungen) in der Form $x(t) + u(t)$ zu schreiben, wobei $u(t)$ die von einer Anfangsabweichung herrührende Störung ist. Wir haben für die Störung die Gleichung

$$u' = f(u, t) = F[x(t) + u, t] - F[x(t), t].$$

Selbst dann, wenn die Funktion $F(x, t)$ wohlbekannt ist, wird man häufig nicht wissen, wie die Funktion $f(u, t)$ von der Variablen t abhängt,

[1] Editors Note: "Strukturstabilität" refers to an asymptotic property of a class of differential equations and is not equivalent to "systèmes grossiers" ("structural stability") as defined by Andronov and Pontryagin.

weil man die ungestörte Bewegung $x(t)$ nicht genau kennt. Die Differentialgleichung für $u(t)$ denkt man sich deshalb nicht für eine bestimmte Lösung $x(t)$ gebildet; vielmehr muß man darin eine Lösungsschar mit mehr oder weniger allgemeinen Charakteristiken, z. B. mit einheitlichen Schranken, in Betracht ziehen. Damit ergibt sich ein weiteres Problem der Strukturstabilität, nämlich innerhalb einer Klasse von Differentialgleichungen, die die triviale Lösung zulassen, deren Stabilität zu bestimmen.

Ein allgemeines, naheliegendes Prinzip für die Lösung des Problems der Strukturstabilität besteht darin, daß man die Beurteilung der betrachteten Gleichungen auf die einer solchen zurückführt, die sich leichter auswerten läßt, z. B. durch Konstruktion einer einfachen *Ljapunow*schen Funktion. Diese als Maßstab gewählte Gleichung mag zur betrachteten Klasse gehören oder nicht, wir werden sie als *Klassenrepräsentanten* bezeichnen. Physikalisch gesehen, vergleichen wir die entsprechenden dynamischen Systeme mit einem Modell, das ein gewünschtes asymptotisches Verhalten unter ungünstigeren dynamischen Bedingungen besitzt.

Eine Frage von praktischem Interesse besteht darin, für eine gegebene Klasse von Differentialgleichungen einen passenden Repräsentanten zu finden. Ein einfaches Verfahren, das man bei Differentialgleichungen zweiter Ordnung und vielleicht auch in anderen Fällen anwenden kann, ist das der *stückweisen Linearisierung:* Man ersetzt die zu behandelnden Gleichungen durch ein stückweise lineares Vergleichssystem. Das Verfahren wird am besten an einem Beispiel erläutert.

Wir betrachten die verallgemeinerte *Linéard*sche Gleichung

$$x'' + f(x)\,x' + g(x) = e(t) \tag{1}$$

oder das äquivalente System

$$\left.\begin{aligned} x' = -F(x) + y + E(t), \qquad y' = -g(x) \\ \left[F(x) = \int_0^x f(\xi)\,d\xi,\ E(t) = \int_0^t e(\tau)\,d\tau\right] \end{aligned}\right\} \tag{2}$$

und setzen folgende Bedingungen voraus, die eine Klasse von Gleichungen bestimmen:

$$\left.\begin{aligned} |\,E(t)\,| \leqq M \qquad |\,e(t)\,| \leqq M \\ f(x) \geqq 2\gamma > 0 \\ 0 < k^2 \leqq \frac{g(x_1) - g(x_2)}{x_1 - x_2} \leqq (\tau k)^2, \qquad \tau > \frac{\gamma}{k} \end{aligned}\right\}; \tag{3}$$

f, g, e mögen stetige Funktionen sein.

Für die Störungen, die wir mit $\{u(t), v(t)\}$ bezeichnen, erhalten wir das System

$$\left.\begin{aligned} u' &= -\varphi(u, t)\, u + v, \qquad v' = -\psi(u, t)\, u \\ &\Big[\varphi(u, t) = \frac{F[x(t) + u] - F[x(t)]}{u} \geqq 2\gamma, \\ k^2 &\leqq \psi(u, t) = \frac{g[x(t) + u] - g[x(t)]}{u} \leqq (\tau k)^2\Big]. \end{aligned}\right\} \tag{4}$$

Für die Gleichungen (2) soll das Beschränktheitsproblem und für die Gleichungen (4) das Stabilitätsproblem gelöst werden. Zu dem Zweck tragen wir für die Funktionen $F(x)$ und $g(x)$ geeignete stückweise lineare Funktionen und für die Funktionen $E(t)$, $\varphi(u, t)$ und $\psi(u, t)$ stückweise konstante (notwendig unstetige) Funktionen ein:

$$\left.\begin{aligned} x' &= -\gamma\tau^{-2}[(\tau^2 + 1) - (\tau^2 - 1)\operatorname{sgn} xy]\, x + y \\ &\quad + \frac{M}{2}[(\tau^2 + 1)\operatorname{sgn} x - (\tau^2 - 1)\operatorname{sgn} y] \\ y' &= -\frac{k^2}{2}[(\tau^2 + 1) - (\tau^2 - 1)\operatorname{sgn} xy]\, x \end{aligned}\right\} \tag{5}$$

bzw.

$$\left.\begin{aligned} u' &= -\gamma\tau^{-2}[(\tau^2 + 1) - (\tau^2 - 1)\operatorname{sgn} uv]\, u + v \\ v' &= -\frac{k^2}{2}[(\tau^2 + 1) - (\tau^2 - 1)\operatorname{sgn} uv]\, u. \end{aligned}\right\} \tag{6}$$

Um die Vergleichssysteme (5) und (6) herzuleiten, betrachtet man die Phasenbahnen in der xy-Ebene oder die Störkurven in der uv-Ebene, die den Lösungen von (2) bzw. (4) entsprechen, und schätzt die Richtungsfelder dieser Trajektorien ab. Dabei entstehen die Gleichungen (5) und (6), deren Lösungskurven (die sogenannten Vergleichskurven) durch die erwähnten Trajektorien von außen nach innen geschnitten werden. Sind die Vergleichskurven in der Nähe des unendlich fernen Punktes der xy-Ebene bzw. in der Umgebung des Ursprunges der uv-Ebene divergente bzw. Spiralen, so ist die künftige Beschränkheit der Lösungen $\{x(t), y(t)\}$ und die asymptotische Stabilität jeder Lösung, also offenbar auch die asymptotische Stabilität im Ganzen gewährleistet.

Zeigt man die Beschränktheit auf anderem Wege unter geringeren Einschränkungen, so erkennt man, daß die Bedingungen (3) für die Funktionen $f(x)$ und $g(x)$ nur zwischen den gemeinsamen Schranken aller Lösungen $x(t)$ von (1) erfüllt sein müssen.

Die Systeme (5) und (6) kommen auf die einheitliche Form (5), wenn wir darin setzen:

$$x = a, \qquad y = b, \qquad M = \mu$$

und für (6) vereinbaren:

$$a = u, \qquad b = v, \qquad \mu = 0.$$

Das ab-System mit dem Parameter μ ist offenbar bezüglich des Ursprunges der ab-Ebene symmetrisch. Um seine Lösungen zu studieren, genügt es, wenn man den Kurvenverlauf in der oberen Halbebene verfolgt.

Zu einem beliebigen positiven Wert a konstruieren wir den Bogen, der vom Punkt $(-a, 0)$ durch den zweiten Quadranten zum Punkt $(0, b_2(a))$ der positiven b-Halbachse hinläuft, und den Bogen, der vom Punkt $(a, 0)$ durch den ersten Quadranten zum Punkt $(0, b_1(a))$ der positiven b-Halbachse führt. Mit der Bezeichnung $\epsilon = \mu/a$ berechnen wir

$$\frac{b_1(a) - b_2(a)}{a} = -(1 + \tau^2)\,\epsilon + d(\epsilon); \tag{7}$$

$d(\epsilon)$ ist eine stetige Funktion, die für alle $\epsilon \geqq 0$ durch die folgenden Formeln definiert ist:

$$d(\epsilon) = \sqrt{\epsilon^2 - 2\gamma\tau^{-2}\epsilon + k^2}\,\exp\left(\frac{\gamma\tau_1}{\sqrt{\tau^4k^2 - \gamma^2}}\right) - \tau\sqrt{\tau^2\epsilon^2 - 2\gamma\epsilon + k^2}\,\exp\left(-\frac{\gamma\tau_2}{\sqrt{\tau^2k^2 - \gamma^2}}\right);$$

$$\tau_1 = \operatorname{arc\,ctg}\left(\frac{\tau^2\epsilon - \gamma}{\sqrt{\tau^4k^2 - \gamma^2}}\right), \qquad \tau_2 = \operatorname{arc\,ctg}\left(\frac{\gamma - \tau^2\epsilon}{\sqrt{\tau^2k^2 - \gamma^2}}\right).$$

Wir wollen annehmen:

$$d(0) > 0. \tag{8}$$

Eine hinreichende Bedingung hierfür ist

$$(\tau^2k)\log\tau \leqq \frac{\gamma\pi}{2}.$$

Aus $\tau > \gamma/k$ ergibt sich jetzt eine Einschränkung für den Parameter τ, z. B.

$$\tau\log\tau \leqq \frac{\pi}{2}.$$

Aus (7) und (8) schließen wir für den Fall $\epsilon = 0$ ($\mu = 0$, *d. h.* für das System (6)):

$$\begin{aligned} b_1(a) - b_2(a) &> 0, \qquad a > 0; \\ b_1(0) - b_2(0) &= 0 \end{aligned} \tag{9}$$

und für den Fall $\epsilon > 0$ (z. B. für das System (5)):

$$b_1(a) - b_2(a) > 0$$

für hinreichend große Werte a ($\epsilon \to 0$), jedoch

$$b_1(0) - b_2(0) = -\tau^2\mu\left[1 + \exp\left(-\frac{\gamma\pi}{\sqrt{\tau^2k^2 - \gamma^2}}\right)\right] < 0.$$

Im ersten Fall nähern sich alle ab-Kurven spiralenförmig dem Ursprung; im zweiten Fall existiert eine positive Zahl a_μ mit der Eigenschaft:

$$b_1(a_\mu) = b_2(a_\mu); \qquad b_1(a) > b_2(a) \quad \text{fur} \quad a > a_\mu.$$

Dem Wert a_μ entspricht ein Zyklus C_μ, während zu den Werten $a > a_\mu$ Spiralen gehören, die sich auf dem Zyklus aufwickeln. Für $\mu = M$ erhalten wir das System (5) und erkennen, daß alle Lösungen $\{x(t), y(t)\}$ der Gleichungen (2) den Zyklus C_M von außen nach innen überschreiten.

Demzufolge sind die Bedingungen (3) zusammen mit der Bedingung (8) hinreichend für Beschränktheit und asymptotische Stabilität im Ganzen.

Auf Grund passender stückweise linearer Vergleichssysteme können wir auch die Übergangserscheinungen beurteilen.

Der Einfachheit halber wählen wir eine lineare Dämpfung, $f(x) = 2\gamma$. Wir verwenden nun das der *Liénard*schen Gleichung (1) äquivalente System

$$x' = y, \qquad y' = -g(x) - 2\gamma y + e(t). \tag{10}$$

Die Störungsgleichungen sind

$$u' = v, \qquad v' = -\psi(u, t)\, u - 2\gamma v. \tag{11}$$

Die gleichen Bedingungen wie oben, aber $\gamma < k$ statt $\gamma < \tau k$, mögen gelten. Folglich ist die Beschränktheit und asymptotische Stabilität im Ganzen gewährleistet.

Um die Übergangslösungen, d. h. den Zeitverlauf der Funktion $u(t)$ abzuschätzen, benötigen wir zwei Vergleichssysteme für (11): Ein System, daß in gewissem Sinne besseren oszillatorischen Eigenschaften entspricht, so daß die Störungen langsamer abklingen; und ein anderes

System mit schlechterem Schwingungscharakter, so daß die Störungen schneller verschwinden. Diese Systeme, die die betrachtete Klasse (11) repräsentieren, sind

$$\left.\begin{aligned} u' &= v, \qquad v' = -\frac{k^2}{2}[(\tau^2+1)-(\tau^2-1)\operatorname{sgn} uv]\,u - 2\gamma v.\\ u' &= v, \qquad v' = -\frac{k^2}{2}[(\tau^2+1)+(\tau^2-1)\operatorname{sgn} uv]\,u - 2\gamma v.\end{aligned}\right\} \tag{12}$$

Wenn man die Störungen $\{u(t), v(t)\}$ wieder durch Trajektorien in der uv-Ebene darstellt und diese mit den Lösungskurven von (12) vergleicht, erfährt man, daß sie die erste Kurvenschar von außen nach innen und die zweite im entgegengesetzten Sinne durchdringen.

Die beiden Vergleichskurvenscharen bestehen aus konvergenten Spiralen, wenn wir annehmen

$$\tau \exp\left(-\frac{\gamma\pi}{\sqrt{\tau^2k^2-\gamma^2}}\right) = e^{-\delta}, \qquad \delta > 0. \tag{13}$$

Jetzt betrachten wir irgendeine Störkurve für $t \geqq 0$; die Folge der sukzessiven Nullstellen von $v(t) = u'(t)$ sei $\{t_n\}$. Für die Werte $u_n = u(t_n)$, die Störungsextrema, haben wir die Abschätzung

$$|\,u_n/u_{n-1}\,| < e^{-\delta},$$

so daß

$$|\,u_n/u_1\,| < e^{-(n-1)\delta}.$$

$v = v_k^{\pm}(u, \bar{u})$ sei ein Bogen des Systems

$$u' = v, \qquad v' = -k^2u - 2\gamma v, \tag{14}$$

der vom Punkt $(\bar{u}, 0)$ durch die obere $(+)$ oder untere $(-)$ Halbebene zur v-Achse läuft. Sei ferner $v_{n-1}(u)$ der Störkurvenbogen zwischen den Punkten $(u_{n-1}, 0)$ und $(u_n, 0)$; wir setzen $\operatorname{sgn} v_{n-1} = \epsilon_{n-1}$.

Dann schließen wir aus der Anordnung der Störkurven und der Vergleichskurven, daß die Bögen

$$v_k^{\epsilon_{n-1}}(u, u_{n-1}) \qquad \text{und} \qquad v_k^{\epsilon_{n-1}}(u, u_n)$$

im Innern von $v_{n-1}(u)$ liegen. Daraus ergibt sich:

$$\begin{aligned} T = \frac{\pi}{\sqrt{k^2-\gamma^2}} &= \int_{u_{n-1}}^{0} [du/v_k^{\epsilon_{n-1}}(u, u_{n-1})] + \int_0^{u_n} [du/v_k^{\epsilon_{n-1}}(u, u_n)] > t_n - t_{n-1}\\ &= \int_{u_{n-1}}^{u_n} [du/v_{n-1}(u)].\end{aligned}$$

Somit haben wir

$$\exp[-(n-1)\delta] < \exp[-(t_n - t_1)\delta/T] < \kappa \exp(-\delta t_n/T) \qquad (\kappa = e^\delta),$$

und wegen

$$|u(t)| \leqq |u_n|; \qquad t \geqq t_n$$

erhalten wir die Ungleichung

$$|u(t)| \leqq \kappa \operatorname*{Max}_{t\geqq 0} |u(t)| \exp(-\delta t_n/T); \qquad t_n \leqq t < t_{n+1}.$$

Die Majorante für $|u(t)|$ ist eine Treppenkurve, die exponentiell abnimmt.

Wir bemerken, daß unsere Resultate über das asymptotische Verhalten der Lösungen von (2) numerische Auswertungen gestatten.

Während sich das Verfahren der stückweisen Linearisierung nur bei Systemen niedriger Ordnung eignet, haben wir eine Methode zur Lösung von Problemen der Strukturstabilität, die man in komplizierten Fällen anwenden kann, die Methode von *Aiserman*. Wir wollen zeigen, wie sich mit ihr Übergangslösungen abschätzen lassen, und wählen dazu das soeben behandelte einfache Beispiel. Der Gedanke der *Aiserman*schen Methode besteht in einem Vergleich der betrachteten Klasse von nichtlinearen Systemen mit einem passenden linearen System, dessen Parameter konstant sind.

Zuerst behandeln wir also die linearisierten Gleichungen (14) und bilden für sie eine *Ljapunow*sche Funktion, nämlich die positiv-definite quadratische Form

$$W(u, v) = Au^2 + 2Buv + Cv^2.$$

Zu dem Zweck gehen wir von der totalen zeitlichen Ableitung mit Hilfe der Gleichungen (14) aus und wählen sie in der Form

$$W' = -4\gamma k^2(u^2 + v^2).$$

Wir berechnen

$$A = k^2(1 + k^2) + 4\gamma^2,$$

$$B = 2\gamma,$$

$$C = 1 + k^2$$

und die Eigenwerte der Koeffizientenmatrix

$$\lambda = 2qk^2(1 - \sqrt{1 - 1/q}), \qquad \mu = 2qk^2(1 + \sqrt{1 - 1/q})$$

$$[4qk^2 = (1 + k^2)^2 + 4\gamma^2].$$

Dann haben wir die Abschätzung

$$\lambda \rho^2 \leqq W(u, v) \leqq \mu \rho^2,$$

wenn wir $\rho^2 = u^2 + v^2$ setzen.

Bezeichnen wir in (11)

$$\psi(u, t) = k^2[1 + \sigma(u, t)],$$

so erhalten wir die Ableitung von $W(u, v)$ vermöge (11)

$$W' = -4\gamma k^2(1 + \sigma)\, u^2 - 2k^2(1 + k^2)\, \sigma u v - 4\gamma k^2 v^2.$$

Sie ist negativ-definit, wenn wir verlangen

$$W' \leqq -4\gamma k^2 \vartheta \rho^2 \qquad (0 < \vartheta < 1).$$

Das bedeutet:

$$\sigma \leqq \frac{8\gamma(1 - \vartheta)}{(1 + k^2)^2} (\gamma + k\sqrt{q}).$$

Diese Bedingung sichert die exponentielle Stabilität der Null-Lösung von (11):

$$\rho(t) \leqq \kappa \rho_0 \exp(-t/T);$$

$$\rho_0 = \rho(0), \qquad \kappa = \sqrt{q} + \sqrt{q - 1}, \qquad T = \frac{q}{\gamma\vartheta}(1 + \sqrt{1 - 1/q}).$$

Für das Störungsmaß $\rho(t)$ wird eine stetig exponentiell abnehmende Majorante angegeben.

Der Begriff der exponentiellen Stabilität kann für praktische Probleme, etwa in der Regelungstheorie, von Bedeutung sein. Deshalb verdient die angedeutete Methode, die man in mancher Hinsicht weiterentwickeln kann, Beachtung.

Eine andere allgemeine Methode zur Lösung des Stabilitätsproblems ist das wohlbekannte Verfahren der Linearisierung. Um zum Schluß noch zu zeigen, daß man es auf Strukturfragen ausdehnen kann, betrachten wir die am Anfang erwähnte Differentialgleichung für die Störung

$$u' = f(u, t).$$

Die zugehörige linearisierte Gleichung ist

$$u' = \frac{\partial f}{\partial u} \cdot u = \left(\frac{\partial F}{\partial x}\right)_{/x(t),t} \cdot u.$$

Haben wir für sie die Stabilitätsfrage beantwortet, so müssen wir entscheiden, inwieweit unsere Resultate im nichtlinearen Fall Gültigkeit haben. Dabei wirkt wieder der Umstand erschwerend, daß man den Zeitverlauf der zu untersuchenden ungestörten Lösung $x(t)$ als unbekannt ansehen muß. Ein Anhaltspunkt ergibt sich aber in wichtigen Fällen daraus, daß die Ausgangsgleichung $x' = F(x, t)$, die das dynamische System beschreibt, eine relative Schranke zuläßt. Dann gibt es im Phasenraum ein gewisses beschränktes Gebiet B derart, daß alle von dort ausgehenden Lösungen für immer in B bleiben. Wir betrachten die Menge S der zur Zeit $t = 0$ in B entspringenden Lösungen und die über S gebildete Klasse von Störungsgleichungen. Mit Hilfe einer von *Opial* vorgeschlagenen Idee läßt sich folgender Satz beweisen:

Wenn die über S definierte Klasse der linearisierten Störungsgleichungen gleichmäßig asymptotisch oder exponentiell stabil ist, gilt das auch für die ursprünglichen Störungsgleichungen.

Beim Studium der betreffenden Klasse von linearisierten Gleichungen im Hinblick auf gleichmäßige Stabilitätseigenschaften eignen sich verschiedene Verfahren zur Bildung einer *Ljapunow*schen Funktion, z. B. das von *Aiserman.*

Periodic Solutions of Hyperbolic Partial Differential Equations

LAMBERTO CESARI
Mathematics Department
University of Michigan, Ann Arbor, Michigan

In the present paper we take into consideration a canonic system of hyperbolic partial differential equations (with fixed characteristics) of the form

$$u_{xy} = f(x, y, u, u_x, u_y), \qquad -\infty < x < +\infty, \qquad -a \leq y \leq a, \tag{0.1}$$

or

$$u_{ixy} = f_i(x, y, u, u_x, u_y), \qquad i = 1, 2, \ldots, n, \tag{0.2}$$

where $u = (u_1, \ldots, u_n)$ is an unknown vector function of the two independent variables x, y, where u_x, u_y, u_{xy} are the vector functions of the partial derivatives with respect to x, to y, and to x, y, and where $f(x, y, z, p, q) = (f_1, f_2, \ldots, f_n)$ denotes a continuous vector function of its arguments for (x, y) varying in a strip $A = [-\infty < x < +\infty, -a \leq y \leq a]$, and all z, p, q, and f is periodic in x of some period T:

$$f(x + T, y, z, p, q) = f(x, y, z, p, q). \tag{0.3}$$

Under usual smoothness conditions, it is well known [7] that, given any two vector functions $u_0(x)$, $-\infty < x < +\infty$, $v_0(y)$, $-a \leq y \leq a$, $v_0(0) = 0$, there exists a unique solution $U(x, y)$ in A satisfying (0.1) and such that

$$U(x, 0) = u_0(x), \qquad U(0, y) = u_0(0) + v_0(y).$$

If we know that $u_0(x)$ is periodic in x of period T, then $U(x, y)$ is periodic in x of period T only on the x-axis but not necessarily in the strip A. The question arises whether it is possible to choose $v_0(y)$ in such a way that $U(x, y)$ is periodic in x of period T for all y, $-a \leq y \leq a$, at least for a sufficiently small. Actually, the periodicity condition $U(x + T, y) = U(x, y)$ reduces to the integral equation in the unknown $v_0(y)$:

$$M(y) = \int_0^T f(x, y, U, U_x, U_y)\, dx = 0, \qquad -a \leq y \leq a,$$

where U is thought of as depending on $v_0(y)$ only. In the present paper we consider the following question which may be thought of as the inverse of the one mentioned above:

Given $u_0(x)$ periodic of period T, and $v_0(y)$, $-a \leq y \leq a$, $v_0(0) = 0$, determine a vector function $U(x, y)$, $-\infty < x < +\infty$, $-a \leq y \leq a$, and a vector function $m(y)$, $-a \leq y \leq a$, such that

$$U_{xy} = f(x, y, U, U_x, U_y) - m(y),$$
$$U(x + T, y) = U(x, y), \tag{0.4}$$
$$U(x, 0) = u_0(x), \qquad U(0, y) = u_0(0) + v_0(y),$$

for all $-\infty < x < +\infty$, $-a \leq y \leq a$.

We will suppose below that f is a continuous function of its arguments satisfying

$$| f(x, y, z, p_1, q_1) - f(x, y, z, p_2, q_2) | \leq M | p_1 - p_2 | + b | q_1 - q_2 |.$$

Then we will prove (Theorem I) that, given $u_0(x)$, $v_0(y)$ of class C^1 as above, if

$$2bT < 1, \qquad Ma < 1, \tag{0.5}$$

there is at least one pair $U(x, y)$, $m(y)$ *satisfying all requirements* (0.4).

Conditions (0.5) can be restated by saying that the problem (0.4) has certainly a solution provided b is not too large with respect to T (that is, f varies not too rapidly with respect to q), and a is small (that is, the strip A is sufficiently narrow).

We will then prove (Theorem II) that, if in addition f satisfies a relation of the form

$$| f(x, y, z_1, p_1, q_1) - f(x, y, z_2, p_2, q_2) |$$
$$\leq M_0 | z_1 - z_2 | + M | p_1 - p_2 | + b | q_1 - q_2 |,$$

then *the pair* $U(x, v)$, $m(v)$ *satisfying* (0.4) *is unique*. Finally (Theorem III) we will prove that, under the same hypotheses, *the pair* $U(x, y)$, $m(y)$ *satisfying* (0.4) *depends continuously on the functions* $u_0(x)$, $v_0(y)$.

The problem of the existence of a solution of the equation $U_{xy} = f$ in A, U periodic in x of period T, is now led to the question of functional analysis as to whether one can determine $v_0(y)$ in such a way that $m(y) = 0$ (or equivalently $M(y) = 0$).

In successive papers we shall apply the present considerations to classes of hyperbolic partial differential equations with fixed or not fixed characteristics, and we shall give practical criteria for the existence of periodic solutions.

1. We first state and prove the existence theorem for the problem (0.4). The proof is based on the Schauder fixed point theorem, and is given in the §§ 2-6 below. The proof is more difficult than the corresponding proof for the Darboux problem, since the requirements for m make the usual estimates inapplicable.

Below $u = (u_1, ..., u_n)$, $v = (v_1, ..., v_n)$, $f = (f_1, ..., f_n)$, and so on, are real n-vectors, $n \geq 1$, and we shall denote by $| u |$, etc., the Euclidean norm of u (the absolute value for $n = 1$).

Theorem I (Existence). If a, $T > 0$, and N, N_1, N_2, L, M, b, M_1, M_2, $M_3 \geq 0$ are constants, if A and R are the sets

$$A = [0 \leq x \leq T, - a \leq y \leq a],$$

$$R = [0 \leq x \leq T, - a \leq y \leq a, | z | \leq M_1, | p | \leq M_2, | q | \leq M_3],$$

if

$$M_1 \geq N + 2^{-1}N_1T + N_2a + LTa,$$

$$M_2 \geq N_1 + 2La, \qquad M_3 \geq N_2 + LT,$$

if $u_0(x)$, $0 \leq x \leq T$, $v_0(y)$, $- a \leq y \leq a$, $v_0(0) = 0$, are vector functions which are continuous with $u_0'(x)$, $v_0'(y)$, if $f(x, y, z, p, q)$, $(x, y, z, p, q) \in R$, is continuous in R, and

$$u_0(T) = u_0(0), \qquad | u_0(0) | \leq N, \qquad | u_0(x_1) - u_0(x_2) | \leq N_1 | x_1 - x_2 |, \tag{1.1}$$

$$v_0(0) = 0, \qquad | v_0(y_1) - v_0(y_2) | \leq N_2 | y_1 - y_2 |, \tag{1.2}$$

$$f(T, y, z, p, q) = f(0, y, z, p, q), \qquad | f(x, y, z, p, q) | \leq L, \tag{1.3}$$

$$| f(x, y, z, p_1, q_1) - f(x, y, z, p_2, q_2) \leq M | p_1 - p_2 | + b | q_1 - q_2 |, \tag{1.4}$$

then for

$$2Tb < 1, \qquad Ma < 1, \tag{1.5}$$

there exists a vector function $\phi(x, y)$, $(x, y) \in A$, continuous in A together with ϕ_x, ϕ_y, ϕ_{xy}, and a continuous vector function $m(y)$, $- a \leq y \leq a$, such that

$$\phi(x, 0) = u_0(x), \tag{1.6}$$

$$\phi(0, y) = \phi(T, y) = u_0(0) + v(y), \tag{1.7}$$

$$m(y) = T^{-1} \int_0^T f(x, y, \phi(x, y), \phi_x(x, y), \phi_y(x, y)) \, dx, \tag{1.8}$$

$$\phi_{xy}(x, y) = f(x, y, \phi(x, y), \phi_x(x, y), \phi_y(x, y)) - m(y), \tag{1.9}$$

for all $0 \leq x \leq T$, $-a \leq y \leq a$. Thus, by extending both $\phi(x, y)$ and $f(x, y, z, p, q)$ for all $-\infty < x < +\infty$, $|y| \leq a$, $|z| \leq M_1$, $|p| \leq M_2$, $|q| \leq M_3$, by means of the periodicity of period T in x, equation (1.9) is satisfied in the strip $-\infty < x < +\infty$, $-a \leq y \leq a$.

2. (Proof of theorem I). Let us first prove that every vector function $\phi(x, y)$, $(x, y) \in A$, satisfying

$$\begin{gathered} \phi(x, 0) = u_0(x), \qquad \phi(0, y) = \phi(T, y) = u_0(0) + v_0(y), \\ |\phi(x_1, y_1) - \phi(x_1, y_2) - \phi(x_1, y_2) + \phi(x_2, y_2)| \leq 2L\,|x_1 - x_2|\,|y_1 - y_2|, \end{gathered} \tag{2.1}$$

also satisfies the relations:

$$\begin{gathered} |\phi(x, y)| \leq M_1, \qquad |\phi(x_1, y) - \phi(x_2, y)| \leq M_2\,|x_1 - x_2|, \\ |\phi(x, y_1) - \phi(x, y_2)| \leq M_3\,|y_1 - y_2|. \end{gathered} \tag{2.2}$$

Indeed we have, for $0 \leq x \leq T$, $-a \leq y \leq a$,

$$|\phi(x_1, y) - \phi(x_2, y) - \phi(x_1, 0) + \phi(x_2, 0)| \leq 2L\,|x_1 - x_2|\,|y|,$$

where

$$|\phi(x_1, 0) - \phi(x_2, 0)| = |u_0(x_1) - u_0(x_2)| \leq N_1\,|x_1 - x_2|,$$

and hence

$$|\phi(x_1, y) - \phi(x_2, y)| \leq (N_1 + 2La)\,|x_1 - x_2| \leq M_2\,|x_1 - x_2|.$$

Analogously, we have

$$|\phi(x, y_1) - \phi(x, y_2) - \phi(0, y_1) + \phi(0, y_2)| \leq 2Lx\,|y_1 - y_2|,$$

$$|\phi(x, y_1) - \phi(x, y_2) - \phi(T, y_1) + \phi(T, y_2)| \leq 2L(T - x)\,|y_1 - y_2|,$$

where

$$|\phi(0, y_1) - \phi(0, y_2)| = |\phi(T, y_1) - \phi(T, y_2| \leq N_2\,|y_1 - y_2|,$$

and hence

$$\begin{aligned} |\phi(x, y_1) - \phi(x, y_2)| &\leq N_2\,|y_1 - y_2| + 2L\,|y_1 - y_2| \min(x, T - x) \\ &\leq (N_2 + LT)\,|y_1 - y_2| \leq M_3\,|y_1 - y_2|. \end{aligned}$$

Finally,

$$|\phi(x, y)| \leq |\phi(0, 0)| + |\phi(0, y) - \phi(0, 0)| + |\phi(x, y) - \phi(0, y)|,$$

$$|\phi(x, y)| \leq |\phi(T, 0)| + |\phi(T, y) - \phi(T, 0)| + |\phi(x, y) - \phi(T, y)|,$$

$$\begin{aligned} |\phi(x, y)| &\leq N + N_2\,|y| + (N_1 + 2La) \min(x, T - x) \\ &\leq N + N_2 a + N_1 T/2 + LTa \leq M_1. \end{aligned}$$

Thus, we have proved that relations (2.1) imply (2.2). Also, the vector functions $\phi(x, y)$ satisfying (2.1) are all Lipschitzian in A, and hence have partial derivatives ϕ_x, ϕ_y a.e. in A satisfying $|\phi_x| \leq M_2$, $|\phi_y| \leq M_3$ a.e. in A.

3. The vector function $f(x, y, z, p, q)$ is continuous in R. Hence, there are continuous monotone functions $\omega_1(\alpha)$, $\omega_2(\beta)$, $\omega_3(\gamma)$, in $0 \leq \alpha \leq T$, $0 \leq \beta \leq 2a$, $0 \leq \gamma \leq 2M$, respectively, such that $\omega_1(0) = \omega_2(0) = \omega_3(0) = 0$, and

$$\begin{aligned} |f(x_1, y, z, p, q) - f(x_2, y, z, p, q)| &\leq \omega_1(|x_1 - x_2|), \\ |f(x, y_1, z, p, q) - f(x, y_2, z, p, q)| &\leq \omega_2(|y_1 - y_2|), \\ |f(x, y, z_1, p, q) - f(x, y, z_2, p, q)| &\leq \omega_3(|z_1 - z_2|), \end{aligned} \tag{3.1}$$

for all $0 \leq x, x_1, x_2 \leq T$; $-a \leq y, y_1, y_2 \leq a$; $|z|, |z_1|, |z_2| \leq M_1$; $|p| \leq M_2$; $|q| \leq M_3$. It is not restrictive to suppose that $\omega_1, \omega_2, \omega_3$ are actually continuous and monotone in $[0, +\infty)$.

The vector functions $u_0'(x)$, $v_0'(y)$ are continuous in $[0, T]$, and $[-a, a]$ respectively. Hence, there are continuous monotone functions $\omega_4(\alpha)$, $0 \leq \alpha \leq T$, $\omega_5(\beta)$, $0 \leq \beta \leq 2a$, $\omega_4(0) = \omega_5(0) = 0$, such that

$$|u_0'(x_1) - u_0'(x_2)| \leq \omega_4(|x_1 - x_2|), \quad |v_0'(y_1) - v_0'(y_2)| \leq \omega_5(|y_1 - y_2|). \tag{3.2}$$

Let

$$\eta_1(\beta) = (1 - 2Tb)^{-1} [\omega_5(\beta) + 2T\omega_2(\beta) + 2T\omega_3(M_3\beta) + 4LMT\beta], \tag{3.3}$$

$$\eta_2(\alpha) = (1 - aM)^{-1} [\omega_4(\alpha) + a\omega_1(\alpha) + a\omega_3(M_2\alpha) + 2Lab\alpha)]. \tag{3.4}$$

Both $\eta_1(\alpha)$, $0 \leq \alpha \leq T$, $\eta_2(\beta)$, $0 \leq \beta \leq 2a$, are continuous monotone functions with $\eta_1(0) = \eta_2(0) = 0$.

4. Let E be the linear space of all vector function $\phi(x, y)$, $(x, y)\ A$, continuous in A together with their partial derivatives ϕ_x, ϕ_y, with norm $\|\phi\| = \max|\phi| + \max|\phi_x| + \max|\phi_y|$, where max is taken in A.

Let K be the subset of E made up of all vector functions $\phi(x, y) \in E$ satisfying relations (2.1) and in addition

$$\begin{aligned} |\phi_x(x_1, y) - \phi_x(x_2, y)| &\leq \eta_2(|x_1 - x_2|), \\ |\phi_x(x, y_1) - \phi_x(x, y_2)| &\leq 2L|y_1 - y_2|, \\ |\phi_y(x_1, y) - \phi_y(x_2, y)| &\leq 2L|x_1 - x_2|, \\ |\phi_y(x, y_1) - \phi_y(x, y_2)| &\leq \eta_1(|y_1 - y_2|). \end{aligned} \tag{4.1}$$

Then the vector functions $\phi \in K$ satisfy relations (2.2) and, then, $|\phi| \leq M_1$, $|\phi_x| \leq M_2$, $|\phi_y| \leq M_3$ everywhere in A. As a consequence, the vector function

$$F(x, y) = f(x, y, \phi(x, y), \phi_x(x, y), \phi_y(x, y)), \qquad (x, y) \in A,$$

is defined everywhere in A and is continuous in A. Then, the vector function

$$m(y) = T^{-1} \int_0^T f(x, y, \phi(x, y), \phi_x(x, y), \phi_y(x, y))\, dx \tag{4.2}$$

is continuous in $[-a. a]$, and the vector function

$$\psi(x, y) = u_0(x) + v_0(y) + \int_0^x \int_0^y [f(x, y, \phi(x, y), \phi_x(x, y), \phi_y(x, y)) - m(y)]\, dxdy \tag{4.3}$$

is continuous in A together with its partial derivatives

$$\psi_x(x, y) = u_0'(x) + \int_0^y [f(x, y, \phi(x, y), \phi_x(x, y), \phi_y(x, y)) - m(y)]\, dy, \tag{4.4}$$

$$\psi_y(x, y) = v_0'(y) + \int_0^x [f(x, y, \phi(x, y), \phi_x(x, y), \phi_y(x, y)) - m(y)]\, dx. \tag{4.5}$$

Relations (4.3), (4.4), (4.5) define a map $\mathscr{T} : \psi = \mathscr{T}\phi$, and we have just proved that $\mathscr{T} : K \to E$.

5. Note that $|m(y)| \leq L$ and that

$$\psi(x, 0) = u_0(x), \qquad \psi(0, y) = \psi(T, y) = u_0(0) + v_0(y), \tag{5.1}$$

$$\begin{aligned} &|\psi(x_1, y_1) - \psi(x_1, y_2) - \psi(x_2, y_1) + \psi(x_2, y_2)| \\ &= \left| \int_{x_1}^{x_2} \int_{y_1}^{y_2} [f(\ldots) - m(y)]\, dx\, dy \right| \leq 2L\, |x_1 - x_2|\, |y_1 - y_2|. \end{aligned} \tag{5.2}$$

In other words, $\psi = \mathscr{T}\phi$ for $\phi \in K$, satisfies relations (2.1) and, hence, relations (2.2), as proved in § 2. Also, we have

$$|\psi_x(x, y_1) - \psi_x(x, y_2)| = \left| \int_{y_1}^{y_2} [f(x, y, \phi(x, y), \phi_x(x, y), \phi_y(x, y)) - m(y)]\, dy \right|,$$

$$|\psi_y(x_1, y) - \psi_y(x_2, y)| = \left| \int_{x_1}^{x_2} f(x, y, \phi(x, y), \phi_x(x, y), \phi_y(x, y))\, dx \right|,$$

and hence

$$| \psi_x(x, y_1) - \psi_x(x, y_2) | \leq 2L \, | y_1 - y_2 |, \tag{5.3}$$

$$| \psi_y(x_1, y) - \psi_y(x_2, y) | \leq 2L \, | x_1 - x_2 |. \tag{5.4}$$

We have further

$$\begin{aligned}
| m(y_1) - m(y_2) | &= \Big| T^{-1} \int_0^T [f(x, y_1, \phi(x, y_1), \phi_x(x, y_1), \phi_y(x, y_1)) \\
&\qquad - f(x, y_2, \phi(x, y_2), \phi_x(x, y_2), \phi_y(x, y_2))] \, dx \Big| \\
&\leq \omega_2(| y_1 - y_2 |) + \omega_3(| \phi(x, y_1) - \phi(x, y_2) |) \\
&\qquad + M \, | \phi_x(x, y_1) - \phi_x(x, y_2) | + b \, | \phi_y(x, y_1) - \phi_y(x, y_2) | \\
&\leq \omega_2(| y_1 - y_2 |) + \omega_3(M_3 \, | y_1 - y_2 |) + 2LM \, | y_1 - y_2 | \\
&\qquad + b\eta_1(| y_1 - y_2 |),
\end{aligned}$$

and hence

$$\begin{aligned}
&| \psi_y(x, y_1) - \psi_y(x, y_2) | \\
&= | v_0'(y_1) - v_0'(y_2) + \int_0^x [f(x, y_1, \phi(x, y_1), \phi_x(x, y_1), \phi_y(x, y_1)) - m(y_1) \\
&\qquad - f(x, y_2, \phi(x, y_2), \phi_x(x, y_2), \phi_y(x, y_2)) + m(y_2)] \, dx | \\
&\leq \omega_5(| y_1 - y_2 |) + T[\omega_2(| y_1 - y_2 |) + \omega_3(| \phi(x, y_1) - \phi(x, y_2) |) \\
&\qquad + M \, | \phi_x(x, y_1) - \phi_x(x, y_2) | + b \, | \phi_y(x, y_1) - \phi_y(x, y_2) |] \\
&\qquad + T \, | m(y_1) - m(y_2) | \\
&\leq \omega_5(| y_1 - y_2 |) + 2T\omega_2(| y_1 - y_2 |) + 2T\omega_3(M_3 \, | y_1 - y_2 |) \\
&\qquad + 4LMT \, | y_1 - y_2 | + 2Tb\eta_1(| y_1 - y_2 |), \\
&= (1 - 2Tb) \, \eta_1(| y_1 - y_2 |) + 2Tb\eta_1(| y_1 - y_2 |) = \eta_1(| y_1 - y_2 |).
\end{aligned} \tag{5.5}$$

Also, we have

$$\begin{aligned}
&| \psi_x(x_1, y) - \psi_x(x_2, y) | \\
&= \Big| u_0'(x_1) - u_0'(x_2) + \int_0^y [f(x_1, y, \phi(x_1, y), \phi_x(x_1, y), \phi_y(x_1, y)) \\
&\qquad - f(x_2, y, \phi(x_2, y), \phi_x(x_2, y), \phi_y(x_2, y))] \, dy \Big| \\
&\leq \omega_4(| x_1 - x_2 |) + a[\omega_1(| x_1 - x_2 |) + \omega_3(| \phi(x_1, y) - \phi(x_2, y) |) \\
&\qquad + M \, | \phi_x(x_1, y) - \phi_x(x_2, y) | + b \, | \phi_y(x_1, y) - \phi_y(x_2, y) |] \\
&= \omega_4(| x_1 - x_2 |) + a\omega_1(| x_1 - x_2 |) + a\omega_3(M_2 \, | x_1 - x_2 |) \\
&\qquad + 2Lab \, | x_1 - x_2 | + aM\eta_2(| x_1 - x_2 |) \\
&= (1 - aM) \, \eta_2(| x_1 - x_2 |) + aM\eta_2(| x_1 - x_2 |) = \eta_2(| x_1 - x_2 |).
\end{aligned} \tag{5.6}$$

Relations (5.1), (5.2), (5.3), (5.4), (5.5), (5.6) show that $\psi = \mathscr{T}\phi$ for $\phi \in K$ satisfies all relations (2.1) and (4.1). Thus $\psi \in K$, and $\mathscr{T} : K \to K$.

6. The transformation $\mathscr{T} : K \to K$, $K \subset E$, is continuous in K with respect to the norm $\| \phi \|$ of E. Indeed, for two vector functions $\phi_j \in K$, $j = 1, 2$, we have, $\psi_j = \mathscr{T}\phi_j$, $m_j = m_{\phi_j}(y)$, $j = 1, 2$, and

$$| m_1(y) - m_2(y) | = \Big| T^{-1} \int_0^T [f(x, y, \phi_1(x, y), \phi_{1x}(x, y), \phi_{1y}(x, y)) \\ - f(x, y, \phi_2(x, y), \phi_{2x}(x, y), \phi_{2y}(x, y)] \, dx \Big| \\ \leq \omega_3(\| \phi_1 - \phi_2 \|) + M \| \phi_1 - \phi_2 \| + b \| \phi_1 - \phi_2 \|.$$

We deduce that

$$| \psi_1(x, y) - \psi_2(x, y) | = \Big| \int_0^x \int_0^y [f(...) - f(...) - m_1(y) + m_2(y)] \, dx \, dy \Big| \\ \leq 2aT(\omega_3 (\| \phi_1 - \phi_2 \|) + (M + b) \| \phi_1 - \phi_2 \|).$$

Analogously, we have

$$| \psi_{1x}(x, y) - \psi_{2x}(x, y) | \leq 2a(\omega_3 (\| \phi_1 - \phi_2 \|) + (M + b) \| \phi_1 - \phi_2 \|),$$
$$| \psi_{1y}(x, y) - \psi_{2y}(x, y) | \leq 2T(\omega_3 (\| \phi_1 - \phi_2 \|) + (M + b) \| \phi_1 - \phi_2 \|),$$

and also

$$\| \psi_1 - \psi_2 \| \leq 2(aT + a + T) (\omega_3(\| \phi_1 - \phi_2 \|) + (M + b) \| \phi_1 - \phi_2 \|),$$

i.e. $\| \psi_1 - \psi_2 \| \to 0$ as $\| \phi_1 - \phi_2 \| \to 0$ uniformly in K.

Finally, the set K is obviously convex, closed and compact with respect to the norm $\| \phi \|$ of E. By Schauder's fixed point theorem we conclude that there is an element $\phi(x, y) \in K$ such that

$$\phi(x, y) = u_0(x) + v_0(y) + \int_0^x \int_0^y [f(x, y, \phi(x, y), \phi_x(x, y), \phi_y(x, y)) - m(y)] \, dx \, dy,$$

$$m(y) = m_\phi(y) = T^{-1} \int_0^T f(x, y, \phi(x, y), \phi_x(x, y), \phi_y(x, y)) \, dx,$$

for all $0 \leq x \leq T$, $-a \leq y \leq a$. Obviously $\phi_x, \phi_y, \phi_{xy}$ exist everywhere in A, are continuous in A, and everywhere in A we have

$$\phi_{xy} = f(x, y, \phi, \phi_x, \phi_y) - m(y).$$

Thereby theorem I is proved.

7. Remark 1. The conditions of theorem I do not assure uniqueness, as the following example shows. Take $T = 1$, $a = 1$, $u_C(x) = 0$, $v_0(y) = 0$, $n = 1$, and

$$f(x, y, z, p, q) = | z |^{1/2} \sin 2\pi x,$$

for $0 \leq x \leq 1$, and all y, z, p, q. Then, the equation $u_{xy} = | u |^{1/2} \sin 2\pi x$, besides the trivial solution $\phi_1(x, y) = 0$, has also the solution $\phi_2(x, y) = (16\pi^2)^{-1} | y | y \sin^4 \pi x$, $0 \leq x \leq 1$, $-1 \leq y \leq 1$, and both satisfy the boundary condition. We have here $m_1(y) = m_2(y) = 0$, $-1 \leq y \leq 1$. Note that we may take $N = N_1 = N_2 = 0$, $L = 1$, $M_1 = 1$, $M_2 = 3$, $M_3 = 2$, $M = 0$, $b = 0$, and we have chosen $a = 1$. All conditions of theorem I are satisfied.

Remark 2. The conditions listed in Theorem I compare well with the usual ones of the existence theorems for the Darboux problem. J. P. Schauder [*11a*] supposes that f satisfies a Hölder condition in x, y, z and a Lipschitz condition in p, q, P. Hartman and A. Wintner [*6*] assume, as we do here, that f is continuous in x, y, z and Lipschitzian in p, q. A. Alexiewicz and W. Orlicz [*1*] give a new proof under the same hypotheses, by approaching f by means of functions which are Lipschitzian in z, p, q. The condition $| f(z) | \leq L$ is required by all authors, in the whole field, i.e., for all z, p, q. Since we are interested in the case where f is a polynomial in z, p, q (e.g., Van der Pol—like equations) it appears as a relaxation to state the bound L for a given bounded set R. The condition that f varies not too rapidly with respect to q seems peculiar of the present difficult problem (0.4). The condition that the strip A be sufficiently narrow is less demanding since we could repeat the reasoning in successive narrow strips below and above A.

8. Theorem II. Under the hypotheses of Theorem I, and $\omega_3(\gamma) = M_0 | \gamma |$ for some constant $M_0 \geq 0$, there is only one vector function $\phi(x, y)$ continuous in A with ϕ_x, ϕ_y, ϕ_{xy}, and one vector function $m(y)$ continuous in $[-a, a]$ satisfying (1.6-1.9).

Proof. Suppose that there are two distinct vector functions $\phi_1(x, y)$, $\phi_2(x, y)$, $(x, y) \in A$, and corresponding functions $m_1(y)$, $m_2(y)$, $-a \leq y \leq a$, as above, both pairs (ϕ_j, m_j), $j = 1, 2$, satisfying (1.6-1.9). The functions ϕ_1, ϕ_2 must be distinct (otherwise $\phi_1 = \phi_2$ and then $m_1 = m_2$), and we may suppose, to simplify notations, that they are distinct in the upper strip $0 \leq x \leq T$, $0 \leq y \leq a$. There must be a minimum number ξ, $0 \leq \xi < a$, such that $\phi_1 = \phi_2$ for all $0 \leq x \leq T$, $0 \leq y \leq \xi$, while $\phi_1 \neq \phi_2$ in any strip $0 \leq x \leq T$, $0 \leq y \leq \xi + \epsilon$,

$\epsilon > 0$. Hence $\phi_1 \neq \phi_2$ in any strip $0 \leq x \leq T$, $\xi \leq y \leq \xi + c$, $c > 0$, $\xi + c \leq a$, while

$$\phi_1(x, \xi) = \phi_2(x, \xi), \qquad \phi_{1x}(x, \xi) = \phi_{2x}(x, \xi), \qquad 0 \leq x \leq T,$$
$$m_1(\xi) = m_2(\xi).$$

We shall take below $c > 0$ sufficiently small, as the reasoning requires. Let

$$\alpha = \max | \phi_1(x, y) - \phi_2(x, y) |,$$
$$\beta = \max | \phi_{1x}(x, y) - \phi_{2x}(x, y) |,$$
$$\gamma = \max | \phi_{1x}(x, y) - \phi_{2y}(x, y) |,$$
$$\delta = \max | m_1(y) - m_2(y) |,$$

the max being taken for all $0 \leq x \leq T$, $\xi \leq y \leq \xi + c$.

Obviously $\alpha > 0$, and then necessarily $\beta > 0$, $\gamma > 0$, while we can say only that $\delta \geq 0$. In the inequalities below we shall always suppose $0 \leq x \leq T$, $\xi \leq y \leq \xi + c$. We have

$$| m_1(y) - m_2(y) | = \Big| T^{-1} \int_0^T [f(x, y, \phi_1(x, y), \phi_{1x}(x, y), \phi_{1y}(x, y)) \\ - f(x, y, \phi_2(x, y), \phi_{2x}(x, y), \phi_{2y}(x, y))] \, dx \Big|.$$

Since

$$| f(x, y, z_1, p_1, q_1) - f(x, y, z_2, p_2, q_2) | \\ \leq M_0 | z_1 - z_2 | + M | p_1 - p_2 | + b | q_1 - q_2 |,$$

we have, using the definitions of α, β, γ,

$$| m_1(y) - m_2(y) | \leq M_0\alpha + M\beta + b\gamma,$$

and hence $\delta \leq M_0\alpha + M\beta + b\gamma$.

Analogously, we have

$$| \phi_1(x, y) - \phi_2(x, y) | = \Big| \int_0^x \int_\xi^y [f(x, y, \phi_1(x, y), \phi_{1x}(x, y), \phi_{1y}(x, y)) \\ - f(x, y, \phi_2(x, y), \phi_{2x}(x, y), \phi_{2y}(x, y)) - m_1(y) + m_2(y)] \, dx \, dy \Big| \\ \leq Tc(M_0\alpha + M\beta + b\gamma + \delta),$$

$$| \phi_{1x}(x, y) - \phi_{2x}(x, y) | = \Big| \int_\xi^y [\ldots] \, dy \Big| \leq c(M_0\alpha + M\beta + b\gamma + \delta),$$

$$| \phi_{1y}(x, y) - \phi_{2y}(x, y) | = \Big| \int_0^x [\ldots] \, dx \Big| \leq T(M_0\alpha + M\beta + b\gamma + \delta).$$

This argument has proved that

$$\alpha \leq TM_0c\alpha + TMc\beta + Tbc\gamma + Tc\delta.$$

$$\beta \leq M_0c\alpha + Mc\beta + bc\gamma + c\delta,$$

$$\gamma \leq TM_0\alpha + TM\beta + Tb\gamma + T\delta,$$

$$\delta \leq M_0\alpha + M\beta + b\gamma.$$

We have now, by eliminating δ,

$$\alpha \leq 2TM_0c\alpha + 2TMc\beta + 2Tbc\gamma,$$

$$\beta \leq 2M_0c\alpha + 2Mc\beta + 2bc\gamma,$$

$$\gamma \leq 2TM_0\alpha + 2TM\beta + 2Tb\gamma,$$

where $2Tb < 1$. If we take $(1 - 2Tb)^{-1} = k$, we have

$$\gamma \leq 2kTM_0\alpha + 2kTM\beta,$$

and then, by using the constants $k' = 1 + 2Tbk$,

$$\alpha \leq 2TM_0k'c\alpha + 2TMk'c\beta,$$

$$\beta \leq 2M_0k'c\alpha + 2Mk'c\beta.$$

We shall now make the first restriction on c, namely $2Mk'c \leq \frac{1}{2}$. Then the second relation gives $(\frac{1}{2})\,\beta \leq 2M_0k'c\alpha$, i.e., $\beta \leq 4M_0k'c\alpha$, and finally, from the first relation, we have

$$\alpha \leq 2TM_0k'c\alpha + 2TMk'c \cdot 4M_0k'c\alpha,$$

$$\alpha \leq 2TM_0k'(1 + 4k'Ma)\, c\alpha.$$

We shall now make on c the further restriction that

$$2TM_0k'(1 + 4k'Ma)\, c \leq \tfrac{1}{2}.$$

Then we obtain $\alpha < \alpha/2$, a contradiction since $\alpha > 0$. Thereby theorem II is proved.

9. Let $[\phi(x, y), m(y)]$ be any solution of (1.6-1.9), where ϕ is a continuous vector function in A with $\phi_x, \phi_y, \phi_{xy}$. Hence $m(y)$ is continuous in $[-a, a]$, and we have

$$\phi(x, y) = u_0(x) + v_0(y) + \int_0^x \int_0^y [f(x, y, \phi(x, y), \phi_x(x, y), \phi_y(x, y)) - m(y)]\, dx\, dy,$$

$$\phi_x(x, y) = u_0'(x) + \int_0^y [f(x, y, \phi, \phi_x, \phi_y) - m(y)]\, dy,$$

$$\phi_y(x, y) = v_0'(y) + \int_0^x [f(x, y, \phi, \phi_x, \phi_y) - m(y)]\, dx,$$

$$m(y) = T^{-1} \int_0^T f(x, y, \phi, \phi_x, \phi_y)\, dx,$$

for all $(x, y) \in A$.

For any y, $-a \leq y \leq a$, let $q(x)$ denote the function $q(x) = \phi_y(x, y)$ of x, $0 \leq x \leq T$. Then $q(x)$ is a continuous function in $[0, T]$ with its first derivative $q'(x) = \phi_{xy}(x, y)$. Analogously, let $u(x) = \phi(x, y)$, $u'(x) = \phi_x(x, y)$, $v = v_0'(y)$, $m = m(y)$, and let $F(x, q)$ be the function of the two independent variables x and q:

$$F = F^{(\phi, y)}(x, q) = f(x, u(x), u'(x), q) = f(x, y, \phi(x, y), \phi_x(x, y), q).$$

Then $F(x, q)$ is continuous in the interval $0 \leq x \leq T$, $|q| \leq M_3$, and Lipschitzian in q with constant b, and $2bT < 1$. Also, the pair m, $q(x)$, $0 \leq x \leq T$, is a solution of the integral relations

$$q(x) = v + \int_0^x [F(x, q(x)) - m]\, dx, \qquad m = T^{-1} \int_0^T F(x, q(x))\, dx, \tag{9.1}$$

where the first one can obviously be replaced by the ordinary differential equation with initial condition

$$q'(x) = F(x, q(x)) - m, \qquad 0 \leq x \leq T, \qquad q(0) = v. \tag{9.2}$$

Nevertheless, the second relation (9.1) shows that we are dealing with a problem in the large concerning the whole interval $[0, T]$. We shall prove below the following uniqueness statement:

(*a*) Both $q(x)$, $0 \leq x \leq T$, and m are uniquely determined by the constant v and the function $F(x, q)$.

The statement is similar to the one of the usual theory of ordinary differential equations, but the proof is quite different.

Proof of (a). Suppose that there are two continuous solutions of (9.1), (9.2), say $q_1(x)$, $q_2(x)$, $0 \leq x \leq T$, and relative constants m_1, m_2. Then we have

$$q_1(0) = q_2(0) = v = q_1(T) = q_2(T),$$

$$m_j = T^{-1}\int_0^T F(x, q_j(x))\,dx, \qquad q_j'(x) = F(x, q_j(x)) - m_j, \qquad j = 1, 2. \tag{9.3}$$

Let $\mu = \max |q_1(x) - q_2(x)|$, where max is taken for $0 \leq x \leq T$. If $q_1(x) \neq q_2(x)$, then $\mu > 0$. By (9.3) we deduce first

$$\begin{aligned} |m_1 - m_2| &= |T^{-1}\int_0^T [F(x, q_1(x)) - F(x, q_2(x))]\,dx| \\ &\leq T^{-1}\int_0^T b\,|q_1(x) - q_2(x)|\,dx \leq b\mu. \end{aligned} \tag{9.4}$$

Since $q_1(0) = q_2(0)$, by (9.3) and (9.4) we deduce

$$\begin{aligned} |q_1(x) - q_2(x)| &= \left|\int_0^x [q_1'(s) - q_2'(s)]\,ds\right| \\ &= \left|\int_0^x [F(s, q_1(s)) - F(s, q_2(s)) - m_1 + m_2]\,ds\right| \\ &\leq \int_0^x [b\,|q_1(s) - q_2(s)| + b\mu]\,ds \\ &\leq 2bT\mu. \end{aligned} \tag{9.5}$$

For some x in $[0, T]$ we have $|q_1(x) - q_2(x)| = \mu$, and hence, by (9.5), we deduce $\mu \leq 2bT\mu$, and finally $\mu < \mu$ since $2bT < 1$ and $\mu > 0$, a contradiction. This proves that the solution $[q(x), m)$ of (9.1) is unique (for given v and F). Thereby statement (a) is proved. This implies, for instance, that the initial data $u_0(x)$, $v = v'(0)$, determine completely $q_0(x) = \phi_y(x, 0)$.

As an example of the statement, take $F(x, q) = \sin x + q$, $T = 2\pi$, $n = 1$. Then, for an arbitrary v, let us determine a pair $[q(x), m]$ such that

$$q'(x) = \sin x + q(x) - m, \qquad q(0) = v,$$

$$m = T^{-1}\int_0^T [\sin x + q(x)]\,dx.$$

The equation is linear and, by straightforward integration, we have

$$q(x) = Ce^x - \tfrac{1}{2}\cos x - \tfrac{1}{2}\sin x + m.$$

The initial condition $q(x) = v$ yields $C = v + \frac{1}{2} - m$, and hence

$$q(x) = (v + \tfrac{1}{2} - m)\, e^{x} - \tfrac{1}{2}\cos x - \tfrac{1}{2}\sin x + m.$$

The condition for m simply assures that $q(x)$ is periodic, hence $v + \frac{1}{2} - m = 0$, and finally

$$m = v + \tfrac{1}{2}, \qquad q(x) = -\tfrac{1}{2}\cos x - \tfrac{1}{2}\sin x + v + \tfrac{1}{2}.$$

Thus, actually, both m and $q(x)$ are univocally determined by v and F and $dm/dv = 1$.

It may well occur that m is independent of v. For instance, this occurs if $F(x, q)$ does not depend on q. Indeed, we have then

$$q'(x) = F(x) - m, \qquad q(0) = v,$$

$$m = T^{-1}\int_0^T F(x)\, dx, \text{ independent of } v.$$

(b) Both $q(x)$, $0 \le x \le T$, and m are continuous functions of v.

Proof of (b). Note that $m = T^{-1}\int_0^T F(x, q(x))\, dx$, and, therefore, if $q(x)$ depends continuously on v (in the uniform topology), then m too depends continuously on v. Note that $|F| \le L$, $|m| \le L$, $|q'(x)| \le 2L$, hence the functions $q(x)$ are uniformly Lipschitzian with constant $2L$, and hence uniformly continuous in $[0, T]$. Finally the relation $q'(x) = F(x, q(x)) - m$ shows that also the derivatives $q'(x)$ are uniformly continuous in $[0, T]$.

If v_1, v_2 denote two distinct vectors v, and $m_1, m_2, q_1(x), q_2(x)$ the corresponding vectors m and $q(x)$, let $d = |v_1 - v_2|$, and $\mu = \max |q_1(x) - q_2(x)|$, where the maximum is taken for $0 \le x \le T$. Then, we have successively, as in (*a*),

$$|m_1 - m_2| = |T^{-1}\int_0^T [F(x, q_1(x)) - F(x, q_2(x))]\, dx| \le b\mu,$$

$$|q_1(x) - q_2(x)| = |v_1 - v_2 + \int_0^x [F(s, q_1(s)) - F(s, q_2(s)) - m_1 + m_2]\, ds$$

$$\le d + 2bT\mu.$$

For some x in $[0, T]$ we have $|q_1(x) - q_2(x)| = \mu$, and hence

$$\mu \le d + 2bT\mu,$$

$$\mu \le (1 - 2bT)^{-1}\, d.$$

Thus $q(x)$ depends continuously on v, and, as a consequence, also m depends continuously on v. Thereby statement (b) is proved.

We shall use the notations $q(x, v)$, $m(v)$ to emphasize the dependence of q on x and v, and of m on v. Also, we shall use the notations

$$q' = \frac{\partial q}{\partial x}, \qquad q_v = \frac{\partial q}{\partial v}, \qquad m_v = \frac{dm}{dv},$$

whenever these derivatives exist.

(c) If $n = 1$, and F is of class C^1 in x and q, then the derivatives $q'(x, v)$, $q_v(x, v)$, $q_v'(x, v)$, $m_v(v)$ exist, are continuous in their arguments, and

$$\frac{dm}{dv} = \left[1 - \exp\left(-\int_0^T F_q dx\right)\right]\left[\int_0^T dx \exp\left(-\int_0^x F_q dx\right)\right]^{-1}. \qquad (9.6)$$

Proof of (c). First let us denote by $Q(x) = Q(x, v, m)$ the solution $Q(x)$ of the differential problem

$$q' = F(x, q) - m, \qquad q(0) = v,$$

when m and v are arbitrary independent parameters. Then

$$q(x, v) = Q(x, v, m(v)).$$

By standard theorems on differential equations, the partial derivatives Q', Q_v, Q_m, Q_v', Q_m' exist, are continuous and

$$Q_m'(x) = F_q(x, Q(x))\, Q_m(x) - 1, \qquad Q_m(0) = 0$$

$$Q_v'(x) = F_q(x, Q(x))\, Q_v(x), \qquad Q_v(0) = 1,$$

and hence, by straightforward integration,

$$Q_v(x) = \exp\left[\int_0^x F_q(x, Q(x))\, dx\right],$$

$$Q_m(x) = -\exp\left[\int_0^x F_q(x, Q(x))\, dx\right] \int_0^x dx \exp\left[-\int_0^x F_q(x, Q(x))\, dx\right].$$

Now we have

$$m(v) = T^{-1} \int_0^T F(x, q(x, v))\, dx$$

$$= T^{-1} \int_0^T F(x, Q(x, v, m(v)))\, dx,$$

and, to an increment Δv of v there corresponds an increment Δm of m, so that

$$\Delta m = T^{-1}\int_0^T [F(x, Q(x, v+\Delta v, m+\Delta m)) - F(x, Q(x, v, m))]\,dx$$

$$= T^{-1}\int_0^T F_q(x, Q+\theta\Delta Q)\,[Q(x, v+\Delta v, m+\Delta m) - Q(x, v, m)]\,dx$$

$$= T^{-1}\int_0^T F_q(x, Q+\theta\Delta Q)\,[Q_v(x, v+\theta'\Delta v, m+\theta'\Delta m)\,\Delta v$$
$$+ Q_m(x, v+\theta'\Delta v, m+\theta'\Delta m)\,\Delta m]\,dx$$

where

$$\Delta Q = Q(x, v+\Delta v, m+\Delta m) - Q(x, v, m),$$

where $\theta = \theta(x)$, $\theta' = \theta'(x)$, $0 < \theta < 1$, $0 < \theta' < 1$, and we know already that $\Delta m \to 0$ as $\Delta v \to 0$, and hence $\Delta Q \to 0$, and Q_r, Q_m are continuous functions of their arguments. Thus, we have

$$\Delta m\left[1 - T^{-1}\int_0^T F_q(x, Q+\theta\Delta Q)\,Q_m(x, v+\theta'\Delta v, m+\theta'\Delta m)\,dx\right]$$
$$= \Delta v\cdot T^{-1}\int_0^T F_q(x, Q+\theta\Delta Q)\,Q_v(x, v+\theta'\Delta v, m+\theta'\Delta m)\,dx. \qquad (9.7)$$

As $\Delta v \to 0$, the coefficient of Δm approaches

$$1 - T^{-1}\int_0^T F_q(x, Q)\,Q_m(x, v, m)\,dx$$
$$= 1 + T^{-1}\int_0^T F_q dx \exp\left(\int_0^x F_q dx\right)\int_0^x dx \exp\left(-\int_0^x F_q dx\right)$$
$$= T^{-1}\exp\left(\int_0^T F_q dx\right)\int_0^T dx \exp\left(-\int_0^x F_q dx\right),$$

where the last equality is proved by integrating by parts. Thus the coefficient of Δm in (9.10) has a limit as $\Delta v \to 0$ which is certainly $\neq 0$. The coefficient of Δv in (9.10), as $\Delta v \to 0$, approaches

$$T^{-1}\int_0^T F_q Q_v dx = T^{-1}\int_0^T F_q \exp\left(\int_0^x F_q dx\right) = T^{-1}\left[\exp\left(\int_0^T F_q dx\right) - 1\right].$$

Thus (9.7) implies that dm/dv exists, and is given by (9.6). Hence, dm/dv is a continuous function of v. Finally, the relation $q(x, v) = Q(x, v, m(v))$ shows that the derivatives q', q_v, q'_v exist and are continuous functions of x and v. Thereby, statement (c) is proved.

10. Theorem III (Continuous dependence upon the boundary values). Under the conditions of Theorems I and II ($\omega_3(\gamma) = M_0 \mid \gamma \mid$), the unique solution $\phi(x, y)$, $m(y)$ of (1.6-1.9) depends continuously on $u_0(x)$ and $v_0(y)$.

Proof. We shall prove this theorem for a strip

$$A' = [0 \leq x \leq T, \qquad 0 \leq y \leq c],$$

of width c sufficiently small, $c = a/k$, k an integer. By repeating this reasoning in each strip $(0 \leq x \leq T, nc \leq y \leq (n + 1)\,c$, $n = 1, 2, \ldots, k - 1$, $n = -1, -2, \ldots, -k$, we will be able to conclude the continuous dependence of $\phi(x, y)$, $(x, y) \in A$, on $u_0(x)$, $v_0(y)$.

Let $u_{01}(x)$, $u_{02}(x)$, $0 \leq x \leq T$, $v_{01}(y)$, $v_{02}(y)$, $0 \leq y \leq c$, be two functions as in Theorems I and II, let $\phi_1(x, y)$, $\phi_2(x, y)$, $(x, y) \in A$, be the relative solutions of (1.6-1.9), and $m_1(y)$, $m_2(y)$, $-a \leq y \leq a$, the corresponding functions m. As in the proof of Theorem II, let

$$\epsilon = \max \mid u_{01}(x) - u_{02}(x) \mid + \max \mid u'_{01}(x) - u'_{02}(x) \mid$$
$$+ \max \mid v_{01}(y) - v_{02}(y) \mid + \max \mid v'_{01}(y) - v'_{02}(y) \mid,$$

and

$$\alpha = \max \mid \phi_1(x, y) - \phi_2(x, y) \mid,$$
$$\beta = \max \mid \phi_{1x}(x, y) - \phi_{2x}(x, y) \mid,$$
$$\gamma = \max \mid \phi_{1y}(x, y) - \phi_{2y}(x, y) \mid,$$
$$\delta = \max \mid m_1(y) - m_2(y) \mid,$$
$$\eta = \alpha + \beta + \gamma,$$

where max is taken for $0 \leq x \leq T, 0 \leq y \leq c$.

As in the proof of Theorem II we have

$$\mid m_1(y) - m_2(y) \mid \leq M_0\alpha + M\beta + b\gamma.$$

Analogously, we have

$$\phi_1(x, y) - \phi_2(x, y) \mid = \mid u_{01}(x) - u_{02}(x) + v_{01}(y) - v_{02}(y)$$
$$+ \int_0^x \int_0^y [f(x, y, \phi_1(x, y), \phi_{1x}(x, y), \phi_{1y}(x, y))$$
$$- f(x, y, \phi_2(x, y), \phi_{2x}(x, y), \phi_{2y}(x, y)) - m_1(y) + m_2(y)]\, dx\, dy$$
$$\leq 2\epsilon + Tc(M_0\alpha + M\beta + b\gamma + \delta),$$

and

$$|\phi_{1x}(x, y) - \phi_{2x}(x, y)| \leq \epsilon + c(M_0\alpha + M\beta + b\gamma + \delta),$$
$$|\phi_{1y}(x, y) - \phi_{2y}(x, y)| \leq \epsilon + T(M_0\alpha + M\beta + b\gamma + \delta).$$

Thus

$$\alpha \leq 2\epsilon + TM_0c\alpha + TMc\beta + Tbc\gamma + Tc\delta,$$
$$\beta \leq \epsilon + M_0c\alpha + Mc\beta + bc\gamma + c\delta,$$
$$\gamma \leq \epsilon + TM_0 + TM\beta + Tb\gamma + T\delta,$$
$$\delta \leq M_0\alpha + M\beta + b\gamma,$$

and, as in the proof of Theorem II, we have successively

$$\alpha \leqslant 2\epsilon + 2TM_0c\alpha + 2TMc\beta + 2Tbc\gamma,$$
$$\beta \leqslant \epsilon + 2M_0c\alpha + 2Mc\beta + 2bc\gamma,$$
$$\gamma \leqslant \epsilon + 2TM_0\alpha + 2TM\beta + 2Tb\gamma,$$

and then, for $k = (1 - 2Tb)^{-1}$, $k' = 1 + 2kTb$, we have

$$\gamma \leqslant k\epsilon + 2kTM_0\alpha + 2kTM\beta,$$
$$\alpha \leqslant (2 + 2Tbc)\,\epsilon + 2TM_0k'c\alpha + 2TMk'c\beta,$$
$$\beta \leqslant (1 + 2bc)\,\epsilon + 2M_0k'c\alpha + 2Mk'c\beta.$$

The same restrictions already used,

$$2Mk'c \leqslant \tfrac{1}{2}, \qquad 2TM_0k'(1 + 4k'Ma)\,c \leqslant \tfrac{1}{2},$$

give successively,

$$\beta \leqslant 2(1 + 2bc)\,\epsilon + 4M_0k'c\alpha,$$
$$\alpha \leqslant 2(1 + Tbc + 2TMk'c(1 + 2bc))\,\epsilon + 2TM_0k'(1 + 4k'Ma)c\alpha,$$

and finally

$$\alpha \leqslant 4(1 + Tbc + TMk'c(1 + 2bc))\,\epsilon.$$

Thus, we have $\eta = \alpha + \beta + \gamma \leqslant K\epsilon$, where K is a constant. Hence $\eta \to 0$ as $\epsilon \to 0$. Since we can repeat this reasoning in the successive strips in which we have divided A, the continuity of the dependence of $\phi(x, y)$, $\phi_x(x, y)$, $\phi_y(x, y)$, $(x, y) \in A$, on $u_0(x)$, $u_0'(x)$, $0 \leqslant x \leqslant T$, $v_0(y)$, $v_0'(y)$, $-a \leq y \leq a$, is proved.

11. In the previous § we have proved that, given $u_0(x)$, $0 \leqslant x \leqslant T$, periodic of period T, and $v_0(y)$, $-a \leqslant y \leqslant a$, both as in Theorem III, then there is a unique function $m(y)$, $-a \leqslant y \leqslant a$, and a unique vector function $\phi(x, y)$, $(x, y) \in A$, periodic in x of period T, such that

$$\phi_{xy}(x, y) = f(x, y, \phi, \phi_x, \phi_y) - m(y), \qquad m(y) = T^{-1} \int_0^T f(x, y, \phi, \phi_x, \phi_y)\, dx.$$

In addition m and ϕ depend continuously upon u_0, v_0, u_0', v_0'. The question now arises as to whether it is possible to choose $u_0(x)$ and $v_0(y)$ in such a way that $m(y)$, $-a \leqslant y \leqslant a$, has preassigned values, or that $m(y) = 0$.

This problem has some formal analogies with the one which has been met in the study we have mentioned [3, 5] of periodic solutions of ordinary nonlinear differential equations. This problem will be discussed in a successive paper. Below, we mention two situations where $m(y) = 0$.

(a) Suppose that f, u_0, etc. are defined for all $-\infty < x < +\infty$ and periodic of period T in x. Suppose that $f(-x, y, u, -p, q) = -f(x, y, u, p, q)$ and that $u_0(-x) = u_0(x)$. Then $m(y) = 0$ for all y.

To prove (a), let us restrict K to contain only functions $\phi(x, y)$ with $\phi(-x, y) = \phi(x, y)$. Then $\phi_x(-x, y) = -\phi_x(x, y)$, $\phi_y(-x, y) = \phi_y(x, y)$. If $F(x, y) = f(x, y, \phi(x, y), \phi_x(x, y), \phi_y(x, y))$, then we have, for $\phi \in K$,

$$\begin{aligned} F(-x, y) &= f(-x, y, \phi(-x, y), \phi_x(-x, y), \phi_y(-x, y)) \\ &= f(-x, y, \phi(x, y), -\phi_x(x, y), \phi_y(x, y)) = -F(x, y). \end{aligned}$$

Hence,

$$m(y) = T^{-1} \int_0^T F(x, y)\, dx = 0.$$

Now we have, for $\phi \in K$,

$$\psi(x, y) = u_0(x) + v_0(y) + \int_0^x \int_0^y F(x, y)\, dx\, dy,$$

$$\psi_x(x, y) = u_0'(x) + \int_0^y F(x, y)\, dy,$$

$$\psi_y(x, y) = v_0'(y) + \int_0^x F(x, y)\, dx,$$

and hence

$$\psi(-x, y) = \psi(x, y), \qquad \psi_x(-x, y) = -\psi_x(x, y), \qquad \psi_y(-x, y) = \psi_y(x, y).$$

Thus, $\mathscr{T} : K \to K$, and the proof of Theorem I remains unchanged.

(b) Hyperbolic systems of the form

$$\begin{aligned} u_{ix} &= f_i(x, y, u), \qquad i = 1, \dots, \mu, \\ u_{iy} &= f_i(x, y, u), \qquad i = \mu + 1, \dots, n, \end{aligned} \tag{11.1}$$

for some n and $1 < \mu < n$, and where, as usual, $u = u(x, y) = (u_1, \dots, u_n)$, are often studied. Systems (11.1) are then reduced to systems (0.1) by partial differentiation with respect to x, or y, and one obtains,

$$u_{ixy} = F_i(x, y, u, u_x, u_y), \qquad i = 1, \dots, n, \tag{11.2}$$

where

$$F_i = f_{iy} + \sum_{j=1}^{n} f_{iu_j} u_{jy}, \qquad F_i = f_{ix} + \sum_{j=1}^{n} f_{iu_j} u_{jx}, \tag{11.3}$$

for $i = 1, \dots, \mu$, and for $i = \mu + 1, \dots, n$, respectively. We shall suppose that system (11.2) satisfies the conditions of Theorem I. If $m(y) = (m_1, \dots, m_n)$ is the corresponding vector function $m(y)$, we state that

$$m_i(y) = 0, \qquad i = \mu + 1, \dots, n, \tag{11.4}$$

for all y.

To prove (11.4) let us consider again the transformation $\mathscr{T}$, and suppose that $\phi = (\phi_1, \dots, \phi_n) \in K$, $\psi = (\psi_1, \dots, \psi_n) \in K$, $\psi = \mathscr{T}\phi$. For $i = \mu + 1, \dots, n$, we have

$$m_i(y) = T^{-1} \int_0^T F_i(x, y, \phi, \phi_x, \phi_y)\, dx = T^{-1} \int_0^T (d/dx) f_i(x, y, \phi)\, dx = 0,$$

the last integral being identically zero because of the periodicity in x of period T.

Considerations analogous to the ones above hold, by a convenient rewording, for systems of the form

$$\sum_{k=1}^{n} a_{ik}(x, y, u)\, u_{kx} - b_i(x, y, u) = 0, \qquad i = 1, \dots, \mu,$$

$$\sum_{k=1}^{n} b_{ik}(x, y, u)\, u_{ky} - b_i(x, y, u) = 0, \qquad i = \mu + 1, \dots, n,$$

studied by H. Lewy [*9*], K. O. Friedrichs [*4*], P. Hartman and A. Wintner [*6*], and others.

12. We shall need the following statement

(a) Under the conditions of Theorems I, II, III, if $n = 1$, if $f(x, y, z, p, q)$ is of class C^1 and $u_0(x)$, $v_0(y)$ of class C^2, then $m(y)$ is of class C^1, and $\phi(x, y)$ of class C^2.

Proof of (a). For every $(x, y) \in A$ we have

$$\phi(x, y) = u_0(x) + v_0(y) + \int_0^x \int_0^y [f(x, y, \phi, \phi_x, \phi_y) - m(y)]\, dx\, dy, \tag{12.1}$$

$$\phi_x(x, y) = u_0'(x) + \int_0^y [f(x, y, \phi, \phi_x, \phi_y) - m(y)]\, dy, \tag{12.2}$$

$$\phi_y(x, y) = v_0'(y) + \int_0^x [f(x, y, \phi, \phi_x, \phi_y) - m(y)]\, dx, \tag{12.3}$$

$$m(y) = T^{-1} \int_0^T f(x, y, \phi, \phi_x, \phi_y)\, dx, \tag{12.4}$$

$$\phi_{xy}(x, y) = f(x, y, \phi, \phi_x, \phi_y) - m(y). \tag{12.5}$$

All these functions are continuous, and we gave in § 5 the modulus of continuity of ϕ, ϕ_x, ϕ_y, m. In the present situation, f is of class C^1 and so are u_0, u_0', v_0, v_0'. Hence, all these functions are Lipschitzian, the functions ω_j of § 5 can be taken as linear, and hence ϕ, ϕ_x, ϕ_y are Lipschitzian, and so are ϕ_{xy}, and m. As a consequence, $m(y)$ is an absolutely continuous function of y, whose derivative $m'(y)$ exists for almost all y, and is bounded, and L-integrable. Also, $\phi_y(\bar{x}, y)$ is absolutely continuous in y for every $\bar{x}$, and $\phi_{yy}(\bar{x}, y)$ exists for almost all y and every $\bar{x}$, and analogously, $\phi_x(x, \bar{y})$ is absolutely continuous in x for every $\bar{y}$, and $\phi_{xx}(x, \bar{y})$ exists for almost all x and every $\bar{y}$.

Actually, the lower and upper partial derivatives $\underline{\phi}_{xx}$, $\bar{\phi}_{xx}$, $\underline{\phi}_{yy}$, $\bar{\phi}_{yy}$ are B-measurable [10, p. 170], hence the set Ω_1 where $\bar{\phi}_{xx} > \underline{\phi}_{xx}$ is B-measurable, and since Ω_1 intersects every line $y = \bar{y}$ in a set of linear measure zero, Ω_1 is of measure zero in A. The same for the set Ω_2 where $\bar{\phi}_{yy} > \underline{\phi}_{yy}$. Finally, the set $\Omega = \Omega_1 \cup \Omega_2$ has measure zero in A, and ϕ_{xx}, ϕ_{yy} both exist everywhere in $A \setminus \Omega$. The function f is of class C^1, hence differentiable everywhere. The derivatives ϕ_x, ϕ_y, ϕ_{xy} exist everywhere in A. At every point $(x, y) \in A \setminus \Omega$, we can apply the elementary reasoning for the existence of the derivatives of composite functions to

$$F(x, y) = f(x, y, \phi(x, y), \phi_x(x, y), \phi_y(x, y)).$$

We then prove that the partial derivatives F_x, F_y exist at every point $(x, y) \in A \setminus \Omega$, and that

$$F_x = f_x + f_z\phi_x + f_p\phi_{xx} + f_q\phi_{xy}, \qquad F_y = f_y + f_z\phi_y + f_p\phi_{xy} + f_q\phi_{yy}.$$

Also F_x, F_y are B-measurable, and bounded, say $|F_x|, |F_y| \leq M$, hence L-integrable in A. Hence, $F_y(x, \bar{y})$ is an L-integrable, function of x for all $\bar{y} \in [-a, a]$ but for a set I_0 of measure zero of numbers $\bar{y}$, $-a \leq \bar{y} \leq a$.

By Fubini's theorem, $F_1(y) = \int_0^T F_y(x, y)\,dx$ is L-integrable in $[-a, a]$, and a.e. the derivative of its indefinite integral, say, for all $\bar{y}$ but a set of measure zero. Analogously, for every rational r, $0 \leq r \leq T$, $F_r(y) = \int_0^r F_y(x, y)\,dx$ is L-integrable in $[-a, a]$, and the derivative of its indefinite integral for every $y = \bar{y}$ of $[-a, a]$ but a set I_r of measure zero. Let I be the union of I_0 and of all I_r, r rational, $0 \leq r \leq T$. Then I has measure zero. Let us prove that, for every $y = \bar{y} \in [-a, a] \setminus I$, and for every real c, $0 \leq c \leq T$, $F_c(y) = \int_0^c F_y(x, y)\,dx$, $a \geq y \leq a$, is the derivative of its indefinite integral. Indeed, given $\eta > 0$ arbitrary, take a rational r, $0 \leq r \leq T$, with $M\,|c - r| < \eta$. Also take $\delta > 0$, such that for $y = \bar{y} + h$, $|h| < \delta$, we have

$$\left| h^{-1} \int_y^{y+h} dy \int_0^r F_y dx - \int_0^r F_y dx \right| < \eta.$$

Then for $|h| < \delta$, we have, by Fubini's theorem,

$$\left| h^{-1} \int_y^{y+h} dy \int_0^c F_y dx - \int_0^c F_y dx \right|$$

$$\leq \left| h^{-1} \int_y^{y+h} \int_r^c F_y dx\, dy \right| + \left| h^{-1} \int_y^{y+h} dy \int_0^r F_y dy - \int_0^r F_y dy \right| + \left| \int_r^c F_y dy \right|$$

$$\leq M\,|c - r| + \eta + M\,|c - r| \leq 3\eta.$$

This proves that, for $y = \bar{y} \in [-a, a] \setminus I$, and every c, $F_c(y) = \int_0^c F_y\,dy$ is the derivative of its indefinite integral. Analogously, there is a set $I' \subset [0, T]$ of measure zero, such that, for every $x = \bar{x} \in [0, T] \setminus I'$, and every d, $-a \leq y \leq a$, $F_d(x) = \int_0^d F_x\,dx$ is the derivative of its indefinite integral.

By a Tonelli's theorem on differentiation under the sign of L-integral [*12*] we conclude that, for all $\bar{y} \in [-a, a] \setminus I$, we have

$$\begin{aligned} m'(\bar{y}) &= T^{-1} \int_0^T (f_y + f_z \phi_y + f_p \phi_{xy} + f_q \phi_{yy})\,dx \\ &= T^{-1} \int_0^T [A(x, \bar{y}) + B(x, \bar{y})\,\phi_{yy}(x, \bar{y})]\,dx. \qquad (12.6) \end{aligned}$$

Also, for the same $\bar{y}$, and every x, we have by (12.3)

$$\phi_{yy}(x, \bar{y}) = v_0''(y) + \int_0^x [A(x, \bar{y}) + B(x, \bar{y})\, \phi_{yy}(x, \bar{y}) - m'(\bar{y})]\, dx, \tag{12.7}$$

where the integrand is a bounded L-integrable function x. Thus $\phi_{yy}(x, \bar{y})$ is absolutely continuous in x, even Lipschitzian (with constant independent of $\bar{y}$), and $\phi_{yy}(x, \bar{y})$, $0 \leq x \leq T$, satisfies, in the sense of Carathéodory [2], the linear differential equation with initial condition

$$\begin{aligned} (d/dx)\, \phi_{yy}(x, \bar{y}) &= [A(x, \bar{y}) - m'(\bar{y})] + B(x, \bar{y})\, \phi_{yy}(x, \bar{y}), \\ \phi_{yy}(0, \bar{y}) &= v_0''(\bar{y}). \end{aligned} \tag{12.8}$$

This proves that for the same $\bar{y}$ and all $0 \leq x \leq T$, we have

$$\begin{aligned} \phi_{yy}(x, \bar{y}) &= \exp\left[\int_0^x B(x, \bar{y})\, dx\right] \\ &\times \left[v_0''(\bar{y}) + \int_0^x [A(x, \bar{y}) - m'(\bar{y})] \exp\left[-\int_0^x B(x, \bar{y})\, dx\right] dx\right]. \end{aligned} \tag{12.9}$$

By replacing (12.9) in (12.6) and some computations we obtain

$$\begin{aligned} & m'(\bar{y}) \cdot \int_0^T dx \exp\left[-\int_0^x B(x, \bar{y})\, dx\right] \\ &= \left[1 - \exp\left(-\int_0^T B(x, \bar{y})\, dx\right)\right] v_0''(\bar{y}) + \int_0^T A(x, \bar{y})\, dx \exp\left[-\int_0^x B(x, \bar{y})\, dx\right]. \end{aligned} \tag{12.10}$$

The function $v''(y)$ is continuous, the functions $A(x, y)$, $B(x, y)$ are given in terms of ϕ, ϕ_x, ϕ_{xy}, f, f_y, f_z, f_p, f_q, all continuous in their arguments, and then $A(x, y)$, $B(x, y)$ are continuous functions of x, y in A. Relation (12.10) shows that $m'(\bar{y})$ coincides for almost all $\bar{y}$ with a continuous function of y. Thus $m(y)$ has continuous derivative $m'(y)$ given by (12.10) for *all* $y = \bar{y}$. By (12.9) now we deduce that $\phi_{yy}(x, y)$ also coincides with a continuous function of x, y, in A for almost all $y = \bar{y}$. By usual arguments we conclude that $\phi_{yy}(x, y)$ exists for all $(x, y) \in A$ and is a continuous function in A. Analogously, for almost all $\bar{x}$, we can differentiate under the sign of integral in (12.2) and we have, for all $-a \leq \bar{x} \leq a$,

$$\begin{aligned} \phi_{xx}(\bar{x}, y) &= u_0''(x) + \int_0^y [f_x + f_z\phi_x + f_p\phi_{xx} + f_q\phi_{xy}]\, dy \\ &= u_0''(x) + \int_0^y [C(\bar{x}, y) + D(\bar{x}, y)\, \phi_{xx}(\bar{x}, y)]\, dy, \end{aligned} \tag{12.11}$$

where the integrand is a bounded L-integrable function of y, $-a \leq y \leq a$. Thus $\phi_{xx}(\bar{x}, y)$ is absolutely continuous in y, and even Lipschitzian with constant independent of $\bar{x}$, satisfying, in the sense of Carathéodory, the linear differential equation with initial condition

$$\begin{aligned} (d/dy)\, \phi_{xx}(\bar{x}, y) &= C(\bar{x}, y) + D(\bar{x}, y)\, \phi_{xx}(\bar{x}, y), \\ \phi_{xx}(\bar{x}, 0) &= u_0''(\bar{x}). \end{aligned} \tag{12.12}$$

This proves that, for the same $\bar{x}$ and all $0 \leq x \leq T$, we have

$$\begin{aligned} \phi_{xx}(\bar{x}, y) = \exp\left(\int_0^y D(\bar{x}, y)\, dy\right) \\ \times \left\{u_0''(\bar{x}) + \int_0^y \left[C(\bar{x}, y) \exp\left(-\int_0^y D(\bar{x}, y)\, dy\right)\right] dy\right\}. \end{aligned} \tag{12.13}$$

The function $u_0''(x)$ is continuous, the functions $C(x, y)$, $D(x, y)$ are given in terms of ϕ, ϕ_x, ϕ_{xy}, f, f_x, f_z, f_p, f_q all continuous in their arguments, and then $C(x, y)$, $D(x, y)$ are continuous functions of x, y in A. Relation (12.13) shows that $\phi_{xx}(\bar{x}, y)$ coincides, for almost all $\bar{x}$, with a continuous function in A. By usual arguments we conclude that $\phi_{xx}(x, y)$ exists for all (x, y) in A, and is a continuous function in A.

13. Remark. The uniqueness statement (a) of § 9 concerning the differential problem (9.1), (9.2), has an analogue which deserves attention.

Given a real vector function $F(x, q)$ as in § 9, and a real vector $a = (a_1, a_2, \ldots, a_n)$, let us consider a vector function $q(x)$, periodic in x of period T, satisfying the differential problem (in the large):

$$\begin{gathered} q'(x) = F(x, q(x)) - m, \qquad 0 \leq x \leq T, \\ T^{-1} \int_0^T q(x)\, dx = a, \\ T^{-1} \int_0^T F(x, q(x))\, dx = m. \end{gathered} \tag{13.1}$$

The following uniqueness statement holds:

(a′) Both $q(x)$, $0 \leq x \leq T$, and m, satisfying (13.1), are uniquely determined by the constant a and the function $F(x, q)$.

The proof is the same as the one given for statement (a) in § 9.

In addition to statement (a′), also theorems of continuous dependence upon the constant vector v, or the constant vector a, can be given. The proofs are similar to (b) and (c).

REFERENCES

[*1*] Alexiewicz, A. and Orlicz, W., Some remarks on the existence and uniqueness of solutions of the hyperbolic equation $z_{xy} = f(x, y, z, z_x, z_y)$. *Studia Math.* **15**, 201-215 (1956).

[*2*] Carathéodory, C., "Vorlesungen über reelle Funktionen." Teubner, Leipzig, 1918.

[*3*] (a) Cesari, L., Existence theorems for periodic solutions of nonlinear Lipschitzian differential systems and fixed point theorems. *In* "Contributions to the Theory of Nonlinear Oscillations" (S. Lefschetz, ed.), Vol. 5, pp. 115-172. Princeton Univ. Press, Princeton, New Jersey, 1960.

(b) Cesari, L., Existence theorems for periodic solutions of nonlinear differential systems. *Bol. soc. mat. mex.* pp. 24-41 (1960).

(c) Cesari, L., Periodic solutions of nonlinear differential systems. *Proc. Symposium Polytech. Inst. Brooklyn* **10**, 545-560 (1960).

(d) Cesari, L., Functional Analysis and Periodic Solutions of Nonlinear Differential Equations. Wisconsin Math. Research Center Report No. 173 (1960).

[*4*] Friedrichs, K. C., Nonlinear hyperbolic differential equations for functions of two independent variables. *Am. J. Math.* **70**, 555-589 (1946).

[*5*] (a) Hale, J. K., On the stability of periodic solutions of weakly nonlinear periodic and autonomous differential systems. *In* "Contributions to the Theory of Nonlinear Oscillations" (S. Lefschetz, ed.), Vol. 5, pp. 91-114, Princeton Univ. Press, Princeton, New Jersey, 1960.

(b) Hale, J. K., and Gambill, R. A., Subharmonic and ultraharmonic solutions for weakly nonlinear systems. *J. rational Mech. and Anal.* **5**, 353-394 (1956).

[*6*] Hartman, P. and Wintner, A., On hyperbolic partial differential equations. *Am. J. Math.* **74**, 834-864 (1952).

[*7*] Kamke, E., "Differentialgleichungen reeller Funktionen." Akad. Verlagsges., Leipzig, 1930.

[*8*] Lefschetz, S., "Lectures on Differential Equations." Princeton Univ. Press, Princeton, New Jersey, 1948.

[*9*] Lewy, H., Über Anfangswertprobleme einer hyperbolischen nichtlinearen partiellen Differentialgleichung. *Math. Ann. (Leipzig)* **98**, 179-191 (1927).

[*10*] Saks, S., "Theory of the Integral." Warszawa, 1937.

[*11*] (a) Schauder, J. P., Zur Theorie stetiger Abbildungen in Funktionalräumen. *Math. Z.* **26**, 47-65, 417-431 (1927).

(b) Schauder, J. P., Der Fixpunktsatz in Funktionalräumen. *Studia Math.* **2**, 171-190 (1930).

[*12*] Tonelli, L., Su la continuità e la derivabilità di un integrale rispetto ad un parametro. *Rend. Accad. Lincei* **19**, 84-89 (1910); also, Tonelli, L., *Opere Scelte* **1**, 166-172 (1960).

Coupling in Certain Classes of Weakly Nonlinear Vibrating Systems[1]

P. R. SETHNA
Department of Aeronautics and Engineering Mechanics
University of Minnesota, Minneapolis, Minnesota

Introduction

Vibrations of dynamical systems of multiple-degrees of freedom for motions in the neighborhood of stable static equilibrium have been studied for a long time. In the classical theory [*7*] the analysis is restricted to values of the coordinates in the immediate neighborhood of an equilibrium configuration. The differential equations describing the motion are in general linear with constant coefficients and due to the relative simplicity of these equations a considerable amount of information is available for motions of this type. A problem that naturally arises is the one where the motions are considered to be in a sufficiently large neighborhood of the equilibrium configuration to make the first nonlinear terms of the Taylor expansion of the differential equations of motions significant. This problem can often be formulated as a system of ordinary differential equations which are linear with constant coefficients and with small nonlinear perturbation terms. Among the several methods of analysis that are available for such systems, the "Method of Averaging" of Bogoliuboff and Mitropol'skiy [*1*] has proved very useful especially if the solutions are not restricted to periodic solutions.

A restricted class of dynamical systems for motions of the type mentioned above will be studied here. Included in the analysis will be nonlinear effects arising from both the kinetic and potential energies of the system, and furthermore the system will be assumed to be acted upon by dissipative forces and forces explicitly dependent on time. The analysis will be concerned with both periodic and nonperiodic solutions.

Russian authors, notably Bogoliuboff and Mitropol'skiy [*1*], [*3*], have dealt with this general type of problem. They, however, discuss for the

[1] The author wishes to express his appreciation of the support received from the Aeronautical Systems Division, Air Force Systems Command, during the preparation of this work.

most part what are called "monofrequent oscillations" [3, p. 143], that is, oscillations that are predominantly near one or the other of the linear modes of motion. Such oscillations arise naturally and are of considerable interest in certain types of nonautonomous systems and certain regenerative autonomous systems. It is felt, however, that for the types of systems studied here more general solutions should be investigated. The problem will therefore be studied, wherever possible, for arbitrary initial conditions. It will be shown that the "averaged equations" have various types of coupling which can be classified into four distinct types. Solutions of the averaged equations will be studied. In several cases they are periodic with periods that are much longer than the periods of any of the linear modes. These long-period solutions can be either unisolated or isolated and are of interest both from mathematical as well as engineering viewpoints.

Nature of the Systems under Consideration

Consider a holonomic dynamical system with N degrees of freedom with generalized coordinates η_i, $i = 1, 2, \ldots, N$. The constraints of the system will be assumed to be independent of time, so that the kinetic energy T can be expressed thus:

$$T = \frac{1}{2} \sum_{i,j=1}^{N} m_{ij} \dot{\eta}_i \dot{\eta}_j; \tag{1}$$

where $m_{ij} = m_{ij}(\eta_1, \eta_2, \ldots, \eta_N)$ and $m_{ij} = m_{ji}$. A dot above a symbol will mean derivative with respect to time. Let some of the forces of the system be conservative and derivable from a potential function

$$V = V(\eta_1, \eta_2, \ldots, \eta_N). \tag{2}$$

The coordinates will be assumed to be measured from a stable static equilibrium configuration so that

$$\left(\frac{\partial V}{\partial \eta_i}\right)_{\eta_1=\eta_2=\ldots=\eta_N=0} \equiv V_{,i}^{0} = 0, \qquad i = 1, 2, \ldots, N. \tag{3}$$

We introduce a damping function

$$D = \frac{1}{2}\sum_{i,j=1}^{N} d_{ij}\dot{\eta}_i\dot{\eta}_j + \frac{1}{3}\sum_{i,j,k=1}^{N} d_{ijk}\dot{\eta}_i\dot{\eta}_j\dot{\eta}_k + \frac{1}{4}\sum_{i,j,k,l=1}^{N} d_{ijkl}\dot{\eta}_i\dot{\eta}_j\dot{\eta}_k\dot{\eta}_l \tag{4}$$

where the quantities d_{ij}, d_{ijk}, d_{ijkl} are completely symmetric and functions of the coordinates.

The Lagrange's Equations of motion for the system will be taken as

$$\frac{d}{dt}\left(\frac{\partial T}{\partial \dot{\eta}_i}\right) - \frac{\partial T}{\partial \eta_i} + \frac{\partial V}{\partial \eta_i} + \frac{\partial D}{\partial \dot{\eta}_i} = \bar{Q}_i(t) \tag{5}$$

where $\bar{Q}_i(t)$ are external forces acting on the system.

For systems of this type it is possible to solve equations (5) for the second derivatives of the coordinates [7, p. 39] and we get

$$\ddot{\eta}_r + \sum_{p=1}^{N} M_{pr} \left\{ \sum_{i,j=1}^{N} [ij, p]\, \dot{\eta}_i \dot{\eta}_j + \frac{\partial V}{\partial \eta_p} + \frac{\partial D}{\partial \dot{\eta}_p} - \bar{Q}_p \right\} = 0 \tag{6}$$

where M_{pr} is a matrix related to m_{pr} in (1) and $[i\,j, p]$ are the Christoffel symbols of the first kind [6].

For motion in the neighborhood of static equilibrium, the classical theory of linear vibrations is obtained by expanding each term in (6) in a Taylor series and by retaining in the analyses only the linear terms in the coordinates and their derivatives. The resulting equations are linear with constant coefficients. Here again motions in the neighborhood of static stable equilibrium will be considered but included in the analysis will be the first significant nonlinear term in the Taylor expansion of the terms in (6). It can be shown that unless cubic terms are retained in this expansion, except for certain special cases, the physically interesting results are not obtained in the first approximation of the analytical procedure which will be employed here. Terms up through the cubic terms will therefore be included and the quadric terms will be assumed to be zero for simplicity.

Let

$$\eta_i = \epsilon^{1/2} \xi_i, \qquad i = 1, 2, \ldots, N, \qquad 0 < \epsilon \ll 1 \tag{7}$$

and

$$\bar{Q}_i(t) = \epsilon^{3/2} Q(t), \qquad i = 1, 2, \ldots, N. \tag{8}$$

Furthermore, in order to remove the coupling in (6) due to linear terms arising from T and V, let

$$\xi_i = \sum_{j=1}^{N} b_{ij} \zeta_j, \qquad i = 1, 2, \ldots, N, \tag{9}$$

where b_{ij} is the ith component of the jth eigenvector of the system

$$\sum_{i=1}^{N} m_{ri}^{0} \ddot{\xi}_i + \sum_{i=1}^{N} V_{,ri}^{0} \xi_i = 0, \qquad r = 1, 2, \ldots, N. \tag{10}$$

The notation in (10) is similar to that employed in (3).

Let

$$\sum_{i=1}^{N} b_{in} Q_i(t) = P_n \cos \nu t \tag{11}$$

where ν is a real number. Substituting (7-11) in (6) and retaining linear and cubic terms in the Taylor expansion we get

$$\ddot{\zeta}_n + \omega_n^2 \zeta_n = \epsilon \Big[\sum_{p=1}^{N} D_{np} \dot{\zeta}_p + \sum_{p,q,r=1}^{N} (A_{npqr} \zeta_p \zeta_q \zeta_r$$

$$+ B_{npqr} \zeta_p \zeta_q \dot{\zeta}_r + C_{npqr} \zeta_p \dot{\zeta}_q \dot{\zeta}_r + D_{npqr} \dot{\zeta}_p \dot{\zeta}_q \dot{\zeta}_r) + P_n \cos \nu t \Big] \tag{12}$$

$$n = 1, 2, \ldots, N,$$

where ω_i, $i = 1, 2, \ldots, N$ are roots of the polynomial in ω:

$$\det (V^0_{,ij} - \omega^2 m^0_{ij}) = 0. \tag{13}$$

It will be assumed that the roots of (13) are distinct.

Equation (12) is of the form

$$\ddot{\zeta}_n + \omega_n^2 \zeta_n = \epsilon f_n(\zeta_i, \dot{\zeta}_i) + \epsilon P_n \cos \nu t. \tag{14}$$

For reasons that will become apparent later we write (14) thus:

$$\ddot{\zeta}_n + s_n^2 \nu^2 \zeta_n = \epsilon[\epsilon^{-1}(s_n^2 \nu^2 - \omega_n^2)\, \zeta_n + f_n(\zeta_i, \dot{\zeta}_i) + P_n \cos \nu t] \tag{15}$$

where

$$s_n = \frac{p_n}{q_n}, \tag{16}$$

$$\left| \frac{p_n^2}{q_n^2} \nu^2 - \omega_n^2 \right| < \epsilon \tag{17}$$

and p_n and q_n, $n = 1, 2, \ldots N$ are relatively primed integers satisfying (17) but otherwise arbitrary at this stage of the analysis.

Reduction of the System to Normal Form and the Method of Averaging

Now (15) will be transformed to a form suitable for the use of the Method of Averaging of Bogoliuboff and Mitropol'skiy [*1*]. For this, new variables $A_n(t)$, $\Psi_n(t)$, $n = 1, 2, \ldots, N$ are introduced thus:

$$\zeta_n = A_n(t) \cos [s_n \nu t + \Psi_n(t)], \quad \dot{\zeta}_n = - s_n \nu A_n(t) \sin [s_n \nu t + \Psi_n(t)] \tag{18}$$

Now let

$$\theta_n = s_n \nu t + \Psi_n(t)$$

Then $f_n(\zeta_j, \dot{\zeta}_j)$ is periodic in θ_n thus

$$f_n = \sum_{m_1, m_2, \ldots, m_N = -R_n}^{+R_n} g_{n, m_1, m_2, \ldots m_N}(A_1, A_2, \ldots, A_N) \exp\left(i \sum_{r=1}^{N} m_r \theta_r\right) \tag{19}$$

where $g_{n, m_1, m_2, \ldots, m_N}(A_1, A_2, \ldots, A_N)$ are coefficients of the Fourier polynomial expansion of f_n. Substituting (17-19) in (15) we get

$$\dot{A}_n = -\frac{\epsilon}{s_n \nu}\left[\epsilon^{-1}(s_n^2 \nu^2 - \omega_n^2)\frac{A_n}{2}\sin 2\theta_n + \frac{P_n}{2}\sin(\theta_n + \nu t) + \frac{P_n}{2}\sin(\theta_n - \nu t)\right]$$

$$- \frac{\epsilon}{2 i s_n \nu} \sum_{m_1, m_2, \ldots, m_N = -R_n}^{+R_n} g_{n, m_1, \ldots, m_N}(A_j)$$

$$\times \left\{\exp\left[i\left(\sum_{r=1}^{N} m_r \theta_r + \theta_n\right)\right] - \exp\left[i\left(\sum_{r=1}^{N} m_r \theta_r - \theta_n\right)\right]\right\}, \quad n = 1, 2, \ldots, N. \tag{20}$$

$$A_n \dot{\Psi}_n = -\frac{\epsilon}{s_n \nu}\left[\epsilon^{-1}(s_n^2 \nu^2 - \omega_n^2)\frac{A_n}{2}(1 + \cos 2\theta_n) + \frac{P_n}{2}\cos(\theta_n + \nu t) + \frac{P_n}{2}\cos(\theta_n - \nu t)\right]$$

$$- \frac{\epsilon}{2 s_n \nu} \sum_{m_1, m_2, \ldots, m_N = -R_n}^{+R_n} g_{n, m_1, \ldots, m_N}(A_j)$$

$$\times \left\{\exp\left[i\left(\sum_{r=1}^{N} m_r \theta_r + \theta_n\right)\right] + \exp\left[i\left(\sum_{r=1}^{N} m_r \theta_r - \theta_n\right)\right]\right\}, \quad n = 1, 2, \ldots, N. \tag{21}$$

Equations (20) and (21) are now in the desired form

$$\dot{A}_n = \epsilon F_n(A_1, A_2, \ldots, A_N, \Psi_1, \Psi_2, \ldots, \Psi_N, t) \tag{22}$$

$$\dot{\Psi}_n = \epsilon G_n(A_1, A_2, \ldots, A_N, \Psi_1, \Psi_2, \ldots, \Psi_N, t) \tag{23}$$

Following the Method of Averaging, we introduce new variables

$$A_n = a_n(t) + \epsilon \tilde{F}_n(a_1, a_2, \ldots, a_n, \psi_1, \psi_2, \ldots, \psi_n, t) \tag{24}$$

$$\Psi_n = \psi_n(t) + \epsilon \tilde{G}_n(a_1, a_2, \ldots, a_n, \psi_1, \psi_2, \ldots, \psi_n, t) \tag{25}$$

where $\tilde{F}_n$, $\tilde{G}_n$, are related to F_n and G_n in (22) and (23) [2]. It will not be necessary to refer to $\tilde{F}_n$ and $\tilde{G}_n$ in the rest of the discussion. The variables a_n and ψ_n, $n = 1, 2, ..., N$ satisfy equations of the first approximation:

$$\dot{a}_n = \epsilon \underset{t}{M}(F_n); \qquad \dot{\psi}_n = \epsilon \underset{t}{M}(G_n), \qquad n = 1, 2, ..., N \tag{26}$$

where the operator $\underset{t}{M}$ operating on a function Q is defined thus:

$$\underset{t}{M}(Q) = \lim_{T\to\infty} \frac{1}{T} \int_0^T Q\, dt.$$

The integration is to be performed with respect to the explicitly occurring t in the function Q. Thus the equations of the first approximation for (20) and (21) are

$$\begin{aligned}\dot{a}_n = -\frac{\epsilon}{2is_n\nu}\Big\{&\sum\nolimits^{*} g_{n,m_1,\dots,m_N}(a_j)\exp\Big[i\Big(\sum_r m_r\psi_r + \psi_n\Big)\Big]\\ &-\sum\nolimits^{**} g_{n,m_1,\dots,m_N}(a_j)\exp\Big[i\Big(\sum_r m_r\psi_r - \psi_n\Big)\Big]\Big\} - \delta_{1n}\frac{\epsilon P_1}{2\nu}\sin\psi_1\end{aligned} \tag{27}$$

$$\begin{aligned}a_n\dot{\psi}_n = -\frac{\epsilon}{2s_n\nu}\Big\{&\epsilon^{-1}(s_n^2\nu^2 - \omega_n^2)\frac{a_n}{2} + \sum\nolimits^{*} g_{n,m_1,\dots,m_N}(a_j)\exp\Big[i\Big(\sum_r m_r\psi_r + \psi_n\Big)\Big]\\ &+\sum\nolimits^{**} g_{n,m_1,\dots,m_N}(a_j)\exp\Big[i\Big(\sum_r m_r\psi_r - \psi_n\Big)\Big]\Big\} - \delta_{1n}\frac{\epsilon P_1}{2\nu}\cos\psi_1\end{aligned} \tag{28}$$

where $\delta_{1n} = 1$ when $n = 1$ and $\delta_{1n} = 0$ when $n \neq 1$ and Σ^* and Σ^{**} mean respectively summed over those values of $m_1, m_2, ..., m_N$ such that

$$m_1\frac{p_1}{q_1} + m_2\frac{p_2}{q_2} + ... + m_{n-1}\frac{p_{n-1}}{q_{n-1}} + (m_{n+1})\frac{p_n}{q_n} + m_{n+1}\frac{p_{n+1}}{q_{n+1}} + ... + m_N\frac{p_N}{q_N} = 0 \tag{29}$$

and

$$\begin{aligned}m_1\frac{p_1}{q_1} + m_2\frac{p_2}{q_2} + ... + m_{n-1}\frac{p_{n-1}}{q_{n-1}} &+ (m_n - 1)\frac{p_n}{q_n}\\ &+ m_{n+1}\frac{p_{n+1}}{q_{n+1}} + ... + m_N\frac{p_N}{q_N} = 0.\end{aligned} \tag{30}$$

Equations (27) and (28) are written for the case when the external excitation frequency ν is in the neighborhood of one of the linear natural frequencies, in this case near ω_1. If this frequency ν is not in the neighborhood of any of the linear natural frequencies, the equations of the first approximation (27) and (28) will have the last term absent in each equation. A system which has the external frequency in the neighborhood of one of the linear natural frequencies will be said to have "external resonance". In addition to "external resonance" a system may or may not exhibit what will be called "internal resonance".

In a given system the manner in which (29) and (30) are satisfied determines whether internal resonance is present or absent. If every term in (27) and (28) is obtained by satisfying (29) and (30) respectively for arbitrary values p_i and q_i, $i = 1, 2, \ldots, N$ except for the restrictions imposed on them by inequalities (16), then we have no internal resonance. Otherwise the system is said to have internal resonance.

Thus the class of systems discussed here divides into four distinct sub-classes.

A. Those systems possessing no internal or external resonance or briefly non-resonant systems.

B. Those systems possessing internal resonance but no external resonance.

C. Those systems possessing external resonance but no internal resonance.

D. Those systems possessing internal as well as external resonance.

The averaged equations of the first approximation in each of the four sub-classes mentioned above have in general solutions which can be discussed under the following three headings:

1. Special solutions
2. Periodic solutions
3. General solutions

Among the special solutions are those for which a_i and ψ_i, $i = 1, 2, \ldots, N$ are constants. These are often referred to as steady state solutions. Another kind of special solutions of great interest are the mono-frequent solutions. They occur in systems without internal resonance and are solutions in which only two of the $2N$ variables a_i, ψ_i, $i = 1, 2, \ldots N$ are not of order ϵ. These solutions, when they occur, occur for initial conditions in the neighborhood of one of the linear principal modes of motion. The periodic solutions when they occur may or may not be isolated. General solutions of the averaged equations are found only in rare cases.

Classes of Systems

We will now discuss briefly each of the four sub-classes. Since monofrequent and steady state solutions are widely discussed in the literature [*1-4*] the emphasis will be placed on periodic and general solutions [*5*].

A. Non-Resonant Systems [*5*]

If every term in (27) and (28) is obtained by satisfying (29) and (30) respectively by taking

$$m_i = 0; \qquad i = 1, 2, ..., (n-1), (n+1), ..., N; \qquad m_n = -1$$

and (31)

$$m_i = 0; \qquad i = 1, 2, ..., (n-1), (n+1), ..., N; \qquad m_n = +1$$

and if ν is not near one of the linear natural frequencies, we have the non-resonant case. The equations of the first approximation (27) and (28) become in this case

$$\mathring{a}_n = \epsilon \left[\frac{D_{nn}}{2} a_n + \sum_{\substack{p=1 \\ p\neq n}}^{N} \tfrac{1}{4}(B_{nppn} + \omega_p^2 D_{nppn})\, a_p^2 a_n + \tfrac{1}{8}(B_{nnnn} + 3\omega_n^2 D_{nnnn})\, a_n^3\right] \tag{32}$$

$$a_n\dot{\psi}_n = -\frac{\epsilon}{\omega_n}\left[\sum_{\substack{p=1 \\ p\neq n}}^{N} \tfrac{1}{4}(A_{nnpp} + A_{npnp} + A_{nppn} + \omega_p^2 C_{nnpp})\, a_p^2 a_n + \tfrac{1}{8}(3A_{nnnn} + \omega_n^2 C_{nnnn})\right] \qquad n = 1, 2, ..., N$$

which can be written in the form

$$\mathring{a}_n = -\epsilon\left(C_n + \sum_{p=1}^{N} h_{np} a_p^2\right) a_n \tag{34}$$

$$a_n\dot{\psi}_n = -\epsilon \sum_{p=1}^{N} k_{np} a_p^2 a_n \qquad n = 1, 2, ..., N. \tag{35}$$

It is seen that equations (34) are decoupled from equations (35) and thus the order of the system of coupled differential equations to be integrated is reduced from the original $2N$ to N. The integration of (35) is then reduced to quadratures once (34) are integrated. This fact of reduction of order of the coupled system makes the study of systems of two degrees of freedom in the non-resonant case considerably simpler than in the general case. Solutions can be studied in the (a_1, a_2) phase plane and if periodic solutions occur their existence can be simply and rigorously determined.

It is not possible to integrate the general case of (34) in terms of elementary functions. It is possible, however, to make some qualitative statements about the behavior of the solutions. Multiplying both sides of (34) by a_n and letting $a_n^2 = x_n$, $n = 1, 2, \ldots N$ we get

$$\dot{x}_n = -2\epsilon \left(C_n + \sum_{p=1}^{N} h_{np}x_p\right) x_n. \tag{36}$$

Since we are interested only in real values of a_i, $i = 1, 2, \ldots N$ the solutions of (36) that are of interest are those for which $x_n(t) \geqq 0$, $n = 1, 2, \ldots, N$. It is not difficult to show the following properties of the solutions of (36) [5]:

(a) If $x_n(0) > 0$ then $x_n(t) > 0$ for $0 < t < \infty$, $n = 1, 2, \ldots N$.

(b) If $C_n > 0$; $h_{np} > 0$; $x_n(0) > 0$; all $n, p = 1, 2, \ldots N$ then $x_n(t) \to 0$ monotonically as $t \to \infty$.

(c) If $C_n > 0$; $\sum_{n,p=1}^{N} h_{np}x_nx_p > 0$; $x_n(0) > 0$; $x_n(t) \to 0$ as $t \to \infty$.

Conditions (b) and (c) are sufficient for the absence of limit cycles in the system.

If $h_{np} = 0$; $n, p = 1, 2, \ldots N$, (34) and (35) can be integrated explicitly and solutions in terms of the principal coordinates corresponding to the linear part of the system can be given. They are

$$\zeta_n = a_{n0}e^{-\epsilon cnt} \cos\left[\omega_n t - \sum_{p=1}^{N} k_{np}\frac{a_{p0}^2}{2c_p}(1 - e^{-\epsilon 2cpt}) + \psi_{n0}\right] \quad n = 1, 2, \ldots, N \tag{37}$$

where a_{n0} and ψ_{n0} are arbitrary constants depending on initial conditions.

To discuss resonant systems of types B, C and D it is necessary to study special cases of (27) and (28). In order to simplify the discussion we shall restrict the nature of the systems under consideration. We will assume that the systems are of only two degrees of freedom ($N = 2$), the quantities m_{ij} in (1) are constants, and the damping function is of the special form

$$D = \tfrac{1}{2} \sum_{i,j=1}^{N} d_{ij}\dot{\eta}_i\dot{\eta}_j$$

where d_{ij} are constants. Thus the nonlinear effects will be restricted to those arising from the potential energy.

B. Systems with Internal Resonance but No External Resonance

It is not difficult to show that for two degrees of freedom systems with the restrictions as stated above, internal resonance occurs in only two cases. It occurs when the linear natural frequencies are such that $|\omega_1 - \omega_2| = O(\epsilon)$ and $|3\omega_1 - \omega_2| = O(\epsilon)$. The averaged equations of the first approximation for the former case are more complicated than those for the latter case. We will here give the equations of the first approximation when $|3\omega_1 - \omega_2| = O(\epsilon)$. These equations are:

$$\dot{a}_1 = \frac{\epsilon D_1}{2} a_1 - \frac{3}{8} \frac{\epsilon}{\omega_1} A_2 a_1^2 a_2 \sin (3\psi_1 - \psi_2) \tag{38}$$

$$a_1 \dot{\psi}_1 = -\left(\frac{3}{8}\right) \frac{\epsilon}{\omega_1} [A_1 a_1^3 + 2A_3 a_1 a_2^2 + A_2 a_1^2 a_2 \cos (3\psi_1 - \psi_2)] \tag{39}$$

$$\dot{a}_2 = \frac{\epsilon D_2}{2} a_2 + \frac{\epsilon}{24\omega_1} A_2 a_1^3 \sin (3\psi_1 - \psi_2) \tag{40}$$

$$a_2 \dot{\psi}_2 = -\frac{\epsilon}{8\omega_1} \left[\frac{4}{3} \epsilon^{-1}(9\omega_1^2 - \omega_2^2) a_2 + 2A_3 a_1^2 a_2 + A_4 a_2^3 + \tfrac{1}{3} A_2 a_1^3 \cos (3\psi_1 - \psi_2)\right] \tag{41}$$

where the following notation is used:

$$A_1 = A_{1111}; \qquad A_2 = A_{1112}; \qquad A_3 = A_{1122}; \qquad A_4 = A_{2222};$$
$$D_1 = D_{11}; \qquad D_2 = D_{22}.$$

It is to be noted that if $A_2 = 0$ in spite of the fact that

$$|3\omega_1 - \omega_2| = 0(\epsilon)$$

the solutions are the same as those for the non-resonant case. If $D_1 = D_2 = 0$ (38-41) can be integrated by elementary means. It can be shown that the amplitudes a_1 and a_2 are periodic functions of time with periods that can be very large compared to the lowest period of the corresponding linear modes. These periodic solutions are not isolated. They vary continuously with initial conditions and with the values of the system parameters. The above results will be given in detail in a forthcoming paper.

C. System with External Resonance and No Internal Resonance

By inspection of (27) and (28) it is not difficult to show that the set of coupled averaged equations to be integrated in this case are of order $N + 1$ for an N degree of freedom system.

Again we restrict the analysis to systems with two degrees of freedom and restrict the non-linear terms to those arising from the potential energy. The averaged equations of the first approximation are:

$$\dot{a}_1 = \frac{\epsilon D_1}{2} a_1 - \frac{\epsilon P_1}{2} \sin \psi_1 \tag{42}$$

$$a_1 \dot{\psi}_1 = -\frac{\epsilon}{\nu} \left[\epsilon^{-1}(\nu^2 - \omega_1^2) \frac{a_1}{2} + \frac{3}{8} A_1 a_1^3 + \tfrac{3}{4} A_3 a_1 a_2^2 + \frac{p_1}{2} \cos \psi_1 \right] \tag{43}$$

$$\dot{a}_2 = \frac{\epsilon D_2}{2} a_2 \tag{44}$$

$$a_2 \dot{\psi}_2 = -\frac{\epsilon}{\omega_2} \left[\frac{3}{4} A_3 a_1^2 a_2 + \frac{3}{8} A_4 a_2^3 \right] \tag{45}$$

The external frequence ν is taken in the neighborhood of ω_1. We note that equation (45) is decoupled from (42), (43) and (44). Special steady state solutions with $\dot{a}_1 = \dot{a}_2 = \dot{\psi}_1 = \dot{\psi}_2 = 0$ are easily obtained and the relationship between the amplitude a_{10} and the excitation frequency ν is the same as that for Duffing's Equation:

$$\nu^2 = \omega_1^2 + \tfrac{3}{4} \epsilon A_1 a_{10}^2 + \frac{\epsilon P_1}{a_{10}}, \qquad a_{10} \neq 0$$

where a_{10} is the steady state value of a_1. If $D_1 = D_2 = 0$ it is possible to integrate (42-45) by elementary means, and it can be shown that the amplitude a_1 is a periodic function of time with a period that can be very much longer than $2\pi/\omega_1$. These periodic solutions are again not isolated. Such solutions, however, are to be expected since they also occur in linear undamped systems.

D. Systems with Internal and External Resonance

In this, as in the case B, the order of the coupled set of equations for a system with N degrees of freedom ($N \geqq 2$) can be anywhere between four and $2N$. Again, we restrict the analysis to systems with two degrees of freedom and with non-linear terms arising only from the potential energy. In this case, as in the case with internal resonance and without external resonance, we have only two cases. One case occurs when $| \omega_1 - \omega_2 | = O(\epsilon)$ and the other case occurs when $| 3\omega_1 - \omega_2 | = O(\epsilon)$. Following are the averaged equations of the first approximation when $| 3\omega_1 - \omega_2 | = O(\epsilon)$:

$$\dot{a}_1 = \frac{\epsilon D_1}{2} a_1 - \frac{3}{8} \frac{\epsilon}{\nu} A_2 a_1^2 a_2 \sin (3\psi_1 - \psi_2) - \frac{\epsilon P_1}{2\nu} \sin \psi_1 \tag{46}$$

$$a_1\dot{\psi}_1 = -\frac{\epsilon}{2\nu}[\epsilon^{-1}(\nu^2 - \omega_1^2)\, a_1 + \tfrac{3}{4} A_1 a_1^3 + \tfrac{3}{2} A_3 a_1 a_2^2 + \tfrac{3}{4} A_2 a_1^2 a_2 \cos(3\psi_1 - \psi_2) + P_1 \cos\psi_1] \quad (47)$$

$$\dot{a}_2 = \frac{\epsilon D_2}{2} a_2 + \frac{\epsilon}{24\nu} A_2 a_1^3 \sin(3\psi_1 - \psi_2) \quad (48)$$

$$a_2\dot{\psi}_2 = -\frac{\epsilon}{6\nu}[\epsilon^{-1}(9\nu^2 - \omega_2^2)\, a_2 + \tfrac{3}{2} A_3 a_1^2 a_2 + \tfrac{3}{4} A_2 a_2^3 + \tfrac{1}{4} A_2 a_1^3 \cos(3\psi_1 - \psi_2)]. \quad (49)$$

Special "steady state" solutions with $\dot{a}_1 = \dot{a}_2 = \dot{\psi}_1 = \dot{\psi}_2 = 0$ can be easily obtained, though the computations of the roots of the resulting algebraic equations can get complicated. Steady state solutions of this type were obtained and checked against experimental results [*2*, *4*] when $\omega_2 = 3\omega_1$.

If ω_2 is in the neighborhood of $3\omega_1$, without being equal to $3\omega_1$, experimental results indicate that as ν deviates from $\omega_2/3$ the amplitudes a_1 and a_2 develop periodic motions of long periods. These periodic motions are isolated. That is, the same periodic motions occur for arbitrary initial conditions as $t \to \infty$. Further study of this problem is in progress.

Remarks

The remarks that have been made above will often be applicable to systems with more general nonlinear terms. The most restrictive condition in the discussion has been the requirement on the external time dependent forces that they have a single frequency. When these forces have more than one frequency, the values of which may or may not be commensurate, the analysis is considerably more complicated. If the external time dependent forces are not periodic, in a great many cases, the Method of Averaging is not applicable. If the system of equations has time dependent coefficients, again the analysis gets much more complicated.

The accuracy of the method of analysis used here has been checked in many instances by comparing the results with those from an analog type computer. In most cases the accuracy obtained from the first approximation is very good even for values of ϵ as high as 0.5. In general, it has been observed that the accuracy obtained from the first approximation is good if the solutions obtained from the second approximation do not introduce a qualitative change in the nature of the solution. In many cases the error introduced by taking only a few terms of the Taylor series expansion of the differential equation is greater than that introduced by the method of averaging employed to solve them.

REFERENCES

[*1*] Bogoliuboff, N. N., and Mitropol'skiy, Yu. A., "Asimptoticheskie Methody v teorii Nelieninykh Kolebanii." Gos. Izdat. Fiz.-Mat. Lit., Moscow, 1958.

[*2*] Chow, S. T., and Sethna, P. R., Steady-State Damped Vibrations and Stability of a Class of Nonlinear Discrete Systems. WADC Tech. Report No. 59-543 (1960).

[*3*] Mitropol'skiy, Yu. A., "Nonstationary Process in Nonlinear Oscillatory Systems" (Nestatsionarnye Protesessy v Nelineinykh Koleoatel'nykh Sistemakh). Akad. Nauk Ukr. S.S.R., Kiev, 1955; English transl., Air Technical Intelligence, ATIC-270579, F-TS-9085/v.

[*4*] Sethna, P. R., Steady-state undamped vibrations of a class of nonlinear discrete systems. *J. appl. Mech. (Trans. ASME, Ser. E)* **27**, 187-195 (1960).

[*5*] Sethna, P. R., Transients in certain autonomous multiple-degree-of-freedom nonlinear vibrating systems. ASME Paper No. 61.06.176. Submitted to *J. appl. Mech.*

[*6*] Sokolnikoff, I. S. "Tensor Analysis," p. 78. Wiley, New York, 1951.

[*7*] Whittaker, E. T., "A Treatise on the Analytical Dynamics of Particles and Rigid Bodies," Chapter VII. Dover Publications, New York, 1944.

Perturbation Theory for Almost Periodic Solutions for Undamped Nonlinear Differential Equations

JÜRGEN MOSER
Institute of Mathematical Sciences
New York University, New York, New York

I. A Theorem on Area Preserving Mappings

In recent years A. N. Kolmogorov [*7*, *8*] and his student V. I. Arnold [*1*, *2*] announced several new results concerning the existence of almost periodic solutions of nonlinear differential equations. These results have solved several outstanding problems which have been open since Poincaré's work at the beginning of the century [*13*]. I want to report on these theorems and present some new results of mine which are based on a new method which furnishes the construction of solutions for nonlinear problems. Although this method has many other applications in analysis (partial differential equations, differential geometry, etc.), I will illustrate it only with a few examples which show the strength of the new method clearly.

Consider an annulus mapping defined in the annulus $y_1 \leq y \leq y_2$ where y denotes the radial coordinate of polar coordinates and x (mod 2π) the angular variable. The mapping we are going to discuss is assumed to be close to the following "twist mapping" M_0

$$\begin{aligned} x_1 &= x + a(y) \\ y_1 &= y \end{aligned}$$

where $a(y)$ is a monotone function and, say, $a'(y) \geq 1$.

The mapping M we are going to study is a perturbation of M_0:

$$\begin{aligned} x_1 &= x + a(y) + f(x, y) \\ y_1 &= y + g(x, y) \end{aligned} \tag{1}$$

where the functions f, g are, of course, of period 2π in x. The smallness of the deviation is measured in terms of a derivative norm: For any function $h(x, y)$ which is r times continuously differentiable we set

$$| h |_r = \sup_{\rho \leq r;\, x,y} | D^\rho h(x, y) |$$

where D^p denotes any partial derivative of order p. We will require that the quantity

$$|f|_r + |g|_r,$$

which we also denote by $|M - M_0|_r$, is sufficiently small, for some r to be specified later.

Furthermore, the mapping is assumed to map the annulus one-to-one and area preserving onto itself. The problem consists in finding invariant curves of the mapping M for sufficiently small values of $|M - M_0|_r$. For M_0 obviously every circle $y =$ const represents such an invariant curve, which under M_0 is rotated by the angle $a(y)$. Therefore, one can consider our question as a continuation problem or perturbation problem for the invariant curves.

Assuming for the moment that such an invariant curve C of M exists, we can label it by its rotation number: The mapping M restricted to C can be considered as a circle mapping of the type which A. Denjoy investigated [*4*]. If x_n is the x coordinate of the image point under the nth iterate M^n of M the limit

$$\lim_{n\to\infty} \frac{x_n}{n} = \omega \tag{2}$$

exists and is called the rotation number. For M_0 the rotation number agrees with the rotation angle.

We first want to point out a difficulty of this problem which has its analytical manifestation in the so called small divisors. If we draw our attention to a circle for which

$$a(y) = \frac{m}{n} 2\pi$$

we see that the whole circle consists of fixed points of M_0^n. To find a continuation of such a circle would suggest to find a curve of fixed points of M^n. However, it is well known that in general only a finite number of fixed points of M^n will exist (in general $2n$). This phenomenon corresponds to the phenomenon of phase-locking in ordinary differential equations. One can say, that in general the curves with rational $\omega/2\pi$ cannot be continued, but break up into finitely many points.

Therefore, in the following we will concentrate on curves for which $\omega/2\pi$ is irrational. In fact, we will require that this number can only be badly approximated by rationals, e.g.

$$\left| \omega - \frac{m}{n} 2\pi \right| \geq \gamma n^{-5/2} \tag{3}$$

for all integers m, and positive integers n. The numbers ω so excluded have—as is easily seen—a density which is small if the constant γ is small.

It is remarkable that the circles with a rotation number satisfying (3) can be continued for sufficiently small $| M - M_0 |_r$. We formulate this statement as follows:

Theorem I. Let ω be a number satisfying (3) and

$$a(y_1) + \gamma < \omega < a(y_2) - \gamma$$

and assume M to be area preserving and $a' \geq 1$. Then there exists an integer and a positive ϵ_0 depending on M_0 and γ such that for

$$| M - M_0 |_r = \epsilon < \epsilon_0$$

there exists an invariant curve $C(\omega)$ of M with rotation number ω. Moreover, $C(\omega)$ can be parametrised by

$$x = \theta + u(\theta)$$

$$y = y_0 + v(\theta)$$

where

$$| u |_1 + | v |_1 \to 0 \quad \text{as} \quad \epsilon \to 0$$

and the mapping M induced on $C(\omega)$ is given by

$$\theta_1 = \theta + \omega.$$

We do not specify the best values for r since we need as many as 333 derivatives. Moreover, if r increases the number of derivatives of u, v increases at the same rate. The main point of theorem 1 is that finitely many derivatives suffice for the existence of invariant curves. On the other hand it is easily seen by examples that the smallness of $| M - M_0 |_0$ does not suffice for the existence of invariant curves. This is rather surprising because the problem seems of topological nature but the statement requires bounds on derivatives.

II. Applications

The theorem can be sharpened in several ways[1] so as to be useful for stability investigations of periodic solutions. As is well known, the

[1] e.g. One can drop the assumption that both boundaries of the annulus are mapped onto itself.

stability problem of periodic solutions can be reduced to the study of a mapping near a fixed point (see Levi-Civita [*9*]). If this mapping is two dimensional (as in Levi-Civita's work) then one can take as annulus a circular neighborhood of the fixed point from which the fixed point itself is deleted. As a special application, we mention the stability problem of area preserving mappings near an elliptic fixed point. Let u, v be rectangular local coordinates near the fixed point $u = v = 0$ of an area perserving mapping M

$$u_1 = u \cos \alpha - v \sin \alpha + f_2$$

$$v_1 = u \sin \alpha + v \cos \alpha + g_2$$

where f_2, g_2 are assumed to vanish with their first derivatives at the fixed point. Moreover, M is assumed to have $r \geq 5$ derivatives near $u = v = 0$. Birkhoff [*3*] has shown that this mapping can be brought into the form

$$\left.\begin{aligned} u_1 &= u \cos (\alpha + \beta y) - v \sin (\alpha + \beta y) + f_5 \\ u_1 &= u \sin (\alpha + \beta y) + v \cos (\alpha + \beta y) + g_5 \end{aligned}\right. \qquad y = u^2 + v^2 \tag{4}$$

by an appropriate coordinate transformation provided that $\alpha/2\pi$ is not a rational number with denominator 1, 2, 3, 4. Actually, Birkhoff assumed analyticity of the mapping but the proof is trivially extended to the case where only finitely many derivatives exist; f_5, g_5 vanish with their derivatives up to order 4. Applying the above theorem (in a slightly sharper version) to an annulus $\epsilon \leq y \leq 2\epsilon$ one finds: The area preserving mapping (4) is indeed stable near $u = v = 0$ provided the number β is different from zero, and provided f_5, g_5 are sufficiently often differentiable.

In particular, this result proves the stability of the periodic solutions of the first kind of the restricted three body problem (for verification of the assumptions see Moser [*10*, *11*]).

Application to the Circle Mapping

We return to theorem 1 and discuss it for the special case $g = 0$:

$$x_1 = x + \omega + y + f(x, y) \qquad \text{if} \qquad a(y) = \omega + y.^1$$

[1] In this case, we will not assume that the annulus mapping is area-preserving so that our theorem is not directly applicable. But, since the invariant curves are given as circles, the area preserving property is not needed in the proof.

This represents a mapping of a circle onto itself near a rotation depending on a parameter y. Theorem 1 shows that under the assumptions stated, there exists a parameter value y such that this mapping has the rotation number ω which we always assume to satisfy condition (3). Moreover, this mapping can be transformed by

$$x = \theta + u(\theta)$$

into the rotation

$$\theta_1 = \theta + \omega.$$

This is contained in Denjoy's well-known theorem [*4*]. However, here we have the additional statement that u is continuously differentiable, which cannot be proven with the methods of Denjoy. In fact, it is not known whether such a statement is true for any sufficiently smooth circle mapping with a rotation number satisfying (3); in other words, whether such a statement is true if the mapping is not close to a rotation.

Thus, theorem 1 does not only prove the existence of invariant curves, but also provides the parametrization in which the mapping appears as a rotation.

Application to the Duffing Equation

To give a more concrete application of theorem 1 consider Duffing's equation

$$\ddot{x} + \alpha x + \beta x^3 = \epsilon F(x, t) \qquad \alpha, \beta > 0 \tag{5}$$

which is periodic of period 2π in t and otherwise sufficiently often differentiable. For $\epsilon = 0$ the solutions can be described by the energy relation

$$\frac{\dot{x}^2}{2} + \frac{\alpha x^2}{2} + \frac{\beta x^4}{4} = h$$

which in the phase plane represents a family of ovals. All the solutions are periodic. Denoting their frequency by $a(h)/2\pi$ one finds that $a(h)$ is a monotonic increasing function. One has

$$a(0) = 2\pi\sqrt{\alpha}$$

and for $h \rightarrow \infty$ the function increases like $h^{1/4}$.

For $\epsilon \rightarrow 0$ the differential equation (5) depends periodically on t and we have to consider the motion in a three dimensional phase space, with coordinates x, $\dot{x}$, t where t values differing by 2π are identified. The energy surfaces above then appear as tori.

Following the solutions of (5) from $t = 0$ to $t = 2\pi$ one obtains a mapping which is area preserving, and which for $\epsilon = 0$ maps the above ovals onto itself. Moreover, the rotation number $a(y)$ has a nonvanishing derivative so that we can apply theorem 1 to find invariant curves for the mapping corresponding to (5) for sufficiently small ϵ. Let

$$x = U(\theta), \qquad \dot{x} = V(\theta)$$

be the representation of this curve in the normalizing parametrization. We extend the definition of U, V in such a way that

$$U\left(t, \theta + \omega \frac{t}{2\pi}\right), \qquad V\left(t, \theta + \omega \frac{t}{2\pi}\right)$$

describes the solutions with the initial values $U(\theta)$, $V(\theta)$. Since $t = 2\pi$ describes the image point under the mapping, one has

$$U(2\pi, \theta + \omega) = U(\theta + \omega) = U(0, \theta + \omega)$$

and a similar relation for V. This proves that $U(t, \theta)$, $V(t, \theta)$ have period 2π in both variables showing that

$$x = U(t, \theta), \qquad \dot{x} = V(t, \theta)$$

represents an invariant torus for the flow defined by (5). The flow on this torus is given by

$$\dot{\theta} = \omega/2\pi$$

The solutions so obtained can be described by almost periodic solutions with two basic frequencies, namely 1 and $\omega/2\pi$. Therefore, the result can be described as a continuation of almost periodic solutions.

In this form, namely as continuation of almost periodic solution, Kolmogorov stated his theorem which we want to formulate: Consider a Hamiltonian system

$$\dot{x} = H_y(x, y, \epsilon)$$

$$\dot{y} = -H_x(x, y, \epsilon)$$

where $x = (x_1, ..., x_n)$, $y = (y_1, ..., y_n)$ and the x variables are angular variables of period 2π. The Hamiltonian H is assumed to be analytic in all variables and for $\epsilon = 0$ independent of x: $H(x, y, 0) = H(y)$.

Theorem (Kolmogorov). Let $\alpha = (\alpha_1, ..., \alpha_n)$ be a vector satisfying the inequality

$$\left| \sum_{\nu=1}^{n} m_\nu \alpha_\nu \right| \geq \gamma \left(\sum | m_\nu | \right)^{-\delta}$$

for all integers m, with $\Sigma \mid m_\nu \mid > 0$. Let y be so chosen that

$$H_y(y_0) = \alpha \qquad \text{for} \qquad \epsilon = 0.$$

Assuming

$$\det H_{yy} \neq 0 \qquad \text{for} \qquad y = y_0 \quad \text{and} \quad \epsilon = 0$$

there exist analytic functions u, v of period 2π in all variables such that

$$\begin{aligned} x &= \theta + u(\theta) \\ y &= y_0 + v(\theta) \end{aligned} \qquad \theta = (\theta_1, \theta_2, \ldots, \theta_n)$$

represent an invariant torus. The flow on this torus is given by

$$\dot{\theta} = \alpha.$$

III. Remarks Concerning the Proof:

A proof[1] of these results would be out of place here. However, the main points can be made quite clearly. The difficulty in the problems discussed is the occurrence of the small divisors. Any construction of the functions sought will involve the operator

$$Lu = \sum_{k \neq 0} \frac{u_k}{e^{ik\omega} - 1} e^{ikx} \qquad \text{if} \qquad u = \sum_k u_k e^{ikx}$$

which is unbounded. One can, however, easily establish an estimate of the type

$$| Lu |_r \leq c \, | u |_{r+3}$$

under the assumption (3). This means that the rth derivatives of Lu can be estimated in terms of derivatives of u up to order $r + 3$.

In an iterative procedure, one will obtain the $(n + 1)$ approximation u_{n+1} in terms of u_n by application of the above operator and other bounded operations. The point is that at each step, one will lose 3 derivatives. Although usually this phenomenon is symptomatic for divergence, one still can establish convergence under special circumstances; namely, in case one finds an iteration process which converges "accelerated," by which we mean an iteration in which one has an estimate

$$| u_{n+1} - u_n | \leq c \, | u_n - u_{n-1} |^{\kappa} \qquad (\kappa > 1)$$

[1] The proof of Theorem 1 will appear in *Nachr. Akad. Wiss. Göttingen, Math.-physik. Kl. IIa* (1962).

if one ignores the loss of derivatives. In this case, the deviation of the nth approximation from the solution will be of the order ϵ^{κ^n} instead of the order ϵ^n which one is used to from Picard's iteration. A typical example is Newton's approximation for the construction of a zero of a function where one has $\kappa = 2$.

This accelerated convergence phenomenon is enough to counteract the unboundedness effect from the operator L provided the functions involved are analytic. This is essentially the observation of Kolmogorov.

In case of differentiable mappings, this method still would fail since after finitely many iterations all derivatives would be lost. In this case, one can devise a smoothing process like, for instance, the truncation of a Fourier series at the Nth frequency. Such a smoothing operator will depend on a parameter N (like the cut off point N). This parameter has to be made to tend to infinity with the iteration number n in an appropriate way so that convergence can be established.[1]

IV. Ergodic Theory

Finally, we want to point out that the above results have a surprising implication for the applicability of ergodic theory to differential equations. We return to the area preserving mapping discussed in theorem 1. The question whether such a mapping is ergodic or not is a very delicate one and there are no methods available to decide this question for a given mapping. However, there are investigations which study whether ergodicity is more frequent than nonergodicity among all area preserving mappings [*6*, *12*]. It can be shown that the ergodic mappings are indeed more frequent in the Baire category sense. In these investigations, all measure preserving mappings are eligible. On the other hand, if one admits only sufficiently smooth mappings (as they usually arise from differential equations) then according to theorem 1 a full neighborhood of M_0 is not ergodic since any invariant curve separates the annulus into two invariant sets. It is interesting in this connection to read an old investigation of Fermi, in which he tried to establish quasi-ergodicity in a similar situation [*5*].

[1] In this connection see: A new technique for the construction of solutions of nonlinear differential equations, *Proc. Natl. Acad. Sci. U.S.* **47**, 1824-1831 (1961).

REFERENCES

[*1*] Arnold, V. I., *Doklady Akad. Nauk S.S.S.R.* **137**, 255-257 (1961).

[*2*] Arnold, V. I., *Doklady Akad. Nauk S.S.S.R.* **138**, 13-15 (1961).

[*3*] Birkhoff, G. D., "Dynamical Systems," New York, 1927; also *Acta Math. (Stockh.)* **43**, 1-119 (1920).

[*4*] Denjoy, A., *J. pures appl. Math.* [9] **11**, 333-375 (1932).

[*5*] Fermi, E., Beweis, dass ein mechanisches Normalsystem im allgemeinen quasiergodisch ist. *Physik. Z.* **24**, 261-264 (1923).

[*6*] Halmos, P., Approximation theories for measure-preserving transformations. *Trans. AMS* **55**, 1-18 (1944).

[*7*] Kolmogorov, A. N., *Doklady Akad. Nauk S.S.S.R.* **98**, 527-530 (1954).

[*8*] Kolmogorov, A. N., General theory of dynamical systems and classical mechanics. *Proc. intern. Congr. of Math., Amsterdam 1954*, Vol. 1, pp. 315-333 (1957).

[*9*] Levi-Civita, T., *Ann. Mat. pura appl.* [3] **5**, 221 (1901).

[*10*] Moser, J., *Math. Ann. (Heidelberg)* **126**, 325-335 (1953).

[*11*] Moser, J., *Nachr. Akad. Wiss. Göttingen, Math.-physik. Kl. IIa* pp. 87-120 (1955).

[*12*] Oxtoby, J. C., and Ulam, S. M., Measure-preserving homeomorphisms and metrical transitivity. *Ann. Math. (Princeton)* **42**, 87-920 (1941).

[*13*] Poincaré, H., Sur les courbes définies par une équation differentielles. *In* "Œuvres," Vol. 1. Paris, 1892.

Asymptotic System of a Perturbed System[1]

TARO YOSHIZAWA
RIAS[1]*, Baltimore, Maryland and*
Mathematics Department
Nihon University, Tokyo, Japan

In this article we shall discuss relationships between solutions of a perturbed system and solutions of the unperturbed system. It is based on my paper entitled "Asymptotic behavior of solutions of a system of differential equations", to be published in *Contributions to Differential Equations* [*6*]. Relationships between solutions of an unperturbed system and solutions of a perturbed system have been discussed by many authors, especially concerning the stability of solutions of the perturbed system [*2*]. Moreover, recently Hale and Stokes have discussed the behavior of solutions of a perturbed system near manifolds of periodic solutions of the unperturbed system [*3*].

In this paper we shall consider a system of differential equations

$$x' = F(t, x) + G(t, x), \tag{1}$$

where $G(t, x)$ is an integrable perturbation term. This type of system has been discussed by Antosiewicz and several authors [*1*]. Especially relationships between the positive limiting sets of solutions of a perturbed equation and the solutions of the unperturbed equation have been discussed by Markus [*4*] and Opial [*5*] concerning an equation of form

$$x' = H(x) + G(t, x), \tag{2}$$

where $G(t, x)$ is a perturbation term. Assuming the uniqueness of the solution of the Cauchy-problem, they have shown that the positive limiting sets of solutions of a perturbed equation are composed of the solutions of the unperturbed equation.

Now we assume that $F(t, x)$ and $G(t, x)$ of (1) are continuous in $I \times Q$ ($I : 0 \leqq t < \infty, Q :$ an open set in Euclidean n-space R^n) and

[1] This research was partially supported by the United States Air Force through the Air Force Office of Scientific Research of the Air Research and Development Command, under Contract Number AF 49(638)-382. Reproduction in whole or in part is permitted for any purpose of the United States Government.

that if $x(t)$ is continuous and bounded on $t_0 \leqq t < \infty$, that is, $x(t) \subset Q^*$ (Q^*: a compact set in Q), we have

$$\int_{t_0}^{\infty} \| G(s, x(s)) \| \, ds < \infty. \tag{3}$$

Let Ω be a closed set in the space. We assume that $F(t, x)$ satisfies the following two conditions:

(a) $F(t, x)$ tends to a function $H(x)$ for $x \in \Omega$ as $t \to \infty$ and on any compact set in Ω this convergence is uniform. Consequently $H(x)$ is a continuous function on Ω.

(b) Corresponding to each $\epsilon > 0$ and each $y \in \Omega$, there exist positive numbers $\delta(y)$ and $T(y)$ such that if $\| x - y \| < \delta(y)$ and $t \geqq T(y)$, we have

$$\| F(t, x) - F(t, y) \| < \epsilon.$$

In Markus' case, $G(t, x)$ of (2) tends to zero uniformly for each compact set in Q. Therefore, Markus' condition is a special case of our case where $Q = \Omega$. Moreover, in Opial's case, $G(t, x)$ of (2) satisfies (3) and hence his case is also a special case of our case.

Let $x(t; x_0, t_0)$ be a solution of (1) through the point (t_0, x_0) and let Γ^+ be a set in R^n such that

$$\Gamma^+ = \bigcap_{s<\tau} \overline{\bigcup_{s \leqq t} x(t; x_0, t_0)},$$

where the interval $[t_0, \tau)$ is the maximum interval on which the solution $x(t; x_0, t_0)$ is defined. Then Γ^+ is a closed subset in Q and if $x(t; x_0, t_0)$ is bounded, Γ^+ is compact and nonempty.

Under the above assumptions, we can prove the following theorems.

Theorem 1. We assume that $F(t, x)$ and $G(t, x)$ are continuous in $I \times Q$ and that Γ^+ is nonempty. Then we have $\tau = +\infty$.

Proof. Suppose that $\tau < \infty$ and $\omega \in \Gamma^+$. Let B_1, B_2 be two balls centered at ω and lying in Q and $B_1 \subset B_2$. If $\Gamma^+ \subset B_2$, then $\tau = +\infty$, since $x(t)$ does not approach the boundary of Q. Otherwise, $x(t)$ must traverse the set $B_2 - B_1$ an infinite number of times. Since $\| F(t, x) + G(t, x) \|$ is bounded for $t_0 \leqq t \leqq \tau$ and $x \in B_2$, there arises a contradiction. Therefore, we have $\tau = +\infty$.

Theorem 2. We assume that Γ^+ is nonempty and $x(t; x_0, t_0) \to \Omega$ as $t \to \infty$, where Ω is a closed set in the space Q and that $F(t, x)$ satisfies the conditions (a), (b) and $G(t, x)$ satisfies the condition (3). Then Γ^+ is the union of solutions of the equation

$$x' = H(x), \qquad x \in \Omega. \tag{4}$$

Proof. Since Γ^+ is nonempty, $x(t) = x(t; x_0, t_0)$ is defined on $t_0 \leqq t < \infty$. Therefore, if $\omega \in \Gamma^+$, there exists a sequence $\{t_k\}$ such that

$$x(t_k) \to \omega, \qquad t_k \to \infty \quad \text{as} \quad k \to \infty. \tag{5}$$

We can choose a set $S = \{x; \| x - \omega \| < \alpha\}$ such that $\bar{S} \subset Q$, because $\omega \in \Gamma^+ \subset \Omega$. Since $\Omega_1 = \bar{S} \cap \Omega$ is a compact set in R^n, there exists a continuous, bounded function $H^*(x)$ on R^n such that $H^*(x) = H(x)$ on Ω_1. Now we consider a system

$$x' = H^*(x). \tag{6}$$

Since $H^*(x)$ is bounded, there is a $\lambda > 0$ such that $\bar{E} \subset [0, \lambda] \times S$, where E is the set of all the points consisting of the solution curves of (6) for $0 \leqq t \leqq \lambda$, starting from the point $(0, \omega)$. The behavior of the solutions of (6) through (t_k, ω) is the same as one of the solutions through $(0, \omega)$, because (6) is an autonomous system.

Let $x^*(t)$ be a continuous function defined on $t_0 \leqq t < \infty$ such that $x^*(t) \in \bar{S}$ and $x^*(t) = x(t)$ when $x(t) \in \bar{S}$. Now we consider a system

$$x' = H^*(x) + F(t, x^*(t)) - H^*(x^*(t)) + G(t, x^*(t)). \tag{7}$$

As long as $x(t) \in \bar{S}$, $x(t)$ is a solution of (7). By (3), if k is sufficiently large, for a given $\delta > 0$

$$\int_{t_k}^{t_k+\lambda} \| G(s, x^*(s)) \| \, ds < \frac{\delta}{2}. \tag{8}$$

By the condition (b), there are $\delta_1 > 0$ and $T > 0$ such that if $y \in \Omega_1$, $\| x - y \| \leqq \delta_1$ and $t \geqq T$, we have

$$\| F(t, x) - F(t, y) \| < \frac{\delta}{6\lambda}.$$

Since $x(t) \to \Omega$, we can assume that $x^*(t) \to \Omega$. Therefore, for large t we have $x^*(t) \in U(\Omega_1, \delta_1) \cap \bar{S}$, where $U(\Omega_1, \delta_1)$ is a δ_1-neighborhood of Ω_1. For every point $x^*(t)$ there is a point $y^*(t)$ such that

$$d(x^*(t), \Omega_1) = \| x^*(t) - y^*(t) \|,$$

because Ω_1 is a compact set. If k is sufficiently large, we have on $J_k = [t_k, t_k + \lambda]$

$$\| F(t, x^*(t)) - F(t, y^*(t)) \| < \frac{\delta}{6\lambda}, \tag{9}$$

because of $\| x^*(t) - y^*(t) \| \leqq \delta_1$. By the condition (a), for sufficiently large t and $y \in \Omega_1$, we have $\| F(t, y) - H(y) \| < \delta/6\lambda$. Therefore, if k is sufficiently large, we have on J_k

$$\| F(t, y^*(t)) - H^*(y^*(t)) \| = \| F(t, y^*(t)) - H(y^*(t)) \| < \frac{\delta}{6\lambda}. \tag{10}$$

Moreover, since $H^*(x)$ is uniformly continuous on $\bar{S}$, there is a $\delta_2 > 0$ such that if $\| x - y \| < \delta_2$, $\| H^*(x) - H^*(y) \| < \delta/6\lambda$. Therefore, if k is sufficiently large, we have on J_k

$$\| H^*(y^*(t)) - H^*(x^*(t)) \| < \frac{\delta}{6\lambda}. \tag{11}$$

Since $H^*(y^*(t)) = H(y^*(t))$, we have from (9), (10) and (11)

$$\| F(t, x^*(t)) - H^*(x^*(t)) \| < \frac{\delta}{2\lambda}. \tag{12}$$

Therefore, if k is sufficiently large, by (8) and (12), we have

$$\int_{t_k}^{t_k+\lambda} \| F(s, x^*(s)) - H^*(x^*(s)) + G(s, x^*(s)) \| \, ds < \delta.$$

On the other hand, by (5), for sufficiently large k we have $\| x(t_k) - \omega \| < \delta$. Thus, by Theorem 2 in [6], we can see that there is a solution of (7) through $(t_k, x(t_k))$ defined on J_k for sufficiently large k.

The solution $x(t)$ passes through $(t_k, x(t_k))$ and $x(t_k) \in \bar{S}$. If we suppose that $x(t)$ does not lie in $\bar{S}$ for some t on J_k, there is a $\sigma > 0$ such that $x(t) \in \bar{S}$ for $t_k \leqq t \leqq t_k + \sigma$ and $\| x(t_k + \sigma) \| = \alpha$. For $t_k \leqq t \leqq t_k + \sigma$ we have $x(t) = x^*(t)$ and hence $x(t)$ is a solution of (7) through $(t_k, x(t_k))$. Therefore, if k is sufficiently large, $x(t) \subset S$ for $t \in J_k$. Consequently, we can see that if k is sufficiently large, $x(t)$ is a solution of (7) defined on J_k. Therefore, by Theorem 2 in [6], we can see that there is a solution $\varphi_k(t)$ of (6) through (t_k, ω) such that for a given $\epsilon > 0$

$$\| x(t) - \varphi_k(t) \| < \epsilon \qquad \text{for} \qquad t \in J_k. \tag{13}$$

Since $\varphi_k(t)$ is a solution of (6) through (t_k, ω), we have

$$\varphi_k(t) = \omega + \int_{t_k}^{t} H^*(\varphi_k(s)) \, ds \qquad (t_k \leqq t \leqq t_k + \lambda)$$

and if we denote $\varphi_k(t + t_k)$ $(0 \leqq t \leqq \lambda)$ by $\varphi_k(t)$ again, we have

$$\varphi_k(t) = \omega + \int_0^t H^*(\varphi_k(s))\, ds \qquad (0 \leqq t \leqq \lambda).$$

Therefore, for a sequence $\{\epsilon_k\}$ approaching zero as $k \to \infty$, there exist solutions $\varphi_k(t)$ of (6) such that

$$\begin{gathered} \varphi_k(t) = \omega + \int_0^t H^*(\varphi(s))\, ds \qquad (0 \leqq t \leqq \lambda) \\ \varphi_k(t) \subset U(\Omega, \epsilon_k) \cap \bar{S}. \end{gathered} \tag{14}$$

Since the sequence of functions $\{\varphi_k(t)\}$ is uniformly bounded and equicontinuous, we can select a uniformly convergent subsequence. Let $\varphi(t)$ be its limit function and then by (14) we have

$$\varphi(t) = \omega + \int_0^t H^*(\varphi(s))\, ds \qquad (0 \leqq t \leqq \lambda)$$

and

$$\varphi(t) \subset \Omega \cap \bar{S} = \Omega_1,$$

because $\epsilon_k \to 0$ as $k \to \infty$. On the other hand, $H^*(\varphi(t)) = H(\varphi(t))$ and hence we have

$$\varphi(t) = \omega + \int_0^t H(\varphi(s))\, ds \qquad (0 \leqq t \leqq \lambda),$$

that is, $\varphi(t)$ is a solution of (4) passing through $(0, \omega)$. For any $\sigma > 0$ such that $0 < \sigma \leqq \lambda$, $\varphi(\sigma)$ belongs to Γ^+, because by (13) and the uniform convergence of $\{\varphi_k(t)\}$, there exists a sequence $\{t_k + \sigma\}$ such that $t_k + \sigma \to \infty$, $x(t_k + \sigma) \to \varphi(\sigma)$ as $k \to \infty$. Therefore, $\varphi(t)$ is a solution of (4) which passes through $(0, \omega)$ and remains in Γ^+ for $0 \leqq t \leqq \lambda$. The theorem is proved.

Remark. If a solution $x(t; x_0, t_0)$ of (1) is bounded for all $t \geqq t_0$, then its positive limiting set Γ^+ is a nonempty, compact set and $x(t; x_0, t_0) \to \Gamma^+$ as $t \to \infty$. Therefore, we can prolong the solution $\varphi(t)$ of (4) up to any t, and hence we can obtain a solution of (4) which passes through $(0, \omega)$ and remains in Γ^+ for all t. When Γ^+ has this property, we have said in the paper [*6*] that Γ^+ is a semi-invariant set of (4). If every solution of (4) is unique for Cauchy-problem, the semi-invariant set is an invariant set of (4).

The above result can be applied to discussing the asymptotic behavior of solutions of (1) [*6*].

REFERENCES

[*1*] Antosiewicz, H. A., Stable systems of differential equations with integrable perturbation term. *J. London math. Soc.* **31**, 208-212 (1956).

[*2*] Cesari, L., "Asymptotic Behavior and Stability Problems in Ordinary Differential Equations." Springer, Berlin, 1959.

[*3*] Hale, J. K., and Stokes, A. P., Behavior of solutions near integral manifolds. *Arch. rational Mech. Anal.* **6**, 133-170 (1960).

[*4*] Markus, L., Asymptotically autonomous differential systems. *In* "Contributions to the Theory of Nonlinear Oscillations (S. Lefschetz, ed.), Vol. 3, pp. 17-29. Princeton Univ. Press, Princeton, New Jersey, 1956.

[*5*] Opial, Z., Sur la dépendance des solutions d'un système d'équations différentielles de leurs seconds membres. Application aux systèmes presque autonomes. *Ann. polon. Math.* **8**, 75-89 (1960).

[*6*] Yoshizawa, T., Asymptotic behavior of solutions of a system of differential equations. *Contribs. Differential Equations* to be published.

On Some New Long Periodic Solutions of The Plane Restricted Three Body Problem

C. C. CONLEY
New York University, New York, New York

In this paper the well known three body problem is discussed, i.e. the following situation: Two bodies of arbitrary positive mass move in circles about their center of mass under the influence of gravity, and we study the motion of a third body of mass zero, so that it does not affect the motion of the other two.

In particular, we assume this zero mass point moves in the plane determined by the circles and close to one of the positive mass points, which corresponds to lunar orbits. We are interested in the existence of infinitely many periodic solutions of arbitrarily long period for large negative energy.

For convenience, we normalize the gravitational constant, the distance between the two positive mass-points, and the sum of the masses to be one; in particular we take the positive masses to be μ, $1 - \mu$, $0 < \mu < 1$. We study the motion of the zero mass-point in a coordinate system rotating so that the positive mass-points are fixed in the rotating system and assume the zero mass-point moves close to the body of mass μ which is to lie at the origin of the rotating system.

Birkhoff [*1a*] discusses the case where $\mu = 1$. Following H. Poincaré, he constructs an annulus map, fixed points of which correspond to periodic solutions of long period. The existence of fixed points is insured by the Birkhoff-Poincaré fixed point theorem (Theorem 2 below). Having this construction for $\mu = 1$, he is able, by a perturbation argument to extend it for small values of $1 - \mu$. In our treatment of the problem, μ is not treated as a parameter, and we obtain the solutions for all values of $0 < \mu < 1$. In particular, if μ is close to zero we have the case of lunar theory.

Let position coordinates in the rotating plane be given by x_1, x_2, and the conjugate momentum coordinates by y_1, y_2. If we let

$$x = x_1 + ix_1; \qquad y = y_1 + iy_2$$

the equations of motion are then given by

$$\dot{x} = H_y; \qquad \dot{y} = -H_x$$

where the Hamiltonian function H is given by[1]

$$H = \tfrac{1}{2} | y |^2 + i(x\bar{y} - \bar{x}y) - \frac{2\mu}{| x |} - \frac{2(1-\mu)}{| 1 + x |} - (1-\mu)(x + \bar{x}).$$

We know that H is a constant along solution paths, and the value of this constant is called the energy of the system. We study the equations on the energy surface $H = - C$ where C is a large positive constant.

The limit situation when the zero mass-point coincides with the body of mass μ (i.e. when $x = 0$) corresponds to a singularity in the equations of motion. An appropriate transformation, as given by Levi-Civita for example regularizes this singularity. Namely, if we let

$$x = \tfrac{1}{2} z^2; \qquad y = \frac{w}{\bar{z}}$$

we are led to a new Hamiltonian system with Hamiltonian function given by

$$K = | z |^2 (H + C)$$

where x, y in H are replaced by their expression in terms of z, w.

Solutions of the new equations,

$$\dot{z} = K_w \qquad \dot{w} = - K_z \tag{1}$$

correspond to solutions of the three body problem only on the energy surface $K = 0$; i.e. $H = - C$.

We notice that the new system of equations has an equilibrium solution; $z \equiv w \equiv 0$. Also, for large values of C, the energy surface $K = 0$ lies close to this equilibrium point so that in effect we have exchanged a study near a singularity for a study near an equilibrium.

To facilitate this study a series of transformations is introduced which lead to equations in the form[2] ($\epsilon^2 = 1/C$)

$$\dot{\xi}_1 = i \left[\xi_1 + \frac{\epsilon}{8} | \xi_1 | \xi_1 + 0_5(\xi) \right]$$

$$\dot{\xi}_2 = i \left[\xi_2 - \frac{\epsilon}{8} | \xi_2 | \xi_2 + 0_5(\xi) \right]$$

with an integral

$$G(\xi) = | \xi_1 |^2 + | \xi_2 |^2 + 0_4(\xi)$$

[1] See for example Wintner [*11*].

[2] $0_n(\xi)$ stands for terms of order n or more in ξ, $\bar{\xi}$.

where we are interested in solutions on the integral surface

$$G = 4\mu\epsilon^2.$$

We now recall that a theorem of Liapounov guarantees the existence of periodic solutions near an equilibrium point under conditions on the linearized equations. This theorem states that corresponding to each pair of purely imaginary eigenvalues $i\lambda$, $-i\lambda$ there exists a one parameter family of periodic solutions provided no other eigenvalue is an integer multiple of $i\lambda$.

In our case we have the double pair of eigenvalues $(i, -i)$ so that the theorem does not apply. However, using techniques similar to those of Cesari and Hale [B, pp. 125-136], a theorem is proved containing the following as a special case:

Theorem 1. Let there be given the system of equations

$$\dot{\xi}_1 = i\xi_1 + a\,|\,\xi_1\,|^2\,\xi_1 + b\,|\,\xi_2\,|^2\,\xi_2 + 0_5(\xi)$$

$$\dot{\xi}_2 = i\xi_2 + c\,|\,\xi_1\,|^2\,\xi_2 + d\,|\,\xi_2\,|^2\,\xi_2 + 0_5(\xi)$$

which admit the integral

$$G = |\,\xi_1\,|^2 + |\,\xi_2\,|^2 + 0_4(\xi)$$

then under the condition that $a \neq c$, there exists a one parameter family of periodic solutions given in the form

$$\xi_1 = \eta + \phi_1(\eta, \bar{\eta})$$

$$\xi_2 = 0 + \phi_2(\eta, \bar{\eta})$$

where

$$\eta = e^{i\alpha t}$$

α is a real function of $(\eta\bar{\eta}) = |\,\eta\,|^2$, and ϕ_1, ϕ_2 are of at least 2nd order in η, $\bar{\eta}$.

The condition $a \neq c$ can be described by saying that to the next highest order the natural frequencies differ.

A similar result holds, of course, when $b \neq d$. In our case we have $a = -d = \epsilon/8$ and $b = c = 0$ so that we obtain the existence of two families of periodic solutions in a neighborhood of the equilibrium. It follows that for large enough C (small enough ϵ) there will be one solution from each family on the integral surface $G = 4\mu\epsilon^2$. These are the well known Hill's solutions in the case when μ is small. They are analogous to the Moulton solutions of the genuine three body problem which can, in fact, be derived in a similar way.

To obtain the solution of long period, we construct, in the energy surface, an annular surface of section bounded by the pair of circular solutions. This is done by first transforming coordinates from ξ_k to η_k in such a way that the circular solutions are given by $\eta_k \equiv 0$ ($k = 1$ or 2). Letting $\eta_k = \rho_k \exp(i\theta_k)$, the annulus can be described as that set of points in the integral surface where $\psi = \theta_1 + \theta_2 \equiv 0$.

Since for small ϵ, $\dot{\psi}$ is close to 2, we see that the annulus is indeed a surface of section, and that the flow defines a mapping of the annulus into itself. The fixed points of iterates of this mapping clearly correrpond to periodic solutions of the equations. The existence of fixed points is guaranteed by means of the:

Theorem 2. (Birkhoff-Poincaré Fixed Point Theorem). Let T be a mapping of an annulus into itself which takes boundaries into themselves, and satisfies the following conditions:

(1) The rotation numbers of the boundary maps differ from each other.

(2) The map takes no open set properly interior to itself.

It follows that the set of fixed points of iterates of T is infinite.

In fact, there are an infinite number of integers n for which T^n has a fixed point P, and $T^m P \neq P$ for $0 < m < n$.

Condition (2) of the theorem follows in our case from the Hamiltonian character of the equations. In particular, that a four dimensional volume element is preserved. From this we conclude that a 3 dimensional volume element is preserved in each energy surface, and condition (2) follows easily.

Condition (1) is shown to be satisfied by means of a lengthy computation the result of which is that if β_1, β_2 are the rotation numbers of the boundary maps, then

$$|\beta_1 - \beta_2| = \tfrac{3}{2}\mu^2\epsilon^6 + 0(\mu^3\epsilon^8).$$

The conclusions drawn are embodied in the following theorem:

Theorem 3. For the restricted plane three body problem as described above, there exists a large negative constant E_0 independent of the mass ratio of the two positive masses such that in each energy surface with energy less than E_0 there are infinitely many periodic solutions of arbitrarily long period.

We remark that in the rotating coordinate system these solutions look like ellipses precessing about the mass μ. They are periodic solutions in the rotating system but not necessarily in the inertial system.

In conclusion I would like to thank Professor J. Moser who suggested this problem and without whose aid it would not have been completed.

REFERENCES

[*1*] Birkhoff, G. D., "Collected Mathematical Papers," in 3 volumes. Am. Math. Soc., Providence, Rhode Island, 1950:
(*a*) The restricted problem of three bodies. Vol. I, p. 752.
(*b*) Sur le problème restreint des trois corps (I). Vol. II, p. 466.
(*c*) Sur le problème restreint des trois corps (II). Vol. II, p. 668.
(*d*) Poincaré's geometric theorem. Vol. I, p. 673.
(*e*) An extension of Poincaré's last geometric theorem. Vol. II, p. 252.
(*f*) Dynamical systems with two degrees of freedom. Vol. II, see particularly p. 286ff.

[*2*] Birkhoff, G. D., and Lewis, D. C., On Periodic Motions Near a Given Periodic Motion of a Dynamical System.

[*3*] Cesari, L., "Asymptotic Behavior and Stability Problems in Ordinary Differential Equations." Springer, Berlin, 1959.

[*4*] Hill, G. W., Researches in lunar theory. *Am. J. Math.* **1** (1878).

[*5*] Levi-Civita, T., Sur la régularisation du problème des trois corps. *Acta Math. (Stockh.)* **42**, 99-144 (1920).

[*6*] Moser, J., On a generalization of a theorem of A. Lyapounov. *Communs. pure appl. Math.* May (1958).

[*7*] Moulton, F. R., A class of periodic solutions of the problem of three bodies with applications to lunar theory. *Trans. AMS.* pp. 537-577 (1906).

[*8*] Nemickii, V. V., and Stepanov, V. V., "Qualitative Theory of Differential Equations." Princeton Univ. Press, Princeton, New Jersey, 1960.

[*9*] Poincaré, H., Sur un théorème de géométrie. *Rend. Circ. mat. Palermo* **33**, 375-407 (1912).

[*10*] Siegel, C. L., "Vorlesungen über Himmelsmechanik." Springer, Berlin, 1956.

[*11*] Wintner, A., "Analytical Foundations of Celestial Mechanics." Princeton Univ. Press, Princeton, New Jersey, 1941.

Continuous Groups and Autosynartetic Solutions

ROBERT W. BASS,
Aeronca Manufacturing Corporation
Baltimore, Maryland
DANIEL C. LEWIS
Johns Hopkins University
Baltimore, Maryland.

1. Introduction

We refer to the concept of an autosynartetic solution of a system of ordinary differential equations,

$$dx/dt = f(t, x), \tag{1.1}$$

under a transformation S of the form $s = P(t, x)$, $y = k(t, x)$, as developed in a previous paper [3]. In these formulas, t, s, and P are scalars, while x, f, y, and k are n-vectors. The functions f, P, k are supposed to be of class C' in a suitable region. The system (1.1) is assumed to be invariant under S. Briefly this means that, if $x(t)$ is an arbitrary solution of (1.1), there exists another solution $y(t)$ such that

$$y[P(t, x(t))] \equiv k[t, x(t)].$$

This solution $y(t)$ is said to be synartetic to $x(t)$ under S; and, if furthermore $y(t) \equiv x(t)$, then $x(t)$ is said to be autosynartetic under S.

Part of the previous paper considered the case in which S was imbedded in a continuous group of transformations S_λ, which we shall write in the form,

$$s = P(t, x, \lambda), \qquad y = k(t, x, \lambda), \tag{1.2}$$

where λ is the parameter of the group. We shall consider here only the case where λ is a real scalar, and we may suppose that $S = S_{\lambda_0}$. It was shown that, if (1.1) is invariant under every transformation of such a group, then the autosynartetic solutions of (1.1) under S have certain properties analogous to corresponding properties of periodic solutions of autonomous systems. For instance, such an autosynartetic solution is necessarily degenerate in the sense defined in [3].

The purpose of the present paper is to develop this order of ideas a bit further. The Lie theory of the application of continuous transformation groups to differential equations is used to reduce the problem of integrating (1.1) near a given autosynartetic solution to a similar problem involving a system whose order is $n - 1$ (instead of n) followed by a quadrature. This result is analogous to well known results on the introduction of a so-called "angular variable" in the autonomous periodic case. It is also proved that the appropriate autosynartetic solution of the reduced system has degeneracy exactly one less than the degeneracy of the corresponding autosynartetic solution originally given for the system (1.1).

Our results include a proof of the fact that if (1.1) is invariant under a one-parameter family of transformations S_λ given by (1.2), then (1.1) is, in general, through an easily available transformation of the form $\tau = \Pi(t, x)$, $\xi = \Xi(t, x)$, equivalent to an *autonomous* system $d\xi/d\tau = \varphi(\xi)$, where ξ, Ξ, φ are n-vectors, while τ and Π are scalars. Furthermore, any solution of (1.1) which is autosynartetic under S_{λ_0} corresponds to a periodic solution of $d\xi/d\tau = \varphi(\xi)$ with a period which is a simple function of λ_0. In fact the period is precisely equal to λ_0, if the parameter λ is chosen as an additive parameter, so that $S_{\lambda'+\lambda''} = S_{\lambda'}S_{\lambda''}$.

Finally the preparatory Sections 2 and 3 contain a number of routine theorems in the general development of the theory of autosynartetic solutions which are not dependent on the imbedment of S in a group S_λ.

2. Invariance of the Property of Autosynarteticity

In the sequel we shall have several occasions to transform a system (1.1) into another system, say

$$d\xi/d\tau = \varphi(\tau, \xi), \tag{2.1}$$

of the same type, by means of a transformation T of the form,

$$\tau = \Pi(t, x), \qquad \xi = \Xi(t, x), \tag{2.2}$$

where, as before, ξ, φ, Ξ are n-vectors, while τ and Π are scalars, and φ, Π, and Ξ are of Class C'. We shall also assume that the Jacobian of T is not zero, so that T has an inverse.

Theorem 2.1. If (1.1) is transformed into (2.1) by the transformation T given by (2.2), and if (1.1) is invariant under S, then the system (2.1) is invariant under $\Sigma = TST^{-1}$, and a solution x of (1.1) is autosynartetic under S if and only if the corresponding solution $\xi = Tx$ of (2.1) is autosynartetic under Σ. This "corresponding solution" ξ is here meant to be the solution of (2.1) related to $x(t)$ by the functional equation,

$$\xi(\Pi(t, x(t))) \equiv \Xi(t, x(t)).$$

The proof of this theorem follows obvious lines and will be omitted.

3. Invariance of the Property of Degeneracy

Let us now consider the variational system of (1.1)

$$dx^*/dt = A(t)\, x^* \tag{3.1}$$

based on a fixed solution $x(t)$ of (1.1) so that $A(t)$ is the matrix $f_x(t, x(t))$. We also consider the variational system of (2.1)

$$d\xi^*/d\tau = \Gamma(\tau)\, \xi^*, \tag{3.2}$$

based on the solution $\xi = Tx$ of (2.1), so that $\Gamma(\tau)$ is the matrix $\varphi_\xi(\tau, \zeta(\tau))$. We wish to consider a transformation T^* which takes (3.1) into (3.2). This T^*, to be called the variational modification of T with respect to the solution of x of (1.1), will have the form

$$\tau = \Pi(t, x(t)), \qquad \xi^* = B(t)\, x^*, \tag{3.3}$$

where

$$B(t) = \Xi_x(t, x(t)) - \varphi[\Pi(t, x(t)), \Xi(t, x(t))]^0\, \Pi_x(t, x(t)),$$

(here $\varphi^0\Pi_x$ means the matrix the element in whose ith row and jth column is the ith component of φ multiplied by the jth component of Π_x, so that $(\varphi^0\Pi_x)\, x^* = \varphi(\Pi_x x^*)$, where $\Pi_x x^*$ denotes, of course, the scalar product of the vectors Π_x and x^*). We may now cite without proof the following theorems:

Theorem 3.1. The transformation T^* given by (3.3) transforms the linear variational system (3.1) of (1.1) into the linear variational system (3.2) of (2.1). It is hereby assumed that the variational systems (3.1) and (3.2) are based on solutions of (1.1) and (2.1), respectively, that correspond to each other under the transformation T given by (2.2).

Theorem 3.2. Assuming that the Jacobian of T is not zero, the transformation T^* of the preceding theorem (i.e., the variational modification of T) also has a non-vanishing Jacobian and hence has an inverse.

We next consider a further transformation R given by equations of the form

$$\sigma = \Lambda(\tau, \xi), \qquad \zeta = Z(\tau, \xi), \tag{3.4}$$

where σ and Λ are scalars while ζ and Z are n-vectors The functions Λ and Z are of class C'. We also consider the variational modification R^* of R with respect to $\xi = Tx$ as well as the variational modification $(RT)^*$ of RT with respect to x. Then the following theorems may also be established:

Theorem 3.3. $(RT)^* = R^*T^*$.

Theorem 3.4. Denoting by $(T^{-1})^*$ the variational modification of T^{-1} with respect to the solution $\xi = Tx$ of (2.1), $(T^{-1})^* = (T^*)^{-1}$.

Let $x(t)$ be an autosynartetic solution of (1.1) under S and let S^* be the variational modification of S with respect to $x(t)$. Then, by definition, the degeneracy of $x(t)$ is the maximum number of solutions of the vatiational system (3.1) in any set of linearly independent solutions autosynartetic under S^*.

Theorem 3.5. Under the assumption of Theorem 2.1 let us further assume that x is an autosynartetic solution of (1.1) under S and that consequently $\xi = Tx$ is an autosynartetic solution of (2.1) under $\Sigma = TST^{-1}$. Then the degeneracy of x is the same as the degeneracy of ξ.

This, of course, is the principal theorem of the present section. Its proof, which we omit, is a fairly straightforward application of Theorems 3.3 and 3.4.

4. Fundamental Properties of Systems invariant under One-Parameter Groups

From now on a fundamental assumption is that the system (1.1) is invariant, for every value of the real parameter λ, under the transformation S_λ given by (1.2). Furthermore it is assumed that these transformations form a group G. According to a familiar result of the Lie theory, we may, without loss of generality, assume that the parameter λ is *additive*. Known results of this assumption are to the effect that $S_0 = I$, the identity element in G, that $S_\lambda^{-1} = S_{-\lambda}$, and that there exist functions $\psi(s, y)$ and $\varphi(s, y)$ where ψ is a scalar and φ an n-vector, such that the P and k of (1.2) satisfy the following system of autonomous differential equations.

$$\begin{aligned} P_\lambda(t, x, \lambda) &= \psi[P(t, x, \lambda), k(t, x, \lambda)] \\ k_\lambda(t, x, \lambda) &= \varphi[P(t, x, \lambda), k(t, x, \lambda)] \end{aligned} \tag{4.1}$$

We assume that ψ, φ, P, and k are all of class C'.

Theorem 4.1. Let $x(t)$ be the solution of (1.1) such that $x(t_0) = x_0$, where t_0 and x_0 are constant. Let $y(s, \lambda) = S_\lambda x(t)$ be the solution synartetic to $x(t)$ under S_λ. Then $y(s, \lambda)$, considered as a function of λ, satisfies the autonomous system,

$$y_\lambda(s, \lambda) = \varphi[s, y(s, \lambda)] - f[s, y(s, \lambda)]\, \psi[s, y(s, \lambda)]. \tag{4.2}$$

and the initial condition,

$$y(s, 0) = x(s). \tag{4.3}$$

If we take $s = t_0$, we have in particular

$$y(t_0, 0) = x(t_0) = x_0. \tag{4.4}$$

Proof. Since $y = S_\lambda x$, we have the identity

$$y[P(t, x(t), \lambda), \lambda] \equiv k(t, x(t), \lambda). \tag{4.5}$$

We differentiate (4.5) with respect to λ and use (4.1). We also use the fact that $y_s(s, \lambda) = f[s, y(s, \lambda)]$, which expresses the fact that y, considered as a function of its first argument, is a solution of (1.1). In this way we readily find

$$f[P(t, x(t), \lambda), k(t, x(t), \lambda)]\, \psi[P(t, x(t), \lambda), k(t, x(t), \lambda)] + y_\lambda[P(t, x(t), \lambda), \lambda]$$
$$\equiv \varphi[P(t, x(t), \lambda), k(t, x(t), \lambda)].$$

Hence, if we set

$$s = P(t, x(t), \lambda) \tag{4.6}$$

so that (4.5) becomes $y(s, \lambda) = k(t, x(t), \lambda)$, and if we recall that $S_\lambda^{-1} = S_{-\lambda}$, we obtain

$$t = P(s, y(s, \lambda), -\lambda), \qquad x(t) = k(s, y(s, \lambda), -\lambda) \tag{4.7}$$

and

$$f[s, y(s, \lambda)]\, \psi[s, y(s, \lambda)] + y_\lambda[s, \lambda] = \varphi[s, y(s, \lambda)],$$

which is equivalent to (4.2).

Setting $\lambda = 0$ in (4.7) and remembering that $S_0 = I$, we have $t = s$ and $x(t) = y(s, 0) = x(s)$, thus establishing (4.3) and completing the proof of the theorem.

In the sequel we shall always assume that

$$-P_\lambda(t, x, \lambda) + P_x(t, x, \lambda)\, [\varphi(t, x) - f(t, x)\, \psi(t, x)] \neq 0. \tag{4.8}$$

As a consequence of this assumption, notice that, if $y(s, \lambda)$ is known for a particular value of s, $x(t)$ is determined by (4.7). Namely the first equation of (4.7) determines $\lambda = \lambda(t, s)$ so that $\lambda(s, s) = 0$ (since $S_0 = I$), whence the second equation of (4.7) yields

$$x(t) = k[s, y(s, \lambda(t, s)), -\lambda(t, s)]. \tag{4.9}$$

A slight elaboration of this reasoning allows us to establish the following.

Theorem 4.2. Under the assumption (4.8), let $y(t_0, \lambda)$ satisfy the system (4.2) with $s = t_0$ and the initial condition

$$y(t_0, 0) = x_0. \tag{4.10}$$

Let $\lambda(t, t_0)$ be determined implicitly by the equation,

$$t = P(t_0, y(t_0, \lambda), -\lambda). \tag{4.11}$$

Then the vector function

$$x(t) = k[t_0, y(t_0, \lambda(t, t_0)), -\lambda(t, t_0)] \tag{4.12}$$

satisfies (1.1) and the initial condition $x(t_0) = x_0$.

We next cite the following rather interesting theorem.

Theorem 4.3. A necessary and sufficient condition that a solution $x(t)$ of (1.1) be autosynartetic under S_{λ_0} is that the solution $y(s, \lambda)$ of (4.2), for which $y(s, 0) = x(s)$, be periodic in λ with period λ_0. Here s is any fixed number.

For brevity we prove only the necessity of the condition. Namely, if $x(t)$ is autosynartetic under S_{λ_0}, then by definition of autosynarteticity $y(t, \lambda_0) \equiv x(t)$, since it is known from Theorem 4.1 and the uniqueness theorem for the system (4.2) that $y(t, \lambda)$ is synartetic to $x(t)$ under S_λ. Hence $y(s, \lambda_0) = x(s)$. But $y(s, 0) = x(s)$ by hypothesis. Therefore $y(s, \lambda_0) = y(s, 0)$ and, since the system (4.2) is autonomous, $y(s, \lambda)$ must be periodic as required.

Theorems 4.1 and 4.2 show that the integration of (1.1) is equivalent to the integration of the autonomous system

$$d\xi/d\tau = \varphi(t_0, \xi) - f(t_0, \xi)\, \psi(t_0, \xi), \tag{4.13}$$

where t_0 is an arbitrary constant, only one value of which need be considered. We have here changed the notation from y to ξ and from λ to τ because in the next section we wish to discuss the transformation from (1.1) to (4.13) in the notation of Sections 2 and 3.

5. On the Degeneracy of a Periodic Solution of (4.13)

Let us consider the transformation T which carries (1.1) into (4.13). This transformation is not very simple to write down, but a perusal of the preceding section, shows that its inverse T^{-1}, which sends (τ, ξ) into (t, x) is given by the equations

$$t = P(t_0, \xi, -\tau), \qquad x = k(t_0, \xi, -\tau).$$

The transformation $S = S_{\lambda_0}$, sending (t, x) into (s, y) is, of course, $s = P(t, x, \lambda_0)$, $y = k(t, x, \lambda_0)$.

Theorem 5.1. The transformation TST^{-1} taking (τ, ξ) into (σ, η) is given by the formulas $\sigma = \tau - \lambda_0$, $\eta = \xi$.

In the proof we let Σ denote the transformation defined by the last two simple formulas and then prove that $T^{-1}\,\Sigma = ST^{-1}$. The details are straight-forward and will be omitted.

Theorems 2.1 and 5.1 give an alternative proof of Theorem 4.3. Moreover Theorems 3.5 and 5.1 yield immediately the following.

Theorem 5.2. The degeneracy of an autosynartetic solution of (1.1) under S_{λ_0} is the same as the degeneracy of the corresponding periodic solution with period λ_0 of (4.13).

6. On the Reduction of Order of the System (4.13) by Introduction of an Angular Variable

Once the system (1.1) has been reduced to the autonomous system (4.13) we may by known methods introduce an angular variable (cf. for example Lewis [*2*] or Diliberto [*1*]), and write (4.13) in the form

$$d\xi/d\tau = F(\theta, \xi) \qquad d\theta/d\tau = \Theta(\theta, x), \tag{6.1}$$

where $F(\theta, 0) \equiv 0$, $\Theta(\theta, 0) = 2\pi\lambda_0^{-1}$. Here ξ and F are now $(n - 1)$-vectors, while θ and Θ are scalars. Moreover F and Θ are periodic in θ with period 2π. Such a reduction is possible in a neighborhood of a given periodic solution and will thus have significance in the study of (1.1) near the corresponding autosynartetic solution. The corresponding periodic solution of (4.13), when written in the form (6.1), is, of course, $\xi \equiv 0$, $\theta \equiv 2\pi\lambda_0^{-1}\tau \pmod{2\pi}$.

From (6.1) we may eliminate the independent variable τ by writing the following non-autonomous system of order $(n - 1)$.

$$d\xi/d\theta = F(\theta, \xi)/\Theta(\theta, \xi). \tag{6.2}$$

Theorem 6.1. The degeneracy of the periodic solution $\xi \equiv 0$, $\theta = 2\pi\lambda_0^{-1}\tau \pmod{2\pi}$ of period λ_0 of (6.1) is exactly one greater than the degeneracy of the corresponding periodic solution $\xi \equiv 0$ of (6.2).

The proof of this intuitively evident theorem is not entirely trivial but must nevertheless, for lack of space, be omitted here. We discuss, of course, the linearly independent periodic solutions of the variational equations of both (6.1) and (6.2), but this turns out to be sufficient to prove only that $d \leqq \delta + 1$, where d is the degeneracy of (6.1) and δ that of (6.2). To prove that $d \geqq \delta + 1$, we need also to study the adjoint systems, which, as is well known, also determine the degeneracies in question.

Urabe has proved almost this theorem [*4*], but he formulated his results in terms of characteristic exponents instead of in terms of degeneracies. On account of the possibility of different sorts of elementary divisors, his result does not seem to be exactly the same as ours.

Theorem 6.1, taken in conjunction with the results of Sections 2-5, leads at once to all the results mentioned in the Introduction, Section 1.

REFERENCES

[*1*] Diliberto, S. P., and Hufford, G., Perturbation theorems for nonlinear ordinary differential equations. *In* "Contributions to the Theory of Nonlinear Oscilations" (S. Lefschetz, ed.), Vol. 3, pp. 207-236. Princeton Univ. Press, Princeton, New Jersey, 1956 (see especially Theorem 3.1).

[*2*] Lewis, D. C., On the perturbation of a periodic solution when the variational has non-trivial periodic solutions. *J. rational Mech. and Anal.* **4**, 795-815 (1955).

[*3*] Lewis, D. C., Autosynartetic solutions of differential equations. *Am. J. Math.* **83**, 1-32 (1961).

[*4*] Urabe, M., Geometric study of nonlinear autonomous oscillations. *Funk. Ekv.* **1**, 1-84 (1958) (see especially Theorem 4, p. 19).

Autosynartetic Families of Solutions

DANIEL C. LEWIS
Mathematics Department
The Johns Hopkins University, Baltimore, Maryland

We consider a system of ordinary differential equations of the form,

$$dx/dt = f(t, x), \tag{1}$$

where t is the scalar independent variable, where x and f are n-vectors, and where $f(t, x)$ is of class C' in t and x. It is assumed that for every x_0 representing a point on some n-dimensional set M there exists a solution $x(t)$ defined for $-\infty < t < +\infty$, such that $x(0) = x_0$ and $x(t) \in M$ for all t.

We suppose also that the system (1) is invarient under a transformation involving, in general, both the independent variable t and the dependent vector variable x. We write this transformation in the form

$$t' = P(t, x), \qquad x' = h(t, x), \tag{2}$$

where P is a scalar function and h is a vector function, both being of class C' in t and x.

The supposition that (1) is invariant under (2) means that to every solution $x(t)$ of (1) there corresponds a solution $x'(t)$ of (1) such that $x'[P(t, x(t))] \equiv h[t, x(t)]$. In accordance with the terminology introduced in an earlier paper [*2*], $x'(t)$ is said to be synartetic to $x(t)$ under the transformation (2).

We are primarily interested in families of solutions $\{x(t, \alpha), \alpha \in \Sigma\}$, which are closed under the operation of setting up their synartetic solutions. That is, for such a family, we assume that for every $\alpha \in \Sigma$ there exists $\beta \in \Sigma$, such that the solution $x(t, \beta)$ is synartetic to the solution $x(t, \alpha)$ under (2). We call families of solutions of this kind autosynartetic families of solutions. This idea generalizes the notion of an autosynartetic *solution*, in which the index set Σ consists of just one element [*2*]. It also generalizes the notion of a periodic k-manifold which arises when Σ is a k-manifold ($k < n$) and $f(t - 2\pi, x) \equiv f(t, x)$ so that we may take $P(t, x) \equiv t - 2\pi$, $h(t, x) \equiv x$ [*3*].

It is useful to observe that questions related to autosynartetic families of solutions are equivalent to corresponding questions related to invariant sets of a certain transformation $T : x_0 \rightarrow y_0$, which we define as follows: Let x_0 be any point of M. Let $x(t)$ be the solution of (1) such that $x(0) = x_0$. Let $y(t)$ be synartetic to $x(t)$ and then set $y_0 = y(0)$.

This transformation T may be given implicitly in terms of the general solution $\varphi(t, x_0)$ of (1) such that $\varphi(0, x_0) = x_0$. For let $\varphi(t, y_0)$ be synartetic to $\varphi(t, x_0)$. Then, by definition of synarteticity (under (2)) we have the identity, $\varphi[P(t, \varphi(t, x_0)), y_0] \equiv h(t, \varphi(t, x_0))$. Now setting $t = 0$, we get $\varphi[P(0, x_0), y_0] = h(0, x_0)$, which furnishes the desired relationship between x_0 and its image y_0 under T.

Conversely, it follows from [*2*, p. 7, Theorem 3.3] that, when x_0 and y_0 are related by this last formula, the solution $y(t)$ such that $y(0) = y_0$ is synartetic under (2) to the solution $x(t)$ such that $x(0) = x_0$. Thus, synartetic families of solutions correspond to invariant sets under T and conversely. In particular, an invariant point under T corresponds to an autosynartetic solution.

Notice that in the special case $P(t, x) \equiv t - 2\pi$, $h(t, x) \equiv x$, the above formula giving T reduces to $\varphi(-2\pi, y_0) = x_0$, or $y_0 = \varphi(2\pi, x_0)$, which is a transformation often considered in this "periodic" special case.

One kind of problem inviting consideration occurs when $f(t, x)$ depends upon a parameter γ and when, for a particular value of γ, say $\gamma = 0$, we have a known autosynartetic family of solutions. We then wish to inquire whether such an autosynartetic family still exists for $\gamma \neq 0$ but for $|\gamma|$ sufficiently small. In this "perturbution" problem, the transformation T also depends on the parameter γ; and, when $\gamma = 0$, it is known that T admits an invariant set. We wish to inquire whether this invariant set persists in recognizable form for $\gamma \neq 0$. Our problem thus reduces to problems of a type previously considered. We refer particularly to some work of Walter T. Kyner [*1*], which, through the use of the transformation T, may be applied directly to the present problem.

In this investigation, I have generalized somewhat the results of Kyner and adapted them to the present problem. Changing our notation, to conform with Kyner's, we write T, in suitable coordinates, in the form

$$\bar{x} = N(\theta)x + W(x, \theta, \gamma)$$

$$\bar{\theta} = V(x, \theta, \gamma).$$

Here x is an n-vector, θ is an s-vector, and $n + s$ is what was previously called n. $N(\theta)$ is an $n \times n$ matrix, W is an n-vector, and V is an s-vector. All three of these functions, N, W, V are supposed to be of class C^l in their arguments.

We assume that $W(0, \theta, 0) \equiv 0$ and $W_x(0, \theta, 0) \equiv 0$, so that a given s-dimensional invariant manifold, when $\gamma = 0$, is represented simply by $x = 0$. The problem of perturbing this invariant manifold Σ_0 is equivalent to the problem of finding a fixed point of a certain mapping T_γ of the set S of all sufficiently regular n-vector functions $f(\theta)$, defined on Σ_0, into S.

We introduce various metrics for these vector functions:

$$\| f \|_0 = \max_i \max_\theta | f_i(\theta) |, \qquad i = 1, ..., n.$$

$$\| f \|_k = \max [\| f \|_0, \| \partial f/\partial \theta_p \|_0, ..., \| \partial^k f/\partial \theta_p ... \partial \theta_q \|_0], \qquad p, q = 1, ..., s.$$

Evidently the definition of $\| f \|_k$ is effective if f is of class C^k. The coordinates θ used over Σ_0 are not necessarily uniform; different coordinates may be used in a finite number of overlapping neighborhoods covering Σ_0. The definition of the metric, however, refers to a preferred coordinate system in each of these neighborhoods.

For a fixed γ, the transformation $g = T_\gamma(f)$ is given by

$$g(\bar{\theta}) = N(\theta) f(\theta) + W[f(\theta), \theta, \gamma],$$

where θ is a function of $\bar{\theta}$, f, and γ, defined implicitly by

$$\bar{\theta} = V_{f\gamma}(\theta) = V[f(\theta), \theta, \gamma]. \tag{3}$$

To insure that (3) can be solved for θ, we assume that γ and f may be so restricted that the $s \times s$ matrix $\partial\bar{\theta}/\partial\theta$ is non-singular.

For each f and γ with $\| f \|_k$ and $| \gamma |$ sufficiently small, define the *linear* transformation $L_{f\gamma}$ by the formula,

$$[L_{f\gamma} g] (\theta) = g[V_{fr}(\theta)] - N(\theta) g(\theta). \tag{4}$$

For convenience in stating our theorem we refer to hypothesis H_k, which is to the effect that $L_{f\gamma}$ is invertible and that its inverse is uniformly bounded in the metric $\| ... \|_k$. That is $\| L_{f\gamma}^{-1} g \| \leqq A \,.\, \| g \|_k$ for all g of class C^k, at least so long as $\| f \|_k$ and $| \gamma |$ are sufficiently small. A is a positive constant (independent of both g and f).

Theorem. If T_γ is defined as above and if the hypotheses H_k and H_{k+1} are satisfied (for some $k < l$) then there exists a positive number γ_0 such that, if $| \gamma | \leq \gamma_0$, T_γ has a unique fixed point f_γ of class C^k in θ and γ, and its kth derivatives satisfy a uniform Lipschitz condition in θ.

We can give here only a brief outline of the proof.

If f is invariant under T_γ, then

$$f[V_{f\gamma}(\theta)] = N(\theta) f(\theta) + W[f(\theta), \theta, \gamma].$$

Therefore

$$f[V_{f\gamma}(\theta)] - N(\theta) f(\theta) = W[f(\theta), \theta, \gamma].$$

In other words, $[L_{f\gamma} f](\theta) = W[f(\theta), \theta, \gamma]$, so that $f = L_{f\gamma}^{-1} W[f, \gamma]$, where $[W(f, \gamma)](\theta) = W[f(\theta), \theta, \gamma]$.

Thus it is necessary to show the existence of a fixed point of the transformation $f \to g$, given by the formula $g = L_{f\gamma}^{-1} W[f, \gamma]$.

From hypothesis H_k and H_{k+1}, we have at once that

$$\| g \|_k \leqq A \| W(f, \gamma) \|_k \quad \text{and} \quad \| g \|_{k+1} \leqq A \| W(f, \gamma) \|_{k+1},$$

if $\| f \|_k$, $\| f \|_{k+1}$ and $| \gamma |$ are sufficiently small. We next use the Taylor expansions of $W(x, \theta, \gamma)$ with respect to x and γ about the point $x = 0$ and $\gamma = 0$. Remembering that $W_x(0, \theta, 0) \equiv 0$ as well as all the partial derivatives of W with respect to the θ's, when $x = 0$, $\gamma = 0$, we are enabled to choose positive numbers r_1, r_2 and γ_1 so that $\| g \|_k < r_1$ and $\| g \|_{k+1} < r_2$ as long as $\| f \|_k < r_1$, $\| f \|_{k+1} < r_2$, and $| \gamma | < \gamma_1$.

But the convex set S^* of functions f such that $\| f \|_k < r_1$, and $\| f \|_{k+1} < r_2$ is, by the Ascoli-Arzela theorem, conditionally compact in the $\| \ldots \|_k$ metric, although not necessarily in the $\| \ldots \|_{k+1}$ metric. Once we show that, if $f \in \bar{S}^*$, then $g = L_{f\gamma}^{-1} W[f, \gamma] \in \bar{S}^*$, and that f depends *continuously* on g in the sense of the $\| \ldots \|_k$ metric, the Schauder fixed point theorem shows the existence of a fixed point as required by our theorem. Uniqueness and continuity in γ is proved by means of a supplementary argument, which also points the way to an alternative existence proof by a method of successive approximations. We have space here only to settle the important point that if $f \in \bar{S}^*$, then also $g \in \bar{S}^*$.

Let

$$f_m \in S^* \quad \text{and} \quad \lim_{m \to \infty} \| f_m - f \|_k = 0.$$

Define

$$g_m = L_{f_m \gamma}^{-1} W[f_m, \gamma], \tag{5}$$

so that, as above, $g_m \in S^*$. Hence, from the conditional compactness of S^*, we know that there exists $g \in \bar{S}^*$ such that

$$\lim_{m \to \infty} \| g_m - g \|_k = 0,$$

at least if we confine attention to a suitable subsequence, as we henceforth do. It now remains only to prove that $g = L_{f\gamma}^{-1} W[f, \gamma]$ or in other words that

$$[L_{f\gamma} g](\theta) = W[f(\theta), \theta, \gamma]. \tag{6}$$

From (5) we have $[L_{f_m\gamma} g_m](\theta) = W[f_m(\theta), \theta, \gamma]$. By (4) this last equation may be written

$$W[f_m(\theta), \theta, \gamma] + N(\theta) g_m(\theta) = g_m[V(f_m(\theta), \theta, \gamma)]. \tag{7}$$

From the continuity of W, it is clear that the left hand side of (7) approaches a certain limit,

$$g^*(\theta) = W[f(\theta), \theta, \gamma] + N(\theta) g(\theta). \tag{8}$$

We also obviously have

$$\begin{aligned} | g^*(\theta) - g[V(f(\theta), \theta, \gamma)] | \leqq & \; | g^*(\theta) - g_m[V(f_m(\theta), \theta, \gamma)] | \\ & + | g_m[V(f_m(\theta), \theta, \gamma)] - g_m[V(f(\theta), \theta, \gamma)] | \\ & + | g_m[V(f(\theta), \theta, \gamma)] - g[V(f(\theta), \theta, \gamma)] |. \end{aligned} \tag{9}$$

Evidently $| g^*(\theta) - g_m[V(f_m(\theta), \theta, \gamma)] | \to 0$, since the left, and hence the right, hand side of (7) tends to $g^*(\theta)$. Also

$$| g_m[V(f_m(\theta), \theta, \gamma)] - g_m[V(f(\theta), \theta, \gamma)] | \to 0$$

because of the equicontinuity of $\{g_m\}$, the continuity of V, and the fact that $f_m \to f$. Finally

$$| g_m[V(f(\theta), \theta, \gamma)] - g[V(f(\theta), \theta, \gamma)] | \to 0$$

because $g_m \to g$. Hence, from (9), we have $g^*(\theta) = g[V(f(\theta), \theta, \gamma)]$, so that, from (8), we have also

$$W[f(\theta), \theta, \gamma] + N(\theta) g(\theta) = g[V(f(\theta), \theta, \gamma)],$$

which is simply another way of writing (6).

We conclude our outline of the proof of this modified theorem of Kyner, by remarking (with him) that the continuity of the transformation leading from f to g, in the metric $|| \dots ||_k$, is proved by a repetition of the above argument except that we allow $f_m \in \bar{S}^*$ instead of restricting it further to S^* as above.

REFERENCES

[*1*] Kyner, W. T., A fixed point theorem. *In* "Contributions to the Theory of Nonlinear Oscillations" (S. Lefschetz, ed.), Vol. 3, pp. 197-205. Princeton Univ. Press, Princeton, New Jersey, 1956.

[*2*] Lewis, D. C., Autosynartetic solutions of differential equations. *Am. J. Math.* **83**, 1-32 (1961).

[*3*] Lewis, D. C., Poincaré's theorem generalized for the perturbation of invariant manifolds. *Proc. intern. Symposium on Ordinary Differential Equations, Mexico City, 1959*, pp. 205-219 (1961).

Periodic Functions Generated as Solutions of Nonlinear Differential-Difference Equations[1]

G. STEPHEN JONES
RIAS, Baltimore, Maryland

1. Introduction

Consider functional equations of the form

$$f'(x) = -\alpha \sum_{i=1}^{n} \lambda_i f(x - \tau_i) [af(x)^2 + bf(x) + 1], \tag{1}$$

where $0 < \tau_1 < \tau_2 < \ldots < \tau_n$ and α, a, b and the λ_i's are real parameters. Such equations, usually categorized as differential-difference equations with retarded arguments, occur in an impressively wide and varied field of applications. Most prevalent among these are control systems, biological growth behavior, and econometrics. One of the simplest members of this class,

$$f'(x) = -\alpha f(x - 1) [1 + f(x)], \tag{2}$$

even appears in the application of probability methods in the determination of the asymptotic density of prime numbers. For further discussion of these applications the reader is referred to Bellman [*1*], Lord Cherwell [*2*], Cunningham [*3*], and Wright [*11*].

Now in defining what we shall mean by a solution of equation (1), we specify φ to be a bounded and Lebesgue integrable function defined on the interval $(-\tau_n, 0]$. A function f such that $f(x) = \varphi(x)$ for x in $(-\tau_n, 0]$ and satisfying the equation

$$f(x) = f(0) - \alpha \sum_{i=1}^{n} \lambda_i \int_0^x f(t - \tau_i) [af(t)^2 + bf(t) + 1] \, dt, \tag{3}$$

for x in some interval $(0, X)$, $X > 0$, is referred to as a solution of equation (1) on $(0, X)$ corresponding to the initial function φ. Using

[1] This research was partially supported by the United States Air Force through the Air Force Office of Scientific Research of the Air Research and Development Command, under Contract Number AF 49(638)-382. Reproduction in whole or in part is permitted for any purpose of the United States Government.

the method of successive approximation it is easily shown that corresponding to each such initial specification φ, there exists an interval $(0, X)$ on which equation (1) has a unique solution. It follows that a solution is uniquely determined by its specification on any interval of length τ_n within its domain of definition.

In the sequel φ always denotes an initial function of the type described in the previous paragraph and f denotes the corresponding solution of equation (1). The notation $\varphi \sim f$ signifies that f is the solution of (1) corresponding to φ. In addition, we assume throughout that $\alpha > 0$ and that $b^2 - 4a \geqq 0$ which implies the quadratic equation

$$ay^2 + by + 1 = 0 \tag{4}$$

has real roots. For $a \neq 0$ we denote these roots by r_1 and r_2 where $| r_1 | < | r_2 |$. If $a = 0$ then the single root of (4) is denoted r_1. Since we can do so without loss of generality, we specify r_1 to be negative.

Now if we suppose that $r_1 \neq r_2$ and choose an initial function φ such that $\varphi(0) \neq r_j$, $j = 1, 2$, then by continuity there exists a solution f on an interval $(0, X]$ for which $f(x) \neq r_j$ for all x in $(0, X]$. Equation (1) can be rewritten

$$\frac{f'(x)}{(f(x) - r_1)(f(x) - r_2)} = -\alpha a \sum_{i=1}^{n} \lambda_i f(x - \tau_i)$$

for x in $(0, X]$, since $f(x) \neq r_j$, and through obvious manipulations we get

$$\frac{f(x) - r_1}{f(x) - r_2} = \frac{\varphi(0) - r_1}{\varphi(0) - r_2} \exp\left[\alpha a(r_2 - r_1) \sum_{i=1}^{n} \lambda_i \int_0^x f(t - \tau_i)\, dt\right], \tag{5}$$

and

$$f(x) = \frac{r_1(\varphi(0) - r_2) - r_2(\varphi(0) - r_1) \exp\left[\alpha a(r_2 - r_1) \sum_{i=1}^{n} \lambda_i \int_0^x f(t - \tau_i)\, dt\right]}{(\varphi(0) - r_2) - (\varphi(0) - r_1) \exp\left[\alpha a(r_2 - r_1) \sum_{i=1}^{n} \lambda_i \int_0^x f(t - \tau_i)\, dt\right]}. \tag{6}$$

In a similar fashion when $r_1 = r_2$ or $a = 0$ we get

$$f(x) = r_1 + (\varphi(0) - r_1)\left[1 + \alpha a(\varphi(0) - r_1) \sum_{i=1}^{n} \lambda_i \int_0^x f(t - \tau_i)\, dt\right]^{-1} \tag{7}$$

or

$$f(x) = r_1 + (\varphi(0) - r_1) \exp\left[-\alpha b \sum_{i=1}^{n} \lambda_i \int_0^x f(t - \tau_i)\, dt\right] \tag{8}$$

respectively for x in some interval $(0, X]$.

It is our purpose to discuss briefly conditions under which equation (1) can be shown to have non-constant periodic solutions and to indicate a method of proof. Also, we shall describe some of the general properties of these solutions and present three illustrative examples which essentially characterize the behavior of the whole class.

2. The Existence of Periodic Solutions

Now for positive constants A and B let S_{AB} denote the set of continuously differentiable functions defined on the interval $[-\tau_n, 0]$ with the property that φ in S_{AB} implies $\varphi(x) \not\equiv 0$, $0 \leqq \varphi(x) \leqq A$, $\varphi(0) = 0$, and $|\varphi'(x)| \leqq B$. To show that equation (1) with a particular specification of parameters has non-constant periodic solutions, it is sufficient to verify that there exist constants A and B such that the following conditions are satisfied.

Condition 1. φ in S_{AB} and $f \sim \varphi$ imply f can be defined for all $x > 0$.

Condition 2. φ in S_{AB} and $f \sim \varphi$ imply $r_1 < f(x) \leqq A$ and $|f'(x)| \leqq B$ for all $x > 0$.

Condition 3. For φ in S_{AB} the set of zeros of f form an infinite discrete set $\{z_k(\varphi)\}_{k=1}^{\infty}$, $z_{k+1}(\varphi) > z_k(\varphi) > 0$, and there exists a positive constant C, independent of φ, such that $\tau_n \leqq z_{k+1}(\varphi) - z_k(\varphi) \leqq C$ for all k.

Condition 4. There exist $\epsilon_1 > 0$ and an integer p such that φ in S_{AB} and $\max\{|\varphi(x)| : x \text{ in } [-\tau_n, 0]\} = \epsilon_1$ imply that at some point $x_1(\varphi)$ in the interval $(z_{2p}(\varphi) - \tau_n, z_{2p}(\varphi))$, $f(x_1(\varphi)) \geqq \epsilon_1$.

With conditions 1, 2, and 3 satisfied it makes sense to define a mapping $T: S_{AB} \to S_{AB}$ by the formula

$$T(\varphi(x)) = f(x + z_2(\varphi)), \tag{9}$$

for x in $[-\tau_n, 0]$. The continuity of T follows directly from the fact that $z_2(\varphi)$ is bounded and with a little investigation one can show that T is topological as well. Let us define $S_{AB}(\epsilon) \subset S_{AB}$ to be such that φ in $S_{AB}(\epsilon)$ implies $\max\{|\varphi(x)| : x \text{ in } [-\tau_n, 0]\} \geqq \epsilon > 0$. Since T is topological, condition 4 implies that T^p maps $S_{AB}(\epsilon_1)$ into itself. Also, since $S_{AB}(\epsilon_1)$ is a compact convex subset of a Banach space it follows from the Schauder fixed point theorem that there exists φ^* in $S_{AB}(\epsilon)$ such that $T^P(\varphi^*) = \varphi^*$. Finally, since a solution of equation (1) is uniquely determined by its specification on an arbitrary interval of length τ_n, it follows immediately that the existence of the fixed point φ^* implies the existence of a non-constant periodic solution of equation (1).

In some cases we are able to show that T itself has a fixed point in S_{AB} and indications are that this fixed point is unique.

3. General Properties of Solutions

In this section we are primarily concerned with indicating how certain restrictions on the parameters of equation (1) imply some or all of the conditions 1, 2, 3, and 4. First of all by the use of formulas (6), (7), and (8) the following lemma is easily established.

Lemma 1. $\varphi(0) > r_j$, $\varphi(0) = r_j$, and $\varphi(0) < r_j$ imply $f(x) > r_j$, $f(x) = r_j$, and $f(x) < r_j$ respectively for all x in the domain of definition of f.

Let us define E to be the largest open segment of the real line containing zero but not r_1 or r_2. We notice that Lemma 1 indicates that $\varphi(0)$ in E implies $f(x)$ is in E for all x in its domain of definition. In particular, $\varphi(0)$ in E and $r_1 r_2 < 0$ implies $r_1 < f(x) < r_2$ for all x in the domain of f. In fact, we can easily prove the following lemmas.

Lemma 2. If $r_1 r_2 < 0$ then conditions 1 and 2 are satisfied. If $a = 0$ and the λ_i's are nonnegative, then conditions 1 and 2 are satisfied.

Lemma 3. Let $\varphi(0)$ be in E, $r_1 r_2 < 0$ or $a = 0$, and let the λ_i's be nonnegative. If the zeros of f form a finite set, then $f(x) \to 0$ as $x \to \infty$. If $\alpha > 1/\Sigma_{i=1}^{n} \lambda_i$, then the set of zeros of f is not bounded. If $\varphi(0) = 0$ and $\varphi(x)$ of constant sign imply $f(x)$ is of constant sign on $(0, \tau_n)$, and if $\alpha > 1/\Sigma_{i=1}^{n} \lambda_i$, then condition 3 is satisfied.

Lemma 4. Let the λ_i's be nonnegative. If $r_1 r_2 > 0$ and condition 1 is satisfied, then condition 2 is satisfied. If, in addition, $\varphi(0) = 0$ and $\varphi(x)$ of constant sign imply $f(x)$ is of constant sign on $(0, \tau_n)$, and if $\alpha > 1/\Sigma_{i=1}^{n} \lambda_i$, then condition 3 is satisfied.

We now turn our attention to determining types of parametric specifications which imply condition 4. For this purpose we observe that "small" solutions of equation (1) might be expected to have behavior similar to those of the linear differential-difference equation

$$y'(x) = -\alpha \sum_{i=1}^{n} \lambda_i y(x - \tau_i). \tag{10}$$

We easily determine that the transcendental equation

$$s + \alpha \sum_{i=1}^{n} \lambda_i e^{-\tau_i s} = 0 \tag{11}$$

is the characteristic equation of (10). Now let us specify that equation (11) has a unique pair of conjugate complex roots with largest positive real part which we denote by $u \pm iv$. Solutions of equation (10) corresponding to initial functions in S_{AB} may be expanded in terms of the roots of equation (11) and the coefficient corresponding to $u + iv$ is

$$C(\varphi) = \frac{\alpha \sum_{k=1}^{n} \lambda_i \int_{-\tau_i}^{0} \exp\left[-(u+iv)(t+\tau_k)\right] \varphi(t)\, dt}{1 - \alpha \sum_{k=1}^{n} \lambda_k \tau_k \exp\left[-(u+iv)\,\tau_k\right]}.$$

Lemma 5. If there exists a positive constant K such that φ in $S_{AB}(\epsilon)$ implies $| C(\varphi) | \geqq K\epsilon$ for arbitrary ϵ positive and sufficiently small, then condition 4 is satisfies.

Proceeding now with a few general properties of periodic solutions of equation (1) we mention that they are analytic in a strip of the complex plane containing the real line. Also if ξ denotes the period of a particular solution f, then

$$\sum_{i=1}^{n} \lambda_i \int_{x}^{x+\xi} f(t - \tau_i)\, dt = 0,$$

and if

$$\sum_{i=1}^{n} \lambda_i \neq 0, \qquad \text{then} \qquad \int_{x}^{x+\xi} f(t)\, dt = 0.$$

Finally it can be demonstrated that the equations represented by (1) under different parametric specifications fall naturally into two general but distinctive categories determined by the roots of (4). If $r_1 r_2 < 0$ we are considering an equation of category 1, and if $a = 0$ or $r_1 r_2 > 0$ we have a category 2 equation. Periodic solutions of equations in category 1 are bounded between r_1 and r_2 for all $\alpha > 0$ and there exists a bound, independent of α, on their periods. For equations in category 2 periodic solutions are bounded below by r_1 for all $\alpha > 0$ but may become arbitrarily large in the positive direction as $\alpha \to \infty$. Also their periods may tend to infinity as $\alpha \to \infty$.

For detailed discussions of some of the analysis indicated in this section see Jones [*6*], Langer [*8*], and Wright [*9*, *10*].

4. Examples

The differential-difference equation

$$f'(x) = -\alpha f(x-1)\,[1 - f(x)^2] \tag{12}$$

is an interesting example of category 1. Here we observe that $r_1 = -1$, $r_2 = 1$, and formula (6) reduces to

$$f(x) = \frac{(1 + \varphi(0)) \exp\left[-2\alpha \int_0^x f(t-1)\, dt\right] - (1 - \varphi(0))}{(1 + \varphi(0)) \exp\left[-2\alpha \int_0^x f(t-1)\, dt\right] + (1 - \varphi(0))}. \tag{13}$$

If we specify $\varphi(0) = 0$, (13) reduces to the functional relation

$$f(x) = \tanh\left[-\alpha \int_0^x f(t-1)\, dt\right].$$

By lemmas 2 and 3 we have that conditions 1, 2, and 3 are satisfied. Now we observe that equation (11) reduces to

$$s + \alpha e^{-s} = 0 \tag{14}$$

which is known to have a unique pair of conjugate complex roots with largest positive real part for $\alpha > \pi/2$. With this information we are able to demonstrate that condition 4 is satisfied by using lemma 5. Figure 1

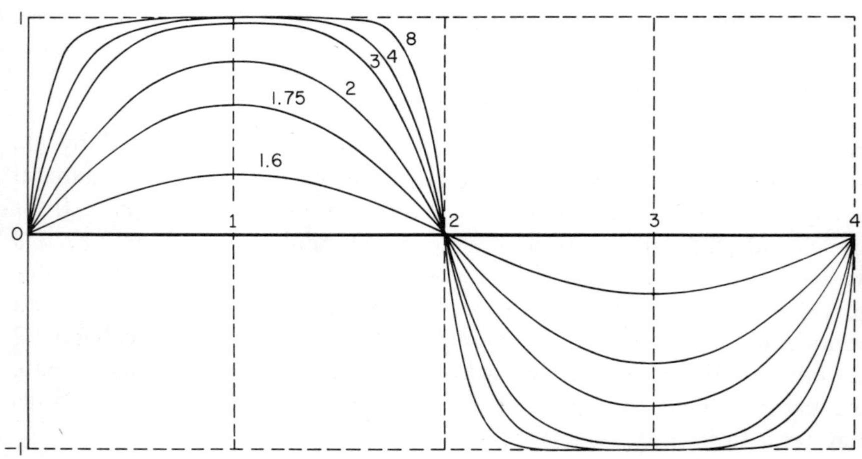

FIG. 1

reveals numerically computed periodic solutions of equation (12) for several values of α. The number appearing above each curve denotes the corresponding value of α used. We notice that these functions are bounded between 1 and -1 and that the squarer forms correspond to the larger values of α. In addition to these properties, one observes that all functions represented apparently have the same period of 4. This independence of period from α, however, and the apparent symmetry of these functions is not characteristic of category 1 but depend rather on the special property that $r_1 = -r_2$. It is also interesting that in this particular case the periodic solution of period 4 is unique for fixed $\alpha \geqq 3$.

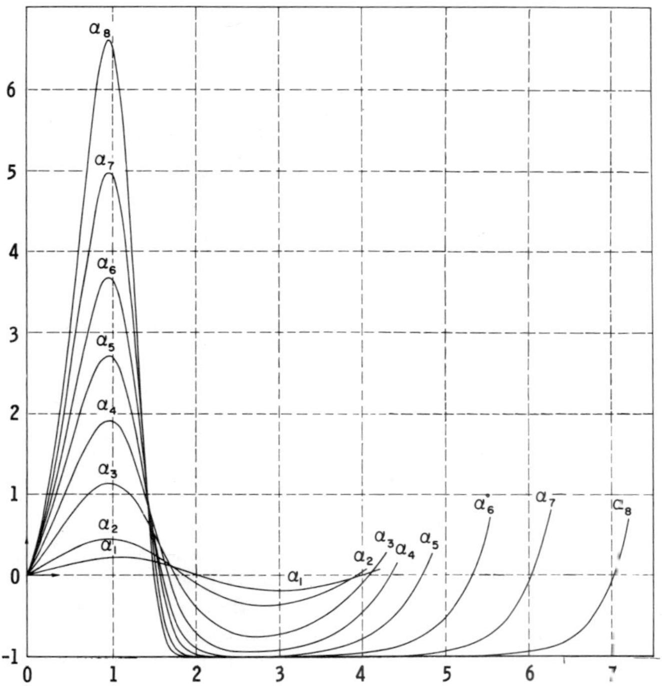

FIG. 2

Equation (2) is an example of category 2 with $a = 0$. As in the previous example our four conditions can be shown to be satisfied for $\alpha > \pi/2$. Its periodic solutions are bounded below by -1 for all $\alpha > 0$ but become arbitrarily large in the positive direction as $\alpha \to \infty$. Also their minimum value tends to -1 and their period to infinity as $\alpha \to \infty$. Figure 2 reveals the form of the periodic solutions of equation (2) for several values of α. $\alpha_1 = 1.58$, $\alpha_2 = 1.6$, $\alpha_3 = 1.75$, $\alpha_4 = 2$, $\alpha_5 = 2.25$, $\alpha_6 = 2.5$, $\alpha_7 = 2.75$, and $\alpha_8 = 3$.

The functional equation

$$f'(x) = -\alpha f(x-1)\,[f(x)+1]^2 \tag{15}$$

is another equation of category 2 whose periodic solutions are similar in form to those of equation (2). However, their existence is assured only for $\pi/2 < \alpha < \beta$ where $2 < \beta < 3$. For α sufficiently large we can not even satisfy condition 1.

For more detailed information on equation (2), the reader is referred to Jones [*5*, *6*], Kakutani and Markus [*7*], and to Wright [*10*].

REFERENCES

[*1*] Bellman, R. E., A Survey of the Mathematical Theory of Time-lag Retarded Control and Hereditary Processes. RAND Corp. Report, 1954.

[*2*] Cherwell, Lord, Number of primes and probability considerations. *Nature* **148**, 436 (1941); **150**, 121 (1942).

[*3*] Cunningham, W. J., A nonlinear differential-difference equation of growth. *Proc. natl. Acad. Sci. U.S.* **40**, 709-713 (1954).

[*4*] Jones, G. S., Asymptotic behavior and periodic solutions of a nonlinear differential-difference equation. *Proc. natl. Acad. Sci. U.S.* **47**, **6**, 879-882 (1961).

[*5*] Jones, G. S., On the nonlinear differential-difference equation $f'(x) = \alpha f(x-1)[1+f(x)]$. *J. math. Anal. and Appl.* **2** (1962).

[*6*] Jones, G. S., The existence of periodic solutions of $f'(x) = -\alpha f(x-1) \times [1+f(x)]$. *J. math. Anal. and Appl.* **2** (1962).

[*7*] Kakutani, S., and Markus, L., On the nonlinear differential-difference equation $y'(t) = [A - By(t-\tau)]\,y(t)$. *In* "Contributions to the Theory of Nonlinear Oscillations" (S. Lefschetz, ed.), Vol. 4, pp. 1-18. Princeton Univ. Press, Princeton, New Jersey, 1958.

[*8*] Langer, R. E., The asymptotic location of the roots of a certain transcendental equation. *Trans. AMS* **31**, 837-844 (1929).

[*9*] Wright, E. M., The linear difference-differential equation with constant coefficients. *Proc. roy. Soc. Edinburgh* **A62**, 387-393 (1949).

[*10*] Wright, E. M., A nonlinear difference-differential equation. *J. reine angew. Math.* **194**, 66-87 (1955).

[*11*] Wright, E. M., A functional equation in the heuristic theory of primes. *Math. Gaz. (London)* **45**, 15-16 (1961).

Uniformly Valid Asymptotic Approximations for Certain Non-Linear Differential Equations[1]

J. D. COLE AND J. KEVORKIAN
Firestone Flight Sciences Laboratory
California Institute of Technology, Pasadena, California

Various physical problems lead to differential equations characterized by the presence of a small force active for a long time. For example, see equation (1) below. It is often useful to be able to construct an approximation which is a good representation of the solution for the complete time interval $0 < t < \infty$. In these circumstances the simple perturbation approximation in terms of the small parameter measuring the perturbing force and one time scale fails due to the cumulative action of the small force. In this note a new method is presented which utilizes an expansion in terms of two or more time scales. This method is applicable when the overall behavior of the solution is sufficiently well known. The physical significance of various time scales, for example, natural period, decay time, etc., is well understood by physicists usually in terms of the solution for a linear damped oscillator. However the various time scales have significance also for non-linear systems since they are formed of combinations of the physical constants (with dimensions) which enter the description of the phenomena. The small parameter, ϵ, of the problem can be expressed as a ratio of two significant time scales. The explicit introduction of several time scales changes the original ordinary differential equation into what is formally a partial differential equation, which can be exploited in various ways. The solution to this equation is expanded in a suitable series, each term of which satisfies an ordinary differential equation. In this note the method is applied to several simple examples. More complicated examples and some discussion of the general theory are presented by Kevorkian [*1*] and a similar idea is studied by Kuzmak [*2*]. It should be mentioned that the class of problems considered here involving small damping can also be solved by the method of Kryloff and Bogoliuboff. The advantage of the present method is its simplicity especially for computing higher approximations and its

[1] This research was carried out with the support of the Air Force Office of Scientific Research, Air Research and Development Command under Contract AF 49(638)-521.

applicability to a large class of problems for which the motion is bounded for large times.

In order to fix the ideas consider an equation of the form

$$\frac{d^2y}{dt^{*2}} + \epsilon f\left(y, \frac{dy}{dt^*}\right) + y = 0 \qquad 0 \leqslant t^* < \infty \tag{1}$$

with

$$y(0) = a, \qquad \frac{dy}{dt^*}(0) = 0,$$

ϵ is the small parameter and f is such that the solution is bounded. t^* is the natural dimensionless time variable for the simple harmonic solutions with $\epsilon = 0$. For this class of problems the partial differential equation is obtained by writing

$$y = F(t^\dagger, \tilde{t}; \epsilon), \tag{2}$$

where

$$t^\dagger = t^*\{1 + \epsilon^2\omega_2 + ...\}$$

$$\tilde{t} = \epsilon t^*.$$

$t^\dagger$ is a variable close to t^* which is introduced to account for a frequency shift due to the generalized damping, as is familiar in the method of Poincaré for periodic solutions. $\tilde{t}$ on the other hand is a slowly varying function of t^* and is used to describe the slow modulation of the oscillations. For these cases the expansion has the form

$$y = F \cong F^{(0)}(t^\dagger, \tilde{t}) + \epsilon F^{(1)}(t^\dagger, \tilde{t}) + \epsilon^2 F^{(2)}(t^\dagger, \tilde{t}) + ... \tag{3}$$

Consider first the classical case of linear damping.

$$\frac{d^2y}{dt^{*2}} + 2\epsilon \frac{dy}{dt^*} + y = 0. \tag{4}$$

The straightforward expansion procedure in (ϵ, t^*) yields an expansion valid only for a finite time interval $0 \leqslant t^* < T$, an initially-valid expansion of the form

$$y \cong h_0(t^*) + \epsilon h_1(t^*) + ... = a \cos t^* + \epsilon a\{\sin t^* - t^* \cos t^*\} + ... \tag{5}$$

The origin of the "secular" term is clearly seen from the exact solution

$$y(t^*; \epsilon) = ae^{-\epsilon t^*} \left[\cos (1 - \epsilon^2)^{1/2} t^* + \frac{\epsilon}{(1 - \epsilon^2)^{1/2}} \sin (1 - \epsilon^2)^{1/2} t^*\right]. \tag{6}$$

The expansion of the exponential function is not valid for large times; also the dependence of frequency on ϵ is manifest. Using the more general expansion (3) produces the following set of approximation equations

$$\frac{\partial^2 F^{(0)}}{\partial t^{\dagger 2}} + F^{(0)} = 0 \tag{7}$$

$$\frac{\partial^2 F^{(1)}}{\partial t^{\dagger 2}} + F^{(1)} = -2\frac{\partial^2 F^{(0)}}{\partial t^\dagger \partial \tilde{t}} - 2\frac{\partial F^{(0)}}{\partial t^\dagger} \tag{8}$$

$$\frac{\partial^2 F^{(2)}}{\partial t^{\dagger 2}} + F^{(2)} = -2\left\{\frac{\partial^2 F^{(1)}}{\partial t^\dagger \partial \tilde{t}} + \frac{\partial^2 F^{(0)}}{\partial \tilde{t}^2} + \omega_2 \frac{\partial^2 F^{(0)}}{\partial t^{\dagger 2}} + \frac{\partial F^{(0)}}{\partial \tilde{t}} + \frac{\partial F^{(1)}}{\partial t^\dagger}\right\}. \tag{9}$$

The initial conditions are

$$F^{(0)}(0, 0) = a, \qquad F^{(1)}(0, 0) = F^{(2)}(0, 0) = \dots 0 \tag{10}$$

$$\frac{\partial F^{(0)}}{\partial t^\dagger}(0, 0) = 0, \qquad \frac{\partial F^{(0)}}{\partial \tilde{t}}(0, 0) + \frac{\partial F^{(1)}}{\partial t^\dagger}(0, 0) = 0,$$

$$\frac{\partial F^{(1)}}{\partial \tilde{t}}(0, 0) + \frac{\partial F^{(2)}}{\partial t^\dagger}(0, 0) + \omega_2 \frac{\partial F^{(0)}}{\partial t^\dagger}(0, 0) = 0, \qquad \text{etc.} \tag{11}$$

The solution of (7) is

$$F^{(0)}(t^\dagger, \tilde{t}) = A^{(0)}(\tilde{t}) \sin t^\dagger + B^{(0)}(\tilde{t}) \cos t^\dagger. \tag{12}$$

The functions $A^{(0)}$, $B^{(0)}$ are determined by considerations of boundedness on the $F^{(1)}$ solution. Eq. (8) reads

$$\frac{\partial^2 F^{(1)}}{\partial t^{\dagger 2}} + F^{(1)} = -2\left[\frac{dA^{(0)}}{d\tilde{t}} + A^{(0)}\right] \cos t^\dagger - 2\left[\frac{dB^{(0)}}{d\tilde{t}} + B^{(0)}\right] \sin t^\dagger. \tag{13}$$

It is clear that to prevent occurrence of secular terms of the form $t^\dagger \cos t^\dagger$ in $F^{(1)}$, unbounded on the $t^\dagger$ scale,

$$\frac{dA^{(0)}}{d\tilde{t}} + A^{(0)} = 0 \qquad \text{and} \qquad A^{(0)}(\tilde{t}) = \alpha^{(0)} e^{-t} \tag{14}$$

$$\frac{dB^{(0)}}{d\tilde{t}} + B^{(0)} = 0 \qquad\qquad B^{(0)}(\tilde{t}) = \beta^{(0)} e^{-t}. \tag{15}$$

Taking account of the initial conditions $\alpha^{(0)} = 0$, $\beta^{(0)} = a$. Thus the uniformly valid first approximation is

$$F^{(0)}(t^\dagger, \tilde{t}) = a e^{-t} \cos t^\dagger. \tag{16}$$

It is clear that this contains the first term of the initially valid expansion if ($\epsilon \to 0$, t^* fixed). It is instructive also to consider the next term. Now

$$F^{(1)}(t^\dagger, \tilde{t}) = A^{(1)}(\tilde{t}) \sin t^\dagger + B^{(1)}(\tilde{t}) \cos t^\dagger \tag{17}$$

Considerations of the forcing terms in Eq. (9) now fix $A^{(1)}$ and $B^{(1)}$ as well as the frequency shift ω_2. Eq. (9) reads

$$\begin{aligned}\frac{\partial^2 F^{(2)}}{\partial t^{\dagger 2}} + F^{(2)} = &- 2\left[\frac{dA^{(1)}}{d\tilde{t}} + A^{(1)} - (\tfrac{1}{2} + \omega_2)\, ae^{-\tilde{t}}\right] \cos t^\dagger \\ &+ 2\left[\frac{dB^{(1)}}{d\tilde{t}} + B^{(1)}\right] \sin t^\dagger.\end{aligned} \tag{18}$$

Again, boundedness of $F^{(2)}$ with respect to $t^\dagger$ means that

$$\frac{dA^{(1)}}{d\tilde{t}} + A^{(1)} = (\tfrac{1}{2} + \omega_2)\, e^{-\tilde{t}} \tag{19}$$

$$\frac{dB^{(1)}}{d\tilde{t}} + B^{(1)} = 0. \tag{20}$$

Taking account of the initial conditions,

$$F^{(1)} = a[1 + (\tfrac{1}{2} + \omega_2)\, \tilde{t}]\, e^{-\tilde{t}} \sin t^\dagger. \tag{21}$$

Now, a stronger boundedness condition on the solution is needed to fix the frequency shift ω_2. We require that

$$y/B^{(0)}(\tilde{t}) = y/ae^{-\tilde{t}} \text{ is bounded for all } \tilde{t}, t^\dagger.$$

Thus $\omega_2 = -\frac{1}{2}$, which is easily seen to be correct from the exact solution (6). For this simple case the stronger boundedness condition can easily be verified by a direct calculation. The first two terms of the uniformly valid approximations are thus

$$y \simeq ae^{-\tilde{t}} \cos t^\dagger + \epsilon ae^{-\tilde{t}} \sin t^\dagger. \tag{22}$$

No new features appear in continuing the process to higher terms and all the significant features of the method in its applications to non-linear problems of Eq. (1) are illustrated in this example. As a concrete example of a non-linear case consider a mechanical interpretation for a Van der Pol equation with small damping

$$\begin{gathered} m\frac{d^2\bar{y}}{dt^2} - \left[\alpha\frac{d\bar{y}}{dt} - \frac{\beta}{3}\left(\frac{d\bar{y}}{dt}\right)^3\right] + k\bar{y} = 0 \\ \bar{y}(0) = A \qquad \frac{d\bar{y}}{dt}(0) = 0 \end{gathered} \tag{23}$$

where m = mass, k = spring constant, α, β = damping coefficients. Length and time scales can be calculated depending on the physical constants

$$L = \text{characteristic length} = \frac{\alpha m}{\beta k}$$

$$T_1 = \text{characteristic time for oscillations} = (m/k)^{1/2}$$

$$T_2 = \text{characteristic damping time} = m/\alpha$$

Appropriate dimensionless variables are ($y = \bar{y}/L$, $t^* = t/T_1$) and the small parameter of the problem $\epsilon = T_1/T_2$. In these variables the problem reads:

$$\frac{d^2y}{dt^{*2}} - \epsilon\left[\frac{dy}{dt^*} - \frac{1}{3}\left(\frac{dy}{dt^*}\right)^3\right] + y = 0 \tag{24}$$

$$y(0) = a = \frac{A\beta k}{\alpha m} \qquad \frac{dy}{dt^*}(0) = 0.$$

Applying the expansion (3) it is found, as before, that

$$F^{(0)} = A^{(0)}(\tilde{t}) \sin t^\dagger + B^{(0)}(\tilde{t}) \cos t^\dagger \tag{25}$$

$$\begin{aligned}\frac{\partial^2 F^{(1)}}{\partial t^{\dagger 2}} + F^{(1)} &= \cos t^\dagger \left[-2\frac{dA^{(0)}}{d\tilde{t}} - \frac{A^{(0)3}}{4} - \frac{A^{(0)}B^{(0)2}}{4} + A^{(0)}\right] \\ &+ \sin t^\dagger \left[2\frac{dB^{(0)}}{d\tilde{t}} + \frac{B^{(0)3}}{4} + \frac{A^{(0)2}B^{(0)}}{4} - B^{(0)}\right] \\ &+ \cos 3t^\dagger[...] + \sin 3t^\dagger[...].\end{aligned} \tag{26}$$

Equating to zero the coefficients of $\sin t^\dagger$ and $\cos t^\dagger$ in the right-hand side of (26) yields the differential equation for $A^{(0)}$ and $B^{(0)}$ and taking account of the initial conditions (10) and (11) gives the uniformly valid first approximation:

$$y \simeq F^{(0)} = \frac{2a \cos t^\dagger}{(a^2 - (a^2 - 4)\, e^{-\tilde{t}})^{1/2}}\,. \tag{27}$$

The approach to the limit cycle with amplitude 2 is clearly shown, as well as the periodic solution for $a = 2$.

Finally, we summarize the results obtained by applying the method to a somewhat more complicated example, the Mathieu equation

$$\frac{d^2y}{dt^{*2}} + [\delta + \epsilon \cos t^*]\, y = 0. \tag{28}$$

The equation is linear but the presence of the small forcing term again modulates the free oscillation solutions of $\epsilon = 0$. Stoker [3], by studying the periodic solutions of (28), defines the boundary curves of regions of stability[1] in the (δ, ϵ) plane. These transitional curves intersect $\epsilon = 0$ at $\delta_c = n^2/4$, $n = 0, 1, 2, \ldots$. Through each of these points except $n = 0$ pass left ${}^{(n)}\delta_{(\epsilon)}$ and right $\delta_{(\epsilon)}^{(n)}$ transitional curves. At the origin only ${}^{(n)}\delta$ exists and the region to the left has unstable solutions. As each of the transitional curves is crossed the stability of the solution changes. A two variable expansion for the solutions of the Mathieu equation is constructed here by representing

$$\delta = \delta_c + \sum_{i=1}^{n} \delta_i \epsilon^i, \qquad \delta_i \text{ arbitrary.} \tag{29}$$

On the n-parameter family of curves

$$\delta = \delta_c + \sum_{i=1}^{n} \delta_i \epsilon_i$$

the solutions of Mathieu's equation can be uniformly represented by an asymptotic expansion in terms of the two variables t^*, and $t_\eta = \eta(\epsilon)\, t^*$ for some appropriate $\eta(\epsilon) = 0(\epsilon)$ as $\epsilon \to 0$. The particular η depends on the critical point δ_c.

It can be shown that for each critical point δ_c there is only one order class of function $\eta_c(\epsilon)$ for which the expansion for F has the following properties, and we will determine η by demonstrating these properties.

(i) For values of δ_i associated with the stable region F will be a bounded function of the two variables t^* and t_η.

(ii) For values of δ_i associated with an unstable region F will be a bounded function of t^*, but an unbounded function of t_η.

The expansion for F must conform with the results of the general theory in the following respects.

(i) On the transition curves F must admit a periodic solution for the appropriate initial conditions, and the general initial value problem must be unstable.

(ii) Since F is presumed to represent y uniformly, the basic stability, and instability, of the solution must be exhibited by the leading term in the expansion.

[1] Stability means boundedness for all $t^* > 0$ under arbitrary initial conditions.

Taking initial conditions

$$y(0) = b, \qquad \frac{dy}{dt^*}(0) = a. \tag{30}$$

Some of the results obtained are the following.

For $n = 0$, the appropriate expansion is

$$= F(t^*, \tilde{t}) = \frac{1}{\epsilon} F^{(0)}(t^*, \tilde{t}) + F^{(1)}(t^*, \tilde{t}) + \ldots, \qquad \text{where} \qquad \tilde{t} = \epsilon t, \tag{31}$$

$$\delta = \delta_2 \epsilon^2 + \ldots \tag{32}$$

and

$$\begin{aligned} F^{(0)} = B^{(0)}(\tilde{t}) &= \frac{a}{(\delta_2 + \frac{1}{2})^{1/2}} \sin (\delta_2 + \tfrac{1}{2})^{1/2}\, \tilde{t} \qquad && \delta_2 > -\tfrac{1}{2} \\ &= \frac{a}{[-(\delta_2 + \frac{1}{2})]^{1/2}} \sinh [-(\delta_2 + \tfrac{1}{2})]^{1/2}\, \tilde{t} && \delta_2 < -\tfrac{1}{2}. \end{aligned} \tag{33}$$

In order to find this solution it is necessary to proceed two steps ahead and consider the forcing terms of the equation for $F^{(2)}$.

For $n = 2$, the appropriate expansion is

$$y = F = F^{(0)}(t^*, \tilde{t}) + \epsilon F^{(1)}(t^*, \tilde{t}) + \ldots, \qquad \text{where} \qquad \tilde{t} = \epsilon^2 t, \tag{34}$$

$$\delta = 1 + \epsilon^2 \delta_2 + \ldots \tag{35}$$

where

$$F^{(0)} = A^{(0)}(\tilde{t}) \sin t^* + B^{(0)}(\tilde{t}) \cos t^*. \tag{36}$$

The solutions are stable if $\delta_2 > 5/12$ or $\delta_2 < -1/12$ and

$$\begin{aligned} A^{(0)}(\tilde{t}) &= b \left(\frac{12\delta_2 - 5}{12\delta_2 + 1}\right)^{1/2} \sin \left[\left(\delta_2 - \frac{5}{12}\right)\left(\delta_2 + \frac{1}{12}\right)\right]^{1/2} \tilde{t} \\ &\quad + a \cos \left[\left(\delta_2 - \frac{5}{12}\right)\left(\delta_2 + \frac{1}{12}\right)\right]^{1/2} \tilde{t} \\ B^{(0)}(\tilde{t}) &= b \cos \left[\left(\delta_2 - \frac{5}{12}\right)\left(\delta_2 + \frac{1}{12}\right)\right]^{1/2} \tilde{t} \\ &\quad - a \left(\frac{12\delta_2 + 1}{12\delta_2 - 5}\right)^{1/2} \sin \left[\left(\delta_2 - \frac{5}{12}\right)\left(\delta_2 + \frac{1}{12}\right)\right]^{1/2} \tilde{t}. \end{aligned} \tag{37}$$

The solutions are unstable if $-1/12 < \delta_2 < 5/12$ and

$$
\begin{aligned}
A^{(0)}(\tilde{t}) &= \frac{[b\sqrt{r} + a\sqrt{s}]^2 e^{2\sqrt{rs}\,\tilde{t}} + (sa^2 - rb^2)}{2\sqrt{s}\,(b\sqrt{r} + a\sqrt{s})\, e^{\sqrt{rs}\,\tilde{t}}} \\
B^{(0)}(\tilde{t}) &= \frac{[b\sqrt{r} + a\sqrt{s}]^2 e^{2\sqrt{rs}\,\tilde{t}} - (sa^2 - rb^2)}{2\sqrt{r}\,(b\sqrt{r} + a\sqrt{s})\, e^{\sqrt{rs}\,\tilde{t}}}
\end{aligned}
\tag{38}
$$

where $r = 5/12 - \delta_2$, and $s = \delta_2 + 1/12$.

These solutions are valid in some neighborhood of the transitional curves and clearly show the change from stability to instability as well as the slow modulation of the solution.

It is hoped that these examples illustrate the spirit of the method and the type of results to be expected. It has also been seen however that various special conditions may have to be imposed to guarantee the unique construction of the successive approximation terms. Theoretical considerations concerning the precise meaning of the approximation must be left to future publications.

REFERENCES

[*1*] Kevorkian, J. K., The Uniformly Valid Asymptotic Approximations to the Solutions of Certain Non-linear Ordinary Differential Equations. Ph.D. Thesis, California Inst. Technol. (1961).

[*2*] Kuzmak, G. N., Asymptotic solutions of non-linear second order differential equations with variable coefficients. *Priklad. Mat. i Mekh.* **23**, No. 3, 730 (1960).

[*3*] Stoker, J. J., "Non-linear Vibrations in Mechanical and Electrical Systems." Interscience, New York, 1950.

New Directions of Research in the Theory of Differential Equations

RICHARD BELLMAN
The RAND Corporation, Santa Monica, California

1. Introduction

If the language of science is mathematics, much of its prose and poetry is occupied by the theory of differential equations. Consequently, as new shapes loom on the scientific horizon, this theory is progressively enriched. In the following pages we wish to indicate some of the effects of a number of recent scientific and technological developments such as the theory of control processes, the study of timelags, hereditary processes and nonlocal physical theories, nuclear reactors and planetary atmospheres, and digital computers on the field of differential equations.

Each advance permits us to survey both the modern and the classic in a new light and reveals novel and unsuspected interconnections and simplifications.

2. Feedback Control

The classical theory of differential equations is concerned with vector differential equations of the form

$$\frac{dx}{dt} = g(x, t), \qquad x(0) = c. \tag{1}$$

The basic questions are those connected with existence and uniqueness of the solution, stability and asymptotic behavior, and approximation to the solution [*3*, *24*, *25*].

Conceptually, the foregoing equation represents the law of transformation of the vector $x(t)$ defining the state of a physical system S. All aspects that have been mentioned above are concerned with the *descriptive* aspects of the underlying physical process.

If the system in its native form does not operate in a most desirable fashion, we attempt to modify or *control* its behavior. One way to do this is to introduce a forcing function, dependent in general upon time, $y(t)$

into the system. The equation corresponding to (1) now has the form

$$\frac{dx}{dt} = h(x, t; y(t)), \qquad x(0) = c. \tag{2}$$

The function $y(t)$ is to be chosen, subject to various constraints, so as to force $x(t)$ to follow some prescribed behavior. This problem can be considered to lie within the domain of the calculus of variations—but, in the main, unprofitably (cf. the discussion in ref. [*6*], Chapters I and V). Generally, different devices are required.

One of the fundamental ideas of modern technology (with, as usual, a history stretching back to dateless antiquity [*1*] is that of *feedback control.* In place of considering the control function $y(t)$ to be merely a function of time, we regard it as a function of the current state of the system, i.e., $y = y(x)$. This interpretation permits of most important industrial and engineering applications and simultaneously allows an entirely new formulation of the classical calculus of variations, control processes and all types of multistage decision processes by means of the theory of dynamic programming [*5*, *6*]. Prior to a brief discussion of this, let us indicate some of the novel problems that arise in the modern study of control processes.

3. "Bang-Bang" Control

Consider the linear vector differential equation

$$\frac{dx}{dt} = Ax + y, \qquad x(0) = c, \tag{1}$$

where A is a constant matrix and $y = y(t)$ or $y(x)$ (depending upon which interpretation is most convenient at the moment) is to be chosen, subject to constraints such as

$$|y_i(t)| \leq m_i, \qquad i = 1, 2, ..., N, \qquad t \geq 0, \tag{2}$$

or

$$y_i(t) = \pm m_i, \qquad i = 1, 2, ..., N, \qquad t \geq 0, \tag{3}$$

so as to reduce x to 0, the null vector, in minimum time. This problem can be treated by interesting combinations of classical and modern abstract analysis and a very satisfying theoretical solution obtained [*15*, *23*]. At the present time, however, no completely satisfactory computational algorithm for actually obtaining the solution exists.

4. Constraints

The problem described above is one of a large class of problems arising in modern control theory in which there are simple and natural restrictions upon the type of control which can be exerted upon a system.

The problem of minimizing a functional such as

$$J(x, y) = \int_0^T [(x, Bx) + (y, Cy)]\, dt \tag{1}$$

over all functions $y(t)$ where x and y are related by means of (3.1) is routine since the Euler equation is linear. If we add constraints such as (3.2) or (3.3), we no longer possess an explicit analytic solution, and an outstanding challenge is to find a routine numerical solution to questions of this nature.

The minimization of a general functional such as

$$J(x, y) = \int_0^T g(x, y)\, dt \tag{2}$$

where x and y are related by means of the differential equation

$$\frac{dx}{dt} = h(x, y), \qquad x(0) = c, \tag{3}$$

leads to a nonlinear differential equation (the Euler equation) with two-point boundary conditions. We shall present several techniques for treating these difficult problems below. If we impose constraints upon x and y, the problem of obtaining analytic and computational solutions becomes quite formidable, although a number of techniques do exist [*6*, *11*, *21*].

Particularly what is desired in the consideration of these questions are simple characterizations of the structure of the solution in terms of the nature of $g(x, y)$, $h(x, y)$ and the character of the constraints. Some results of this nature may be found in Bellman [*6*, *16*].

Many of these variational problems subject to constraints lead to the study of mixed systems of differential equations and differential inequalities. Very little has been done in this area. For a brief discussion of some known results and further references, see Beckenbach and Bellman [*2*].

5. Successive Approximations

The method of successive approximations is one of the basic techniques of analysis. Without its help much of the theory of differential equations would be rather sterile and existential. It is rather odd then, in the face of this, to observe that so little application of this versatile technique has been made to the study of variational problems.

Consider the problem of minimizing the functional

$$J(u) = \int_0^T g(u, u')\,dt, \qquad u(0) = c; \tag{1}$$

where $u(t)$ is a scalar function. Let u_0 be an initial approximation, and write

$$g(u, u') = g(u_0, u_0') + (u - u_0)\,g_{u_0} + (u' - u_0')\,g_{u_0'} + \ldots, \tag{2}$$

the Taylor series expansion of $g(u, u')$ around $u = u_0$, $u' = u_0'$. Let $k(u, u'; u_0, u_0')$ represent the sum of the constant, linear and quadratic terms in the foregoing expansion, and

$$J(u) = \int_0^T k(u, u'; u_0, u_0')\,dt, \qquad u(0) = c. \tag{3}$$

Let u_1 be the function minimizing $J_2(u)$. Generally, let the sequence of functions $\{u_n\}$ be generated recursively by the minimization of

$$J_{n-1}(u) = \int_0^T k(u, u'; u_{n-1}, u_{n-1}')\,dt, \qquad u(0) = c. \tag{4}$$

The minimizing function is denoted by u_n.

Some important questions associated with this process are:

(a) Under what conditions upon $g(u, u')$ does the sequence converge to a solution of the original minimization problem?

(b) What is the rate of convergence? (5)

(c) Can one devise an approximation method possessing the monotonicity property $J(u_n) \leq J(u_{n-1})$?

Some of these questions can be attacked by the techniques of dynamic programming and quasilinearization discussed below. The questions are of even greater interest if we add constraints such as

$$|u| \leq m_1, \qquad |u'| \leq m_2. \tag{6}$$

6. Dynamic Programming

A new approach to the analytic and computational treatment of variational problems is afforded by the theory of dynamic programming [5, 6]. Regarding the minimization problem of (5.1) as a multistage decision process, the principle of optimality [5, 6], readily yields the nonlinear partial differential equation

$$\frac{\partial f}{\partial T} = \min_{v} \left[g(c, v) + v \frac{\partial f}{\partial c} \right], \qquad f(c, 0) = 0, \tag{1}$$

for the function $f(c, T)$ defined by

$$f(c, T) = \min_{u} \int_{0}^{T} g(u, u')\, dt, \qquad u(0) = c. \tag{2}$$

If we add the constraint

$$k(u, u') \leq 0, \tag{3}$$

then (1) becomes

$$\frac{\partial f}{\partial T} = \min_{k(c,v) \leq 0} \left[g(c, v) + v \frac{\partial f}{\partial c} \right]. \tag{4}$$

As indicated in [4, 6], this equation can be used to determine the structure of the optimal policy, given certain simple structural features of $g(c, v)$ and $k(c, v)$.

A number of interesting questions arise in connection with the study of the existence and uniqueness of solutions of (1) and (4), and in connection with the use of successive approximations in function space and policy space [5, 6]. Very little has been done in this direction. Starting with the defining equations, (1) or (4), it should be possible to derive basic results of the calculus of variations from an extended theory of semigroups. The present theory of semigroups of operations deals with operator equations of the form

$$\frac{\partial f}{\partial T} = Af + b, \tag{5}$$

where A is an operator. Many of the results can be extended to the case of equations of the form

$$\frac{\partial f}{\partial T} = \max_{v} \left[A(v) f + b(v) \right], \tag{6}$$

where $A(v)$ is an operator dependent upon v. All of this is closely related to the theory of quasilinearization which we shall discuss below.

7. Stochastic and Adaptive Control Processes

Whatever the difficulties and unexpected hurdles we face in dealing with deterministic control processes, they compound when we introduce various features of uncertainty. It would take us too far afield here to discuss the many intriguing and novel questions that arise in the study of stochastic and adaptive control processes (the interested reader may refer to ref. [*6*]). Let us merely state that we encounter a most interesting blend of modern probability theory and the classical theory of differential equations in this new and rapidly developing area.

8. Quasilinearization

Linear equations are the only equations over which we can pretend to have any mastery. Hence, when forced to face the existence of nonlinear equations we attempt by one artifice or another to reduce them to linear equations. There are various ways of accomplishing this, each with its individual advantages and disadvantages. One method which combines a number of advantages is the technique of quasilinearization.

To illustrate the method, consider the Riccati equation

$$\frac{du}{dt} = u^2 + a(t), \qquad u(0) = c. \tag{1}$$

This is a purely deterministic equation, having nothing to do with control processes. Suppose we use the simple identity

$$u^2 = \max_v \, [2uv - v^2]. \tag{2}$$

Then (1) can be written

$$\frac{du}{dt} = \max_v \, [2uv - v^2 + a(t)], \qquad u(0) = c, \tag{3}$$

which bears a great resemblance to the equations of § 6 obtained by means of dynamic programming. The function v can then be thought of as a *policy*, which means that we can use the method of successive approximations not only in the usual function space, but also in policy space, and thus obtain monotone convergence (the details will be found in Bellman [*4, 5*]), Kalaba, [*22*].

Suppose that we consider the linearized equation

$$\frac{dw}{dt} = 2wv - v^2 + a(t), \qquad w(0) = c, \tag{4}$$

which can be solved by means of quadratures for any fixed function $v(t)$. Is it true that

$$u = \max_v w, \tag{5}$$

where w is now a function determined by (4)? It is indeed true, and similar results hold for many of the important equations of mathematical physics [*2*, *4*, *22*]. We shall indicate an application of the technique in § 12 below.

9. Time Lag and Retarded Control

In the first few sections of the paper we supposed that the equations describing the system S had the form of ordinary differential equations. This is equivalent to the assumption that the rate of change of $x(t)$ is dependent only upon the state of the system at time t. In many cases, particularly in control processes where a nonzero time is expended in exerting control, this is not a permissible premise.

Taking this into account, we encounter equations of the form

$$\frac{dx}{dt} = g(x(t), x(t - t_1), ..., x(t - t_k)), \tag{1}$$

and, generally, equations of the form

$$\frac{dx}{dt} = g\left(\int_{-\infty}^{t} K(t, s)\, x(s)\, ds\right). \tag{2}$$

The problems of existence, uniqueness and asymptotic behavior are now very much more complex and interesting. A good deal has been done in this direction (see for example Bellman and Cooke [*10*]), but the surface has barely been scratched.

Of particular interest is the study of control processes involving time lags and of variational problems, in general, involving nonlocal formulations. These have become of great physical interest in recent years.

10. Invariant Imbedding

Many of the significant equations of mathematical physics are of boundary value type, as opposed to initial value type. Equivalently, we might say that the classical integral equations arising from the usual formulations are of Fredholm type as distinct from Volterra type. This fact introduces a number of analytic complications, as far as establishing existence and uniqueness theorems is concerned, and also as far as computational solutions are concerned.

The theory of dynamic programming offers us a way of treating variational problems as initial value problems despite the fact that the usual procedure yields a differential equation with two-point boundary conditions. Nevertheless, by emphasizing the multistage aspect of the process, this formidable difficulty can be avoided.

The same idea can be applied to the study of large classes of physical processes. Employed systematically by Ambarzumian and Chandrasekhar in the study of radiative transfer, the concepts have been systematically extended to many different areas, such as neutron transport, scattering, wave propagation, plasma physics, by Bellman, Kalaba, Wing, Vasudevan, Ueno, *et al.* [*12*, *13*, *17-19*], and, independently, to interesting classes of scattering and electromagnetic problems by Redheffer [*27*].

This approach introduces initial value equations and provides a systematic means of establishing the existence and uniqueness of solutions of many equations which classical tèchniques do not handle easily or completely [*20*].

The essential feature of these investigations is the recognition of the existence of many alternative mathematical formulations of the same physical process. Some formulations are well suited for one type of investigation, some for another. The study of the relations between the various formulations is itself of interest [*19*].

11. Digital Computers

The advent of the digital computer has so vastly enlarged the scope of mathematics in general, and analysis in particular, as to defy description. Only the passage of twenty-five or fifty years will permit us to observe the numerous offshoots of the powerful stimulus of the electronic calculator. Here, we wish only to mention a few of the interesting and significant new problems that have arisen. We shall not discuss the questions of stability and accuracy since they have received a great amount of attention to date and are sufficiently will recognized by now.

Instead we shall focus on certain aspects of the questions of storage and time which as yet have received little notice.

12. Two-Point Boundary-Value Problems

Given a system of nonlinear ordinary differential equations with two-point boundary conditions, we have several ways of obtaining numerical solutions. We can follow the usual trial-and-error method, improved by various types of techniques of successive approximations; we car use the theory of dynamic programming to treat the original variational problem if these equations are the Euler equations of such a problem; we can use invariant imbedding to treat the associated physical process directly; or, we can use quasilinearization.

To illustrate this last technique, consider the scalar equation

$$u'' = g(u), \qquad u(0) = u(1) = 0. \tag{1}$$

Write

$$u_n'' = g(u_n) + (u_n - u_{n-1})\, g'(u_{n-1}), \qquad u_n(0) = u_n(1) = 0. \tag{2}$$

Starting with an initial choice of $u_0(t)$, $0 \leq t \leq 1$, we can compute each of the functions $u_n(t)$ as the solution of a *linear* equation with two-point boundary conditions, a routine matter. The advantage of the foregoing technique over the usual techniques lies in quadratic convergence, i.e.

$$| u_n - u_{n-1} | = 0(| u_{n-1} - u_{n-2} |^2), \tag{3}$$

and, occasionally, in monotone convergence (for details, see Kalaba [*22*]).

This method does have one drawback, however. To compute u_n, the values of u_{n-1} must be stored. Although of little significance in this case, in more complex problems this requirement could prevent any use of the method due to the quite limited storage capacity of modern computers. As it turns out, there is a way out of this difficulty.

This, then, is one of the endless problems raised by the presence of the computer. Given an algorithm which has certain desired analytic characteristics, how do we carry out the calculations in a reasonable time with a computer of limited storage? There is no systematic way of tackling questions of this nature. What is important about these problems is that they afford great opportunity and motivation for ingenuity. Thus, a device, the digital computer, which some feared would dry up analysis by providing automatic numerical solution of any equation has actually been responsible for a tremendous surge of interest in analysis and a revival of the emphasis upon elegant techniques and particular solutions. All of this is a very welcome counterballast to the axiomatic sterility and unmotivated asceticism of the Bourbaki school which is exerting such an unhealthy influence upon our graduate schools and young mathematicians.

13. Successive Approximations without Storage

Let us now discuss several ways of applying successive approximations without storage. Consider the problem posed in § 12, and let us examine the calculation of u_1. The equation is

$$u_1'' = g(u_0) + (u_1 - u_0)\, g'(u_0), \qquad u_1(0) = u_1(1) = 0. \tag{1}$$

Since the equation is linear in u_1, we can readily determine the value

$u_1'(0)$. Hence, the function $u_1(t)$ is determined by the same differential equation and the initial values $u_1(0)$, $u_1'(0)$. Turn now to the determination u_2 from the equation

$$u_2'' = g(u_1) + (u_2 - u_1)\, g'(u_1), \qquad u_2(0) = u_2(1) = 0. \tag{2}$$

To find the missing value $u_2'(0)$ we must solve the linear equation in (2) subject to two different sets of initial values. In place of storing the values of u_1, we adjoin the equation in (1) with the initial conditions $u_1(0)$, $u_1'(0)$. Having determined $u_2'(0)$, we determine u_3, without storage, by adjoining the two equations for u_1 and u_2 with initial values. Continuing in this way, we determine u_n by solving n simultaneous differential equations with initial values. (Details will be found in ref. [*1*].)

14. Application to Differential-Difference Equations

The foregoing idea can be applied to the computational solution of differential-difference equations. Consider, for example, the equation

$$u'(t) = g(u(t-1)), \qquad t \geq 1, \tag{1}$$

with the initial condition

$$u(t) = v(t), \qquad 0 \leq t \leq 1. \tag{2}$$

This equation can be solved in a straightforward way by storing the values of $u(t)$ in $[k, k+1]$ while computing the values in $[k+1, k+2]$. To obtain the solution numerically without storage, we proceed as follows. Define the sequence of functions

$$u_n(t) = u(t+n), \qquad 0 \leq t \leq 1, \tag{3}$$

$n = 0, 1, 2, \ldots$. Then (1) can be written as an infinite system of ordinary differential equations

$$u_n'(t) = g(u_{n-1}(t)), \qquad 0 \leq t \leq 1, \tag{4}$$

$n = 1, 2, \ldots$. Unfortunately, we do not have the initial values $u_n(0)$, $n = 2, 3, \ldots$. Let us suppose for the moment that $u_0(t) = v(t)$ is a simple function such as a constant, v_1, so that no storage is required for it.

Then, solving the first equation

$$u_1'(t) = g(v_1), \qquad u_1(0) = v_1, \tag{5}$$

as an initial value problem, we readily determine $u_1(1) = u_2(0)$. Let us then add to (5) the second equation, obtaining the initial value system

$$\begin{aligned} u_1'(t) &= g(v_1), & u_1(0) &= v_1, \\ u_2'(t) &= g(u_1(t)), & u_2(0) &= u_1(1) \quad \text{(a known value).} \end{aligned} \tag{6}$$

Solving this system, we determine $u_2(1) = u_3(0)$, adjoint the third equation and continue in this way, with no storage of functional values. (For further details, see ref. [*8*].)

15. Successive Approximations—II

Let us return to the equation

$$u'' = g(u), \qquad u(0) = u(1) = 0, \tag{1}$$

and proceed in the following fashion. Guess a value of $u'(0)$ and integrate (1) as an initial value problem, obtaining in this way a value $u_1(1)$ and $u_1'(1)$. If $u_1(1) = 0$, we have a solution (or the solution) and we stop. If not, use the value $u(1) = 0$, and the previous value of $u_1'(1)$ as initial values starting at $t = 1$ and integrate back to $t = 0$. In this way, we obtain values $u_2(0)$ and $u_2'(0)$. If $u_2(0) = 0$, we stop; if not, we use the initial conditions $u_3(0) = 0$, $u_3'(0) = u_2'(0)$ and integrate out to $t = 1$, and so on. Once again, we are using successive approximations without the requirement of storage.

Does the process converge? Sometimes it does, and sometimes it doesn't. A brief discussion is given in Bellman [*9*]. A major question is that of deriving a modification of the foregoing procedure which converges for a larger class of equations than that discussed in ref. [*9*].

The problem becomes of greater significance when we attempt to obtain the computational solution of equations of the form

$$\begin{aligned} u''(t) &= g(u(t), u(t-1)), & 1 &\le t \le n, \\ u(t) &= v(t), & 0 &\le t \le 1, \\ &= w(t), & n-1 &\le t \le n. \end{aligned} \tag{2}$$

These arise from variational problems associated with functionals of the type

$$J(u) = \int_0^n h(u(t), u(t-1), u'(t))\, dt. \tag{3}$$

See Myskis [*26*] for a brief discussion of these questions.

16. Sequential Computation

The usual methods for the numerical solution of vector differential equations of the form

$$\frac{dx}{dt} = g(x), \qquad x(0) = c, \tag{1}$$

employ fixed step-sizes in t. Clearly, the size of the step at t should depend upon the behavior of x in the neighborhood of t. How should this be done in some optimal fashion? For a very brief discussion, see ref. [*6*]. Practically nothing has been done in this important area.

17. Unconventional Numerical Techniques

The usual approach to the numerical solution of an equation such as

$$u_t = uu_x, \qquad u(x, 0) = g(x), \tag{1}$$

is by way of characteristics, or by means of a finite difference scheme such as

$$\frac{u(x, t + \Delta) - u(x, t - \Delta)}{2\Delta} = u\left(\frac{u(x + \Delta_1, t) - u(x - \Delta_1, t)}{2\Delta_1}\right), \tag{2}$$

where Δ and Δ_1 are both small quantities.

There are, however, many finite difference schemes which reduce to (1) in the limit. For example,

$$u(x, t + \Delta) = u(x + u\Delta, t), \qquad t = 0, \Delta, 2\Delta, \ldots, \tag{3}$$

is one which works very much better [*14*] than any of the usual schemes for solving (1). Very little has been done in the directions of a systematic study of wide classes of difference schemes. For example, to solve

$$f_T = g(c, T) + f_c^2, \qquad f(c, 0) = h(c) \tag{4}$$

we could use quasilinearization and write

$$f_T = \max_v \, [g(c, T) - v^2 + 2vf_c]. \tag{5}$$

This equation can then be considered to be the limiting form of

$$\begin{aligned} f(c, T + \Delta) &= \max_v \, [(g(c, T) - v^2)\, \Delta + f(c + v\Delta, T)], \\ f(c, 0) &= h(c), \end{aligned} \tag{6}$$

an equation typical of those encountered in dynamic programming [*6*].

Calculations based upon equations of this type work out quite well [*4*], but there has been no work done on the relative efficacy of this method as opposed to the usual methods of solving (4).

REFERENCES

[*1*] Bateman, H., The control of an elastic fluid. *Bull. AMS* **51**, 601-646 (1945).
[*2*] Beckenbach, E. F., and Bellman, R., "Inequalities." Springer, Berlin, 1961.
[*3*] Bellman, R., "Stability Theory of Differential Equations." McGraw-Hill, New York, 1953.
[*4*] Bellman, R., Functional equations in the theory of dynamic programming, V. Positivity and quasilinearity. *Proc. natl. Acad. Sci. U.S.* **41**, 743-746 (1955).
[*5*] Bellman, R., "Dynamic Programming." Princeton Univ. Press, Princeton, New Jersey, 1957.
[*6*] Bellman, R., "Adaptive Control Processes: A Guided Tour." Princeton Univ. Press, Princeton, New Jersey, 1961.
[*7*] Bellman, R., Successive approximations and computer storage problems in ordinary differential equations. *Communs. Assoc. computing Machinery* **4**, 222-223 (1961).
[*8*] Bellman, R., On the computational solution of differential-difference equations. *J. math. Anal. and Appl.* **2**, 108-110 (1961).
[*9*] Bellman, R., On the iterative solution of two-point boundary value problems. *Boll. Unione mat. ital.* **16**, 145-149 (1961).
[*10*] Bellman, R., and Cooke, K. L., "Differential-difference Equations." Academic Press, New York, in press, 1962.
[*11*] Bellman, R., and Dreyfus, S., "Applied Dynamic Programming." Princeton Univ. Press, Princeton, New Jersey, in press, 1962.
[*12*] Bellman, R., and Kalaba, R., On the principle of invariant imbedding and propagation through inhomogeneous media. *Proc. natl. Acad. Sci. U.S.* **42**, 629-632 (1956).
[*13*] Bellman, R., and Kalaba, R., On the principle of invariant imbedding and diffuse reflection from cylindrical regions. *Proc. natl. Acad. Sci. U.S.* **43**, 514-517 (1957).
[*14*] Bellman, R., Cherry, I., and Wing, G. M., A note on the numerical integration of a class of nonlinear hyperbolic equations. *Quart. appl. Math.* **16**, 181-183 (1958).
[*15*] Bellman, R., Glicksberg, I., and Gross, O., On the 'bang-bang' control problem. *Quart. appl. Math.* **14**, 11-18 (1956).
[*16*] Bellman, R., Glicksberg, I., and Gross, O., Some Aspects of the Mathematical Theory of Control Processes. RAND Corp. Report No. R-313 (1958).
[*17*] Bellman, R., Kalaba, R., and Prestrud, M., On a new computational solution of time-dependent transport processes, I. One-dimensional case. *Proc. natl. Acad. Sci. U.S.* **47**, 1072-1074 (1961).
[*18*] Bellman, R., Kalaba, R., and Vasudevan, R., An invariant imbedding approach to dielectric breakdown. To be published.
[*19*] Bellman, R., Kalaba, R., and Wing, G. M., Invariant imbedding and mathematical physics. I. Particle processes. *J. Math. Phys.* **1**, 280-308 (1960).

[*20*] Bellman, R., Kalaba, R., and Wing, G. M., Invariant imbedding, conservation relations, and nonlinear equations with two-point boundary values. *Proc. natl. Acad. Sci. U.S.* **46**, 1258-1260 (1960).

[*21*] Dreyfus, S., The numerical solution of variational problems. *J. math. Anal. and Appl.* to be published.

[*22*] Kalaba, R., On nonlinear differential equations, the maximum operation, and monotone convergence. *J. Math. and Mech.* **8**, 519-574 (1959).

[*23*] La Salle, J. P., On time optimal control systems. *Proc. natl. Acad. Sci. U.S.* **45**, 573-577 (1959).

[*24*] LaSalle, J. P., and Lefschetz, S., "Stability by Liapunov's Direct Method with Applications." Academic Press, New York, 1961.

[*25*] Lefschetz, S., "Differential Equations: Geometric Theory." Interscience, New York, 1957.

[*26*] Myskis, A. D., "Linear Differential Equations with Retarded Argument." Akad. Nauk S.S.S.R., Moscow, 1951; German transl., Deut. Verlag Wiss., Berlin, 1955.

[*27*] Redheffer, R., The Mycielski-Paszkowski diffusion problem. *J. Math. and Mech.* **9**, 607-622 (1960).

Some Aspects of Quasilinearization

ROBERT KALABA
The RAND Corporation, Santa Monica, California

1. Introduction to Quasilinearization

Early in the development of the theory of dynamic programming it was recognized that functional equations of the form

$$L_1(u) = \max_q [L_2(u, x, q) + a(x, q)], \tag{1}$$

where L_1 and L_2 are linear operators on u, a scalar function of the vector x, and q is a decision vector, an element of a space Q, would play a significant role [*4*]. In 1955 Bellman pointed out that many of the classical nonlinear equations of mathematical physics and engineering could be reduced to the form of equation (1). The list of such equations is now known to be very large and includes the Riccati equation [*2*, *3*], the eikonal equation of optics [*11*], various forms of the Hamilton-Jacobi equation of mechanics [*8*, *10*], and the Emden-Fowler equation of astrophysics and nuclear physics [*1*] as well as various nonlinear diffusion and potential equations [*6*].

A detailed discussion of a variety of special cases of equation (1.1) and of the general equation itself is given in [*9*]. Among the equations discussed are the initial value problem

$$u' = f(u, x), \qquad u(0) = c, \tag{2}$$

the boundary value problem

$$u'' = f(u, x), \qquad u(0) = u(b) = 0, \tag{3}$$

and the nonlinear potential and diffusion equations

$$u_{xx} + u_{yy} = f(u, x, y), \tag{4}$$

$$u_t - u_{xx} = f(u, x, y). \tag{5}$$

Let us drop the maximum operator in equation (1) and consider the associated linear equation for $w(x, q)$

$$L_1(w) = L_2(w, x, q) + a(x, q). \tag{6}$$

Then, under certain circumstances, the solution of the original equation (1) may be represented in the form

$$u(x) = \max_q w(x, q). \tag{7}$$

This is the essence of quasilinearization. The representation in equation (7) leads to the production of lower bounds for the solution of equation (1). In addition it leads to a method of successive approximations for the solution of nonlinear boundary value problems, the convergence being both monotone and quadratic. Consequently the technique, abstractly equivalent to the Newton-Raphson method, is of interest in the computational solution of these equations.

The aim of this paper is to provide a sketch of some of the principal ideas and applications associated with the notion of quasilinearization, and it is intended to be quite self-contained. We shall first show how equations of the form of equation (1) arise from application of the principle of optimality to a simple problem in the calculus of variations. Next we shall examine the Riccati equation. Section 4 is devoted to a discussion of the nonlinear boundary value problem arising from minimization of the integral

$$\int_0^1 [ke^y + \tfrac{1}{2}(y'')^2]\, dx,$$

subject to the conditions

$$y(0) = y'(0) = 0,$$

$$y \quad \text{at} \quad x = 1 \quad \text{is free.}$$

In particular some interesting numerical results are presented for, as was indicated above, one of the major applications is to the solution of nonlinear boundary value problems, many of which arise from variational contexts.

Section 5 contains some remarks on the computational solution of boundary value problems for nonlinear partial differential equations. Finally, a discussion of some additional matters is given.

2. A Variational Problem

Equations of the form of equation (1.1) occur naturally in analysis. To show this we consider a classical variational problem from the functional equation viewpoint of dynamic programming.

We consider the problem of maximizing the integral

$$\int_0^T F(x, y)\, dt = J[y], \tag{1}$$

where

$$\frac{dx}{dt} = G(x, y), \qquad x(0) = c. \tag{2}$$

Many problems in the theory of automatic control are of this type. First we observe that the maximal value is solely a function of c, the initial state, and T, the duration of the process,

$$\max_{y(t)} J[y] = f(c, T). \tag{3}$$

Next we use Bellman's principle of optimality [*4*] to see that

$$f(c, T) = \max_{Y} [F(c, Y)\,\Delta + f(c + G(c, Y)\,\Delta, T - \Delta) + o(\Delta)], \tag{4}$$

where $Y = y(0)$. Of course, the maximizing value of Y depends on c and T. Then, under suitable regularity assumptions, a passage to the limit yields the nonlinear differential equation

$$f_T = \max_{Y} [F(c, Y) + G(c, Y) f_c]. \tag{5}$$

Furthermore, as an initial condition, we have

$$f(c, 0) = 0. \tag{6}$$

We assume that these equations have a unique solution in some region for which $T \geq 0$. Equation (5) is of the form which we wish to study.

Next let us consider the *linear* differential equation which we obtain by dropping the maximum operator in equation (5):

$$w_T = F(c, Y) + G(c, Y)\, w_c, \tag{7}$$

$$w(c, 0) = 0. \tag{8}$$

We regard Y as a given function of c and T. In view of equations (5) and (7) we see that

$$(f - w)_T = G(c, Y)\,(f - w)_c + p, \tag{9}$$

where

$$p \geq 0. \tag{10}$$

In addition for $T = 0$ we find

$$f(c, 0) - w(c, 0) = 0. \tag{11}$$

But equations (9) and (10) imply that for some region for which $T \geq 0$

$$f(c, T) - w(c, T) \geq 0, \tag{12}$$

$$f(c, T) \geq w(c, T). \tag{13}$$

On the other hand, equality will hold for an appropriate choice of the function Y in equation (7) which determines the function w. This leads to the representation

$$f(c, T) = \max_{Y} w. \tag{14}$$

We have now represented the solution of the nonlinear equation (5) in terms of a maximization over Y of solutions of the associated linear equation (7).

The purpose of this section has been to show that equations of the form of equation (1.1) occur naturally in analysis. Rather than continue along these lines, let us now turn to a study of one of the classical equations, that of Riccati, to illustrate our methodology.

3. The Riccati Equation and a Generalization

Let us consider the Riccati equation

$$\frac{du}{dx} = u^2 + g(x), \qquad 0 \leq x \leq b, \tag{1}$$

$$u(0) = c, \tag{2}$$

where $g(x)$ is continuous, and b is sufficiently small to guarantee the existence of a solution. Since

$$u^2 \geq 2vu - v^2, \tag{3}$$

with equality holding for

$$v \equiv u, \tag{4}$$

we may rewrite equation (1) in the form

$$\frac{du}{dx} = \max_{v} [2vu - v^2 + g]. \tag{5}$$

This shows that the Riccati equation can be transformed into an equation having the form of equation (1.1).

To proceed with the analysis we introduce the associated linear equation

$$\frac{dw}{dx} = 2vw - v^2 + g, \qquad w(0) = c, \tag{6}$$

which we obtain by dropping the maximum operator in equation (5). To indicate that w depends on x and the choice of the function $v = v(x)$ we write

$$w = w[x; v]. \tag{7}$$

Our first objective is to show that u may be represented in the form

$$u = \max_v w[x; v], \tag{8}$$

where the maximization is over all functions $v = v(x)$, $0 \leq x \leq b$ for which the integrals in equation (15) below exist.

Equation (5) may be written

$$\frac{du}{dx} = 2vu - v^2 + p + g, \tag{9}$$

where p is a nonnegative function on the interval $[0, b]$. It follows that the difference function $z = u - w$ satisfies the equation

$$\frac{dz}{dx} = 2vz + p, \qquad z(0) = 0, \tag{10}$$

for which the solution is

$$z(x) = \int_0^x p(s) \exp\left[2 \int_s^x v(t)\, dt\right] ds. \tag{11}$$

The nonnegativity of this integral shows that

$$z(x) \geq 0, \qquad 0 \leq x \leq b. \tag{12}$$

In addition we know that if we select

$$v(x) = u(x), \tag{13}$$

then the solution of equation (6) is

$$w[x; u(x)] = u(x). \tag{14}$$

This completes the proof of the representation of the solution of the Riccati equation (1) in the form of equation (8). By solving equation (6) for w we find the interesting representation for the solution of the Riccati equation

$$u(x) = \max_v \left\{ c \exp\left[2\int_0^x v(s)\, ds\right] + \int_0^x [g(s) - v^2(s)] \cdot \exp\left\{2\int_s^x v(t)\, dt\right\} ds \right\},$$
$$0 \leq x \leq b. \tag{15}$$

It follows, of course, that uniform lower bounds for u can be obtained by evaluating the right-hand side of equation (15) for simple admissible functions $v = v(x)$.

As a slight digression let us note that we can use the representation formula just derived to obtain a representation for the solution of the linear homogeneous equation of second order

$$\frac{d^2y}{dx^2} + g(x)y = 0, \qquad 0 \leq x \leq b, \tag{16}$$

$$y(0) = y_0 \neq 0, \qquad y'(0) = y_1. \tag{17}$$

We use the well-known transformation

$$-\frac{y'}{y} = u \tag{18}$$

to rewrite equation (16) as

$$\frac{du}{dx} = u^2 + g(x), \tag{19}$$

$$u(0) = -y_1/y_0 = c. \tag{20}$$

Then application of formula (15) yields the result

$$\frac{y(x)}{y_0} = \min_v \exp\left\{-\int_0^x \left[c \exp\left\{2\int_0^s v(t)\, dt\right\}\right.\right.$$
$$\left.\left. + \int_0^s (g(t) - v^2(t)) \exp\left\{2\int_t^s v(t_1)\, dt_1\right\} dt\right] ds\right\}, \qquad 0 \leq x \leq b, \tag{21}$$

for b sufficiently small.

We now turn to the problem of constructing monotone sequences of functions which converge to u, the solution of equation (1). Choose an arbitrary initial function $v = v_0(x)$ and determine the solution of the associated linear equation (6),

$$\frac{du_0}{dx} = 2v_0u_0 - v_0^2 + g, \qquad u_0(0) = c. \tag{22}$$

Then determine an improved choice of the function v, call it v_1, as the function which maximizes the expression

$$2u_0v - v^2 + g.$$

This maximizing function is

$$v = v_1 = u_0. \tag{23}$$

Next compute the function $u_1(x)$ as the solution of the equation

$$\frac{du_1}{dx} = 2u_0u_1 - u_0^2 + g, \qquad u_1(0) = 0. \tag{24}$$

Then determine an improved choice of the function v, $v = v_2(x)$, as the function which maximizes

$$2u_1v - v^2 + g.$$

This maximizing function is

$$v = v_2 = u_1. \tag{25}$$

Continuing in this way we are led to consider the sequence of functions $\{u_n\}$ defined by the linear equations

$$\begin{aligned} \frac{du_0}{dx} &= 2v_0u_0 - v_0^2 + g, \qquad u_0(0) = c, \\ \frac{du_{n+1}}{dx} &= 2u_nu_{n+1} - u_n^2 + g, \qquad u_{n+1}(0) = c, \qquad n \geq 0. \end{aligned} \tag{26}$$

This type of reasoning is familiar in dynamic programming and is known as approximation in policy space [*4*].

The recurrence relations of equation (26) may be viewed as an extension of Newton's method, for the right hand sides of the latter equations may be considered to arise from expanding $f(u) = u^2 + g$ around $u = u_n$ and keeping only the linear terms, $(u_n^2 + g) + (u_{n+1} - u_n)\, 2u_n$. In the more familiar Picard scheme of successive approximations only the first term, $u_n^2 + g$, is retained. As we shall see, in some cases the Newtonian scheme has marked computational advantages.

The sequence of functions $u_n(x)$ tends monotonically to $u(x)$, and, in addition, the convergence is quadratic. Consider the equations

$$\frac{du_{k+1}}{dx} = 2u_ku_{k+1} - u_k^2 + g, \qquad u_{k+1}(0) = c, \tag{27}$$

$$\frac{du_k}{dx} = 2u_{k-1}u_k - u_{k-1}^2 + g, \qquad u_k(0) = c, \qquad k = 0, 1, \ldots. \tag{28}$$

The maximum of the right-hand side of equation (28) is assumed when $u_{k-1} = u_k$, which implies that

$$\frac{du_k}{dx} \leq 2u_k^2 - u_k^2 + g. \tag{29}$$

Consequently, we have

$$d(u_{k+1} - u_k)/dx \geq 2u_k(u_{k+1} - u_k), \qquad u_{k+1}(0) - u_k(0) = 0, \tag{30}$$

which, in view of equations (10) and (12), implies that

$$u_{k+1}(x) \geq u_k(x), \qquad 0 \leq x \leq b. \tag{31}$$

This is the desired monotonicity result.

To demonstrate the quadratic nature of the convergence, which is of great importance from the computational viewpoint, we note that

$$\frac{d(u - u_n)}{dx} = u^2 - [u_{n-1}^2 + 2u_{n-1}(u_n - u_{n-1})]. \tag{32}$$

This can be rewritten to yield

$$\frac{d(u - u_n)}{dx} = u^2 - u_{n-1}^2 - (u - u_{n-1})\, 2u_{n-1} - (u_n - u)\, 2u_{n-1}. \tag{33}$$

Since

$$u(0) - u_n(0) = 0, \tag{34}$$

upon integrating from 0 to x, we find

$$u(x) - u_n(x) = \int_0^x [(u - u_{n-1})^2 + (u - u_n)\, 2u_{n-1}]\, ds, \qquad 0 \leq x \leq b. \tag{35}$$

Upon introducing the norm

$$\| u - u_n \| = \max_{0 \leq x \leq b} | u(x) - u_n(x) |, \tag{36}$$

we find

$$\| u - u_n \| \leq b[\| u - u_{n-1} \|^2 + \| u - u_n \| m], \tag{37}$$

where

$$| 2u_{n-1} | \leq m, \qquad 0 \leq x \leq b. \tag{38}$$

(A region of uniform boundedness for the sequence $\{u_n\}$ is readily established.) Thus

$$\| u - u_n \| \leq \frac{b}{1 - bm} \| u - u_{n-1} \|^2, \tag{39}$$

which demonstrates the quadratic convergence on a sufficiently small interval $[0, b]$.

Similar considerations can be carried out for a generalization of the Riccati equation

$$u' = f(u, x), \qquad u(0) = c, \tag{40}$$

under the assumption that f is a convex function of u for all x [9]. For in this case we can use the well-known property of convex functions that

$$f(u, x) = \max_{v} \left[f(v, x) + (u - v) \frac{\partial f}{\partial v} \right]. \tag{41}$$

Thus equation (40) can be written

$$u' = \max_{v} \left[f(v, x) + (u - v) \frac{\partial f}{\partial v} \right], \tag{42}$$

a form for which our formalism is applicable (see ref. [9] for details).

4. A Numerical Experiment with a Nonlinear Boundary Value Problem

Consider the following variational problem. Determine the function $y = y(x)$ on the interval $[0, 1]$ so as to minimize the integral

$$J[y] = \int_0^1 [ke^y + \tfrac{1}{2}(y'')^2]\, dx. \tag{1}$$

We assume that at $x = 0$ the function y satisfies the conditions

$$y(0) = y'(0) = 0, \tag{2}$$

and at $x = 1$ it satisfies the natural boundary conditions

$$y''(1) = y'''(1) = 0. \tag{3}$$

The Euler equation for this problem is

$$y^{\mathrm{IV}} + ke^y = 0. \tag{4}$$

Consequently we wish to find an efficient computational procedure for solving the nonlinear equation (4) subject to the boundary conditions (2) and (3). Our aim is to convert this boundary value problem into a rapidly converging sequence of initial value problems.

The procedure which we shall use is the following. We first select an initial approximation $v_0(x) \equiv 0$. Then the higher approximations are determined by solving the linear recurrence relations

$$u_0^{\mathrm{IV}} = -ke^{v_0} - k(u_0 - v_0)\, e^{v_0}, \tag{5}$$

$$u_{n+1}^{\mathrm{IV}} = -ke^{u_n} - k(u_{n+1} - u_n)\, e^{u_n}, \tag{6}$$

the boundary conditions being

$$\begin{aligned} &u_{n+1}(0) = u_{n+1}'(0) = 0, \\ &u_{n+1}''(1) = u_{n+1}'''(1) = 0, \qquad n = 0, 1, 2, \ldots. \end{aligned} \tag{7}$$

The computational determination of the function $u_{n+1}(x)$, given $u_n(x)$, is quite simple. First the particular solution of equation (6), $p(x)$, subject to the initial conditions $p(0) = p'(0) = p''(0) = p'''(0) = 0$ is computed and stored. Then the homogeneous solution, $w(x)$, satisfying the initial conditions $w(0) = w'(0) = w''(0) = 0$, $w'''(0) = 1$, is computed and stored. Next the homogeneous solution, $z(x)$, satisfying the initial conditions $z(0) = z'(0) = z'''(0) = 0$, $z''(0) = 1$ is computed and stored. Consequently the desired function $u_{n+1}(x)$ can be expressed in the form

$$u_{n+1}(x) = p(x) + Aw(x) + Bz(x). \tag{8}$$

The initial conditions $u_{n+1}(0) = u_{n+1}'(0) = 0$ are automatically satisfied. The constants A and B are determined by using the computed values of $p(1)$, $w(1)$ and $z(1)$ in conjunction with the remaining conditions

$$u_{n+1}''(1) = u_{n+1}'''(1) = 0. \tag{9}$$

This requires merely the solution of two linear equations in two unknowns. This completes the computational determination of $u_{n+1}(x)$. Notice that only the solution of initial value problems is required, along with the solution of some linear algebraic equations.

Now let us look at some computational results for the case $k = 6$, showing the rapidity of convergence. The entire calculation consumed only a few seconds on the IBM 7090. An Adams-Moulton integration procedure yielding about six significant figures was employed. The programming was very capably handled by Dr. Bella Kotkin.

TABLE 1
SOME COMPUTATIONAL RESULTS ($v_0(x) \equiv 0$)

x	$u_0(x)$	$u_1(x)$	$u_2(x)$	$u_3(x)$
0.0	0.0	0.0	0.0	0.0
0.20312	-0.375572×10^{-1}	-0.394922×10^{-1}	-0.394976×10^{-1}	-0.394976×10^{-1}
0.40625	−0.128544	−0.135598	−0.135617	−0.135617
0.60156	−0.242775	−0.256792	−0.256831	−0.256831
0.8469	−0.373308	−0.395724	−0.395786	−0.395786
1.0	−0.501782	−0.532645	−0.532730	−0.532730

When k becomes sufficiently negative the minimum problem can fail to have a solution, and, of course, the scheme breaks down. Apparently this happens for k lying between -4 and -5.

Similar techniques have been used successfully by Collatz [7], Hestenes [6], and others.

In applying the quasilinearization technique to higher order equations or to systems of equations, especially where high accuracy is required, problems due to the limited high-speed storage of modern computers (about 32,000 words, at best) arise. A powerful technique for avoiding them is discussed in the paper by Bellman [5].

5. Computational Solution of Some Nonlinear Partial Differential Equations

In Part III of ref. [9] treatments of both the nonlinear potential equation

$$\Delta u = f(u), \tag{1}$$

and the nonlinear diffusion, equation

$$u_t - u_{xx} = f(u), \tag{2}$$

under various boundary conditions, are given using the quasilinearization technique. In [6] a report is given concerning the results of numerical experiments which were conducted to compare the relative merits of the Picard approximation scheme versus the Newtonian scheme. This advantage of a quadratically convergent scheme over a linearly convergent scheme is not so clear in the case of partial differential equations. There are many reasons for this, but among the most prominent is the difficulty in solving the associated linear equations to a sufficiently high degree of accuracy. Nevertheless, we did find that in those cases for which the solution has large gradients, the higher order method was superior.

6. Discussion

Though the wide applicability of the notion of quasilinearization is by now apparent, let us mention several additional ones which are discussed in refs. [*2*] and [*9*]. In the first place, systems of equations can be treated, provided that an inequality between vectors implies that relation between components. In addition some of P. Lax's theory of weak solutions of nonlinear first order partial differential equations can be derived [*13*]. Also various nonlinear integral and differential-integral equations are amenable to treatment. Much remains to be done.

REFERENCES

[*1*] Bellman, R., "Stability Theory of Differential Equations." McGraw-Hill, New York, 1953.

[*2*] Bellman, R., Functional equations in the theory of dynamic programming, V. Positivity and quasi-linearity. *Proc. natl. Acad. Sci. U.S.* **41**, 743-748 (1955).

[*3*] Bellman, R., Functional equations in the theory of dynamic programming, II. Nonlinear differential equations. *Proc. natl. Acad. Sci. U.S.* **41**, 482-485 (1955).

[*4*] Bellman, R., "Dynamic Programming." Princeton Univ. Press, Princeton, New Jersey, 1957.

[*5*] Bellman, R., Successive approximations and computer storage problems in ordinary differential equations. *Communs. Assoc. Computing Machinery.* **4**, 222-223 (1961).

[*6*] Bellman, R., Juncosa, M., and Kalaba, R., "Some Numerical Experiments using Newton's method for nonlinear parabolic and elliptic boundary-value problems. *Communs. Assoc. Computing Machinery.* **4**, 187-191 (1961).

[*7*] Collatz, L., Monotonie und Extremalprinzipien beim Newtonschen Verfahren. *Numer. Math.* **3**, 99-106 (1961).

[*8*] Dreyfus, S., Dynamic programming and the calculus of variations. *J. math. Anal. and Appl.* **1**, 228-239 (1960).

[*9*] Kalaba, R., On nonlinear differential equations, the maximum operation, and monotone convergence. *J. Math. and Mech.* **8**, 519-574 (1959).

[*10*] Kalaba, R., Dynamic programming and the variational principles of classical and statistical mechanics. *Proc. 5th midwestern Conf. on Solid Mechanics*, Plenum Press, New York, 1961.

[*11*] Kalaba, R., Dynamic programming, Fermat's principle, and the Eikonal equation. *J. opt. Soc. Am.* **51**, 1150-1151 (1961).

[*12*] Kalaba, R., Computational considerations for some deterministic and adaptive control processes. *In* "Optimization Techniques," (G. Leitmann, ed.). Academic Press, New York, in press.

[*13*] Lax, P., Initial Value Problems for Nonlinear Hyperbolic Equations. Univ. of Kansas Tech. Report No. 14 (1955).

On the Differential Equations of Electrical Circuits and the Global Nature of the Solutions

JÜRGEN MOSER
Institute of Mathematical Sciences
New York University, New York, New York, and
IBM Research Center, Yorktown Heights, New York

I. The Differential Equations

1. The purpose of this paper is to describe some general properties of the differential equations occurring in the theory of electrical circuits. Most of the theories which one finds in text books on circuit theory deal with linear equations and the Laplace transform. However, in recent times many engineering problems have led to differential equations which cannot be properly approximated by linear circuits. A typical situation are circuits having several steady state solutions, which is obviously a nonlinear phenomenon. In fact, this investigation was initiated by a study of tunnel diode circuits and was performed at the IBM Research Center, Yorktown Heights, New York.

2. We consider an electrical circuit consisting of a graph with several branches. Each branch may contain a resistor and battery or a condensor or an inductor. Let i_ν denote the current in the νth branch, and v_ν the voltage (both measured with a proper sign according to an orientation of each branch). Consider a branch containing a resistor. The nature of the resistor is characterized by a function $v_\nu = v_\nu(i_\nu)$ relating the current and the voltage. In the linear case this relation is given by Ohm's law but we will permit more general, even nonmonotonic functions $v_\nu(i_\nu)$.

Our first result concerns the form of the differential equations. Let $\rho = 1, 2, ..., r$ denote the label of those branches containing inductances and i_ρ $(\rho = 1, 2, ..., r)$ the currents in those branches. Similarly let $\sigma = r + 1, ..., r + s$ denote the branches containing capacitances and v_σ the voltage across the capacitance. We want to assume that the branches with $1 \leq \rho \leq r$ do not contain all the branches issuing from any node (since the i_ν would then be related variables by Kirchhoff's law). This assumption, however, is no restriction of generality since for those branches the resistance was neglected and the assumption is satisfied by supplying an extra branch with a resistor.

Similarly, we assume that the branches $\sigma = r + 1, \ldots, r + s$ do not contain a cycle, since otherwise the v_σ would be related. On the other hand, we will assume that the i_ρ $(\rho = 1, \ldots, r)$, v_σ $(\sigma = r + 1, \ldots, r + s)$ determine the currents and voltage in all branches via the Kirchhoff law.

This amounts to the assumption that the static problem can be solved if i_ρ $(1 \leq \rho \leq r$, v_σ $(r < \sigma \leq r + s)$ are given parameters. The dynamic behavior is given by the following differential equations.

Statement. The differential equations of an *RLC* circuit can be written in the form

$$L_\rho \frac{di_\rho}{dt} = \frac{\partial P}{\partial i_\rho} \qquad \rho = 1, \ldots, r$$

$$C_\sigma \frac{dv_\sigma}{dt} = -\frac{\partial P}{\partial v_\sigma} \qquad \sigma = r + 1, \ldots, r + s.$$

The content of this statement is that the right hand sides of the differential equations satisfy certain integrability conditions so that they can be obtained as derivatives of one function P. We will not prove this statement here (see [*4*]) but mention that it is a consequence of Kirchhoff's laws and the law describing the dynamics of inductors and condensors.

The function P has, as is to be expected, a physical significance. It is related to Raleigh's dissipation function and we remark that such a function has been considered by Cherry [*1*] and Millar [*3*]. There similar functions were introduced rather formally and called contents and co-content. Even earlier, Duffin [*2*] used a similar concept. In a note of Weyl [*5*], which deals only with linear electrical circuits without condensors and inductors, a concept (namely, the inner product) is vaguely reminiscent of the function P. However, the differential equations of the statement seem to be new.

3. We want to describe how to construct the function P for a given circuit. Such a construction immediately gives the differential equations. The function can be written as

$$P = \int \sum_{\nu > r+s} v_\nu(i_\nu)\, di_\nu + \sum_{\sigma=r+1}^{r+s} v_\sigma I_\sigma.$$

The first term denotes the sum of all "potentials" of the resistors and represents a line integral which is independent of the path. The second term constitutes contributions from the capacitances and $I_\sigma(i, v)$ are the currents in the capacitances for prescribed i_ρ, v_σ. To get the differential equations of the statement, one has to express P in terms of i_ρ, v_σ.

We give an example of a circuit:

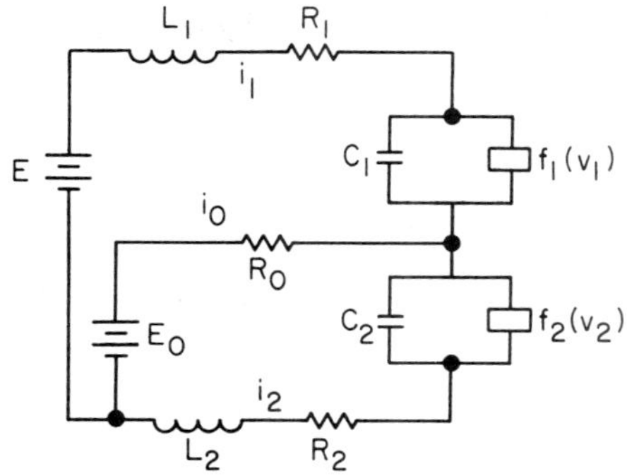

FIG. 1

and the construction of P. The squares indicate resistor elements with a nonlinear characteristic given by $f_1(v_1)$, $f_2(v_2)$. We are not consistent with the above notation in which case we would have to introduce nine currents and voltages. Most of these quantities are related trivially and it suffices to introduce i_1, i_2, $i_0 = i_1 - i_2$, v_1, v_2. Denoting the currents in the capacitances by I_1, I_2 we have

$$I_k = i_k - f_k(v_k) \text{ for } k = 1, 2$$

and

$$P = E_0 i_0 + E_1 i_1 - \frac{R_1 i_1^2}{2} - \frac{R_2 i_2^2}{2} - \frac{R_0 i_0^2}{2} - \sum_{k=1}^{2} \int v_k df_k(v_k)$$

$$- \sum_{k=1,2} i_k v_k + \sum f_k(v_k) \cdot v_k$$

$$= \sum_{k=0}^{1} E_k i_k - \frac{1}{2} \sum_{k=0}^{2} R_k i_k^2 - \sum_{k=1,2} \left(i_k v_k - \int f_k dv_k \right).$$

Replacing i_0 by $i_1 - i_2$ one gets P as a function of i_1, i_2, v_1, v_2.

In general P has the form

$$P = - A(i) + (i, \gamma v) + B(v)$$

where A depends on the currents i_ρ only, B on the v_σ only, and the coupling term is given by a matrix γ which contains $+ 1$, $- 1$ and 0 only. It reflects some property of the graph. The matrix

$$- \left(\frac{\partial^2 P}{\partial i_\rho \, \partial i_{\rho'}} \right) = - P_{ii}$$

is a resistance matrix which for linear circuits is positive definite. P_{vv} is an admittance which for linear circuits is positive definite. We will not make the assumption that P_{vv} is positive definite but do require that the lowest eigenvalues of A and B are positive if $|i|$, $|v|$ are large enough.

II. Non-Oscillating Circuits

1. Mathematically speaking one has a system of differential equations of the form

$$- J\dot{x} = \frac{\partial P}{\partial x}, \qquad P = P(x) \tag{1}$$

where $x = (i, v)$ is an $r + s$ dimensional vector and

$$J = \begin{pmatrix} -L & 0 \\ 0 & C \end{pmatrix};$$

is an indefinite diagonal matrix. One proves easily: A necessary and sufficient condition for a transformation $x \to y = f(x)$ to map the above differential equation into one of the same type

$$- J\dot{y} = \frac{\partial P}{\partial y}$$

is that the differential

$$(dx, Jdx) = -(di, Ldi) + (dv, Cdv)\ [1]$$

be invariant. Therefore we associate this differential form with the above system.

We note that the critical points of P are exactly the steady state solutions (equilibria) of (1). We want to discuss some conditions under which all solutions approach these equilibria; in particular, this would exclude periodic solutions.

2. In order to find such a criterion, we differentiate P along the solutions and find

$$\frac{d}{dt} P = (P_x, \dot{x}) = -(\dot{x}, J\dot{x})$$

[1] This differential form is not meant in the sense of Cartan's calculus but with a commutative multiplication.

which takes both signs. If, however, the circuit does not contain any inductor we have

$$\frac{d}{dt} P = -(\dot{x}, C\dot{x}) \qquad \text{since then} \qquad J = C$$

and obviously P would decrease, except at the equilibria. Following Liapounov's ideas one can derive from this—if $P \to \infty$ as $|x| \to \infty$— that every solution tends to one of the equilibrium solutions. In fact, it is known that oscillations cannot occur if $L = 0$ or $C = 0$.

It is intuitively clear that a similar result should hold if L is sufficiently small. How large can L become without causing oscillations? This question can be discussed by the following device, which is a way of constructing a Liapunov function.

3. For this purpose we ask whether one can describe the system (1) by another pair $\hat{J}$, $\hat{P}$ in place of J, P. Obviously this requires

$$\hat{P}_x = \hat{J} J^{-1} P_x \tag{2}$$

and we aim at finding a pair for which

$$(dx, \hat{J} dx)$$

is positive definite.

We note first if (J_1, P_1), (J_2, P_2) are two pairs describing (1) so are

$$(\alpha J_1 + \beta J_2, \alpha P_1 + \beta P_2)$$

which gives considerable freedom in the construction of other pairs.

To find one non-trivial pair besides J, P we observe: If A is any constant symmetric matrix and we set

$$\hat{P} = \tfrac{1}{2}(P_x, AP_x)$$

then

$$\hat{P}_x = (P_{xx} A) P_x$$

so that $(\hat{J}, \hat{P})$ is a pair if we set

$$\hat{J} = P_{xx} A J.$$

By the linear superposition we obtain the more general pairs

$$J_0 = (I + P_{xx} A) J, \qquad P_0 = P + \tfrac{1}{2}(P_x, AP_x) \tag{3}$$

where A ranges over all constant symmetric matrices.

As an example we assume that $R = - P_{ii}$ is a constant positive definite matrix and choose

$$A = +2 \begin{pmatrix} R^{-1} & 0 \\ 0 & 0 \end{pmatrix}.$$

Then

$$J_0 = \begin{pmatrix} I + 2P_{ii}R^{-1} & 0 \\ 2P_{iv}R^{-1} & I \end{pmatrix} J$$

$$= \begin{pmatrix} -I & 0 \\ 2\gamma^T R^{-1} & I \end{pmatrix} \begin{pmatrix} -L & 0 \\ 0 & C \end{pmatrix} = \begin{pmatrix} +L & 0 \\ 2\gamma^T R^{-1} L & C \end{pmatrix}.$$

Here we set $P_{ii} = - R$, $P_{vi} = \gamma$, $P_{vv} = Q$.

The form

$$(dx, J_0 dx) = (di, Ldi) + 2(R^{-1}\gamma dv, Ldi) + (dv, Cdv)$$

is positive definite whenever the matrix norm[1]

$$\| L^{1/2} R^{-1} \gamma C^{-1/2} \| < 1. \tag{4}$$

This result gives indeed a criterion for all solutions to tend to the equilibria if in addition $P_0 \to \infty$ as $| x | \to \infty$.

4. Notice that the criterion (4) is free from conditions on $P_{vv} = Q$ which might contain negative resistances, i.e., have negative eigenvalues. For the four dimensional system introduced above, one finds simply for

$$R_1 = R_2; \qquad L_1 = L_2; \qquad C_1 = C_2$$

the condition

$$\frac{L_k}{R_k} < C_k R_k \qquad (k = 1, 2)$$

which is an inequality between two constants of the dimension of time. Notice that the size of R_0 as well as $f_k(v_k)$ is quite irrelevant for this criterion for nonoscillatory behavior.

The expression (4) represents a dimensionless number which has to be less than 1. One checks easily that this condition is sharp. Indeed, there are even linear systems with periodic solutions if this number exceeds 1.

Another obvious choice of A is

$$A = \begin{pmatrix} L^{-1} & 0 \\ 0 & C^{-1} \end{pmatrix}$$

[1] $\| A \|^2 = \sup_y (Ay, Ay)$ under the side condition $(y, y) = 1$.

so that

$$\hat{J} = P_{xx}\begin{pmatrix} -I & 0 \\ 0 & I \end{pmatrix} = \begin{pmatrix} R & -\gamma \\ \gamma^T & Q \end{pmatrix}, \tag{5}$$

$$(dx, \hat{J}dx) = (di, Rdi) + (dv, Qdv). \tag{6}$$

Hence if L, C are constant positive definite matrices and

$$R = -P_{ii}, \qquad Q = P_{vv}$$

are positive definite, then

$$\hat{P} = (P_i, L^{-1}P_i) + (P_v, C^{-1}P_v) \tag{7}$$

is a decreasing function from which one can derive that—under appropriate additional assumptions—all solutions approach equilibria as $t \to \infty$.

Combining the pair $(\hat{J}, \hat{P})$ given in (5), (7) with the original one leads to

$$J_0 = \lambda J + \hat{J} \qquad P_0 = \lambda P + \hat{P}$$

and a Liapunov function results whenever

$$(dx, J_0 dx) = (di, (R - \lambda L)\, di) + (dv, (Q + \lambda C)\, dv) \tag{8}$$

is positive definite for an appropriate parameter λ. This condition means that the sum of the lowest eigenvalues of

$$L^{-1/2}RL^{-1/2} \qquad \text{and} \qquad C^{-1/2}QC^{-1/2}$$

is positive.

III. Time Estimates

In applications it is sometimes of importance to determine or estimate in what time a solution can reach a given vicinity of an equilibrium starting from a given state. The technique described above suggests some simple lower estimates for such time intervals.

Let us ask how long it takes a solution to approach an equilibrium $x^{(2)}$ from a given initial point $x^{(1)}$ in the phase space. We assume we found a pair (J_0, P_0) for which

$$(dx, J_0 dx) = (ds)^2$$

is positive definite and therefore can be used as a length element in the phase space. Assuming, moreover, that P_0 has a local minimum at $x^{(2)}$ we can describe a neighborhood of $x^{(2)}$ by the inequality

$$P_0 \leq p^{(2)} \tag{9}$$

where $p^{(2)}$ is slightly larger than $P_0(x_2)$. Moreover, we set

$$P_0(x^{(1)}) = p^{(1)} \tag{10}$$

and determine a lower estimate for the time it takes a solution to travel from (10) to the set (9).

For this purpose let $x(t)$ $(t^{(1)} \leq t \leq t^{(2)})$ be the solution under consideration and l the length of the curve described by the solution in the metric introduced. Then Schwarz inequality gives

$$l^2 = \left(\int ds\right)^2 = \left(\int_{t(1)}^{t(2)} \sqrt{(\dot{x}, J_0\dot{x})}\, dt\right)^2 \leq \int_{t(1)}^{t(2)} (\dot{x}, J_0\dot{x})\, dt(t^{(2)} - t^{(1)})$$

$$= -(P_0^{(2)} - P_0^{(1)})(t^{(2)} - t^{(1)}) \leq (p^{(1)} - p^{(2)})(t^{(2)} - t^{(1)}).$$

Thus the time $T = t^{(2)} - t^{(1)}$ satisfies

$$T \geq \frac{l^2}{p^{(1)} - p^{(2)}}$$

where l is the length of the solution curve in our metric. Usually l cannot easily be computed but it is easily estimated, for instance, by the length of the shortest curve (not solution) connecting $x^{(1)}$ and $x^{(2)}$.

This simple device is quite useful for lower bounds for the switching times on tunnel diode circuits.

REFERENCES

[*1*] Cherry, C., Some general theorems for non-linear systems possessing reactance. *Phil. Mag.* [*7*] **42**, 1161-1177 (1951).

[*2*] Duffin, R., Nonlinear networks, IIa. *Bull. AMS* **53**, 963-971 (1947).

[*3*] Millar, W., Some general theorems for nonlinear systems possessing resistance. *Phil. Mag.* [7] **42**, 1150-1160 (1951).

[*4*] Moser, J., On Nonlinear Electrical Circuits. IBM Research Report No. RC-458 (1961).

[*5*] Weyl, H., Repartición de corriente en una real conductora. *Rev. mat. hispano-am.* **5**, 153-164 (1923).

Differential — Difference Equations

KENNETH L. COOKE
Mathematics Department
Pomona College, Claremont, California

1. Introduction

Though differential-difference equations were encountered by such early analysts as Euler [*12*], and Poisson [*28*], a systematic development of the theory of such equations was not begun until E. Schmidt published an important paper [*32*] about fifty years ago. The subsequent gradual growth of the field has been replaced, in the last decade or so, by a rapid expansion, in part due to the stimulus of various applications. The subject now offers attractive opportunities for research and for application.

This report is intended as a brief introduction to the study of differential-difference equations. We shall outline some of the main features of the theory as it stands at the present time, suggest one or two areas of application, and mention one or two unsolved problems. We shall not attempt, in the limited space available, to discuss either the theory or the applications with anything approaching completeness. The interested reader can refer to a forthcoming book of Bellman and Cooke [*4*] for additional information.

The following topics are discussed in this paper: In § 2, we discuss one application, to control theory, and mention several others. In §§ 3-8, we present the basic theory, emphasizing the Laplace transform method, of linear equations with constant coefficients, and discuss the expansion of the solutions in series of exponentials. In § 9, we discuss the asymptotic behavior of solutions, and the problem of stability. In § 10, we point out that the series of exponentials can be regarded as generalized Fourier series. §§ 11-12 are devoted to the asymptotic behavior and stability of linear equations with non-constant coefficients. In §§ 13-17, various results on nonlinear equations are quoted.

2. Applications

In order to suggest the manner in which differential-difference equations arise in applications, we shall consider an engineering problem, the problem of controlling the temperature in a reaction tank. Let us say that the tank temperature is to be maintained at a constant value, or as nearly constant as possible, by the operation of a controlling device such as an electric heating coil, or steam passing through heating pipes. Whether the controlling action is automatic or manual, it may well happen that there is a time lag inherent in the action of the controlling device. This lag could, for example, be due to the time required for a thermometer in the tank to reach the tank temperature, or due to the time required for the heating coil to cool down when the current through it is decreased.

The theory of such control systems was outlined about twenty-five years ago by Hartee, Porter, Callender, and Stevenson [*6*, *15*] and, as we shall see, it leads to linear differential-difference equations. Following Hartree *et al.*, we let $u(t)$ denote the difference, at time t, between the actual temperature and the desired temperature. Variation of $u(t)$ may be due to any of the following three causes:

(1) Random disturbances;
(2) Inherent effects due to u being non-zero. For example, if u exceeds zero, heat may flow from the tank to its surroundings;
(3) Operation of the control device.

Accordingly, we shall write

$$u'(t) = D(t) + C(t) - mu(t), \tag{2.1}$$

where $D(t)$ represents the effect of the random disturbances, $C(t)$ that of the control device, and $-mu(t)$ that of the inherent effect. Moreover, let us assume that the value of the control function $C(t)$ depends on the deviation u of temperature from its desired value, and is not affected by the random effects. If there is no time lag in the system, this leads to an assumed law of control of the form

$$C'(t) = -n_1u(t) - n_2u'(t) - n_3u''(t), \tag{2.2}$$

where n_1, n_2, n_3 are constants. For a system with time lag, we replace this equation by

$$C'(t+T) = -n_1u(t) - n_2u'(t) - n_3u''(t), \tag{2.3}$$

where T is the time lag, which we shall suppose to be constant.

Combining (2.1) and (2.3) we obtain the differential-difference equation

$$u''(t) + mu'(t) + n_1u(t-T) + n_2u'(t-T) + n_3u''(t-T) = D'(t), \qquad t > T \tag{2.4}$$

In the papers already cited, this equation was thoroughly investigated, especially with respect to the two prime *desiderata* of control, stability and quick response.

Similar equations arise in the theory of cosmic ray cascades (cf. Ramakrishnan [*31*]), in the theory of elasticity (cf. Volterra [*36*]), in mathematical economics (cf. Kalecki [*20*]), in population studies (cf. Cunningham [*9*]), in the theory of self-sustained oscillations (cf. Minorsky [*24*]), and in a number of other areas.

3. Linear Equations

Much of the theory of linear differential-difference equations of the type we are considering (sometimes called mixed differential-difference equations) can be illustrated with the simple first order equation

$$u'(t) = a(t)\,u(t) + b(t)\,u(t-1) + f(t), \qquad t > 1, \tag{3.1}$$

in which a, b, and f are given functions. An approximate initial condition for this equation is obtained by prescribing the value of $u(t)$ over the interval $0 \leqq t \leqq 1$:

$$u(t) = g(t), \qquad 0 \leqq t \leqq 1, \tag{3.2}$$

for then Equation (3.1) becomes a differential equation for $u(t)$ in the interval $1 < t < 2$. Since $u(1) = g(1)$, this determines the values of $u(t)$ for $1 \leqq t \leqq 2$, assuming that a, b, f, and g are integrable. In the same way, $u(t)$ can be determined successively in the intervals [2, 3], [3, 4], and so on. This *method of continuation* is a basic method in establishing existence and uniqueness of a solution, and also in computing the solution for small values of t. It extends readily to more general equations, such as the system

$$\sum_{i=0}^{m} [A_i y'(t - \omega_i) + B_i y(t - \omega_i)] = f(t), \tag{3.3}$$

where y and f are n-vectors, each A_i and each B_i is an n by n matrix, and each ω_i is a real number.

4. Exponential Solutions

One method of obtaining solutions of equations with constant coefficients is the method of exponential solutions. For example, the function $u(t) = \exp(st)$ is a solution of the equation

$$u'(t) = cu(t-1) \tag{4.1}$$

provided s is a root of the *characteristic equation*

$$s = ce^{-s}. \tag{4.2}$$

By combining such solutions, one can obtain more general solutions in series form.

In general the transcendental characteristic equations which are encountered have infinitely many roots, and there are questions of convergence of series of solutions. Also, it becomes somewhat difficult to determine the coefficients entering into the series from the initial function $g(t)$. One of the best ways of handling this problem is by means of the Laplace transform, as we shall explain below.

5. Laplace Transform

Consider the equation

$$\begin{aligned} u'(t) &= cu(t-1) + f(t), \qquad && t > 1, \\ u(t) &= g(t), && 0 \leqq t \leqq 1. \end{aligned} \tag{5.1}$$

If we assume that the Laplace transforms exist, we find that

$$\int_0^\infty u(t)\, e^{-st}\, dt = \frac{p(s)}{h(s)}, \tag{5.2}$$

where

$$p(s) = g(1)\, e^{-s} + s \int_0^1 g(t)\, e^{-st}\, dt + \int_1^\infty f(t)\, e^{-st}\, dt, \tag{5.3}$$

$$h(s) = s - ce^{-s}. \tag{5.4}$$

Inversion of the Laplace transform is permissible if $g(t)$ is suitably regular, and leads to the formula

$$u(t) = \frac{1}{2\pi i} \int_{(b)} e^{st} \frac{p(s)}{h(s)}\, ds, \tag{5.5}$$

where the integration is over a vertical line Re $(s) = b$, with b sufficiently large. It turns out to be the case that all zeros of $h(s)$ lie in a left half-plane, and that the relation in (5.5) is valid if all these zeros satisfy Re $(s) < b$.

6. Solution by Definite Integrals

The solution obtained above can be represented conveniently in terms of a kernal function defined by the conditions

$$\begin{aligned}
&\text{(a)}\quad k(t) = 0, && t < 0,\\
&\text{(b)}\quad k(0+) = 1, && \\
&\text{(c)}\quad k'(t) = ck(t-1), && t > 0,\\
&\text{(d)}\quad k(t) \text{ is continuous for } t > 0. &&
\end{aligned} \tag{6.1}$$

It is not difficult to see that such a function exists, that its Laplace transform is $(s - ce^{-s})^{-1}$, and that

$$k(t) = \frac{1}{2\pi i} \int_{(b)} \frac{e^{st}}{s - ce^{-s}}\, ds, \qquad t > 0. \tag{6.2}$$

It is now possible to use the Convolution Theorem of transform theory to express the solution in (5.5) in the form

$$u(t) = g(1)\, k(t-1) + c \int_0^1 k(t - t_1 - 1)\, g(t_1)\, dt_1 + \int_1^t k(t - t_1)\, f(t_1)\, dt_1,$$

$$t > 1. \tag{6.3}$$

It can be verified directly that this formula defines a solution of Equation (5.1). This form is useful in various connections to be mentioned later (cf. §§ 11, 13).

7. Series Expansions

The contour integral form of solution, as given in (5.5), leads back to the exponential solutions previously mentioned, if one uses the device of shifting the integration contour to the left. For the function $h(s)$ given in (5.4), it can be shown that the zeros of $h(s)$ can be arranged in a sequence $s_1, s_2, s_2, \ldots$, such that $\text{Re}\,(s_1) \geqq \text{Re}\,(s_2) \geqq \ldots$, and $\text{Re}\,(s_n) \to -\infty$ as $n \to \infty$. The shift of contour to the left leads to the series of residues

$$u(t) = \sum_{r=1}^{\infty} p_r(t)\, e^{s_r t}, \qquad t > 0, \tag{7.1}$$

where $p_r(t)$ is a polynomial of degree less than the multiplicity of s_r. The series can, in fact, be shown to converge uniformly in $t_0 \leqq t \leqq t_0'$ $(t_0 > 0)$, under suitable hypotheses.

In applying this method to equations more general than that in (5.1), care must be exercised, as we shall indicate in the next section.

8. Distribution of Characteristic Roots

More general differential-difference equations lead to more general exponential polynomials. For example, for the equation

$$\sum_{i=0}^{m}\sum_{j=0}^{n} a_{ij}u^{(j)}(t-\omega_i)=f(t), \tag{8.1}$$

where

$$0=\omega_0<\omega_1<\ldots<\omega_m, \tag{8.2}$$

the characteristic exponential polynomial is

$$h(s)=\sum_{i=0}^{m}\sum_{j=0}^{n} a_{ij}s^j e^{-\omega_i s}. \tag{8.3}$$

Once again the Laplace transform method leads to a contour integral representation

$$u(t)=\frac{1}{2\pi i}\int_{(b)} e^{st}\frac{p(s)}{h(s)}\,ds, \qquad t>0, \tag{8.4}$$

for an appropriate function $p(s)$, provided all zeros of $h(s)$ lie in some half-plane to the left of the integration contour. Once again, use of the Residue Theorem suggests that

$$u(t)=\sum p_r(t)\,e^{s_r t}, \tag{8.5}$$

where the sum is over all zeros s_r of $h(s)$, and the $p_r(t)$ are polynomials.

In order to justify such an expansion, it is necessary to study in detail the asymptotic distribution of the zeros of $h(s)$. Work of Pólya [*29*], Langer [*22*], Dickson [*10*], and others, has established that the zeros of $h(s)$ of large modulus fall into a finite number of *chains*. These chains can be of *retarded*, *neutral*, or *advanced* type. The zeros in a retarded or an advanced chain lie asymptotically along a curve with an equation of the form $|s^q e^s| =$ constant. Those in a retarded chain have the property that $\operatorname{Re}(s)\to-\infty$ as $|s|\to\infty$, whereas those in an advanced chain have the property that $\operatorname{Re}(s)\to+\infty$ as $|s|\to\infty$. Zeros in a neutral chain have real parts bounded as $|s|\to\infty$. It is clear that all zeros lie in a left half-plane if, and only if, all chains are of retarded or neutral type. For the equation in (8.1), this will be true if, and only if, $a_{0n}\neq 0$. In such a case, it can be shown that it is possible to close the contour on the left by a sequence of nearly semicircular contours which are uniformly bounded away from zeros of $h(s)$ and such that the annular region between successive contours contains a bounded number of zeros. The series of residues in (8.5) then converges uniformly in an interval $\omega_m - \omega_i \leqq t \leqq T$, for some value of i and any finite T, provided terms corresponding to zeros between successive contours are grouped. It was at first thought that the series without grouping must converge, but Leont'ev [*23*] has shown that it is possible for this series to diverge everywhere.

In case there is an advanced chain of roots, the preceding theory is inapplicable. However, it is still possible to obtain convergent series expansions of essentially the same type as in (8.5), using a modified technique, which was originally suggested by Titchmarsh[*33*], and later developed by Verblunsky [*34*].

9. Asymptotic Behavior and Stability

For equations with only retarded chains of roots, the asymptotic behavior of the solutions can be readily found from the series expansions already described. Indeed in this case the characteristic roots can be put in order of decreasing real parts, and it is not hard to see that for any integer k and any $\epsilon > 0$,

$$u(t) = \sum_{r=1}^{k} p_r(t)\, e^{s_r t} + O(|\, e^{(s_k-\epsilon)t}\,|), \qquad t \to +\infty. \tag{9.1}$$

That is, the asymptotic behavior of $u(t)$ is determined by the zeros of $h(s)$ of largest real part.

In particular, a necessary and sufficient condition in order that all solutions approach zero as $t \to \infty$ is that all zeros of $h(s)$ have negative real parts. Incidentally, it seems intuitively clear that this should be true also even if there are neutral chains of roots, but there is an additional complication here, since it is then possible that all roots satisfy Re $(s) < 0$ but that 0 is an accumulation point of the real parts of the roots. If this is so, the roots cannot be put in order of decreasing real parts, and a formula like that in (9.1) is not available. This critical case has been handled by Hahn [*13*] for the special class of scalar equations of the form

$$\sum_{j=0}^{n} [a_j u^{(j)}(t) + b_j u^{(j)}(t-\omega)] = 0, \tag{9.2}$$

containing only one retardation, but no general treatment has yet been given.

Since the behavior of solutions of functional equations as $t \to \infty$ is of great practical interest, it is important to have a criterion for deciding in each particular case whether or not all zeros of $h(s)$ satisfy Re $(s) < 0$. For ordinary differential equations, the famous criteria of Routh and Hurwitz can be used. For ordinary differential-difference equations, the most general methods are due to Pontryagin [*30*], and to Cebotarev and Meiman [*7*]. From these general theorems, special results can be derived for individual equations of interest. For example, for the equation

$$u'(t) - pu(t) - qu(t-1) = 0, \tag{9.3}$$

which is of interest in economic theory, Pontryagin's theorems can be used to show that the roots satisfy Re $(s) < 0$ if, and only if, $p < 1$ and $p < -q < (a_1^2 + p^2)^{1/2}$, where a_1 is the root of $a = p \tan a$ such that $0 < a < \pi$. (If $p = 0$, $a_1 = \pi/2$). See Bellman and Cooke [*4*] for details. This result, by the way, was originally established by Hayes [*16*] by a different method.

10. Fourier Series

The series expansions described in §§ 7, 8 have an interesting application in deriving representations of prescribed functions in series similar to Fourier series. For example, let a function $g(t)$, which is suitably regular, be given on $0 \leqq t \leqq 1$, and let it be desired to find an expansion of $g(t)$ in a series of exponentials exp $(s_r t)$, where the s_r are zeros of the function

$$h(s) = s - ce^{-s}. \tag{10.1}$$

We consider a differential-difference equation associated with $h(s)$,

$$u'(t) = cu(t-1), \tag{10.2}$$

and take as initial condition

$$u(t) = g(t), \qquad 0 \leqq t \leqq 1. \tag{10.3}$$

The theorem of § 7 then gives an expansion of $u(t)$ for $t > 0$. In particular,

$$g(t) = \sum_{r=1}^{\infty} p_r(t)\, e^{s_r t}, \qquad 0 < t \leqq 1. \tag{10.4}$$

If $c \neq -1/e$, the zeros s_r are all simple, and each $p_r(t)$ is a constant.

To obtain the ordinary Fourier series expansion of a periodic function, we consider the difference equation

$$u(t) = u(t - 1) \tag{10.5}$$

with initial condition as in (10.3). Here $h(s) = 1 - e^{-s}$, and the zeros are $s = 2r\pi i$ ($r = 0, \pm 1, \ldots$). The series expansion yields

$$g(t) = \sum_{r=-\infty}^{\infty} p_r e^{2r\pi i t}, \qquad 0 < t < 1, \tag{10.6}$$

where of course

$$p_r = \int_0^1 g(t_1)\, e^{-2r\pi i t_1}\, dt_1. \tag{10.7}$$

The method can be extended to yield Fourier-like expansions in terms of the zeros of more general exponential polynomials. See Bellman and Cooke [*4*], for rigorous results. Note that these series are obtained without the use of orthogonality of the functions exp $(s_r t)$.

11. Almost-Constant Coefficients

Consider an equation of the form

$$u'(t) + [a_0 + a(t)]\, u(t) + [b_0 + b(t)]\, u(t - \omega) = 0, \tag{11.1}$$

under the assumption that $a(t)$ and $b(t)$ approach zero as $t \to \infty$. It is to be expected that there is a close relation between the asymptotic properties of the solutions of (11.1) and of the solutions of the equation

$$v'(t) + a_0 v(t) + b_0 v(t - \omega) = 0. \tag{11.2}$$

In particular, corresponding to each solution e^{st} of Equation (11.2), where s is a root of

$$s + a_0 + b_0 e^{-\omega s} = 0, \tag{11.3}$$

we would expect to find a solution of Equation (11.1) of the same general asymptotic form. Under suitable restrictions on $a(t)$ and $b(t)$, the existence of such a solution of (11.1) has been established (cf. Bellman and Cooke [*3*]). The simplest case is that in which the root λ of (11.3) of largest real part is real and simple, and in which $b(t) = 0$. In this case, it has been shown that there is a solution of the form

$$u(t) = [1 + 0(1)] \exp\left[\lambda t - c_1 \int_{t_0}^{t} a(t_1)\, dt_1\right], \tag{11.4}$$

where

$$c_1 = (1 - b_0 e^{-\lambda\omega})^{-1}, \tag{11.5}$$

provided either that $| a(t) |$ is integrable over the infinite interval, or that $a(t) = o(1)$, $a'(t) = o(| a(t) |)$, and $a(t)^2$, $| a'(t) |$, and $| a''(t)/a(t) |$ are absolutely integrable.

The proof of this result proceeds along the following lines. We regard the term $- a(t)\, u(t)$ as a forcing term in an inhomogeneous equation with constant coefficients, and use the theory of § 6 in order to replace the differential-difference equation by an integral equation of the form

$$u(t) = c_2 e^{\lambda t} - \int_{t_0}^{t} a(t_1)\, u(t_1)\, k(t - t_1)\, dt_1. \tag{11.6}$$

In the present case,

$$k(t) = c_1 e^{\lambda t} + k_1(t), \qquad k_1(t) = O[e^{(\lambda-\epsilon)t}]. \tag{11.7}$$

If $k_1(t)$ is neglected, we therefore have, on differentiating the equation in (11.6),

$$u'(t) = [\lambda - c_1 a(t)]\, u(t), \qquad u(t_0) = c_2. \tag{11.8}$$

The origin of the expression in (11.4) is now evident. The demonstration that the term containing $k_1(t)$ can be neglected is a matter of some technical difficulty, especially in the case in which $a(t)$ is not absolutely integrable.

Similar, but more involved, arguments enable one to demonstrate the existence of solutions associated with each characteristic root, be it real or complex, simple or multiple. The methods can also be extended to equations more general than that in (11.1). Furthermore, refinements are possible if more information concerning $a(t)$ and $b(t)$ is given. For example, the equation with linear coefficients,

$$\sum_{i=0}^{m} \sum_{j=0}^{n} (a_{ij} t + b_{ij})\, u^{(j)}(t - \omega_i) = f(t), \tag{11.9}$$

has been thoroughly studied by Yates [*40*].

12. Dini-Hukuhara Theory

An important theorem in the stability theory of ordinary differential equations, due to Hukuhara and Dini, has been extended to differential-difference equations by Bellman and Cooke [*2*]. An important feature of the method is the use of an integral operator representation of solutions in which the kernel is the solution of an adjoint equation. For the equation

$$u'(t + h) + a(t)\, u(t) + b(t)\, u(t + h) = w(t), \tag{12.1}$$

the adjoint kernel $v(s, t)$ satisfies

$$- \frac{\partial v(s, t)}{\partial s} + v(s + h, t)\, a(s + h) + v(s, t)\, b(s) = 0, \qquad t > 0, \qquad 0 < s < t - h,$$

$$- \frac{\partial v(s, t)}{\partial s} + v(s, t)\, b(s) = 0, \qquad t - h < s < t, \qquad v(t, t) = 1. \tag{12.2}$$

Then the solution of Equation (12.1) for which $u(t) = 0$ for $0 \leqq t \leqq h$ is given by the equation

$$u(t + h) = \int_0^t v(s, t)\, w(s)\, ds, \qquad t > 0. \tag{12.3}$$

From this formula, it is easily established that if all solutions of

$$u'(t + h) + a(t)\, u(t) + b(t)\, u(t + h) = 0 \tag{12.4}$$

are bounded as $t \to \infty$, if

$$\int^{\infty} [|\, c(t) \,| + |\, d(t) \,|]\, dt < \infty, \tag{12.5}$$

and if $|\, v(s, t) \,|$ is bounded for $t \geqq 0$, $0 \leqq s \leqq t$, then all solutions of

$$u'(t + h) + [a(t) + c(t)]\, u(t) + [b(t) + d(t)]\, u(t + h) = 0 \tag{12.6}$$

are bounded as $t \to \infty$. If $a(t)$ and $b(t)$ are constant, the boundedness of v follows from boundedness of all solutions of Equation (12.4).

Theorems of this type can be established for much more general equations, including certain nonlinear equations.

13. Poincaré-Liapunov Theorem

One important result on nonlinear differential-difference equations which we wish to mention is an analogue of the famous Poincaré-Liapunov stability theorem for ordinary differential equations. We shall illustrate this for the special case of a nonlinear equation of the form

$$u'(t) + b_0 u(t) + b_1 u(t - \omega) = f(u(t), u(t - \omega)), \qquad t > \omega. \tag{13.1}$$

Along with this equation, we consider the linear equation

$$u'(t) + b_0 u(t) + b_1 u(t - \omega) = 0. \tag{13.2}$$

We assume that b_0 and b_1 are constants, and that all roots of the characteristic equation

$$s + b_0 + b_1 e^{-\omega s} = 0 \tag{13.3}$$

have negative real parts, so that every solution of (13.2) approaches zero as $t \to +\infty$. The function $f(u, v)$ is assumed to be continuous in a neighborhood of the origin, $|u| + |v| \leqq c_1$, and to satisfy

$$\lim_{|u|+|v|\to 0} \frac{|f(u, v)|}{|u| + |v|} = 0. \tag{13.4}$$

Under these assumptions, Bellman [*1*] has proved that there exists a $\delta > 0$ such that if

$$\max_{0 \leqq t \leqq \omega} |g(t)| \leqq \delta, \tag{13.5}$$

then the function $g(t)$ can be continued as a solution $u(t)$ of (13.1) over $0 \leqq t < \infty$, and

$$\lim_{t\to\infty} |u(t)| = 0. \tag{13.6}$$

To prove this theorem, one can use the result of § 6 to convert the differential-difference equation to the nonlinear integral equation

$$u(t) = u_0(t) + \int_\omega^t f(u(t_1), u(t_1 - \omega))\, k(t - t_1)\, dt_1, \qquad t > \omega, \tag{13.7}$$

where $u_0(t)$ is a solution of Equation (13.2). Under the stated assumptions, there is a number $c_1 > 0$ such that for $t > 0$,

$$|k(t)| \leqq c_2 e^{-c_1 t},$$

$$|u_0(t)| \leqq c_3 e^{-c_1 t}[\max_{0 \leqq t \leqq \omega} |g(t)|].$$

Using these estimates and (13.4), the proof can now be completed in the same way as for ordinary differential equations.

14. Forced Periodic Solutions

Few results have so far been published concerning periodic solutions of differential-difference equations, and we shall mention only one or two of these. The one to be described here, due to Cooke [*8*], depends on the representation used in the preceding section. Consider the equation

$$u'(t+1) = au(t) + bu(t+1) + f(u(t), u(t+1), \omega t) + cg(\omega t) \qquad (14.1)$$

under the following assumptions:

(a) All roots of the characteristic equation

$$s - b - ae^{-s} = 0 \qquad (14.2)$$

have negative real parts.

(b) a, b, c, and ω are real numbers, and $\omega > 0$.

(c) The function $g(t)$ is real and continuous for all t, has period one, and the average of $g(t)$ over a period is zero.

(d) The function $f(x, y, t)$ is real and continuous for all (x, y, t), has period one in t, and $f(0, 0, t) \equiv 0$. Moreover, to each $\epsilon > 0$ there corresponds a $\Delta > 0$, depending on ϵ but not on t, such that

$$|f(x_1, y_1, t) - f(x_2, y_2, t)| \leqq \epsilon(|x_1 - x_2| + |y_1 - y_2|) \qquad (14.3)$$

for all real t and $|x_1 - x_2| < \Delta$, $|y_1 - y_2| < \Delta$.

Under these conditions, there exists a positive constant δ, depending on a, b, f, and g, but not on c or ω, such that if $|c|(1+\omega)^{-1} < \delta$, then Equation (14.1) has a continuous, periodic solution $p(t)$ of period $1/\omega$. Moreover, there is a positive constant c_1, independent of c and ω, such that

$$|p(t)| \leqq \frac{c_1 |c|}{1+\omega}$$

for all t. The solution $p(t)$ is asymptotically stable.

In proving this theorem, the periodic solution $p(t)$ is constructed by the method of successive approximations as the solution of the integral equation

$$u(t+1) = \int_{-\infty}^{t} k(t-t_1)\,[f(u(t_1), u(t_1+1), \omega t_1) + cg(\omega t_1)]\, dt_1.$$

Asymptotic stability of $p(t)$ follows from the fact that if $x(t)$ is any solution existing for $t \geqq 0$, then $x(t) - p(t)$ satisfies a nonlinear equation to which the Poincaré-Liapunov theorem can be applied.

Other problems are discussed in a paper of Halanaĭ [*14*].

15. The Cherwell-Wright Equation

A particularly interesting nonlinear equation is the equation

$$u'(t) = [a - u(t-1)]\, u(t), \tag{15.1}$$

which takes the form

$$y'(t) = -\, ay(t-1)\,[1 + y(t)] \tag{15.2}$$

under the transformation $u(t) = a + ay(t)$. This equation arises in various applications. For example, see the papers of Cunningham [*9*], deVisme [*35*], and Wright [*37*, *39*].

The equation in (15.1) is one of the few nonlinear differential-difference equations about which a great deal is known. For this knowledge, we are indebted to Wright [*37*], Kakutani and Markus [*19*], and Jones [*17*]. Some of the principal results are as follows.

In the first place, there is a unique solution on $0 \leqq t < \infty$ corresponding to each initial condition

$$u(t) = g(t), \qquad 0 \leqq t \leqq 1, \tag{15.3}$$

if, for example, g is continuous. For $a > 0$ (the case of principal interest and difficulty), if $g(1) < 0$, then $u(t) \to -\infty$. For $a > 0$ and $g(1) > 0$, then either $u(t)$ approaches the constant a monotonically, or else $u(t)$ oscillates boundedly about $u = a$. Both behaviors are possible. In fact if $a > 1/e$ then oscillation occurs, whereas if $0 < a \leqq 1/e$ there are no oscillatory solutions for which the interval between zeros of $u(t) - a$ is at least one for all large t.

As to the nature of the oscillations, the following can be said. If $a \leqq 3/2$, the oscillations damp out to zero, that is $u(t) \to a$ as $t \to \infty$. If $a > \pi/2$, there are bounded oscillatory solutions which are undamped, and in fact there exist nonconstant periodic solutions.

Jones [*18*] has recently extended these results to the more general equation

$$y'(t) = a \sum_{i=1}^{m} k_i y(t - \omega_i)\,[1 + by(t) + cy(t)^2]. \tag{15.4}$$

16. Liapunov Functions

The method of Liapunov functions and its generalizations can be applied successfully to the study of various stability questions for differential-difference equations. We refer the reader to a paper of Krasovskii [*21*] for further information.

17. More General Equations

Some of the results and techniques discussed above have been extended to more general classes of equations. For example Pitt [*26*, *27*] has considered linear equations

$$\sum_{r=0}^{R} \int_{-\infty}^{\infty} u^{(r)}(t - t_1)\, dk_r(t_1) = f(t_1), \tag{17.1}$$

and Brownell [*5*] and Wright [*38*] have examined related nonlinear equations. Myškis [*25*] has treated the system of equations

$$y_i'(t) = \sum_{j=1}^{n} \int_{0}^{\infty} y_j(t - t_1)\, dr_{ij}(t, t_1) + f_i(t), \qquad i = 1, 2, ..., n. \tag{17.2}$$

This system includes an extension of differential-difference equations to equations in which the lags or retardations are variable. A number of papers on such equations have appeared recently in the Russian literature.

Finally, we mention the work of Driver [*11*] on equations in which the lags are themselves known functions of the unknown solution function.

REFERENCES

[*1*] Bellman, R., On the existence and boundedness of solutions of non-linear differential-difference equations. *Ann. Math. (Princeton)* **50**, 347-355 (1949).

[*2*] Bellman, R., and Cooke, K. L., Stability theory and adjoint operators for linear differential-difference equations. *Trans. AMS* **92**, 470-500 (1959).

[*3*] Bellman, R., and Cooke, K. L., Asymptotic behavior of solutions of differential-difference equations. *Mem. AMS* **35** (1959).

[*4*] Bellman, R., and Cooke, K. L., "Differential-Difference Equations." Academic Press, New York, in press.

[*5*] Brownell, F. R., Non-linear delay differential equations. *In* "Contributions to the Theory of Nonlinear Oscillations" (S. Lefschetz, ed.), Vol. 1, pp. 89, 148. Princeton Univ. Press, Princeton, New Jersey, 1950.

[*6*] Callender, A., Hartree, D. R., and Porter, A., Time lag in a control system, I. *Trans. roy. Soc. (London)* **A235**, 415-444 (1936).

[*7*] Cebotarev, N. G., and Meiman, N. N., The Routh-Hurwitz problem for polynomials and entire functions. *Trudy Mat. Inst. Steklov* **26** (1949) (in Russian).

[*8*] Cooke, K. L., Forced periodic solutions of a stable nonlinear differential-difference equation. *Ann. Math. (Princeton)* **61**, 381-387 (1955).

[*9*] Cunningham, W. J., A nonlinear differential-difference equation of growth. *Proc. natl. Acad. Sci. U.S.* **40**, 708-713 (1954).

[*10*] Dickson, D. G., Expansions in series of solutions of linear difference-differential and infinite order differential equations with constant coefficients. *Mem. AMS* **23** (1957).

[*11*] Driver, R. D., This volume, p. 474.

[*12*] Euler, L., Investigatio curvarum quae evolutae sui similes producunt. *Comment. Acad. Sci. imp. Petropol.* **12** (1740).

[*13*] W. Hahn, Über Differential-Differenzengleichungen mit anomalen Lösungen. *Math. Ann. (Leipzig)* **133**, 251-255 (1957).

[*14*] Halanaĭ, A., Some qualitative questions in the theory of differential equations with retarded arguments. *Rev. math. pures. appl.* **2**, 127-144 (1957).

[*15*] Hartree, D. R., Porter, A., Callender, A., and Stevenson, A. B., Time lag in a control system, II. *Proc. roy. Soc. (London)* **A161**, 460-476 (1937).

[*16*] Hayes, N. D., Roots of the transcendental equation associated with a certain difference differential equation. *J. London math. Soc.* **25**, 226-232 (1950).

[*17*] Jones, G. S., Asymptotic behavior and periodic solutions of a nonlinear differential-difference equation. *Proc. natl. Acad. Sci. U.S.* **47**, 879-882 (1961).

[*18*] Jones, G. S., this volume, p. 000.

[*19*] Kakutani, S., and Markus, L., On the non-linear difference-differential equation $y'(t) = [A - By(t - \tau)]\, y(t)$. *In* "Contributions to the Theory of Nonlinear Oscillations" (S. Lefschetz, ed.), Vol. 4, pp. 1-18. Princeton Univ. Press, Princeton, New Jersey, 1958.

[*20*] Kalecki, M., A macrodynamic theory of business cycles. *Econometrica.* **3**, 327-344 (1935).

[*21*] Krasovskiĭ, N. N., On the application of the second method of Liapunov for equations with time retardations. *Priklad. Mat. i Mekh.* **20**, 315-327 (1956) (in Russian).

[*22*] Langer, R. E., The asymptotic location of the roots of a certain transcendental equation. *Trans. AMS* **31**, 837-844 (1929).

[*23*] Leont'ev, A. F., Differential-difference equations. *AMS Translations* [1] No. 78 (1952); original article in *Mat. Sborn.* [N.S.] **24**(66), 347-374 (1949).

[*24*] Minorsky, N., Self-excited mechanical oscillations. *J. appl. Phys.* **19**, 332-338 (1948).

[*25*] Myškis, A. D., "Linear Differential Equations with Retarded Argument." Gos. Izdat. Tekhn.-teoret. Lit., Moscow, 1951; German transl., Deut. Wiss., Verlag Berlin, 1955.

[*26*] Pitt, H. R., On a class of integro-differential equations. *Proc. Cambridge phil. Soc.* **40**, 199-211 (1944).

[*27*] Pitt, H. R., On a class of linear integro-differential equations. *Proc. Cambridge phil. Soc.* **43**, 153-163 (1947).

[*28*] Poisson, S. D., Mémoire sur les équations aux différences mêlée. *J. École polytech. (Paris)* 126-147 (1806).

[*29*] Pólya, G., Über die Nullstellen gewisser ganzer Funktionen. *Math. Z.* **2**, 352-383 (1918).

[*30*] Pontryagin, L. S., On the zeros of some elementary transcendental functions. *AMS Translations* [2] **1**, 95-110 (1955); original article in *Izvest. Akad. Nauk. S.S.S.R., Ser. Mat.* **6**, 115-134 (1942).

[*31*] Ramakrishnan, A., Some simple stochastic processes. *J. roy. statist. Soc.* **13**, 131-141 (1951).

[*32*] Schmidt, E., Über eine Klasse linearer funktionaler Differentialgleichungen. *Math. Ann. (Leipzig)* **70**, 499-524 (1911).

[*33*] Titchmarsh, E. C., Solutions of some functional equations. *J. London math. Soc. 14*, 118-124 (1939).

[*34*] Verblunsky, S., On a class of differential-difference equations. *Proc. London math. Soc.* [3] **6**, 355-365 (1956).

[*35*] deVisme, G. H., The density of prime numbers. *Math Gaz.* **45**, 13-14 (1961).

[*36*] Volterra, E., On elastic continua with hereditary characteristics. *J. appl. Mech.* **18**, 273-279 (1951).

[*37*] Wright, E. M., A non-linear difference-differential Equation. *J. reine angew. Math.* **194**, 66-87 (1955).

[*38*] Wright, E. M., Perturbed functional equations. *Quart. J. Math.* **20**, 155-165 (1949).

[*39*] Wright, E. M., A functional equation in the heuristic theory of primes. *Math. Gaz.* **45**, 15-16 (1961).

[*40*] Yates, B. G., The linear difference-differential equation with linear coefficients. *Trans. AMS* **80**, 281-298 (1955).

Transport Theory, Invariant Imbedding, and Two-Point Boundary Value Problems[1]

G. MILTON WING
Sandia Corporation, Albuquerque, New Mexico

1. Introduction

In this paper we shall show how the consideration of a sequence of simple mathematical models of physical problems leads in a natural manner to results concerning differential, partial differential, and integro-differential equations. More specifically, we apply the method of invariant imbedding to some problems of transport theory and arrive at rather general methods of transforming two-point boundary value problems into initial value problems. These initial value problems are ordinarily of a non-linear nature, even when the original equations are linear. For many purposes, especially those involving actual numerical computation, the non-linearity does not create a great hardship.

In Section 2 a very simple problem of particle transport is studied. Its solution is considered in two different ways, and these two viewpoints lead to a connection between a simple pair of ordinary differential equations with conditions specified at two points and a Riccati equation with a single side condition.

Section 3 consists of a brief discussion, with few details, of three more complicated transport problems which have been studied in the past by both classical and invariant imbedding methods. When the latter method was first employed, each case was handled as a separate and distinct problem, and the need for a more unified attack is noted.

A much more general approach is described in Section 4 and its application to the models of Section 3 is mentioned. This method is applicable to a wide class of problems arising in contexts other than transport theory, but the device is not adequate for dealing with certain types of non-linear two-point boundary value problems which occur in practice. A perturbation scheme which eliminates some of these difficulties is given in Section 5. Section 6 consists of a few remarks on the importance of mathematical model making, as well as a discussion of the significance of the results of earlier sections.

[1] This work was performed under the auspices of the United States Atomic Energy Commission.

2. A Simple Transport Model

We pose the following physical problem. Let point particles be capable of moving to the right or left on a rod—a line segment—extending from 0 to x. All particles have the same speed c. They cannot interact or collide with one another, but they may experience collisions with the material of the rod. Let the probability of such a collision in going a distance Δ be $\sigma\Delta + o(\Delta)$, where σ is a constant, and let the results of such an interaction be the disappearance of the original particle and the appearance of two new particles, one moving to the right and one to the left.

Assume that a time-independent distribution of particles exists in the rod. Let $u(z)$ be the expected density of right-moving particles at z, $0 \leq z \leq x$, and $v(z)$ be the expected density of left-moving particles there. Suppose no particles enter the rod at $z = 0$, while one per second impinges at $z = x$. We seek equations for u and v.

We apply an elementary argument familiar to all physics students. Consider $c\,u(z + \Delta)$, the expected number of particles per second passing $z + \Delta$ to the right—the right flux at $z + \Delta$. This flux will consist in part of particles which come from z without collision in $(z, z + \Delta)$. There will be $(1 - \sigma\Delta + o(\Delta))\; c\,u(z)$ such. Further, certain particles moving to the right at z suffer collisions in $(z, z + \Delta)$ and contribute $(\sigma\Delta + o(\Delta))\; c\,u(z)$ particles to the flux of interest. Finally, of the left-moving particles passing $z + \Delta$ each second, $(\sigma\Delta + o(\Delta))\; c\,v(z + \Delta)$ interact with the rod material in $(z, z + \Delta)$ and produce that number of right-moving particles. All other processes involve at least two collisions. Hence, assuming that u and v are at least bounded we have

$$
\begin{aligned}
cu(z + \Delta) - cu(z) &= - \sigma\Delta cu(z) + \sigma\Delta cu(z) + \sigma\Delta cv(z + \Delta) + o(\Delta) \\
&= \sigma\Delta cv(z + \Delta) + o(\Delta). \qquad (2.1)
\end{aligned}
$$

Assuming continuity and differentiability of the functions we then have

$$\frac{du}{dz} = \sigma v \qquad (2.2a)$$

Similarly

$$-\frac{dv}{dz} = \sigma u \qquad (2.2b)$$

The physical conditions imposed at $z = 0$ and $z = x$ imply

$$u(0) = 0, \qquad cv(x) = 1. \qquad (2.2c)$$

The solution to the system (2.2) is trivial:

$$u(z) = \frac{\sin \sigma z}{c \cos \sigma x},$$
$$v(z) = \frac{\cos \sigma z}{c \cos \sigma x}. \qquad (2.3)$$

It is interesting to note that the functions u and v cease to have meaning if $x = \pi/2\sigma$. This is the *critical length* of the rod. When $x > \pi/2\sigma$ both u and v become negative for certain values of z and hence lose their physical significance. A stationary state cannot exist in a rod for which $x \geq \pi/2\sigma$ (see ref. [*11*]).

We have presented the classical way of solving a transport problem. The method of invariant imbedding utilizes a quite different approach. Consider not a single rod, but a set of rods of arbitrary lengths. We imbed the given rod in this set and ask for $R(x)$, the expected density of particles moving to the right at the right end of the rod. Notice that R is defined *only* at the right end of the rod, and is hence a function of the *length* of the rod.

To obtain an equation for R we study a slightly longer rod, one of length $x + \Delta$, and consider the interval $(0, x)$ as a *sub*-rod lying in the new rod. (See Figure 1.) It is reasonable to expect that a simple relationship exists between $R(x + \Delta)$ and $R(x)$. To discover this we merely keep careful account of the interactions which may occur in $(x, x + \Delta)$.

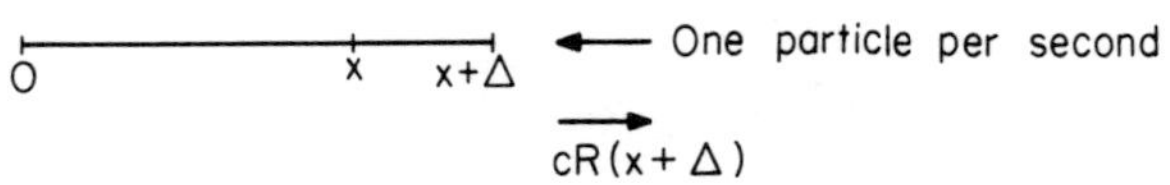

FIG. 1

If the particle impinging at $x + \Delta$ has a collision in $(x, x + \Delta)$, then one particle immediately emerges at $x + \Delta$, another goes into the *sub*-rod $(0, x)$ and acts as a source particle for that *sub*-rod. If there is not a collision of the impinging particle in $(x, x + \Delta)$ then that particle itself becomes a source particle for the *sub*-rod. The sub-rod gives rise to $c\,R(x)$ right-moving particles at x each second for each particle per second impinging at x. Of these right-moving particles some move thru $(x, x + \Delta)$ without incident, while others suffer collisions. Of the particles resulting from such collisions, half move right and out at $x + \Delta$. The others move to the left and provide a *new* source of particles for the sub-rod. Combining these various events we find, under the assumption that $R(x)$ is well-behaved [6, 7],

$$cR(x + \Delta) = \sigma\Delta[1 + cR(x)] \\ + (1 - \sigma\Delta)\,[cR(x)\,(1 - \sigma\Delta) + \sigma\Delta cR(x) + \sigma\Delta c^2R^2(x)] + o(\Delta), \tag{2.4}$$

and hence,

$$c\frac{dR}{dx} = \sigma\{1 + c^2R^2\}. \tag{2.5a}$$

Since no particles emerge from a rod of length zero,

$$R(0) = 0. \tag{2.5b}$$

Equation (2.5a) is a Riccati equation and our transport problem is now posed as an initial value problem. The solution of (2.5) is elementary:

$$cR(x) = \tan \sigma x. \tag{2.6}$$

We note from (2.3) that $u(x) = R(x)$, as expected.

The invariant imbedding formulation has given only the "reflected flux", the number of particles per second emergent from the right end of a rod of length x. However, it has also reduced the two-point boundary value problem (2.2) to an initial value problem, since the condition $u(0) = 0$ of (2.2c) may now be replaced by the condition $c\, u(x) = \tan \sigma x$.

3. Some More Realistic Transport Models

Our transport model of Section 2 is quite unrealistic. Particles are allowed to be of only one energy and of one kind; the rod geometry is a great oversimplification of the usual actual physical situation; the time variable has been suppressed by assuming always the stationary state. In this section we shall briefly mention various models that have been studied, each of which avoids certain of these objections. In each case imbedding results were originally obtained by particle counting methods analogous to that described in obtaining (2.4) although in some cases other derivations are known (see Chandrasekhar [9], Mullikin [10]).

Consider first the time-independent rod geometry but now allow particles to be in one of a finite number, n, of different states—perhaps describing energy or type of particle. Let $c_i\, u_i(z)$, $i = 1, 2, \ldots, n$, be the number of particles in state i crossing z to the right each second; $c_i\, v_i(z)$ is similarly defined for motion to the left. Let the probability of a collision involving a type i particle in $(z, z + \Delta)$ be $\sigma_i\Delta + o(\Delta)$, and let the result of such a collision be F_{ji} particles in state j ($j = 1, 2, ..., n$) moving in the direction of the colliding particle, B_{ji} particles in state j moving in the opposite direction. Again, we allow no interactions between moving particles. Then the equations analogous to (2.2) become matrix equations:

$$\begin{aligned} \frac{dU}{dz} &= \hat{F}U + \hat{B}V, \\ -\frac{dV}{dz} &= \hat{B}U + \hat{F}V. \end{aligned} \tag{3.1}$$

Here U is a column vector of the functions $c_i\, u_i(z)$; similarly for V. The matrix $\hat{F}$ is given by

$$\hat{F} = \begin{pmatrix} \sigma_1(F_{11}-1) & \sigma_2 F_{12} & \cdots & \sigma_n F_{1n} \\ \sigma_1 F_{21} & \sigma_2(F_{22}-1) & \cdots & \cdots \\ \vdots & & & \\ \sigma_1 F_{n1} & \cdots\ \cdots & \cdots & \sigma_n(F_{nn}-1) \end{pmatrix}, \tag{3.2}$$

while

$$\hat{B} = \begin{pmatrix} \sigma_1 B_{11} & \sigma_2 B_{12} & \cdots & \sigma_n B_{1n} \\ \sigma_1 B_{21} & \sigma_2 B_{22} & \cdots & \cdots \\ \vdots & & & \\ \sigma_1 B_{n1} & \cdots & \cdots\ \cdots & \sigma_n B_{nn} \end{pmatrix}. \tag{3.3}$$

(For details, see Bellman *et al.* [*4*] and Wing [*13*]).

If no particles enter the rod from the left, and y_i particles in state i impinge at x each second, then

$$U(0) = 0, \qquad V(x) = Y. \tag{3.4}$$

The equation analogous to (2.5) is (see [*4*, *13*])

$$\frac{d\hat{R}}{dx} = \hat{B} + \hat{R}\hat{F} + \hat{F}\hat{R} + \hat{R}\hat{B}\hat{R}, \tag{3.5}$$

$$\hat{R}(0) = 0,$$

$$\hat{R}_{jk}(x) = c_j u_j(x) \text{ when one particle per second enters at } x \text{ in state } k.$$

(It is clear that the linearity of (3.1) makes it sufficient to consider only this input condition in defining $\hat{R}$.)

Let us return to the model of Section 2 again. Now, however, we shall not demand time-independence, although the rod geometry and the single state of the particle will be retained. Equations (2.2a) and (2.2b) are replaced by

$$\begin{aligned} \frac{\partial u}{\partial z} + \frac{1}{c}\frac{\partial u}{\partial t} &= \sigma v, \\ -\frac{\partial v}{\partial z} + \frac{1}{c}\frac{\partial v}{\partial t} &= \sigma u, \end{aligned} \tag{3.6a}$$

where $u = u(z, t)$ is the density of right-moving particles at z at time t, c is the particle speed, etc. Suppose a single particle enters the system at time $t = 0$ at $z = x$ and that no particles are already in the system at $t = 0$. Then

$$u(0, t) = 0, \qquad t \geq 0;$$

$$cv(x, t) = \delta(t) \text{ (The Dirac } \delta \text{ distribution)}, \qquad t \geq 0;$$

$$u(z, 0) = v(z, 0) = 0, \qquad 0 < z < x. \tag{3.6b}$$

The invariant imbedding equations for this case are [7]

$$\begin{aligned} \frac{\partial R}{\partial x} + \frac{2}{c}\frac{\partial R}{\partial t} &= \frac{\sigma}{c} + \sigma c \int_0^t \frac{\partial R}{\partial s}(x, s)\, R(x, t - s)\, ds, \\ R(x, t) \equiv \int_0^t u(x, s)\, ds, &\qquad R(0, t) = R(x, 0) = 0. \end{aligned} \tag{3.7}$$

These equations have been completely analyzed in ref. [11].

As our final example, we consider a problem of great astrophysical interest, one version of which was originally investigated by Ambarzumian [1]. A slab of material, finite in thickness but infinite in the remaining directions, transports particles. The density of particles, $N(z, \mu)$, depends only upon the distance from the left face of the slab and upon the direction of particle motion, $\mu = \cos\theta$, where θ is measured as shown in Figure 2. The probability of a particle interacting with the medium in going distance Δ is $\sigma\Delta + o(\Delta)$, and the expected result of such a collision is a total of γ particles, uniformly distributed in direction. One particle per second per unit area on $z = x$ impinges in direction μ_0, $-1 \leq \mu_0 < 0$, none enters at $z = 0$. What is the steady-state density of particles reflected at x?

The classical formulation uses the linearized Boltzmann equation:

$$\mu \frac{\partial N}{\partial z}(z, \mu) + \sigma N(z, \mu) = \frac{\gamma\sigma}{2}\int_{-1}^{1} N(z, \mu')\, d\mu',$$

$$N(0, \mu) = 0, \quad 0 < \mu \leq 1, \qquad c\mu N(x, \mu) = \delta(\mu - \mu_0), \qquad -1 \leq \mu < 0. \tag{3.8}$$

We require $N(x, \mu)$, $0 < \mu \leq 1$. This function may be obtained by solving (3.8) and evaluating the result at $z = x$, or by solving the corresponding imbedding equations [2]:

$$\frac{\partial R}{\partial x} = \frac{\sigma\gamma\pi}{|\mu_0|} + \frac{\sigma\gamma}{2|\mu_0|}\int_{-1}^{0} R(x, \mu, \mu')\, d\mu'$$

$$- \sigma R(x, \mu, \mu_0)\left(\frac{1}{\mu} + \frac{1}{|\mu_0|}\right) + \frac{\sigma\gamma}{2}\int_{0}^{1} R(x, \mu', \mu_0)\frac{d\mu'}{\mu'}$$

$$+ \frac{\sigma\gamma}{4\pi}\int_{0}^{1} R(x, \mu', \mu_0)\frac{d\mu'}{\mu'}\int_{-1}^{0} R(x, \mu, \mu'')\, d\mu'',$$

$$R(0, \mu, \mu_0) = 0, \qquad R(x, \mu, \mu_0) = 2\pi N(x, \mu). \tag{3.9}$$

In the Ambarzumian case, $\gamma \leq 1$ and the slab is replaced by a semi-infinite half-space. Since then R must be independent of x, the left side of (3.9) reduces to zero. This results in a great simplification of the problem. A rigorous analytic solution is known (see Chandrasekhar [9]).

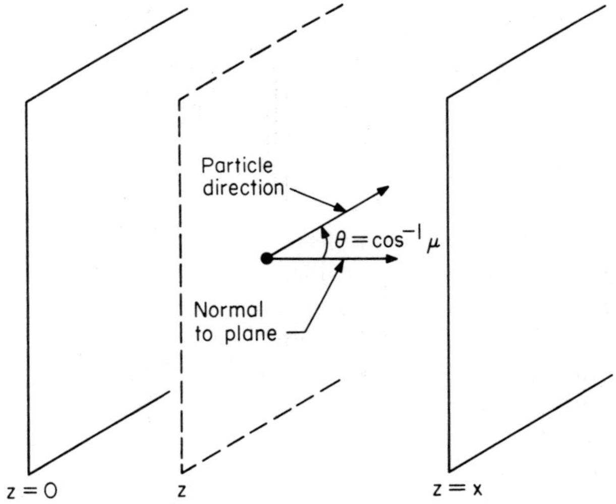

FIG. 2

4. A Generalization of Section 2

Consider the equations

$$\frac{du}{dz} = f(u, v, z),$$

$$-\frac{dv}{dz} = g(u, v, z), \tag{4.1}$$

$$u(0) = 0, \qquad v(x) = y, \qquad 0 \leq z \leq x.$$

If we write the finite difference form

$$u(z + \Delta) - u(z) = f(u, v, z)\Delta + o(\Delta), \tag{4.2a}$$

$$v(z) - v(z + \Delta) = g(u, v, z)\Delta + o(\Delta), \tag{4.2b}$$

then we may think of these equations as describing a time-independent transport process in a rod. (It is convenient to consider $c = 1$.) For example, (4.2a) may be interpreted as saying that the increment in the right-moving flux as particles go from z to $z + \Delta$ is just $f\Delta$, to order Δ. Here f may be a quite arbitrary function describing complicated interactions (including, perhaps, particle-particle collisions) in the interval $(z, z + \Delta)$.

We now define

$$R(x, y) \equiv u(x), \qquad \text{when} \qquad v(x) = y. \tag{4.3}$$

Consider $R(x + \Delta, y)$. If y particles enter each second at $x + \Delta$, then a slightly different number, $\hat{y}$, will enter the sub-rod $(0, x)$ at x each second. We may expect to relate $R(x + \Delta, y)$ and $R(x, \hat{y})$.

This is easily done once we recall that $R(x + \Delta, y) \equiv u(x + \Delta)$ when $v(x + \Delta) = y$, while $R(x, \hat{y}) = u(x)$ when $v(x + \Delta) = y$. From (4.2a)

$$R(x + \Delta, y) - R(x, \hat{y}) = f(u, v, x)\Delta + o(\Delta). \tag{4.4}$$

Further, we note that $v(x) = \hat{y}$ when $v(x + \Delta) = y$. Thus, (4.2b) yields

$$\hat{y} - y = g(u, v, x)\Delta + o(\Delta). \tag{4.5}$$

Hence

$$R(x + \Delta, y) - R(x, y + g\Delta + o(\Delta)) = f(u, v, x)\Delta + o(\Delta), \tag{4.6}$$

leading to

$$\frac{\partial R}{\partial x} - g(R, y, x)\frac{\partial R}{\partial y} = f(R, y, x), \qquad R(0, y) = 0, \tag{4.7}$$

upon introduction of a uniform notation.

Let us compare the results of Section 2 with those obtained above. Comparing (2.2) and (4.1) we find $f = \sigma v$, $g = \sigma u$. Hence (4.7) becomes

$$\frac{\partial R}{\partial x} - \sigma R\frac{\partial R}{\partial y} = \sigma y, \qquad R(0, y) = 0. \tag{4.8}$$

The physical problem of Section 2 suggests the substitution

$$R(x, y) = y\hat{R}(x), \tag{4.9}$$

and (4.8) becomes

$$\frac{d\hat{R}}{dx} - \sigma\hat{R}^2 = \sigma, \qquad \hat{R}(0) = 0. \tag{4.10}$$

This agrees with (2.5). (Recall that $c = 1$.)

The above analysis was first give in Bellman *et al.* [*8*], where it was pointed out that equations (4.1) merely give the characteristic curves associated with (4.7). It is clear, then, that our results here could have been obtained at once by appealing to classical theory. Our new point of view, however, is also applicable to the more complicated problems of Section 3. Each of the results given there, originally obtained by particle counting, can be derived by suitable generalization of the ideas of this section. The standard characteristic arguments are not applicable in treating either the second or third examples of the previous section, and hence the imbedding method seems to lead to truly new results.

The details of these analyses cannot be given here. The interested reader is referred to ref. [*12*]. It is pointed out in that paper that if any non-linearities are introduced in equations such as (3.6a) or (3.8) the treatment given there breaks down completely. That difficulty is removed by a perturbation technique to be described in the next section.

5. A Perturbation Method

In ref. [*3*], Bellman and Kalaba give a new method of deriving (4.7). We outline their result.

Consider (4.1). Let u and v be the solutions of the problem given. Let U and V be the solutions of

$$\begin{aligned} \frac{dU}{dz} &= f(U, V, z), \\ -\frac{dV}{dz} &= g(U, V, z), \end{aligned} \tag{5.1}$$

$$U(0) = 0, \qquad V(x + \Delta) = y, \qquad 0 \leq z \leq x + \Delta.$$

We assume that a valid perturbation expansion exists:

$$\begin{aligned} U(z) &= u(z) + r(z)\Delta + o(\Delta), \\ V(z) &= v(z) + w(z)\Delta + o(\Delta). \end{aligned} \tag{5.2}$$

Then, formal manipulation gives

$$
\begin{aligned}
\frac{dr}{dz} &= f_u r + f_v w, \\
-\frac{dw}{dz} &= g_u r + g_v w, \\
r(0) = 0, \qquad & w(x) = g(u(x), v(x), x).
\end{aligned}
\tag{5.3}
$$

If we differentiate (4.1) with respect to y we find

$$
\begin{aligned}
\frac{d}{dz} u_y &= f_u u_y + f_v v_y, \\
-\frac{d}{dz} v_y &= g_u u_y + g_v v_y, \\
u_y(0) = 0, \qquad & v_y(x) = 1.
\end{aligned}
\tag{5.4}
$$

Noting the similarities between (5.3) and (5.4), and assuming that the uniqueness theorem for systems of linear equations can be invoked, we find

$$
r(z) = g(u(x), v(x), x)\, u_y(z). \tag{5.5}
$$

In the notation (4.3)

$$
R(x + \Delta, y) \equiv U(x + \Delta), \qquad R(x, y) \equiv u(x). \tag{5.6}
$$

But, by (5.1), (5.2), and (5.5),

$$
\begin{aligned}
U(x + \Delta) &= U(x) + \Delta U'(x) + o(\Delta), \\
&= u(x) + [r(x) + f(U, V, x)]\Delta + o(\Delta) \\
&= u(x) + [g(u, v, x)\, u_y(x) + f(u, v, x)]\Delta + o(\Delta).
\end{aligned}
\tag{5.7}
$$

Now, applying (5.6), we readily obtain (4.7).

The advantage of the perturbation method lies in the fact that it formally reduces the problem of finding the invariant imbedding formulation of a given two-point boundary value problem to an almost mechanical level. It may be applied in a straight forward manner to problems such as (3.6) and (3.8), and no great additional difficulties are encountered if non-linear terms are added to these equations. The problem of putting the results on a rigorous basis always remains, of course.

6. Remarks and Summary

We wish to emphasize that while the methods developed in this paper were originally directly connected with transport phenomena, they obviously apply to a wide class of problems—as yet not fully characterized. The result, generally speaking, is that a two-point boundary value problem is replaced by an initial value problem. In some cases this may be quite advantageous. Thus, Chandrasekhar's elegant solution of (3.9) in the case $\partial R/\partial x = 0$ stemmed directly from the imbedding formulation. Bellman, Kalaba, and Prestrud [5] have found numerical computation of (3.9) for the finite slab case relatively straight forward, and have also investigated (3.7) using the Laplace transform and numerical inversion techniques. The author found (3.7) a convenient starting place for a rigorous study of time-dependent transport in a rod [11]. Further possible advantages of the imbedding scheme when applied to numerical calculations are now under investigation at Sandia Corporation.

It is interesting, too, to speculate on possible generalizations of the method of characteristics provided by invariant imbedding. Since (4.1) can be viewed as characteristic equations associated with the partial differential equation (4.7), can (3.8), for example, be similarly considered, in a *useful* manner, as the characteristic equation of (3.9) ? Such questions have not yet been studied.

We have attempted in this brief discussion to show how the investigation of some relatively simple physical models leads naturally to new and useful methods in the study of differential equations and integro-differential equations of certain kinds. The fact that mathematics is often applicable to physical problems is well known to everyone, and the fact that physical considerations may lead naturally to new mathematics is also well known, though sometimes overlooked. The writer feels that one of the most important roles of the applied mathematician is to make full use of the interplay between physical models and mathematical tools. Ideas generated in this way may eventually be put on a firm and rigorous basis by modern mathematical techniques.

REFERENCES

[1] Ambarzumian, V. A., Diffuse reflection of light by a foggy medium. *Compt. rend. acad. sci. U.R.S.S.* **38**, 229-232 (1943).

[2] Bellman, R., and Kalaba, R., On the principle of invariant imbedding and propagation through inhomogeneous media. *Proc. natl. Acad. Sci. U.S.* **42**, 629-632 (1956).

[*3*] Bellman, R., and Kalaba, R., On the fundamental equations of invariant imbedding, I. *Proc. natl. Acad. Sci. U.S.* **47**, 336-338 (1961).
[*4*] Bellman, R., and Kalaba, R., Transport theory and invariant imbedding. *Proc. Symposia in Appl. Math.* **11**, 206-218 (1961).
[*5*] Bellman, R., Kalaba, R., and Prestrud, M. Invariant Imbedding and Radiative Transfer in Slabs of Finite Thickness. RAND Corp. Report No. R-388-ARPA (1962).
[*6*] Bellman, R., Kalaba, R., and Wing, G. M., On the principle of invariant imbedding and neutron transport theory, I. One-dimensional case. *J. Math. and Mech.* **7**, 149-162 (1958).
[*7*] Bellman, R., Kalaba, R., and Wing, G. M., Invariant imbedding and neutron transport theory, II. Functional equations. *J. Math. and Mech.* **7**, 741-756 (1958).
[*8*] Bellman, R., Kalaba, R., and Wing, G. M., Invariant imbedding and the reduction of two-point boundary value problems to initial value problems. *Proc. natl. Acad. Sci.* **46**, 1646-1649 (1960).
[*9*] Chandrasekhar, S., "Radiative Transfer." Oxford, Univ. Press (Clarendon), London and New York, 1950.
[*10*] Mullikin, T., Principles of Invariance in Transport Theory. RAND Corp. Report No. RM-2761-JPL (1961).
[*11*] Wing, G. M., Solution of a time dependent, one dimensional neutron transport problem. *J. Math. and Mech.* **7**, 757-766 (1958).
[*12*] Wing, G. M., Invariant imbedding and transport theory—a unified approach. *J. math. anal. and Appl.* **2**, 277-292 (1961).
[*13*] Wing, G. M., "An Introduction to Transport Theory." Wiley, New York, in press, 1962.

Numerical Study of Periodic Solutions of the van der Pol Equation[1]

MINORU URABE
RIAS, Baltimore, Maryland, and
Mathematics Department
Hiroshima University, Hiroshima, Japan

1. Introduction

In this paper, we are concerned with the numerical study of periodic solutions of van der Pol's equation

$$\frac{d^2x}{dt^2} - \lambda(1 - x^2)\frac{dx}{dt} + x = 0 \qquad (\lambda > 0). \tag{1}$$

The writer first devised a method for computing a periodic solution of a general autonomous system [*6*, *10*] and, using that method, he computed periodic solutions of (1) for $\lambda = 0\ (0.2)\ 1.0$ [*7*]. But, later, this method was found to be inconvenient for the actual computation of periodic solutions of (1) for large values of λ. This is due to the fact that, for large values of λ, the closed orbits of the system

$$\begin{cases} \dfrac{dx}{dt} = y & (\text{def } X(x, y)), \\[2ex] \dfrac{dy}{dt} = -x + \lambda(1 - x^2)\,y & (\text{def } Y(x, y)) \end{cases} \tag{2}$$

which is evidently equivalent to (1), reveal the singular behavior near the x-axis as is seen in Figure 3.

Hence, noticing the symmetric character of the orbits of (2), the writer devised another simpler method to compute a closed orbit of (2), namely a periodic solution of (1). And, using that method, he computed the periodic solutions of (1) for $\lambda = 2, 3, 4, 5, 6, 8, 10$ assisted by Yanagiwara and Shinohara [*9*], and Yanagiwara computed further the periodic solution for $\lambda = 20$ [*14*].

[1] This research was partially supported by the United States Air Force through the Air Force Office of Scientific Research of the Air Research and Development Command, under Contract Number AF 49(638)-382. Reproduction in whole or in part is permitted for any purpose of the United States Government.

When our papers came off the press, we learned that the periodic solutions of van der Pol's equation have been computed also by Krogdahl [*3*] in a different way. His results coincide well with ours in the amplitude, but, in the period, there are some small differences. However, the writer cannot decide whether or not his results should be preferable to ours, because the details of the computation are not described in his paper.

In the present paper, our new method is sketched and the main results of the computation combined with those obtained by the previous method are mentioned. The results of the computation tell us there exists a maximum point of the amplitude between $\lambda = 3$ and $\lambda = 4$. After this was found, Yanagiwara [*15*] sought the maximum point of the amplitude computing the derivative of the amplitude with respect to λ and he found that the amplitude attains the maximum value 2.0235 for $\lambda = 3.2651$. His method is also mentioned in the present paper.

Remark. Our method is based on the numerical solution of a certain equation by means of Newton's method. Consequently, it guarantees the existence of a true periodic solution besides giving a numerical solution [*1*].

2. The Method of Computing a Closed Orbit of the System (2)

As is seen from the form of (2), in order to look for a closed orbit, we have only to seek for an orbit whose intersections with the y-axis are equidistant from the origin.

Let

$$x = \varphi(t, a), \qquad y = \psi(t, a)$$

be the equation of the orbit C passing through any point $A(a, 0)$ $(a > 0)$ on the x-axis at $t = 0$. Then, for our purpose, we want a root a of the equation

$$F(a) \mathrm{def}\, \psi[T_1(a), a] + \psi[T_2(a), a] = 0, \tag{3}$$

where $T_1(a)$ and $T_2(a)$ are respectively the times required to reach the points B_1 and B_2 on the y-axis from A along the orbit C.

Since the functions $T_i(a)$ $(i = 1, 2)$ are analytic with respect to a, the equation (3) can be solved numerically by Newton's method if the approximate root a_0 and the approximate value of the derivative $F'(a_0)$ are found.

The formula for $F'(a_0)$ can be derived from the formula for the normal variation of the orbit [*4*, *8*].

In fact, since any orbit of (2) crosses the x-axis orthogonally, the normal variations of the orbit C at B_1 and B_2 are given by

$$\frac{\sqrt{X_0^2 + Y_0^2}}{\sqrt{X_i^2 + Y_i^2}} e^{h(T_i)} \delta a + (|O\ \delta a\ |^2) \qquad (i = 1, 2),$$

where (X_0, Y_0), (X_1, Y_1) and (X_2, Y_2) are the values of (X, Y) at A, B_1 and B_2 respectively and

$$h(t) = \int_0^t \left(\frac{\partial X}{\partial x} + \frac{\partial Y}{\partial y}\right)_{\substack{x=\varphi(t, a_0)\\ y=\psi(t, a_0)}} dt = \lambda \int_0^t [1 - \varphi^2(s, a_0)]\, ds. \tag{4}$$

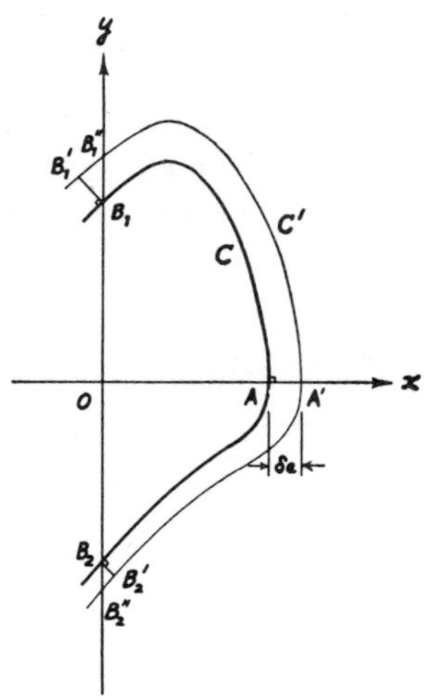

FIG. 1

Then, since the inclinations of C at B_i $(i = 1, 2)$ are Y_i/X_i $(i = 1, 2)$, the components of the above variations of C in the direction of the y-axis are given by

$$\delta\psi[T_i(a_0), a_0] = \frac{\sqrt{X_0^2 + Y_0^2}}{X_i} e^{h(T_i)} \delta a + O(|\ \delta a\ |^2) \qquad (i = 1, 2).$$

From this it readily follows that

$$F'(a_0) = \sqrt{X_0^2 + Y_0^2} \left(\frac{e^{h(T_1)}}{X_1} + \frac{e^{h(T_2)}}{X_2}\right). \tag{5}$$

Now, if a_0 is an approximate root of (3), we may put

$$b \approx \psi[T_1(a_0), a_0] \approx -\psi[T_2(a_0), a_0],$$

consequently we have

$$F'(a_0) \approx \frac{a_0}{b}(e^{h(T_1)} - e^{h(T_2)}).$$

Then, using this approximate value of $F'(a_0)$, the root of (3), namely the initial value of the closed orbit of (2) can be computed by Newton's method as accurately as we desire [5]. This is the method by which we computed the periodic solutions.

3. The Method of Numerical Integration

In our computation, in order to obtain the results as accurately as possible, a somewhat new method of step-by-step numerical integration is used for the system (2).

The formulas used read

$$x_n = x_{n-1} + \frac{k}{720}(11\dot{x}_{n-3} - 74\dot{x}_{n-2} + 456\dot{x}_{n-1} + 346\dot{x}_n - 19\dot{x}_{n+1}),$$

$$x_{n+1} = x_n + \frac{k}{720}(251\dot{x}_{n-4} - 1274\dot{x}_{n-3} + 2616\dot{x}_{n-2} - 2774\dot{x}_{n-1} + 1901\dot{x}_n),$$

where

k is the length of divided intervals;
x_i and $\dot{x}_i$ are the values of $x(t)$
and $\dot{x}(t) = dx(t)/dt$ for $t = ik$.

The first is an integrated expression of Stirling's interpolation formula and the second is an ordinary Adams' extrapolating integration formula.

The computation by means of the above formulas is carried on in the following way.

Let us assume x_i are found up to $i = n - 1$ and we want the value x_n. First, we compute x_n by the second formula and, using the value of x_n, we compute further x_{n+1} by the second formula. Next, by the first formula, we compute the corrected value of x_n using the values of x_n and x_{n+1} obtained already. Using the value of x_n found newly, we compute again x_{n+1} by the second formula. We continue this process until the process ends in a state of oscillation [5] and we take any one set of the values (x_n, x_{n+1}) in a state of oscillation. The value of x_n of this set is adopted as the desired value of x_n and we proceed to the next step.

This method is well suited to the present problem of computing values which are quite good in spite of the large number of steps. This is due to the fact that the truncation errors and the rate of growth of round-off and truncation errors are both smaller than the usual multistep formulas [*13*].

4. The Derivative of the Amplitude with Respect to λ

By the formulas for the normal variation of the orbit [4, 8], the normal variations of the orbit C at B_1 and B_2 corresponding to two variations δa and $\delta\lambda$ are given by

$$\frac{\sqrt{X_0^2+Y_0^2}}{\sqrt{X_i^2+Y_i^2}}\, e^{h(T_i)}\delta a + \frac{e^{h(T_i)}I(T_i)}{\sqrt{X_i^2+Y_i^2}}\,\delta\lambda + o(|\,\delta a\,| + |\,\delta\lambda\,|) \qquad (i = 1, 2),$$

where

$$I(t) = \int_0^t e^{-h(s)}[1 - \varphi^2(s, a_0)]\,\psi^2(s, a_0)\,ds.$$

By making the components of the above variations in the direction of the y-axis as in 2, the desired formula for the derivative is obtained as follows:

$$\frac{da}{d\lambda} = -\frac{e^{h(T_1)}I(T_1) - e^{h(T_2)}I(T_2)}{a[e^{h(T_1)} - e^{h(T_2)}]}, \tag{6}$$

where $a = a(\lambda)$ is the initial value of the closed orbit, namely the amplitude of the oscillation considered as the function of the damping coefficient λ.

TABLE 1

λ	a		ω		h
	Urabe	Krogdahl	Urabe	Krogdahl	Urabe
0	2.000	2.0000	6.283	6.2832	0.000
1	2.009	2.0086	6.687	6.6633	− 1.070
2	2.0199	2.0199	7.6310	7.6299	− 2.3822
3	2.0235	2.0233	8.8613	8.8595	− 3.9409
4	2.0231	2.0230	10.2072	10.2037	− 5.6194
5	2.0216	2.0215	11.6055	11.6124	− 7.3630
6	2.0199	2.0198	13.0550	13.0612	− 9.1316
7	—	2.0181	—	14.5380	—
8	2.0169	2.0167	16.0740	16.0363	− 12.7494
9	—	2.0154	—	17.5501	—
10	2.0145	2.0143	19.1550	19.0763	− 16.3775
20	2.0077	—	34.7103	—	− 34.4544

5. Computed Results and Conclusions

The computed values of the amplitude a, the period $\omega = 2(T_2 - T_1)$ and the characteristic exponent $h = h(\omega)/\omega$ are shown in Table 1 and in Figure 2. In Table 1, the values obtained by Krogdahl [3] are also

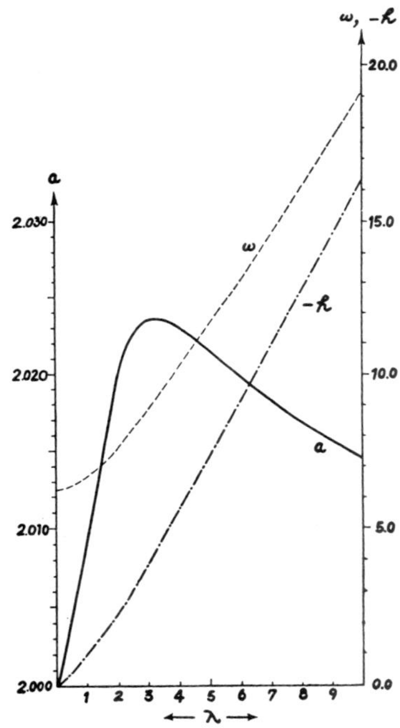

FIG. 2

TABLE 2

a	λ	$\frac{da}{d\lambda} \times 10^5$
	3.0000	+ 85
	3.2400	+ 9
	3.2650	+ 2
	3.2651	+ 0
2.0235	3.2652	− 1
	3.2654	− 4
	3.2662	− 3
	3.2675	− 3
	3.2700	− 8
	3.3000	− 2

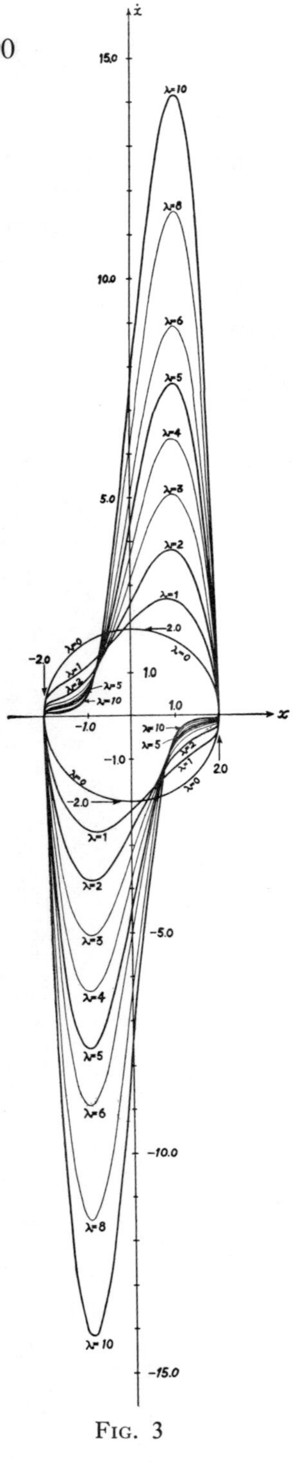
λ=10
λ=8
λ=6
λ=5
λ=4
λ=3
λ=2
λ=1
λ=0
15.0
10.0
5.0
1.0
-1.0
-2.0
2.0
-5.0
-10.0
-15.0

FIG. 3

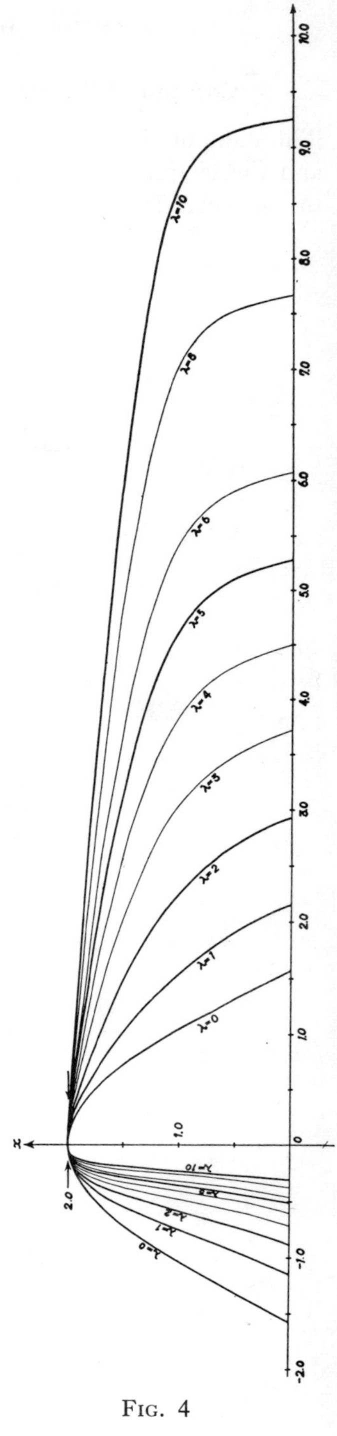
λ=10
λ=8
λ=6
λ=5
λ=4
λ=3
λ=2
λ=1
λ=0
10.0
9.0
8.0
7.0
6.0
5.0
4.0
3.0
2.0
1.0
0
-1.0
-2.0

FIG. 4

shown for reference. Table 2 shows the values of $da/d\lambda$ computed by (6) in order to locate the maximum point of the amplitude.

From Figure 2 and Table 2, the following conclusions are derived:

(1) When λ is increased from 0, the amplitude increases monotonically until $\lambda = 3.2651$ and decreases monotonically after it. The maximum value of the amplitude is 2.0235 and it is attained for $\lambda = 3.2651$;

(2) The period ω increases normally and monotonically with λ;

(3) When λ is increased from 0, the characteristic exponent decreases normally and monotonically. Consequently, the stability of the periodic solution increases very rapidly compared to the variance of λ.

Figures 3 and 4 show the shapes of the closed orbits and of the oscillations.

6. Remarks

For large values of λ, Dorodnicyn [*2*] derived the asymptotic expressions for the amplitude a and the period ω, but the writer observed minor errors in the manipulations leading to the formula for the period [*11*, *12*]. The corrected formula for the period reads

$$\omega = (3 - 2\log 2)\,\lambda + 3\alpha\lambda^{-1/3} - \frac{1}{3}\frac{\log\lambda}{\lambda} + \left(3\log 2 - \log 3 - \frac{3}{2} + b_0 - 2d\right)\frac{1}{\lambda} + O(\lambda^{-4/3}).$$

The Gothic figures show the corrected ones. Table 3 shows the values of a and ω computed in various ways. This table shows that the values

TABLE 3

	a		ω		
	Urabe	Dorodnicyn's Formula	Urabe	Dorodnicyn's Original f.	Dorodnicyn's Corrected f.
10	2.0145	2.0138 (− 0.0007)	19.1550	18.831 (− 0.324)	19.184 (+ 0.029 ... 0.15 %)
20	2.0077	2.0077 (0.0000)	34.7103	34.4925 (− 0.2178)	34.742 (+ 0.032 ... 0.09 %)

computed by the asymptotic expressions of Dorodnicyn are quite accurate even for $\lambda = 10$. Consequently, for values of λ larger than 10, the more accurate values would be gotten by these expressions as is checked for $\lambda = 20$. Then it will be unnecessary to continue the computation of periodic solutions for further values of λ over 10 if greater accuracy is not needed.

REFERENCES

[*1*] Collatz, L., Einige Anwendungen funktionalanalytischer Methoden in der praktischen Analysis. *Z. angew. Math. u. Phys.* **4**, 327-357 (1953).

[*2*] Dorodnicyn, A. A., Asymptotic solution of van der Pol's equation. *Priklad. Mat. i Mekh.* **11**, 313-328 (1947) (in Russian).

[*3*] Krogdahl, W. S., Numerical solutions of van der Pol's equation. *Z. angew. Math. u. Phys.* **11**, 59-63 (1960).

[*4*] Urabe, M., Infinitesimal deformation of cycles. *J. Sci. Hiroshima Univ.* **A18**, 37-53 (1954).

[*5*] Urabe, M., Convergence of numerical iteration in solution of equations. *J. Sci. Hiroshima Univ.* **A19**, 479-489 (1956).

[*6*] Urabe, M., Numerical determination of periodic solution of nonlinear system. *J. Sci. Hiroshima Univ.* **A20**, 125-148 (1957).

[*7*] Urabe, M., Periodic solution of van der Pol's equation with damping coefficients $\lambda = 0$ (0.2) 1.0. *J. Sci. Hiroshima Univ.* **A21**, 193-207 (1958).

[*8*] Urabe, M., Geometric study of nonlinear autonomous oscillations. *Funk. Ekv.* **1**, 1-83 (1958).

[*9*] Urabe, M., Yanagiwara, H., and Shinohara, Y., Periodic solutions of van der Pol's equation with damping coefficient $\lambda = 2 \sim 10$. *J. Sci. Hiroshima Univ.* **A23**, 325-366 (1960).

[*10*] Urabe, M., On a method to compute periodic solutions of the general autonomous system. *J. Sci. Hiroshima Univ.* **A24**, 189-196 (1960).

[*11*] Urabe, M., Remarks on periodic solutions of van der Pol's equation, *J. Sci. Hiroshima Univ.* **A24**, 197-199 (1960).

[*12*] Urabe, M., Periodic solutions of van der Pol's equation with damping coefficient $\lambda = 0 \sim 10$. *IRE Trans. on Circuit Theory* **CT-7**, 382-386 (1960).

[*13*] Urabe, M., Theory of errors in numerical integration of ordinary differential equations. *J. Sci. Hiroshima Univ.* **A-I25**, 3-62 (1961).

[*14*] Yanagiwara, H., A periodic solution of van der Pol's equation with a damping coefficient $\lambda = 20$. *J. Sci. Hiroshima Univ.* **A24**, 201-217 (1960).

[*15*] Yanagiwara, H., Maximum of the amplitude of the periodic solution of van der Pol's equation. *J. Sci. Hiroshima Univ.* **A-I25**, 127-134 (1961).

Finite Elastic Deformation of a Plane Strain Wedgeshaped Radial Crack in a Compressible Cylinder

PAUL J. BLATZ
Guggenheim Aeronautical Laboratory
California Institute of Technology, Pasadena, California

A. Introduction

The problem of the elastic stress-strain distribution around the base of a stationary crack has received much attention [*2*, *4*, *6*]. Inherent in all these analyses is the assumption of linearity which implies that the product of Hookean strains or the product of Hookean strain and rotation or the product of rotations can be neglected with respect to the Hookean strain itself. On the other hand, it is generally recognized that a singularity [*10*] of some sort exists at the base of a crack or indeed, in the neighborhood of any geometrical discontinuity [*7*]. Thus the use of linear theory to predict large or singular stress-strain fields immediately restricts the analyst at best to an approximation, if not an inconsistency. The purpose of this report is to show how the actual elastic stress distribution in a simple geometry is modified by accounting for the terms which are ordinarily dropped in linear theory. It will be assumed that the material under discussion is perfectly elastic and compressible, so that no consideration is given to plastic deformation [*3*].

Because facile application of finite elastic theory is at present restricted to problems in which the displacement field can be specified in some sense, it is necessary to specify first a simple geometry with simple boundary conditions, and secondly, a simple strain-energy function. We choose the perfectly elastic lubricated wedge for which linear solutions have already been presented [*8*]. Since no shear stresses are involved, the complexity of the finite elastic formulation is minimized. In the following, it is assumed that all stress and deformation components are directed along a set of principal coordinate axes.

Secondly, it is necessary to choose a strain energy function which is at once representative of a typically compressible elastomer and at the same time is simple enough to allow for convenient analytical results. In general, the energy stored per unit volume of the deformed isotropic material is at best a function of the invariants [*9*] which are proper to the deformation tensor, isotropy assumed.

$$W = W\{I_1 I_2 I_3\} \tag{1}$$

where

$$\hat{I}_1 = \sum \lambda_i^2 \tag{2}$$

$$I_2 = \sum_{i \neq j} \lambda_i^2 \lambda_j^2 \tag{3}$$

$$I_3 = \pi \lambda_i^2 \tag{4}$$

$$\lambda_i = 1 + \tilde{e}_i \tag{5}$$

$$\tilde{e}_i = u_{i,i}; \qquad i \text{ not summed.} \tag{6}$$

Among the supply of available rubbery materials, there are those [1] whose strain energy function is independent of I_2 and for which $\partial W/\partial I_1$ is a constant. For such materials, it follows that $\partial^2 W/\partial I_1 \partial I_3$ is zero, and therefore $\partial W/\partial I_3$ is independent of I_1. Thus we write:

$$W = c(I_1 - 3) + f(I_3). \tag{7}$$

By comparison with linear theory, it is readily shown that (7) can be expressed in a power series of the form

$$W = \frac{\mu_E}{2}(I_1 - 3) - \left(K_E + \frac{\mu_E}{3}\right) \ln J_3 + (K_E - \tfrac{2}{3}\mu_E)(J_3 - 1) + \frac{c}{2}(J_3 - 1)^2 + \ldots \tag{8}$$

where

$$J_3 = \sqrt{I_3} \tag{9}$$

and where μ_E, K_E are respectively the shear and bulk moduli associated with the diagonalized deformation tensor:

$$E_i = \tilde{e}_i + \tfrac{1}{2}\tilde{e}_i^2 = \frac{\lambda_i^2 - 1}{2}. \tag{10}$$

In the Hooke-Cauchy linear theory, E_i is ordinarily approximated by $\tilde{e}_i$, so that μ_E and K_E are also the ordinary coefficients of the linear theory. In what follows, it is assumed that the compressible elastomer is characterized by a strain energy of the form (8) terminated at the linear term in $(J_3 - 1)$. The constitutive law associated with (8) is then given by:

$$T_i \lambda_i = t_i J_3 = \mu_E(\lambda_i^2 - 1) + (K_E - \tfrac{2}{3}\mu_E)(J_3 - 1). \tag{11}$$

Here T_i is engineering stress (per unit undeformed cross section), and t_i is the true stress.

B. Linear Solution

Figure 1 shows a schematic diagram of the problem at hand. An infinitely long rubbery cylinder of radius a is split lengthwise by a radial crack of depth a. Into this crack is inserted a lubricated wedge of flank angle

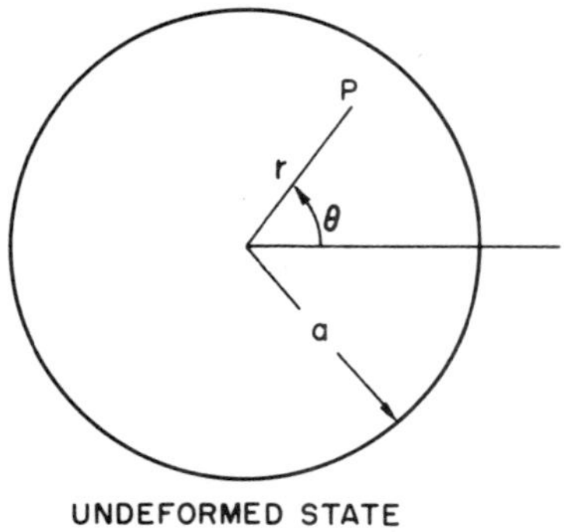

FIG. 1a. Undeformed state

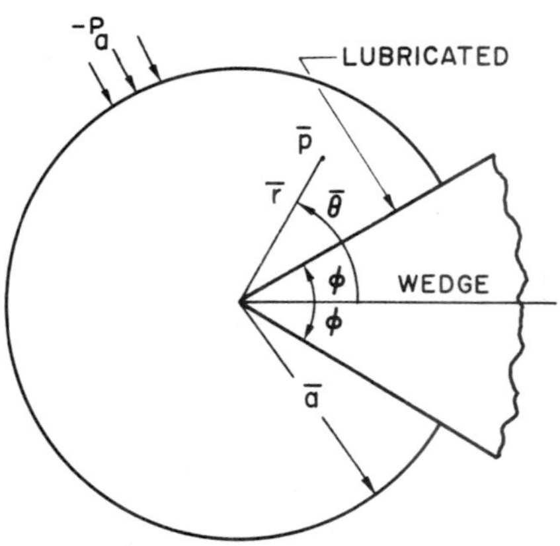

FIG. 1b. Deformed state

2ϕ, also of infinite length, the tip of which is assumed to meet the base of the crack (this statement will be subsequently modified). In order to prevent the rubbery log from slipping off the wedge, a compressive radial stress component P_a is applied to the circumferential periphery of the log. Under these specifications, the displacement field is given as follows:

$$\bar{r} = \bar{r}(r) \tag{12}$$

$$\bar{\theta} = \phi + (\pi - \phi)\frac{\theta}{\pi} = \phi + (1 - \delta)\,\theta \tag{13}$$

$$\bar{z} = z \tag{14}$$

where superscript bars denote the coordinate variables of the deformed body. In linear notation, we have:

$$u = \bar{r} - r \tag{15}$$

$$v = r(\bar{\theta} - \theta) = r(\phi - \delta\theta) \tag{16}$$

$$w = \bar{z} - z = 0 \tag{17}$$

$$\epsilon_r = \frac{du}{dr} \tag{18}$$

$$\epsilon_\theta = \frac{u}{r} - \delta \tag{19}$$

$$\epsilon_z = \epsilon_{r\theta} = 0. \tag{20}$$

After satisfying equilibrium and setting $t_r = -P_a$ at $r = a$, we have:

$$t_r = -p_a - B\left(\frac{1}{r_2} - \frac{1}{a^2}\right) + \frac{\mu_E\delta}{1-\nu}\ln\frac{a}{r} \tag{21}$$

$$t_\theta = -p_a + B\left(\frac{1}{r^2} + \frac{1}{a^2}\right) + \frac{\mu_E\delta}{1-\nu}\left(\ln\frac{a}{r} - 1\right) \tag{22}$$

$$t_z = -2\nu p_a + 2\nu\frac{B}{a^2} + \frac{\nu}{1-\nu}\mu_E\delta\left(2\ln\frac{a}{r} - 1\right) \tag{23}$$

$$2\mu_E\frac{u}{r} = -(1-2\nu)\,p_a + B\left(\frac{1-2\nu}{a^2} + \frac{1}{r^2}\right) + \mu_E\delta\left(\frac{1-2\nu}{1-\nu}\ln\frac{a}{r} + 1\right). \tag{24}$$

At this point, it is tempting to specify that $u = 0$ at $r = 0$, and thereby set $B = 0$. However, this step will be deferred until the significance of what happens at $r = 0$ is clearly understood in the light of the large deformation theory. Following this, equations (21) through (24) will be cast into a form suitable for comparison with the results of the subsequent section.

C. Finite Elastic Solution

Following Rivlin [5], the components of the diagonalized deformation tensor associated with (12), (13), (14) are given by:

$$\lambda_r = \frac{d\bar{r}}{dr} \tag{25}$$

$$\lambda_\theta = \frac{\bar{r}}{r}(1 - \delta) \tag{26}$$

$$\lambda_z = 1. \tag{27}$$

Differentiation of (26) with respect to "r" leads to the relation

$$r\frac{d\lambda_\theta}{dr} = -[\lambda_\theta - (1 - \delta)\lambda_r] \tag{28}$$

with which the equilibrium equation can be rewritten:

$$t_\theta - t_r = \bar{r}\frac{dt_r}{d\bar{r}} = -\frac{\lambda_\theta}{\lambda_r(1 - \delta)}[\lambda_\theta - (1 - \delta)\lambda_r]\frac{dt_r}{d\lambda_\theta}. \tag{29}$$

In terms of its radial and tangential components, (11) becomes

$$t_r = \mu_E\frac{\lambda_r}{\lambda_\theta} + (K_E - \tfrac{2}{3}\mu_E) - \frac{K_E + \dfrac{\mu_E}{3}}{\lambda_r\lambda_\theta} \tag{30}$$

$$t_\theta = \mu_E\frac{\lambda_\theta}{\lambda_r} + (K_E - \tfrac{2}{3}\mu_E) - \frac{K_E + \dfrac{\mu_E}{3}}{\lambda_r\lambda_\theta}. \tag{31}$$

Substituting into (29) yields

$$\frac{d\lambda_r}{d\lambda_\theta} = -\frac{\lambda_r}{\lambda_\theta}\frac{1 + (1 - 2\nu)\lambda_r\lambda_\theta\left[\dfrac{(1 - \delta)\lambda_\theta - \lambda_r}{\lambda_\theta - (1 - \delta)\lambda_r}\right]}{1 + (1 - 2\nu)\lambda_r^2} \tag{32}$$

where

$$1 - 2\nu = \frac{\mu_E}{K_E + \dfrac{\mu_E}{3}}. \tag{33}$$

Along with (32), we need (28) rewritten as:

$$\frac{d\ln r}{d\lambda_\theta} = -\frac{1}{\lambda_\theta - (1 - \delta)\lambda_r}. \tag{34}$$

It is convenient to define:

$$\lambda'_\theta = \lambda_\theta \sqrt{1 - 2\nu} \tag{35}$$

$$\lambda'_r = \lambda_r \sqrt{1 - 2\nu} \tag{36}$$

with which (26), (30), (31), (32), and (34) become:

$$\bar{r} = \frac{r\lambda'_\theta}{(1 - \delta)\sqrt{1 - 2\nu}} \tag{37}$$

$$\frac{2\nu}{1 - 2\nu} - \frac{t_r}{\mu_E} = \frac{1}{\lambda'_r\lambda'_\theta} - \frac{\lambda'_r}{\lambda'_\theta} \tag{38}$$

$$\frac{2\nu}{1 - 2\nu} - \frac{t_\theta}{\mu_E} = \frac{1}{\lambda'_r\lambda'_\theta} - \frac{\lambda'_\theta}{\lambda'_r} \tag{39}$$

$$\frac{d\lambda'_r}{d\lambda'_\theta} = -\frac{\lambda'_r}{\lambda'_\theta}\,\frac{1 + \lambda'_r\lambda'_\theta\left[\dfrac{\lambda'_\theta(1 - \delta) - \lambda'_r}{\lambda'_\theta - (1 - \delta)\lambda'_r}\right]}{1 + \lambda'^2_r} \tag{40}$$

$$\frac{d \ln r}{d\lambda'_\theta} = -\frac{1}{\lambda'_\theta - (1 - \delta)\,\lambda'_r}\,. \tag{41}$$

At the boundary ($r = a$), (38) becomes:

$$\frac{2\nu}{1 - 2\nu} + \frac{p_a}{\mu_E} = 2k = \frac{1}{\lambda'_{ra}\lambda'_{\theta a}} - \frac{\lambda'_{\theta a}}{\lambda'_{ra}}\,. \tag{42}$$

D. Topology of the Phase Plane

In order to investigate the behavior of (40), we look first at the phase plane (λ'_r, λ'_θ) and then seek an asymptotic solution. By definition, λ'_θ is positive; and, λ_r must also be positive to guarantee continuity of the material. Therefore, Figure 2 describes only the upper right (*NE*) quadrant of the phase plane. All integration paths in this quadrant will follow the direction $dr < 0$, i.e. from $r = a$ to $r \to 0$. It follows from (41) that the *NE* quadrant is split by the separatrix

$$\lambda'_r = \frac{\lambda'_\theta}{1 - \delta} \tag{43}$$

into two segments designated as

$$\{\text{I II III}\} \qquad \text{when} \qquad \lambda_r' < \frac{\lambda_\theta'}{1-\delta} \tag{44}$$

$$\text{IV} \qquad \text{when} \qquad \lambda_r' > \frac{\lambda_\theta'}{1-\delta}. \tag{45}$$

There is also a saddle point at $\lambda_r' = \lambda_\theta' = 0$, so that trajectories in {I II III} are directed toward and along the λ_θ' axis, which in turn acts as a separatrix from the *SE* quadrant. Similarly, trajectories in IV are directed toward and along the λ_r' axis, which in turn acts as a separatrix from the *NW* quadrant.

In segment IV, it follows, using (40), that for

$$dr < 0, \qquad d\lambda_\theta' < 0, \qquad \text{and} \qquad d\lambda_r' > 0. \tag{46}$$

The inequalities (46) suggest that an asymptotic solution obtains as $\lambda_\theta' \to 0$, $\lambda_r' \to \infty$, for which (40) becomes:

$$\frac{d\lambda_r'}{d\lambda_\theta'} = -\frac{1}{\lambda_r'\lambda_\theta'} - \frac{1}{1-\delta}. \tag{47}$$

Integration on the assumption that $(\lambda_r'\lambda_\theta' \ll 1 - \delta)$ yields:

$$\lambda_r' = \sqrt{-2 \ln c\lambda_\theta'} \tag{48}$$

which is consistent with this assumption, and with (46).

Substitution into (41) with integration yields:

$$\ln \frac{r}{a} = \frac{\lambda_\theta'}{(1-\delta)\sqrt{-2 \ln c\lambda_\theta'}}. \tag{49}$$

Obviously (49) applies only to the case $r > a$, and thus solutions in segment IV are extraneous to the problem at hand.

Returning to segment {I II III}, we note (Figure 2) that the subsegments are designated by:

I, when $\lambda_r' < \lambda_\theta'(1-\delta)$, in which case, when $dr < 0$, then $d\lambda_\theta' > 0$, and $d\lambda_r' < 0$ (50)

II, when $\lambda_\theta'(1-\delta) < \lambda_r' < \dfrac{\lambda_\theta'}{1-\delta}$, $1 + \lambda_r'\lambda_\theta' \left[\dfrac{\lambda_\theta'(1-\delta) - \lambda_r'}{\lambda_\theta' - (1-\delta)\lambda_r'}\right] < 0$, in which case, when $dr < 0$, then $d\lambda_\theta' > 0$ and $d\lambda_r' < 0$ (51)

III, when $\lambda_\theta'(1-\delta) < \lambda_r' < \dfrac{\lambda_\theta'}{1-\delta}$, and $1 + \lambda_r'\lambda_\theta' \left[\dfrac{\lambda_\theta'(1-\delta) - \lambda_r'}{\lambda_\theta' - \lambda_r'(1-\delta)}\right] < 0$,

in which case, when $dr < 0$, then $d\lambda_\theta' > 0$, $d\lambda_r' > 0$. (52)

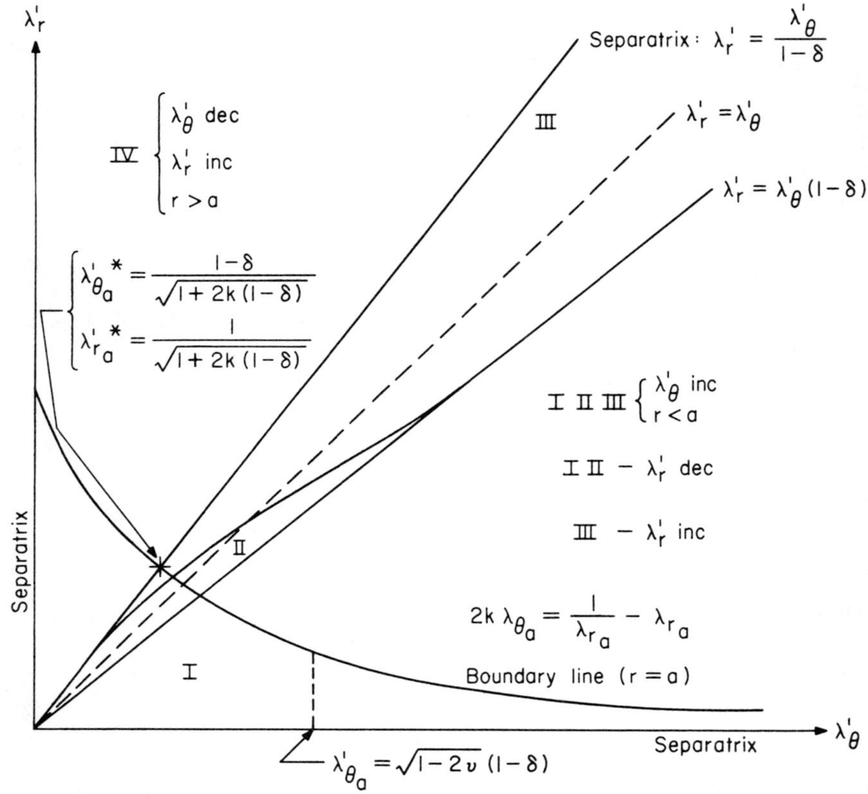

FIG. 2. Phase plane of the equilibrium equation.

It is readily observed that all paths in subsegment III eventually cross over into {I II}, and so the asymptotic behavior of {I II III} is characterized by $\lambda_r' \to 0$, $\lambda_\theta' \to \infty$, for which (40) becomes

$$\frac{d\lambda_r'}{d\lambda_\theta'} = -\frac{\lambda_r'}{\lambda_\theta'} - (1-\delta)\,\lambda_r'^2. \tag{53}$$

Integration yields:

$$\lambda_r' = \frac{1}{(1-\delta)\,\lambda_\theta' \ln c\lambda_\theta'}\,. \tag{54}$$

To the same order of approximation, (41) yields:

$$r\lambda'_{\theta} = \text{constant}. \tag{55}$$

From the definition (37), it follows that the Lagrangian variable must also approach a constant value designated by a_c, a cavity radius, so that (55) becomes

$$r\lambda'_{\theta} = \sqrt{1-2\nu}\,(1-\delta)\,a_c. \tag{56}$$

This striking result is one of the unusual phenomena encountered in nonlinear elasticity and will be termed the "keyhole" effect.

FIX	VARY	RESULT
$\nu, \frac{P_a}{\mu}, \lambda_{\theta_a}=1-\delta$	inc $\delta^* \to \dfrac{1}{1+\dfrac{2\nu}{1-2\nu\left(\dfrac{P_a}{\mu}\right)}}$	$a_c \to 0$
$\nu, \delta, \lambda_{\theta_a}=1-\delta$	dec $\frac{P_a}{\mu} \to \frac{\delta}{1-\delta}\,\frac{2\nu}{1-2\nu}$	$a_c \to 0$
$\nu, \frac{P_a}{\mu}, \delta$	dec $\lambda^*_{\theta_a}$ $\dfrac{1-\delta}{\sqrt{(1-2\nu)\left[1+2k(1-\delta)\right]}}$	$a_c \to 0$
$\frac{P_a}{\mu}, \delta, \lambda_{\theta_a}=1\delta$	inc ν	$a_c \to 0$

FIG. 3. Behavior of cavity radius a_c.

Obviously, when the crack is spread tangentially, the appearance of the keyhole at the base of the crack will produce a shear deformation which will tend to prevent the keyhole formation. In order to anticipate and prevent this effect, we assume that the base of the wedge is also shaped as a keyhole, where the keyhole radius is a predetermined function of the quantities

$$a_c = a_c\left\{\frac{p_a}{\mu_E}, \nu, \delta, \lambda_{\theta a}\right\}. \tag{57}$$

The boundary condition (42) depends only on p_a/μ_E and ν, and the curve for it is depicted in Figure 2. This curve intersects the separatrix at

$$\lambda_{\theta a}'^{*} = \frac{1 - \delta}{\sqrt{1 + 2k(1 - \delta)}}, \qquad \lambda_{ra}'^{*} = \frac{1}{\sqrt{1 + 2k(1 - \delta)}}. \tag{58}$$

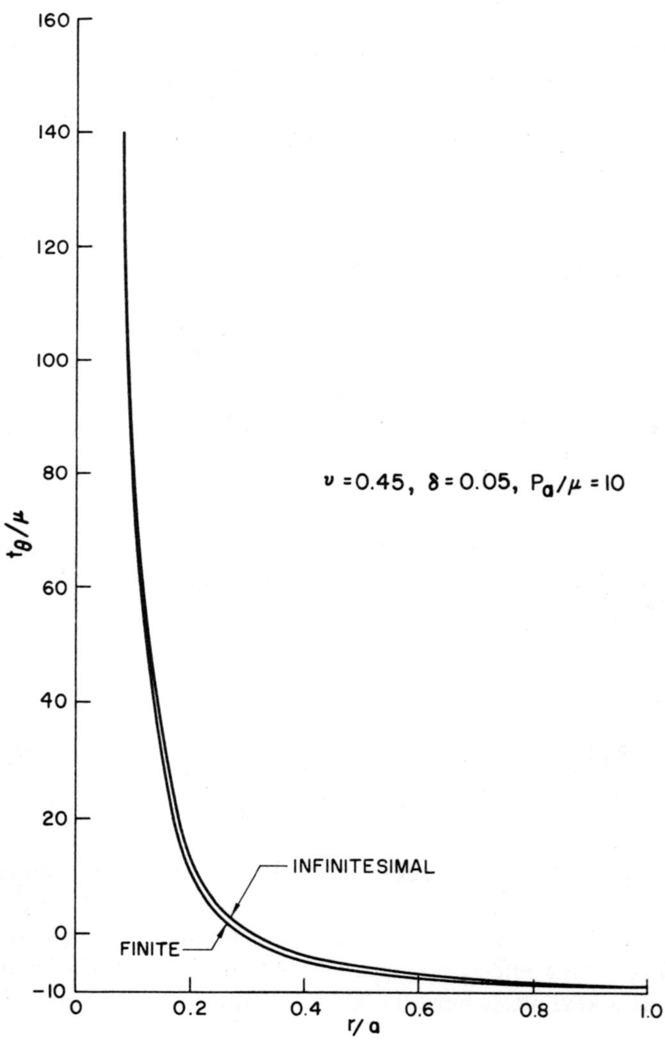

FIG. 4. Tangential stress *vs* radius.

Now observe that at this point the cavity radius must be zero, because

$$\frac{d \ln r}{d\lambda'_\theta} \to -\infty, \qquad \text{as} \qquad (\lambda'_{\theta a} - \lambda'^*_{\theta a}) \to 0. \tag{59}$$

Equation (59) implies that a slight increase in λ'_θ immediately reduces r to zero, and therefore $\bar{r} = a_c$ to zero.

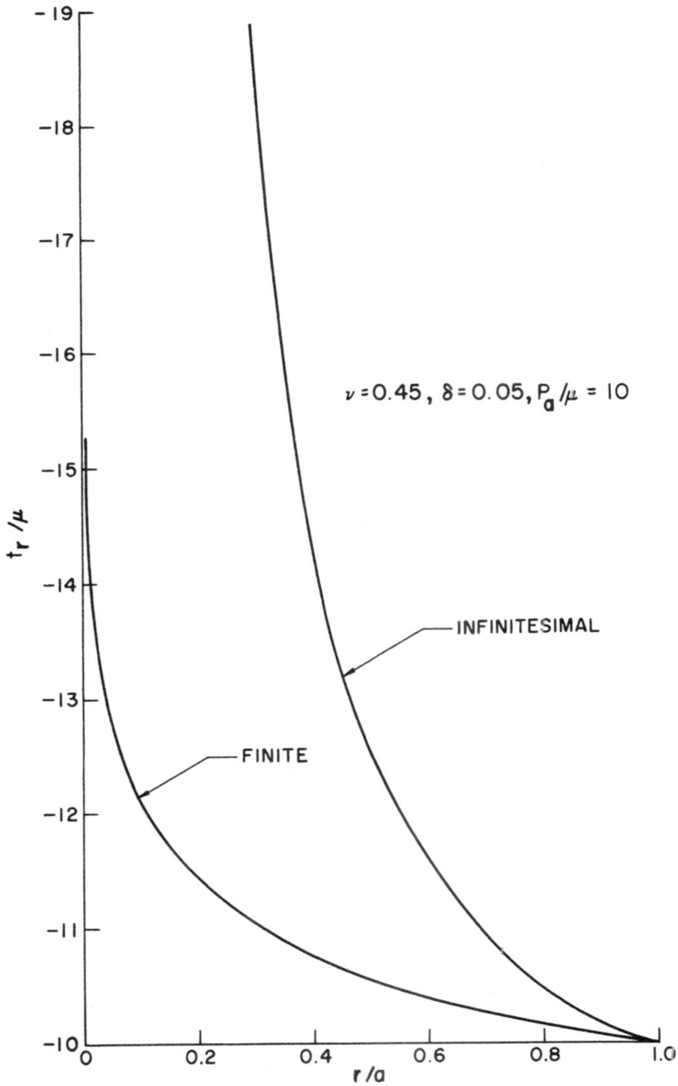

FIG. 5. Radial stress *vs* radius.

Suppose now in addition to applying a compressive stress at $r = a$, the outer boundary is also prevented from displacement after the keyhole-wedge is inserted. This implies that

$$\lambda'_{\theta a} = (1 - \delta)\sqrt{1 - 2\nu}. \tag{60}$$

In general, for given $\{\nu, P_a/\mu_E \text{ and } \delta\}$ there will be some finite a_c. As δ is

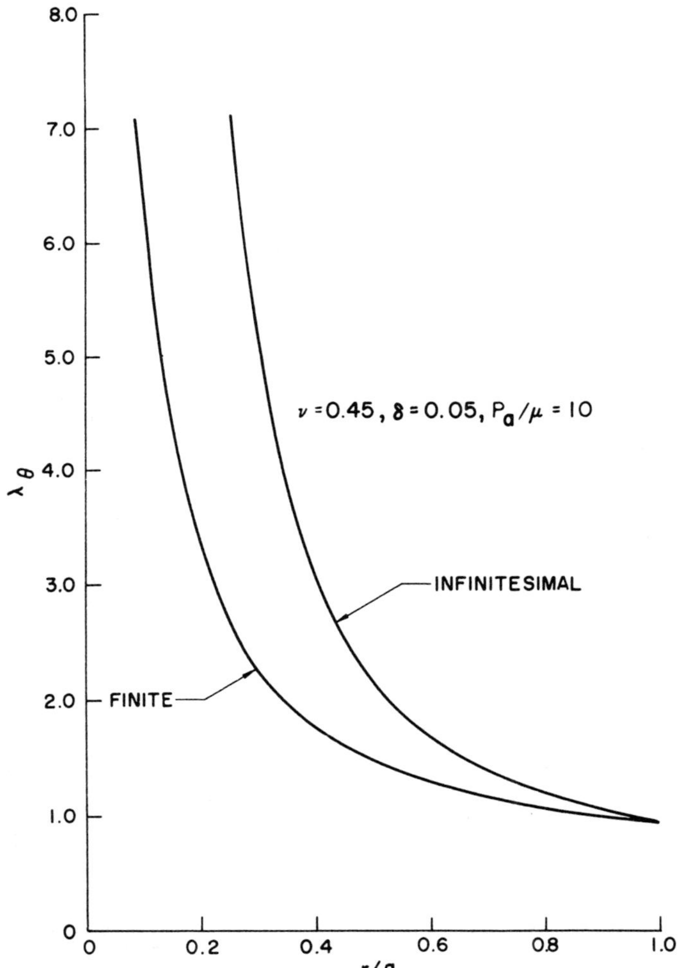

FIG. 6. Tangential deformation ratio *vs* radius.

increased for fixed ν and P_a/μ_E, and maintaining condition (56), a_c will decrease until the separatrix is reached, i.e. when

$$\sqrt{1-2\nu}\,(1-\delta^+) = \frac{1-\delta^+}{\sqrt{1+2k(1-\delta^+)}}\,. \tag{61}$$

This implies that

$$\delta^+ = \frac{1}{1+\dfrac{2\nu}{(1-2\nu)\,p_a/\mu_E}}\,. \tag{62}$$

If now, δ is increased further, either P_a/μ_E must be increased; or, if

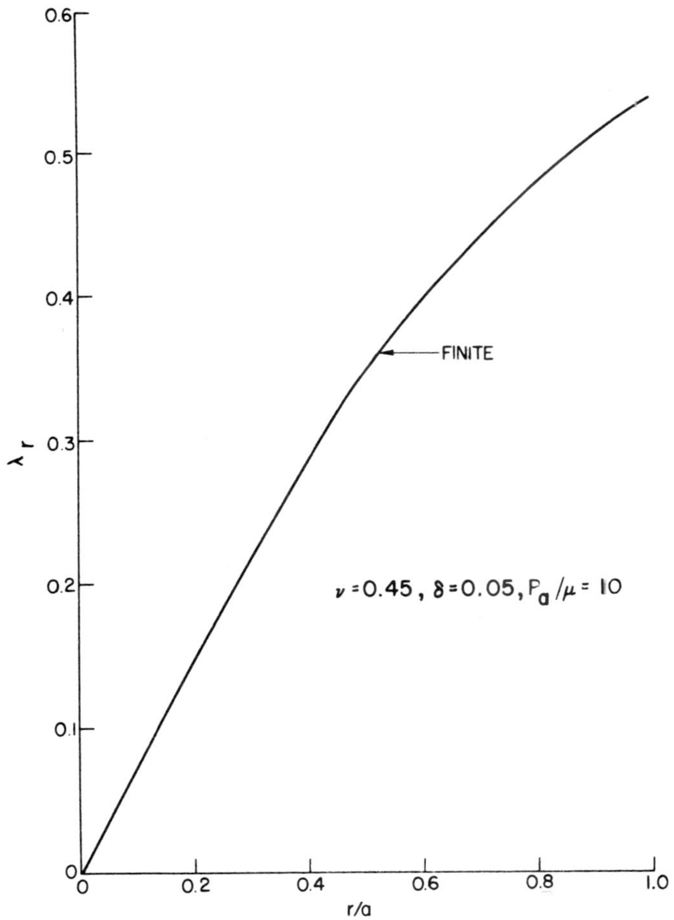

FIG. 7. Radial deformation ratio *vs* radius (Nonlinear theory).

P_a/μ_E is maintained constant, then the outer boundary will displace outwards, with

$$\lambda'_{\theta a} > \sqrt{1-2\nu}\,(1-\delta). \tag{63}$$

Suppose, on the other hand, that $\delta = \delta^+$, and P_a/μ_E is increased, then the outer boundary will displace inwards, with

$$\lambda'_{\theta a} < \sqrt{1-2\nu}\,(1-\delta). \tag{64}$$

From (62), (63) and (64), it follows that the keyhole effect only operates for

$$\frac{p_a}{\mu_E} > \frac{\delta}{1-\delta}\frac{2\nu}{1-2\nu}. \tag{65}$$

The preceding statements are summarized in Figure 3.

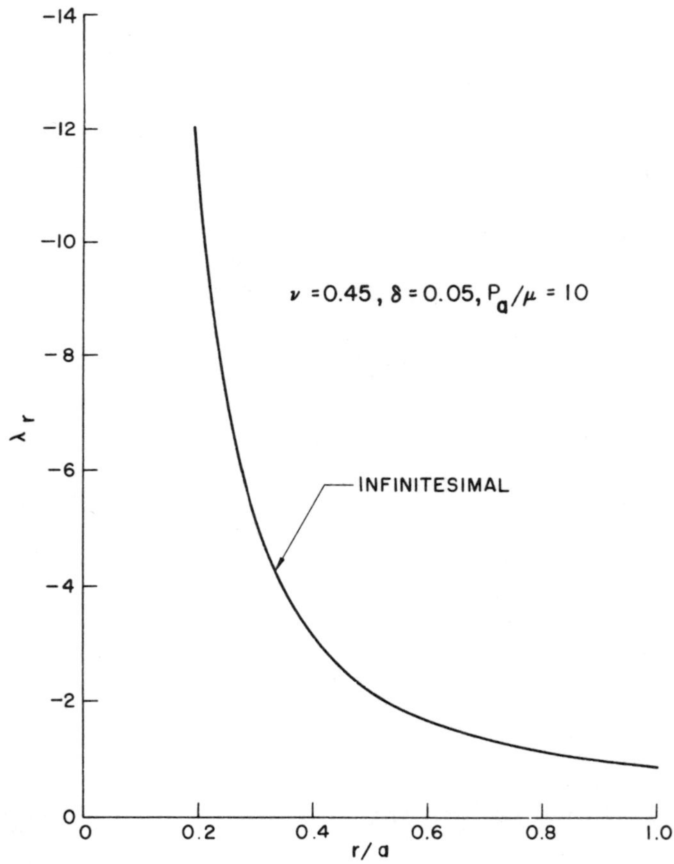

FIG. 8. Radial deformation ratio *vs* radius (Linear theory).

E. Nature of the Singularity

At the base of crack, we have, using (50) and (51)

$$\lambda_\theta \to \frac{(1-\delta)\, a_c}{r} \tag{66}$$

$$\lambda_r \to \frac{r}{(1-2\nu)(1-\delta)^2\, a_c \ln\left[\dfrac{c\sqrt{1-2\nu}\,(1-\delta)\, a_c}{r}\right]} \tag{67}$$

$$\frac{t_\theta}{\mu_E} \to \frac{(1-2\nu)(1-\delta)^3\, a_c^2}{r^2} \ln\left[\frac{c\sqrt{1-2\nu}\,(1-\delta)\, a_c}{r}\right] \tag{68}$$

$$\frac{t_r}{\mu_E} \to -(1-\delta) \ln\left[\frac{c\sqrt{1-2\nu}\,(1-\delta)\, a_c}{r}\right]. \tag{69}$$

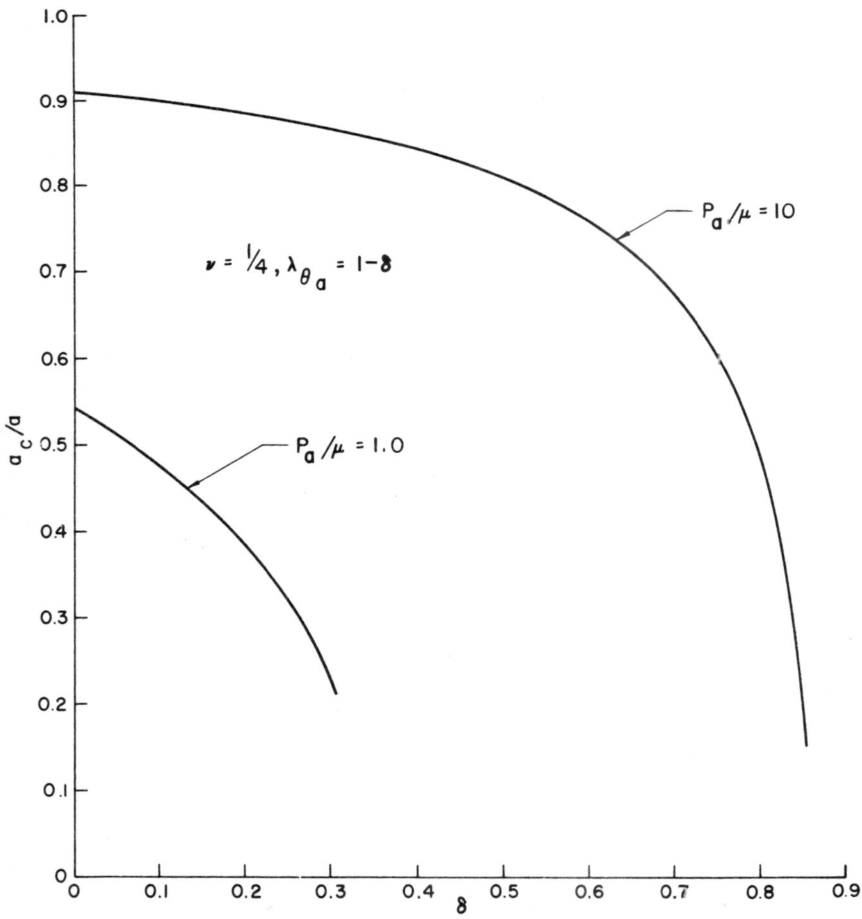

FIG. 9. Cavity radius *vs* half-angle of wedge.

In the infinitesimal case, setting $u = 0$, at $r = a$, we have, from (21), (22), (24):

$$\lambda_\theta \to \frac{1-\delta}{4(1-\nu)}\left[\frac{p_a}{\mu_E}(1-2\nu)-\delta\right]\frac{a^2}{r^2} \tag{70}$$

$$\lambda_r \to \frac{-1}{4(1-\nu)}\left[\frac{p_a}{\mu_E}(1-2\nu)-\delta\right]\frac{a^2}{r^2} \tag{71}$$

$$\frac{t_\theta}{\mu_E} \to \frac{1}{2(1-\nu)}\left[\frac{p_a}{\mu_E}(1-2\nu)-\delta\right]\frac{a^2}{r^2} \tag{72}$$

$$\frac{t_r}{\mu_E} \to \frac{-1}{2(1-\nu)}\left[\frac{p_a}{\mu_E}(1-2\nu)-\delta\right]\frac{a^2}{r^2} \tag{73}$$

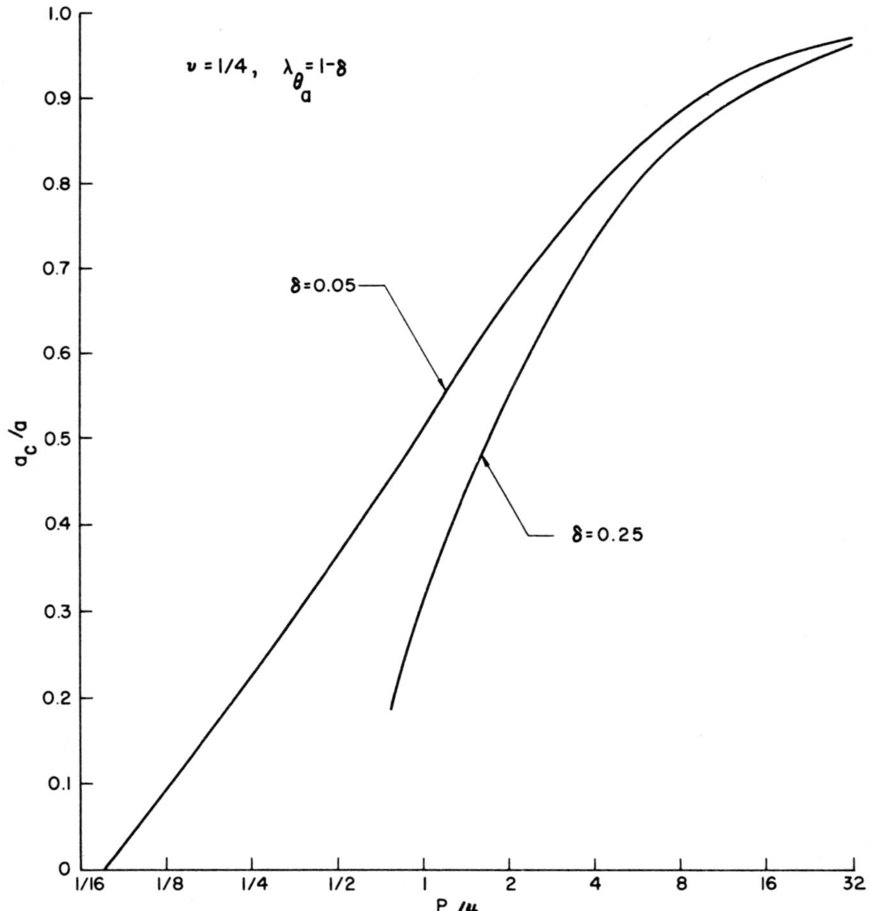

FIG. 10. Cavity radius *vs* boundary pressure.

The only similarity in the two theories arises in the case of the tangential stress which approaches infinity with an inverse square dependence on r. This infinite tangential stress arises because what was originally a point becomes deformed into a sector of a circle. The radial stress in the nonlinear theory is nonsingular as opposed to another inverse square singularity of opposite sign in the linear theory.

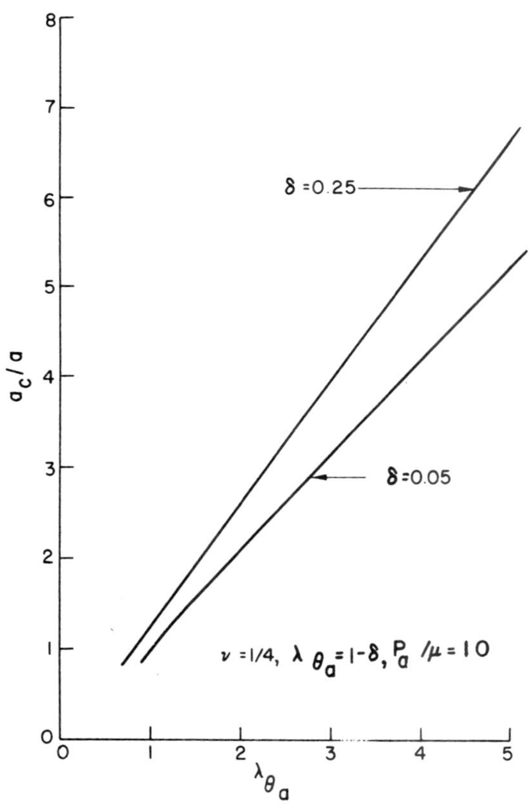

FIG. 11. Cavity radius *vs* boundary displacement.

Turning now to the deformation components, we first note that the linear theory predicts an infinite $1/r$—type displacement at the crack base. This is a striking example of the bound inconsistencies of linear theory. Normally, one avoids this by setting $u = 0$, at $r = 0$, rather than $u = 0$ at $r = a$. We chose the latter in order to effect a direct comparison with the nonlinear theory. The linear theory also predicts a $1/r^2$—type singularity for the radial strain, whereas the nonlinear theory is again nonsingular. Thus a true singularity around the base of a perfectly elastic crack occurs only in the stresses.

Note that, in the linear theory, the sign of all singularities is determined by the coefficient, $[P_a/\mu_E(1 - 2\nu) - \delta]$. If we apply (65), we find that this factor is positive if $2\nu > 1 - \delta$, and vice versa.

F. Numerical Results

Equations (37-39) were integrated on an IBM 704. Figures 4-8, respectively describe the behavior of t_θ, t_r, λ_θ and λ_r as a function of r/a for both linear and nonlinear theories, for the boundary condition $\lambda_{\theta a} = 1 - \delta$. Figures 9-11 describe the behavior of the cavity radius summarized in Figure 3.

The writer wishes to thank Professor M. L. Williams[1] for stimulating discussions on this problem.

REFERENCES

[*1*] Ciferri, A., and Flory, P. J., Stress-strain isothern for polymer networks. *J. appl. Phys.* **30**, 1498 (1959).

[*2*] Inglis, C. E., Stresses in a plate due to the presence of cracks and sharp corners. *Trans. Inst. naval Archit.* **55**, 219 (1913).

[*3*] Murch, S. A., and Nagdhi, P. M., On infinite elastic perfectly plastic wedge. *Proc. 3th U.S. natl. Congr. appl. Mech., June 1958.*

[*4*] Neuber, H., Ein neuer Ansatz sur Lösung räumlicher Probleme der Elastizitätstheorie. *Z. angew. Math. u. Mech.* **14**, 203 (1934).

[*5*] Rivlin, R. S., Large elastic deformations of isotropic materials, VI. *Phil. Trans. roy. Soc. (London)* **A242**, 173 (1949).

[*6*] Sneddon, I. N., The distribution of stress in the neighbourhood of a crack in an elastic solid. *Proc. roy. Soc. (London)* **A 187**, 229 (1946).

[*7*] Sternberg, E., and Eubanks, R. A., On concept of concentraded loads. *J. rational Mech. and Anal.* **4**, 135 (1955).

[*8*] Sternberg, E., and Loiter, W. T., The wedge under a concentrated couple. *J. appl. Mech.* **25**, 575 (1958).

[*9*] Truesdell, C., Mechanical foundations of elasticity. *J. rational Mech. and Anal.* **1**, 132 (1952).

[*10*] Williams, M. L., Stress singularities resulting from various boundary conditions in angular corners of plates in extension. *J. appl. Mech.* **19**, 526 (1952).

[1] Professor, California Institute of Technology, Pasadena, California.

Wave Propagation in Non-Simple Media[1]

B. A. FLEISHMAN[2]
Mathematics Department
Rensselaer Polytechnic Institute, Troy, New York

1. Introduction. Examples

The characteristics of wave propagation in a medium governed by the simple wave equation

$$u_{tt} - u_{xx} = 0 \tag{1}$$

are well known: all wave forms can propagate unchanged, but at a unique speed, unity. Question: What happens to these wave properties when a function of u and/or its derivatives is added to the right-hand side of the above equation? In other words, in a medium governed by the equation

$$u_{tt} - u_{xx} = f(u, u_x, u_t) \tag{2}$$

(which we shall call the "non-simple" wave equation), which wave shapes can propagate at a constant speed and what are these speeds?

As is shown below, the answer, roughly, is this: The roles played by wave form and velocity in the case of the simple wave equation (1) are reversed. For any given function $f(u, u_x, u_t)$ there exist one or more continuous spectra of propagation speeds, while the number of distinct types of wave forms is quite restricted. What happens in the simplest cases is that for each value of the speed, $|c|$, in a certain continuous range of values, there is a one-parameter family of wave profiles, and all members of this two-parameter family of profiles (c is the other parameter) are of the same type.

In (2) the "critical" propagation speed is unity, not only because this is the wave speed in the case $f \equiv 0$, but also for the associated reason that, in general, waves traveling at speeds $|c|$ greater than one (which we call

[1] This research was supported by the United States Air Force through the Air Force Office of Scientific Research of the Air Research and Development Command, under Contract Number AF 49(638)-514. Reproduction in whole or in part is permitted for any purpose of the United States Government.

[2] Presently on leave at the United States Army Mathematics Research Center, University of Wisconsin.

supercritical) have markedly different characteristics from those traveling at speeds less than one (*subcritical*).

The preceding comments hold for linear, as well as nonlinear, functions $f(u, u_x, u_t)$. For example, if $f = \gamma u$ (γ a constant), there can occur the phenomenon of "dispersion": An initial wave form "spreads out" as it propagates due to the different propagation velocities of its various Fourier components. This equation arises in quantum mechanics, where it is known as the Klein-Gordon equation [*3*, p. 137]. Another example of a linear function $f(u, u_x, u_t)$ is the telegrapher's equation, in which $f = \delta u + \eta u_t$ (δ and η constants).

A nonlinear function f was considered by van der Pol [*5*], who studied equation (2) with $f = \epsilon(1 - u^2)\, u_t$, ($\epsilon$ a positive constant), as a model of propagation of electrical impulses along nerve fibers. He found that only "step-type" waves can exist; in particular, no waves of periodic profile can propagate unchanged, at constant speed, in such a medium. As another example of nonlinear f, the author [*2*] considered $f = -\alpha u - \beta u^3$, for all non-zero real values of the parameters α and β. Among the results of this investigation are the following: For fixed positive values of α and β, all supercritical waves have the shape of the cosine type of Jacobian elliptic function; the wave amplitude and velocity may be chosen arbitrarily. On the other hand, not only can there be no periodic subcritical waves, there are no bounded subcritical waves. A final point of interest is that the natures of the sub- and supercritical waves just cited are reversed when the signs of α and β are reversed.

The purpose of this paper is to develop some general properties of wave solutions of (2) for certain classes of function $f(u, u_x, u_t)$; these generalizations include the examples cited above as special cases.

It is assumed that the variables t, x and u and the parameter c are dimensionless and real-valued. Only functions $f(u, u_x, u_t)$ which are continuous in their arguments are considered, and we seek solutions which are defined for all x and t and which, except when $|c| = 1$, are continuous and bounded and have continuous first and second partial derivatives.

2. Reduction to Ordinary Differential Equation

The crux of the analysis is that the problem of finding wave solutions of any non-simple wave equation (2) reduces immediately to that of solving an ordinary differential equation. For suppose that a particular wave propagates with the constant velocity c. The solution of (2) representing this wave must have the form

$$u(x, t) = \phi(x - ct) = \phi(s) \tag{3}$$

where

$$s = x - ct \tag{4}$$

and the function $\phi(s)$ represents the wave profile. Inserting for $u(x, t)$ in (2) the expression given by (3), we have

$$(c^2 - 1)\,\phi'' = f(\phi, \phi', -c\phi') \qquad (' = d/ds) \tag{5}$$

a second-order ordinary differential equation for ϕ as a function of s, with c as a parameter.

If we assume that $|c| \neq 1$ and introduce an additional variable $\psi = d\phi/ds$, equation (5) may be written as a first-order ordinary differential equation

$$\frac{d\psi}{d\phi} = \frac{f(\phi, \psi, -c\psi)}{(c^2 - 1)\psi} = -\frac{\bar{f}(\phi, \psi, -c\psi)}{\psi} \tag{6}$$

where $\bar{f} \equiv -f/(c^2 - 1)$. Given the function $f(u, u_x, u_t)$, for each value of c different from ± 1 equation (6) defines a direction field in a ϕ, ψ-(or phase) plane. If we think of s as a time variable, equation (5), written in the form $\phi'' = -\bar{f}$, has a simple dynamical interpretation. It is the equation of motion of a unit mass subject to the force $-\bar{f}$, where ϕ is the position coordinate of the mass on the line of motion.

The general procedure is now clear. For a given right-hand member f in equation (2), wave solutions of (2) are associated with solutions $\phi(s)$ of (5), or equivalently, when $|c| \neq 1$, with solution $\phi(s)$, $\psi(s)$ and integral curves of (6). In particular, periodic waves correspond to periodic solutions of the ordinary differential equations and thus to closed integral curves in the phase plane. Depending upon the nature of the function $f(u, u_x, u_t)$, one may be able to work directly with equation (5), by quadratures, or else, if this is not possible, one can always obtain qualitative information about the solutions by analyzing the direction field defined by equation (6). In the former category, for instance, are all the examples cited in Section 1.

In sections 3, 4 and 5, assuming $|c| \neq 1$, we present several results concerning wave propagation for certain classes of functions $f(u, u_x, u_t)$, derived by use of one or the other of the procedures just mentioned.

First, however, we comment briefly on the case $|c| = 1$. We then have a singular perturbation problem for equation (5), and to obtain non-trivial results, we must admit for consideration discontinuous solutions. To illustrate, suppose $f = -u + u^3$. It may then be shown (see Fleishman [*2*]) that there exist certain *supercritical* nonperiodic (but bounded) waves of the form

$$\phi(x - ct) = \pm \tanh \left[\left(\frac{1}{2(c^2 - 1)}\right)^{1/2} (x - ct)\right] \qquad (|\, c \,| > 1)$$

Clearly as $|\, c \,|$ approaches 1, these wave profiles assume the forms of step-functions.

3. The Case $f = f(u)$

When the right member $f(u, u_x, u_t)$ of (2) is a function of u alone, exact and simple results may be obtained. In this case equation (5) assumes the form

$$\frac{d^2\phi}{ds^2} = -\bar{f}(\phi). \tag{7}$$

In terms of the dynamical interpretation mentioned above, equation (7) governs a conservative system: the unit mass is acted on only by a restoring force or "spring" force.

Multiplying (7) through by $\psi = d\phi/ds$ and integrating from s_0 to s, we obtain an "energy integral"

$$\frac{\psi^2}{2} + \int_0^{\phi} \bar{f}(\phi')\, d\phi' = \frac{\psi_0^2}{2} + \int_0^{\phi_0} \bar{f}(\phi')\, d\phi' = h. \tag{8}$$

With $\psi = d\phi/ds$ interpreted as velocity, equation (8) is a statement of the law of conservation of energy, with h equal to the total energy of the system.

For given $\bar{f}(u)$ the relation (8) defines a family of curves (energy or integral curves) in the ϕ, ψ-plane. A given initial state (ϕ_0, ψ_0) fixes the value of h, thus defining an integral curve.

The family of solutions $\phi = \phi(s)$ associated with a particular integral curve may be obtained in inverse form by solving (8) for $\psi = d\phi/ds$, separating the variables ϕ and s and then integrating. Furthermore, if the curve is closed and thus corresponds to a periodic solution, an expression for the period is obtained by integrating between appropriate limits. Details may be found in the book by Stoker [*4*, Chapter 2].

As an illustration of the kind of results which follow from the foregoing analysis, we now state some properties of wave solutions of (2) when $f(u)$ is a function of a certain type. If $f(u)$ is an odd function of u, continuous for all u, and such that $uf(u) > 0$ when $u \neq 0$, we shall say that $f(u)$ possesses the property P, written "$f \in P$". For such functions the following result holds.

Suppose that in the non-simple wave equation (2) $f = f(u)$ *and* $f(u) \in P$. *Then all subcritical* ($|c| < 1$) *wave solutions* $u = \phi(x - ct)$ *are periodic, and for each subcritical speed* $|c|$ *there is a one-parameter family of wave forms in which either the amplitude* ϕ_m *or the wave length L may be chosen arbitrarily, while the other is then fixed by the relation*

$$L = 2\sqrt{2(1-c^2)} \int_0^{\phi_m} \left[\int_\phi^{\phi_m} f(\phi')\,d\phi'\right]^{-1/2} d\phi \tag{9}$$

On the other hand there are no periodic or even bounded supercritical ($|c| > 1$) *waves.*

Finally, note that if we introduce a new independent variable $\sigma = s/\sqrt{1-c^2}$ in (7) (assuming c is subcritical), the equation takes the form $d^2\phi/d\sigma^2 = -f(\phi)$. This means that for a given $f(u)$ in equation (2) there exists a "universal" one-parameter family of wave profiles: the set of profiles traveling at any subcritical speed $|c|$ is obtainable from the "universal" set through a "stretching" transformation of the s-axis. Similar considerations hold for supercritical waves.

4. A Similarity Principle When $f = f(u, u_x)$

The statement just above, which says that when $f(u) \in P$ all subcritical waves are periodic, etc., rests on the fact that when $f(u) \in P$ and $|c| < 1$, $\bar{f}(\phi) \in P$. The latter is also true, however, when $-f(u) \in P$ and $|c| > 1$, so that when $-f(u) \in P$ all *supercritical* waves are periodic, all subcritical waves are unbounded, and one may obtain a relation, very similar to (9), between the wave length L, amplitude ϕ_m and (supercritical) wave velocity c.

This exchange of roles of sub- and supercritical waves when the sign of $f(u)$ is changed, does not require $f(u) \in P$, for in general, we see from (6), changing the signs of $c^2 - 1$ and $f(u)$ leaves $\bar{f}(u)$—and therefore the wave profiles—unchanged. Indeed, this so-called "duality" (see ref. [2] for an exact statement) holds also when f depends on u_x as well as on u.

It is not hard to see that the following "similarity principle" generalizes the preceding considerations.

Suppose that two functions of u *and* u_x, f_1 *and* f_2, *have a constant non-zero ratio* λ: $f_2(u, u_x) = \lambda f_1(u, u_x)$. *Then the same set of wave profiles which propagate at speed* $|c_1|$ *when* $f = f_1$ *in equation* (2), *will propagate at speed* $|c_2|$ *when* $f = f_2$, *if and only if* $c_2^2 - 1 = \lambda(c_1^2 - 1)$.

5. Singular Points and Periodic Waves

When the function $f(u, u_x, u_t)$ depends on u_x and/or u_t as well as on u, equation (5) does not, in general, yield an energy integral, and one must rely heavily on phase-plane methods. In this section a known relationship between singular points and periodic solutions is exploited to yield a necessary condition for the existence of periodic waves for (2). Again it is assumed that $c^2 \neq 1$.

As mentioned, equation (6), $d\psi/d\phi = -\bar{f}/\psi$, defines a direction field in the phase plane: at any point (ϕ, ψ) where the numerator and denominator of the right-hand side do not vanish simultaneously, they are the components of a vector which specifies the direction that an integral curve must have when passing through the point. Those points (ϕ, ψ) at which the numerator and denominator vanish simultaneously are called *singular points* of the equation; they correspond to points of equilibrium of the associated dynamical system.

In order, then, that (ϕ, ψ) be a singular point of (6), it is necessary, first of all, that $\psi = 0$; that is, the singular points of (6), if any, must lie on the ϕ-axis. The numerator of the right member of (6) vanishes simultaneously if and only if $\bar{f}(\phi, 0, 0) = 0$, or, since $c^2 \neq 1$, if and only if

$$f(\phi, 0, 0) = 0. \tag{10}$$

Thus, all singular points lie on the ϕ-axis, and their abscissas must satisfy equation (11).

For a detailed discussion and classification of singular points the reader is referred to any one of a number of excellent treatments of the subject, e.g., to Stoker [*4*] or Coddington and Levinson [*1*]. Here we wish merely to recall a well-known result: in order for equation (6) to possess closed integral curves, it is necessary that it possess singular points. Closed integral curves, however, are associated with periodic solutions of equation (5), which, in turn, we have seen, correspond to wave solutions of (2) with periodic profile. These considerations yield the following result.

For a given function $f(u, u_x, u_t)$ *equation* (2) *can have periodic wave solutions with propagation speeds* $|c| \neq 1$ *only if the algebraic equation* $f(u, 0, 0) = 0$ *possesses real roots.*

If $f(u, u_x, u_t)$ is a polynomial, the result can be strengthened. Now if equation (10) holds for some real value of ϕ, either it is satisfied identically or else it has a finite number of discrete roots. In the former case $f(\phi, \psi, -c\psi)$ must possess a factor ψ^m (m a positive integer), in which case the singularities are "removable," i.e., the direction field is structured as if it has no singularities, so that no closed integral curves exist. We have, then, the following result.

Suppose $f(u, u_x, u_t)$ *is a polynomial. Then equation* (2) *can have periodic wave solutions with propagation speeds* $|c| \neq 1$ *only if the equation* $f(u, 0, 0) = 0$ *possesses discrete roots.*

A similar statement can be made for analytic functions $f(u, u_x, u_t)$.

A possible application in reverse of the foregoing results and methods is that if one has reason to believe that a particular physical situation is governed by an equation of the form (2), and one wishes to determine the function f, then observation of the types of waves which actually propagate in the medium may yield information about the nature of this function.

REFERENCES

[*1*] Coddington, E. A., and Levinson, N., "Theory of Ordinary Differential Equations." McGraw-Hill, New York, 1955.

[*2*] Fleishman, B. A., Progressing waves in an infinite nonlinear string. *Proc. AMS* **10**, 329-334 (1959).

[*3*] Morse, P. M., and Feshbach, H., "Methods of Theoretical Physics," Part I. McGraw-Hill, New York, 1953.

[*4*] Stoker, J. J., "Nonlinear Vibrations." Interscience, New York, 1950.

[*5*] van der Pol, B., On a generalization of the nonlinear differential equation $d^2u/dt^2 - \epsilon(1 - u^2)\, du/dt + u = 0$. *Koninkl. Ned. Akad. Wetenschap., Indag. Math.* **19**, 477-480 (1957).

Some Nonlinear Systems of Differential Equations Equivalent to Linear Systems[1]

F. G. DRESSEL AND J. J. GERGEN
Department of Mathematics
Duke University, Durham, North Carolina

1. Introduction

Pinney [*2*] showed that the solution of the nonlinear differential equation $y'' + p(x)\, y + cy^{-3} = 0$ can be written in the form $y = \sqrt{u^2 - v^2}$, where u and v are appropriately selected solutions of the linear equation $u'' + pu = 0$. This result led Thomas [*3*] to pose the problem: What equations of order n have general solutions expressible in the form $y = F(u_1, ..., u_n)$, where $u_1, ..., u_n$ constitute a variable set of solutions of a fixed linear differential equation? Thomas gave the complete answer to this problem when the underlying linear equation is of first order, $y' + py = q$. He also gave the answer for homogeneous second order equations, $u'' + pu' + qu = 0$, when F depends on only one u, or when F is homogeneous of nonzero degree in two u's. Herbst [*1*] showed that these restrictions on the form of F could be removed. He thus gave the complete answer to the problem for linear homogeneous equations of the second order.

In the present paper we consider a similar problem for systems of two and three equations. Our results for the case $n = 2$, given in § 2, closely parallel the results of Herbst and Thomas for linear homogeneous second order equations. Turning next, in § 3, to linear systems in three unknowns, we obtain, except for a special case, information of a negative nature. That is, our results for $n = 2$ cannot be extended to systems in three unknowns within the broad class of solution forms examined here.

2. Two-Dimensional Linear Systems

Let us suppose that the fixed functions $F(y_1, y_2)$, $G_i(X, Y, a_{ij}; W)$, $j,i = 1, 2$, are such that

$$R(y_1, y_2) \equiv y_i F_i(y_1, y_2) \neq 0 \qquad (F_i = \partial F/\partial y_i, y_i F_i = y_1 F_1 + y_2 F_2), \quad (2.1)$$

[1] This research was supported by the United States Air Force through the Office of Scientific Research of the Air Research and Development Command under contract No. AF 49(638)-892.

in the domain under consideration, and that whenever the $a_{ij} = a_{ij}(t)$ are given and $(x_i(t), y_i(t))$, $i = 1, 2$, are solutions of the linear equations

$$dx/dt = a_{11}x + a_{12}y, \qquad dy/dt = a_{21}x + a_{22}y, \tag{2.2}$$

then

$$Y = F(y_1, y_2), \qquad X = x_i F_i \tag{2.3}$$

satisfy

$$dX/dt = G_1(X, Y; a_{ij}; W), \qquad dY/dt = G_2(X, Y; a_{ij}; W), \tag{2.4}$$

where W is the Wronskian of the system (x_i, y_i), $W = x_1y_2 - x_2y_1$. We seek the form of G_1, G_2.

Consider a system (2.2) with solutions (x_i, y_i). On differentiation in (2.3) we obtain from (2.2) and (2.4),

$$dX/dt = G_1(X, Y; a_{ij}; W) = a_{11}X + a_{12}R + a_{21}x_ix_jF_{ij} + a_{22}x_iy_jF_{ij}, \tag{2.5}$$

$$dY/dt = G_2(X, Y; a_{ij}; W) = a_{21}X + a_{22}R. \tag{2.6}$$

Take $a_{22} = 1$, $a_{11} = a_{12} = a_{21} = 0$. Let the initial values of x_i be zero, and the initial values of y_i be y_i^0. Referring to (2.6) we obtain

$$G_2(0, F(y_1^0, y_2^0); a_{ij}; 0) = y_i^0F_i(y_1^0, y_2^0) = R(y_1^0, y_2^0).$$

We infer that R is a function $R = R(F)$ of F. Further, from $R = R(F)$ we obtain

$$y_jF_{ij}(y_1, y_2) = (R' - 1)\,F_i \qquad (i = 1, 2;\, R' = dR/dF). \tag{2.7}$$

which gives

$$x_iy_jF_{ij} = (R' - 1)\,X. \tag{2.8}$$

It remains to consider $x_ix_jF_{ij}$.

Following Thomas [3], we have

$$x_1R = y_1X + F_2W, \qquad x_2R = y_2X - F_1W,$$

which, when used in conjunction with (2.7), give

$$x_ix_jF_{ij} = X^2A(F) + W^2C, \tag{2.9}$$

where

$$R^2C = F_1^2F_{22} - 2F_1F_2F_{12} + F_2^2F_{11}, \qquad A = (R' - 1)/R. \tag{2.10}$$

From (2.10) it follows that $C = C(y_1, y_2)$ is a function of y_1, y_2 alone.

Take $a_{21} = 1$, $a_{11} = a_{12} = a_{22} = 0$, and for each (y_1^0, y_2^0), select (x_1^0, x_2^0) so that

$$X_0 = x_i^0 F_i(y_1^0, y_2^0) = 1, \qquad x_1^0 y_2^0 - x_2^0 y_1^0 = 1, \tag{2.11}$$

which is possible by (2.1). From (2.9) and (2.5) we then obtain

$$G_1(0, F(y_1^0, y_2^0); a_{ij}; 1) = A(F) + C. \tag{2.12}$$

We infer from (2.12) that $C = C(F)$ is a function of F.

From (2.7), (2.8) and (2.10) one sees that F must satisfy the following system of differential equations

$$F_{11} = A(F)F_1^2 + y_2^2 C(F), \qquad F_{12} = AF_1F_2 - y_1y_2C, \qquad F_{22} = AF_2^2 + y_1^2 C,$$
$$y_1F_1 + y_2F_2 = R(F). \tag{2.13}$$

Herbst found that for the system (2.13) to have a solution it is necessary and sufficient that the R, A, C satisfy

$$RC' + (3 - AR)C = 0, \qquad R' - AR = 1. \tag{2.14}$$

Thus we may state the following theorem.

Theorem 1. Suppose that the functions $F(y_1, y_2)$, $G_i(X, Y; a_{ij}; W)$, $i = 1, 2$, are such that $y_iF_i \neq 0$ in the domain under consideration and that whenever the $a_{ij} = a_{ij}(t)$ are given and the (x_i, y_i), $i = 1, 2$, are solutions of the system of linear equations

$$dx/dt = a_{11}x + a_{12}y, \qquad dy/dt = a_{21}x + a_{22}y, \tag{2.15}$$

then

$$X = x_iF_i(y_1, y_2), \qquad Y = F(y_1, y_2) \tag{2.16}$$

satisfy

$$dX/dt = G_1(X, y; a_{ij}; W), \qquad dY/dt = G_2(X, Y; a_{ij}; W), \tag{2.17}$$

with $W = x_1y_2 - x_2y_1$. Then the equations (2.17) have the form

$$dX/dt = a_{11}X + a_{12}R(Y) + a_{21}(A(Y)X^2 + C(Y)W^2) + a_{22}ARX,$$
$$dY/dt = a_{21}X + a_{22}R, \tag{2.18}$$

where R, A, C satisfy $R \neq 0$ and (2.14).

Using the sufficiency part of the Herbst results on the system (2.13) and the fact that for a system of equations (2.15) the Wronskian W of any pair of equations satisfies $W = k \exp(P(t))$, where $P' = a_{11} + a_{22}$ and k is a constant, it is possible to obtain the following further result.

Theorem 2. Given equations of the form

$$dX/dt = a_{11}X + a_{12}R(Y) + a_{21}(A(Y)X^2 + C(Y)\, e^{2P(t)}) + a_{22}ARX,$$

$$dY/dt = a_{21}X + a_{22}R, \qquad P' = a_{11} + a_{22}, \tag{2.19}$$

with $R \neq 0$, and R, A, C satisfying (2.14), and given initial values X_0, Y_0 of X, Y, then there is a solution of the system (2.13) such that

$$X = x_i F_i(y_1, y_2), \qquad Y = F(y_1, y_2),$$

where (x_i, y_i) are properly chosen solutions of the linear system (2.15), have initial values X_0, Y_0, and satisfy (2.19). That is, the solutions of (2.13) provide with the solutions of the linear equations (2.15) a general solution of (2.19) for every set of a_{ij}.

3. Three-Dimensional Linear Equations

Let (x_i, y_i, z_i), $i = 1, 2, 3$, be solutions of the linear equations

$$dx/dt = a_{11}x + a_{12}y + a_{13}z, \qquad dy/dt = a_{21}x + a_{22}y + a_{23}z,$$

$$dz/dt = a_{31}x + a_{32}y + a_{33}z, \tag{3.1}$$

where the coefficients a_{ij} may be functions of t. Let W denote the Wronskian of the solutions (x_i, y_i, z_i),

$$W = \begin{vmatrix} x_1 & x_2 & x_3 \\ y_1 & y_2 & y_3 \\ z_1 & z_2 & z_3 \end{vmatrix}.$$

The results of § 2 suggest that if one takes new variables X, Y, Z of the form

$$X = x_i F_i(y_1, y_2, y_3), \qquad Y = F(y_1, y_2, y_3)$$

$$Z = \alpha_1 z_i F_i + \alpha_2 x_i x_j F_{ij} + \alpha_3 x_i z_j F_{ij} + \alpha_4 x_i y_j F_{ij} \tag{3.2}$$

$$+ \alpha_5 y_i z_j F_{ij} + \alpha_6 z_i z_j F_{ij} + \alpha_7 y_i y_j F_{ij},$$

where the constants α_i are either 0 or 1, then under certain restrictions on the function F these new variables would satisfy nonlinear equations of the type

$$dX/dt = a_{1i}H_i + a_{2i}I_i, \quad dY/dt = a_{2i}G_i, \quad dZ/dt = a_{3i}J_i + a_{3i}K_i + Q, \tag{3.3}$$

where the functions H_i, I_i, J_i and K_i are of the same form as that noted below for G_i.

$$G_i = G_{i1}X + G_{i2}Z + G_{i3}W + G_{i4}X^2 + G_{i5}Z^2 + G_{i6}W^2 + G_{i7}XZ$$
$$+ G_{i8}XW + G_{i9}ZW + G_{i10},$$
$$G_{ik} = G_{ik}(Y) \qquad (i = 1, 2, 3, k = 1, ..., 10). \tag{3.4}$$

In the last equation of (3.3) the function Q contains those terms that result from differentiating $(Z - \alpha_1 z_i F_i)$ with respect to t. The form of Q will not be needed here.

In order to avoid the trivial case, F a constant, and to avoid returning to the linear case (3.1), we assume that

$$F_{ij} \neq 0 \tag{3.5}$$

for some values of i and j.

In case (3.1) takes the special form

$$dx/dt = z, \qquad dy/dt = x, \qquad dz/dt = a_{31}x + a_{32}y + a_{33}z,$$

so that the third order linear equation

$$y''' = a_{33}y'' + a_{31}y' + a_{32}y$$

is under consideration, the general problem posed by Thomas (see Introduction) would prompt one to take $Z = z_i F_i + x_i x_j F_{ij}$. That is the form of Z that results when $\alpha_1 = \alpha_2 = 1$ and all other α_i in (3.2) are put equal to zero. However, it can be shown that for X, Y, Z to be solutions of the nonlinear system (3.3) that α_2, as well as α_3 and α_6, must be zero and the restrictions on F are such that all terms in Z, except the first, are functions of X and Y. These facts and other information on the three-dimensional case are contained in the next two theorems. Proofs of these theorems are quite long and do not appear in this article.

Theorem 3. Suppose that the function $F(y_1, y_2, y_3)$ is such that in the domain under consideration (3.5) holds and that whenever the $a_{ij} = a_{ij}(t)$ are given and the (x_i, y_i, z_i), $i = 1, 2, 3$, are solutions of the system of linear equations (3.1), then X, Y, Z as given by (3.2) satisfy equations of the form (3.3). Then F has the properties that there exist functions R and S such that

$$y_i F_i(y_1, y_2, y_3) = R(F), \qquad F_{ij} = S(F)\, F_i F_j, \tag{3.6}$$

and Z has the representation

$$Z = (\alpha_1 + \alpha_5 S)\, z_i F_i + \alpha_4 RSX + \alpha_7 SR^2.$$

If one takes $Z = \beta_1(Y)\, z_i F_i + \beta_2(Y)\, X + \beta_3(Y)$, it is readily shown that X, Y and this Z satisfy equations of the form (3.3) with only minor restrictions on the functions β_1, β_2, β_3 if F meets the conditions (3.6). We treat here only the special case of this form of Z, namely $Z = z_i F_i$.

Theorem 4. Whenever the $a_{ij} = a_{ij}(t)$ are given, let (x_i, y_i, z_i), $i = 1, 2, 3$, denote solutions of the linear system (3.1). The system of differential equations (3.3) has a general solution expressible in the form

$$X = x_i F_i(y_1, y_2, y_3), \qquad Y = F(y_1, y_2, y_3), \qquad Z = z_i F_i(y_1, y_2, y_3) \tag{3.7}$$

if and only if

$$G_1 = G_2 = H_1 = H_2 = J_1 = J_2 = R(Y),\ G_3 = H_3 = J_3 = Z,\ I_1 = S(Y)X^2,$$
$$I_2 = SRX, \quad I_3 = K_1 = SXZ, \quad K_2 = SRZ, \quad K_3 = SZ^2, \quad Q = 0, \tag{3.8}$$

where the functions S and R satisfy the ordinary differential equation

$$R'(F) = S(F)R + 1 \tag{3.9}$$

and $F(y_1, y_2, y_3)$ is a solution of the system of partial differential equations

$$F_{ij} = S(F)F_i F_j, \qquad F_1 = (R(F) - y_2 F_2 - y_3 F_3)/y_1, \qquad R \neq 0. \tag{3.10}$$

REFERENCES

[1] Herbst, R. T., The equivalence of linear and nonlinear differential equations. *Proc. AMS* **7**, 95-97 (1956).

[2] Pinney, E., The nonlinear differential equation $y'' + p(x)\, y + cy^{-3} = 0$. *Proc. AMS* **1**, 681 (1950).

[3] Thomas, J. M., Equations equivalent to a linear differential equation. *Proc. AMS* **3**, 899-903 (1952).

A Class of Characteristic-Value Problems

RICHARD M. MORONEY
Massachusetts Institute of Technology, Cambridge, Massachusetts

1. The question of interest here is that of the existence of real numbers λ_i such that a solution $x(\lambda, t)$ on $0 \leq t \leq 1$ of

$$x'' + xF(x, \lambda, t) = 0 \tag{1}$$

satisfying the initial conditions

$$x(\lambda, 0) = 0 \qquad x'(\lambda, 0) = 1 \tag{2}$$

also satisfies, when $\lambda = \lambda_i$, the end condition

$$x(\lambda_i, 1) = 0. \tag{3}$$

Throughout this paper all variables are supposed to be real and all functions real-single-valued.

The above problem has been treated by Nehari [*2*] for the case where $F(x, \lambda, t) = G(\lambda^2 x^2, t)$. By taking $F(x, \lambda, t) = \lambda + q(t)$ the classical Sturm-Liouville problem results. (For a detailed treatment of what follows, see ref. [*1*].)

Our goal is to prove a theorem of the following form, i.e. to exhibit a hypothesis 5 for which the following theorem is true:

Theorem. Let D be the set $\{(x, \lambda, t) : -\infty < x < \infty,\ \alpha \leq \lambda < \infty,\ 0 \leq t \leq 1\}$. Let $F(x, \lambda, t)$ be defined on D and satisfy the following hypotheses:

1. $F(x, \lambda, t)$ is continuous in (x, λ, t) on D.
2. For each fixed λ, $F(x, \lambda, t) \in Lip(x)$ uniformly in t.
3. $F(x, \lambda, t) \geq 0$ and $F = 0$ only if $x = 0$.
4. For fixed λ and t

$$x_2 > x_1 \geq 0 \text{ or } x_2 < x_1 \leq 0 \Rightarrow F(x_2, \lambda, t) \geq F(x_1, \lambda, t).$$

5. (Hypothesis describing the behavior of F as a function of λ).

Then

(a) For each $\lambda \in [\alpha, \infty)$ there exists a unique solution of (1) + (2) on $0 \leq t \leq 1$.

(b) There exists a sequence $\{\lambda_i\}$, $\alpha \leq \lambda_i \to \infty$, such that for $\lambda = \lambda_i$ the solution as in (a) satisfies (3).

Remarks 1. *Some* fifth hypothesis is essential, since F could be independent of λ and still satisfy hypotheses 1-4.

2. If F is independent of x, i.e. if (1) is linear, hypothesis 4 is satisfied. Thus this hypothesis may be viewed as "describing" or "restraining" the nonlinearity of (1). Hypothesis 3, admittedly drastic, is the keystone of the theory. As an instance, it assures the existence in the large of solutions of (1) + (2).

3. By lengthening the arguments to follow one may show that hypotheses 3 and 4 need only hold for large λ. Both of these hypotheses allow a pleasant physical interpretation if F in (1) is viewed as a "spring constant".

4. Conditions (2) and (3) can be replaced by more general end conditions, and the number of zeros in (0, 1) of $x(\lambda_i, t)$ can be determined.

5. To save space it will be assumed from here on that F is even in x.

2. Let D be as in the Theorem, and let $F(x, \lambda, t)$ be a function defined on D and satisfying the first four hypotheses of the Theorem. In this section two additional properties which F might possess will be defined. In § 3 it will be shown that a satisfactory hypothesis 5 is that F possess one (either) of these properties.

Let P denote the collection of functions $g(s)$ which are defined and positive for sufficiently large s and tend to zero as $s \to \infty$. For some ρ in (0, 1] define $\check{F}_\rho(x, \lambda)$ and $\hat{F}_\rho(x, \lambda)$ by

$$\check{F}_\rho(x, \lambda) = \sup_{0 \leq t \leq \rho} F(x, \lambda, t) \qquad \hat{F}_\rho(x, \lambda) = \inf_{0 \leq t \leq \rho} F(x, \lambda, t)$$

and, for neatness, drop the ρ subscript.

Definition. $F(x, \lambda, t)$ possesses the *sequential property* if

i) For every $x \neq 0$, $\lim_{\lambda \to \infty} \hat{F}(x, \lambda) = \infty$.

ii) If $g(s) \in P$, and $\lim_{s \to \infty} \hat{F}(g(s), s) \neq \infty$, then $\lim_{s \to \infty} \hat{F}(\bar{g}(s), s) \neq \infty$ where $\bar{g}(s) = g(s)\, \check{F}(g(s), s)$.

Example 1. If $F(x, \lambda, t) = \lambda + q(x, \lambda, t)$ with $q(x, \lambda, t)$ uniformly bounded, then F has the sequential property: condition (ii) above is true because there is no $g \in P$ with $\lim_{s\to\infty} \hat{F}(g(s), s) \neq \infty$.

Example 2. Let $G(u, v)$ be continuous in both variables on

$$\{-\infty < u < \infty, 0 \leq v \leq 1\},$$

and let $\lim_{u\to\infty} G(u, v) = \infty$ uniformly in v. Then $F(x, \lambda, t) = G(\lambda^2 x^2, t)$ has the sequential property.

(Note that

$$\hat{G}(s^2g^2(s)) \nrightarrow \infty \Rightarrow sg(s) \nrightarrow \infty \Rightarrow \check{G}(s^2g^2(s)) \nrightarrow \infty \Rightarrow g(s)\, \check{G}(s^2g^2) = \bar{g}(s) < Mg$$

for some $M \Rightarrow \hat{G}(s^2\bar{g}^2) \nrightarrow \infty$.)

Definition. $F(x, \lambda, t)$ has the *finite property* if for every positive integer n there exist $n + 2$ functions in P: $g(s)$, $C_0(s)$, $C_1(s)$, $\cdots$ $C_n(s)$ such that

i) $\lim_{s\to\infty} g^3(s) \hat{F}(C_j(s), s) = \infty$ $(j = 0, 1, ..., n)$

ii) For all sufficiently-large s and $j = 0, 1, ..., n - 1$

$$20g(s)\, C_{j+1}(s)\, \check{F}(C_{j+1}(s), s) < C_j(s).$$

Example 3. If $F(x, \lambda, t) = \lambda(x^2)^{(3+t)/4}$ then F has the finite property *but not the sequential property.*

1. Given n a positive integer, take $p = (2/7)^{n+2}$ and put $C_n(\lambda) = \lambda^{p-0.5}$. Then take $C_{n-1}(\lambda) = \lambda^{q-0.5}$ where q is the smallest number such that $C_n(\lambda)\, \lambda[C_n(\lambda)]^{3/2} \leq C_{n-1}(\lambda)$. It is easily seen that $q < (2/7)^{n+1}$ and that the other C_j can be defined recursively by a similar scheme. For sufficiently large s and any $g(s) \in P$, $\hat{F}(g(s), s) = sg^2(s)$ and $\check{F}(g(s), s) = sg^{3/2}$. With the above $C_j(s)$ it is easy to choose a g that causes (i) and (ii) to be satisfied.

2. $s^{-0.5} \in P$ violates, for this F, the second condition on functions with the sequential property.

Lemma. If F has the sequential property then it has the finite property.

Sketch of proof. If F has the sequential property (and satisfies the other hypotheses of the Theorem—this is understood throughout this section) it is easy to find a $\varphi(s) \in P$ such that

$$\lim_{s\to\infty} \hat{F}(\varphi(s), s) = \infty;$$

one solves the equation

$$\hat{F}(s, h) = 1/s$$

for $h(s)$ and takes $\varphi(s) = h^{-1}(s)$. Then, given a positive integer n, take $C_0(s) = \varphi(s)$ and solve recursively the equations

$$C_j(s) = C_{j+1}(s)\,\check{F}(C_{j+1}(s), s). \qquad (j = 0, \ldots, n-1).$$

The resulting $\{C_j(s)\}$, and an easily-chosen $g(s)$, satisfy the finite property conditions.

Remark: In view of the Lemma, it is enough to prove the Theorem when hypothesis 5 reads "F has the finite property". Example 3 in conjunction with the Lemma shows that the sequential property need not have entered the discussion at all. It has been retained because of its greater simplicity and because, as will be seen, it is "sharp" in a certain sense.

3. In this section we suppose that hypothesis 5 of the Theorem has been inserted as "5. F has the finite property for some ρ in $(0, 1]$" and now proceed to outline the proof of the Theorem.

1. Define $F(x, \lambda, t)$ for $t > 1$ by $F(x, \lambda, t) = F(x, \lambda, 1)$.

2. For this extended F, every solution of (1) + (2) exists, is unique, and has infinitely many zeros, on $0 \leq t < \infty$.

3. Define $\theta(\lambda, t)$ as usual by

$$\theta(\lambda, t) = \int_0^t \frac{[x'(\lambda, s)]^2 + [x(\lambda, s)]^2\, F(x(\lambda, s), \lambda, s)}{[x(\lambda, s)]^2 + [x'(\lambda, s)]^2}\, ds \tag{4}$$

and note that

$$\theta(\lambda, t) = \tan^{-1}\left[\frac{x(\lambda, t)}{x'(\lambda, t)}\right] \tag{5}$$

$$\theta'(\lambda, t) > 0. \tag{6}$$

4. $\theta(\lambda, 1)$ is continuous in λ, and $x(\lambda, 1) = 0 \Leftrightarrow \theta(\lambda, 1) = n\pi$. Thus the proof will be complete if it is shown that $\theta(\lambda, \rho) \to \infty$, and it is enough to show that for any positive integer n there exists a λ_n such that $\lambda \geq \lambda_n \Rightarrow \theta(\lambda, \rho) \geq n\pi$.

5. Given n, choose λ_n so large that all of the asterisked inequalities below hold when $\lambda \geq \lambda_n$. Let λ be any fixed number $\geq \lambda_n$. In what follows, λ is not written i.e. $x(t) \equiv x(\lambda, t)$, $\theta(t) \equiv \theta(\lambda, t)$ etc.

6. Since $\sin \theta(0) = 0$, follow $x(t)$ to $w > 0$, the smallest number such that $\sin \theta(w) = g$. This is possible because

$$g <^* 1.$$

Also, on $[0, w]$ $\theta' \geq \cos^2 \theta$ hence

$$w <^* 2g.$$

7. $x(w) > C_1$, since otherwise

$$x'(0) \leqslant x'(w) + 2g\, C_1\, \check{F}(C_1)$$

but

$$x/x'(w) = \tan \theta(w) > \sin \theta(w) = g$$

so

$$x'(w) < C_1/g <^* g\, C_1\, \check{F}(C_1)$$

hence

$$x'(0) < 4g\, C_1\, \check{F}(C_1) <^* C_0 <^* 1.$$

8. Follow $x(t)$ until v, the smallest number $> w$ such that $\sin \theta(v) = g$. There exists a $\check{t}$, $w < \check{t} < v$, such that $\sin \theta(t) = 1$. Now $x(v) > C_n$. If not there would be a k, $\check{t} < x < v$, such that $x(k) = C_n$ and $\sin \theta(k) > g$—because

$$x(t) > x(w) > C_1 >^* C_n.$$

Between w and k

$$\theta' \geq \sin^2 \theta F(x) \geq g^2 \hat{F}(C_n)$$

hence, since $\theta(k) - \theta(w) < \pi$,

$$k - \check{t} < k - w < \pi/g^2 \hat{F}(C_n)$$

and

$$x(\check{t}) < x(k) + (k - \check{t}) \mid x'(k) \mid < C_n + \pi C_n/g^3 \hat{F}(C_n) <^* C_1.$$

9. Since $x(v) > C_1$, $v - w < \pi/g^2 \hat{F}(C_1) <^* g$, so $v < 3g$.

10. Let t_1 be the first zero of $x(t)$. As in 5

$$t_1 - v <^* 2g$$

hence

$$t_1 < 5g <^* \rho/n.$$

11. $|x'(t_1)| > x(\check{t})/(t_1 - \check{t}) > C_1/5g$.

12. An inductive argument shows

$$t_j - t_{j-1} < \rho/n$$

$$|x'(t_j)| > C_j/5g$$

hence

$$t_n < \rho.$$

4. The function

$$F(x, \lambda, t) = (\lambda + 2)^{-t} \log (\lambda + 2)$$

has the sequential property for $\rho = 0$ but not for any positive ρ, and it is easily seen that with this F the problem (1) + (2) + (3) has no solution (here $\alpha = 0$). F *does* satisfy all the other hypotheses of the Theorem, so the fifth hypothesis is "sharp" in the sense that if it "just fails" (holds for $\rho = 0$ but no positive ρ) the conclusions of the Theorem may not hold.

5. We consider in this section the following problem P:

$$(P) \qquad \begin{gathered} x'' + x[\lambda + q(x, \lambda, t)] = 0 \\ x(0) = 0 \qquad x(1) = 0, \end{gathered}$$

where $q(x, \lambda, t)$ is supposed to be defined on D and satisfy there the following hypotheses:

1. $q(x, \lambda, t)$ is continuous in (x, λ, t) and bounded.
2. $q(x, \lambda, t) \in \text{Lip}\,(x)$.
3. For sufficiently-large λ and each fixed t

$$x_2 > x_1 \geq 0 \quad \text{or} \quad x_2 \leqslant x_1 \leqslant 0 \Rightarrow q(x_2, \lambda, t) \geq q(x_1, \lambda, t).$$

4. There exists a positive number r such that

$$|x|^{-r} q(x, \lambda, t) \to 0 \quad \text{as} \quad |x| \to 0 \text{ uniformly in } t \text{ and } \lambda.$$

Because of this last hypothesis, the linear Sturm-Liouville problem

$$x'' + x[\lambda + q(t)] = 0$$

$$x(0) = 0 \qquad x(1) = 0$$

is *excluded* but the following Statement, which is true for the linear problem, holds for the nonlinear problem P.

Statement. For every non-negative integer n and every positive number p there exists a λ_{pn} and a solution $x(\lambda_{pn}, t, p)$ of P having n zeros in (0, 1) and such that $x'(\lambda_{pn}, 0, p) = p$. Furthermore, for each n there exists a particular $p = p(n)$ such that the sequence of solutions $\{x(\lambda_{p(n)n}, t, p(n))\}$ admits a sequence $\{y_n(t)\}$ together with which it forms a complete biorthogonal system in $L_2([0, 1])$.

REFERENCES

[*1*] Moroney, R. M., A class of characteristic-value problems. *Trans. AMS* **102**, No. 3, 446-470 (1962).

[*2*] Nehari, Z., Characteristic values associated with a class of nonlinear second-order differential equations. *Acta Math. (Stockh.)* **105**, 141-175 (1961).

Association of Partial Differential Equations with Ordinary Ones and Its Applications[1]

M. Z. v. KRZYWOBLOCKI
Michigan State University, East Lansing, Michigan

Introduction

Dickson in 1924 [*4*] showed that some differential equations can be integrated with the aid of group theory. Birkhoff in 1948 [*2*] suggested that the reduction of independent variables in systems of partial differential equations could be attacked by algebraic methods. Michal undertook the following task: "Given a differential system in the space A of n dimensions, transform this system into the space B of $(n - r)$ dimensions $(1 \leqslant r)$ such that a solution of the system in B determines a solution of it in A. The transformations involved are not to be one-to-one." The resulting theorems involve the operations in the Banach space and some elements of the theory of continuous transformation groups. The Michal work does not provide any usable technique of reducing the number of independent variables. The author of the present work provided such a method of reduction of the number of independent variables from n up to 1, i.e., under suitable conditions this number can be reduced always to one. The application of this technique is demonstrated on various equations of mathematical physics: Laplacian in n-dimensions, Poisson's—Schroedinger—equations, some equations of the viscous flow and boundary layer in four dimensions (space-time) were reduced to ordinary differential equations. The most recent application of this technique is to prove the ergodic theorem of a vibrating membrane. The outlined technique seems to be valuable in the case of partial differential systems with two-curve boundary conditions; in the space of the reduced dimensions this appears as two-point boundary conditions for ordinary differential equations, a problem which is much simpler.

1. Fundamentals of the Method—Michal's Approach

We shall briefly outline the fundamentals of the technique by quoting some results obtained by Michal [*8*, *9*].

[1] The courtesy of the Research Department (Dr. W. Clohessy, Director) of the Martin-Denver Company, in enabling the author to present the paper at the Symposium is gratefully acknowledged.

Michal considers an r-parameter continuous transformation group:

$$\begin{gathered} G_r : \bar{x}^i = f^i(x^1, \dots, x^m ; a^1, \dots, a^r) \equiv f(x ; a), \\ \bar{y} = \varphi(y ; a^1, \dots, a^r) \equiv \varphi(y ; a), \qquad (i = 1, \dots, m, m > 1, r \geqslant 1), \end{gathered} \tag{1.1}$$

where the x's and a's are numerical and y is a variable of class $C^{(n)}$ with n sufficiently large, in a Banach space, B. The group $S_{G_r}(x^i \to \bar{x}^i)$ is assumed to possess $s \geqslant 1$ functionally independent Fréchet differentiable absolute invariants

$$\eta_1(x^1, \dots, x^m), \dots, \eta_s(x^1, \dots, x^m), \tag{1.2}$$

where $s < m$. It is further assumed that G_r possesses a B-valued, Fréchet differentiable, absolute invariant $g(y, x^1, \dots, x^m)$ solvable in y for all admissable y and x's.

Definition 1.1. A B-valued function $y(x^1, \dots, x^m)$ of the numerical variables $(x^1, \dots, x^m)$ is called an invariant function under G_r if, under G_r, y is the same function of the x's as $\bar{y}$ is of the $\bar{x}$'s, i.e. if $y = H(x)$ then $\bar{y} = H(\bar{x})$.

Theorem 1.1.. If B and B_1 are two reflexive Banach spaces (not necessarily distinct, i.e., they may be the same Banach space) and if:

$$\Omega\left(x, y(x), \frac{\partial y}{\partial x}, \dots, \frac{\partial^g y}{\partial x^g}\right), \tag{1.3}$$

is a B_1-valued g-th order differential invariant of invariant B-valued functions $y(x^1, \dots, x^m)$ under the group G_r, then Ω is expressible in the form:

$$\omega\left(\eta, \nu(\eta), \frac{\partial \nu}{\partial \eta}, \dots, \frac{\partial^g \nu}{\partial \eta^g}\right), \tag{1.4}$$

where the corresponding B-valued functions $\nu(\eta_1, \dots, \eta_s)$ exist and are determined by the formula:

$$\nu(\eta_1(x^1, \dots, x^m), \dots, \eta_s(x^1, \dots, x^m)) = g(y(x^1, \dots, x^m), x^1, \dots, x^m). \tag{1.5}$$

An outline of the proof is given in ref. [*9*]. Michal establishes the following partial converse to Theorem 1.1.

Theorem 1.2. If B_1 and B are two Banach spaces (not necessarily distinct) and if $\nu(\eta_1, \dots, \eta_s)$ is an arbitrary B-valued function of the s-invariants $\eta_1, \dots, \eta_s$, (eq. 1.2), then the corresponding B-valued function $y(x^1, ..., x^m)$ determined by equation (1.5) is an invariant function under G_r. In addition, the B_1-valued q'th order differential expression $\omega(\eta, ...)$, (eq. (1.4)), may be expressed as a B_1-valued q'th order differential invariant Ω, (eq. (1.3)), of invariant B-valued functions $y(x^1, ..., x^m)$ under G_r.

Definition 1.2. If $y(x)$ is a solution to the partial differential equation:

$$\Omega\left(x, y(x), \frac{\partial y}{\partial x}, ..., \frac{\partial^g y}{\partial x^g}\right) = 0, \tag{1.6}$$

where Ω is defined in the discussion following equation (1.3) and, in addition $y(x)$ is an invariant function (Def. 1.1) under G_r, then $y(x)$ is said to be an invariant solution of equation (1.6). Theorems 1.1 and 1.2 thus lead to the following theorem concerning partial differential equations.

Theorem 1.3. An invariant solution $y(x^1, ..., x^m)$ of equation (1.6) determines, by means of equation (1.5), a solution $\nu(\eta_1, ..., \eta_s)$ to the corresponding partial differential equation:

$$\omega\left(\eta, \nu(\eta), \frac{\partial \nu}{\partial \eta}, ..., \frac{\partial^g \nu}{\partial \eta^g}\right) = 0, \tag{1.7}$$

where y, $\nu(\eta)$ and $\omega(\eta, \nu(\eta), ...)$ are defined by Theorem 1.1.

Lemma 1. To each solution $\nu(\eta)$ of equation (1.7) corresponds an invariant solution $y(x)$ of equation (1.6). The correspondence is given by equation (1.5).

It is further pointed out that the subject might be related to topological spaces with abstract normed linear coordinates using, for example, the results of ref. [*8*].

2. Morgan's Approach

As was pointed out by Michal, Morgan deals with one-dimensional normed linear spaces and continuous groups in only one numerical parameter (a). Consider the one-parameter group $\Gamma_1(\Gamma_r; r = 1)$ where the x's and y's are numerically valued. The following definition is given by Morgan [*10*].

Definition 2.1. A function Ω of the x's, y's, and the partial derivatives of the y's up to the k-th order is said to be conformally invariant under $\Gamma_1^{E_k}$ (the present paper will drop the superscript E_k) if:

$$\Omega\left(\bar{x}, \bar{y}^1(\bar{x}), \ldots, \bar{y}^n(\bar{x}), \frac{\partial \bar{y}^1}{\partial \bar{x}^1}, \ldots, \frac{\partial^k \bar{y}^n}{\partial (\bar{x}^m)^k}\right)$$

$$= F\left(x, y^1, \ldots, y^n, \ldots, \ldots, \frac{\partial^k y^n}{\partial (x^m)^k}; a\right) \Omega\left(x, y^1(x), \ldots, y^n(x), \frac{\partial y^1}{\partial x^1}, \ldots, \frac{\partial^k y^n}{\partial (x^m)^k}\right), \tag{2.1}$$

under Γ_1 where $F(x, y^1, \ldots, y^n, \ldots, \ldots, \partial^k y^n/\partial(x^m)^k; a)$ is an arbitrary function and Ω is the same function of the original and transformed variables.

If equation (2.1) is satisfied with $F(x, y^1, \ldots, \ldots; a)$ a function of "a" alone, Ω is said to be constant conformally invariant. If $F(x, y^1, \ldots, \ldots; a)$ is equal to one, Ω is said to be absolutely invariant. Since, for the sub-group $S_{r_1}(x^i \to \bar{x}^i)$, the matrix in equation has rank $p = 1$, the group S_{r_1} has $(m - 1)$ functionally independent Fréchet differentiable (in this case we may simply say differentiable) absolute invariants:

$$\eta_1(x^1, \ldots, x^m), \ldots, \eta_{m-1}(x^1, \ldots, x^m). \tag{2.2}$$

Γ_1 has $m + n - 1$ functionally independent absolute invariants. The first $m - 1$ of these invariants may be chosen as the η's (the invariants of the sub-group S_{r_1}) and the other n may be taken as the functionally independent, differentiable, absolute invariants:

$$g_1(y^1, \ldots, y^n, x^1, \ldots, x^m), \ldots, g_n(y^1, \ldots, y^n, x^1, \ldots, x^m). \tag{2.3}$$

Choose a function $h = h(x^1, \ldots, x^m)$ which is functionally independent of the η's (eq. (2.2)), i.e., such that

$$\begin{vmatrix} \dfrac{\partial \eta}{\partial x} \\ \dfrac{\partial h}{\partial x} \end{vmatrix} \neq 0.$$

Since the η's, g's and h are functionally independent, it may then be inferred that:

$$\begin{vmatrix} \dfrac{\partial \eta}{\partial x} & \dfrac{\partial \eta}{\partial y} \\ \dfrac{\partial h}{\partial x} & \dfrac{\partial h}{\partial y} \\ \dfrac{\partial g}{\partial x} & \dfrac{\partial g}{\partial y} \end{vmatrix} \neq 0.$$

However, η is independent of y and therefore $\partial\eta/\partial y^j = 0$ ($j = 1, ..., n$) and, by definition, $\partial h/\partial y^j = 0$ ($j = 1, ..., n$). The determinant in question is, therefore, equal to zero if the determinant $| \partial g/\partial y | = 0$ which is easily verified from the basic properties of matrices and determinants. It therefore follows that one must have:

$$\left| \frac{\partial g}{\partial y} \right| \neq 0, \tag{2.4}$$

i.e., the g's are solvable for the y's in some neighborhood of $(x^1, ..., x^m)$. The η's and g's constitute the $(m + n - 1)$ functionally independent absolute invariants of Γ_1. Consider a system of partial differential equations:

$$\Omega_\lambda \left(x, y^1, ..., y^n, \frac{\partial y^1}{\partial x^1}, ..., \frac{\partial^k y^n}{\partial (x^m)^k}\right) = 0, \qquad (\lambda = 1, ..., f), \tag{2.5}$$

with m independent variables $(x^1, ..., x^m)$ where each of the functions Ω_λ is conformally invariant under Γ_1. Morgan [*10*] proves the following theorem relating to this system of equations.

Theorem 2.1. The invariant solutions (Def. 1.2) of $\Omega_\lambda = 0$ (eq. (2.5)) can be expressed in terms of the solutions of a system of the form:

$$\omega_\lambda \left(\eta, \nu_1, ..., \nu_n, \frac{\partial \nu_1}{\partial \eta_1}, ..., \frac{\partial^k \nu_n}{\partial \eta^k_{m-1}}\right) = 0, \qquad (\lambda = 1, ..., f), \tag{2.6}$$

with "$m - 1$" independent variables $\eta_1, ..., \eta_{m-1}$. The correspondence is determined by n relations of the form:

$$\nu_j(\eta_1, ..., \eta_{m-1}) = g_j(y^1, ..., y^n, x^1, ..., x^m), \qquad (j = 1, ..., n), \tag{2.7}$$

where ν_j's are of class $C^{(t)}$ ($t \geqslant 1$), the η's are the invariants of S_{Γ_1}, (eq. (2.2)), and the g's are the invariants of Γ_1 in the sense of equation (2.3).

Conversely, each solution of $\omega_\lambda = 0$, (eq. (2.6)), determines through equation (2.7) an invariant solution of $\Omega_\lambda = 0$, (eq. (2.5)).

3. Particular Example

Consider the equation:

$$\frac{\partial u}{\partial y} = 0, \qquad u = u(x, y). \tag{3.1}$$

The solution of equation (3.1) is, of course, given by: $u = u(x)$. We now inquire into the invariant solutions of equation (3.1) under:

$$\Gamma_1 : \bar{x} = f^1(x, y; a) = a^j x, \quad \bar{y} = f^2(x, y; a) = a^k y, \quad \bar{u} = f^3(u; a) = a^l u, \tag{3.2}$$

where j, k and l are any real numbers and a ($\neq 0$) is the parameter. From:

$$d\bar{x} = \frac{\partial \bar{x}}{\partial x} dx + \frac{\partial \bar{x}}{\partial y} dy, \qquad d\bar{y} = \frac{\partial \bar{y}}{\partial x} dx + \frac{\partial \bar{y}}{\partial y} dy, \tag{3.3}$$

$$\frac{\partial \bar{u}}{\partial \bar{y}} = \frac{\partial \bar{u}}{\partial x}\frac{\partial x}{\partial \bar{y}} + \frac{\partial \bar{u}}{\partial y}\frac{\partial y}{\partial \bar{y}} = a^{l-k}\frac{\partial u}{\partial y}, \tag{3.4}$$

which shows that equation (3.1) is constant conformally invariant (Def. 2.1) under Γ_1 (eq. (3.2)). Assume that the value of k in equation (3.2) is not equal to zero. The invariants of Γ_1 may be chosen as follows. Set:

$$\eta = xf(y) = \bar{x}f(\bar{y}) = a^j x f(a^k y). \tag{3.5}$$

Equation (3.5) is satisfied for $f(y) = y^{-j/k}$, $f(\bar{y}) = \bar{y}^{-j/k}$, since, in this case:

$$\eta = \bar{x}f(\bar{y}) = \bar{x}\bar{y}^{-j/k} = a^j x(a^k y)^{-j/k} = xy^{-j/k} = xf(y). \tag{3.6}$$

Similarly, by analogy to equations (2.3) and (2.7), choose:

$$g = uh(y) = uy^{-l/k} = \nu(\eta). \tag{3.7}$$

Equations (3.1), (3.4), (3.6) and (3.7) lead to the following equation:

$$\begin{aligned}\frac{\partial u}{\partial y} &= \frac{\partial}{\partial y}(\nu y^{l/k}) = y^{l/k}\left(\frac{l}{k}\nu y^{-1} + \frac{\partial \nu}{\partial y}\right) \\ &= y^{l/k}\left(\frac{l}{k}\nu y^{-1} + \frac{d\nu}{d\eta}\frac{\partial \eta}{\partial y}\right) = y^{l/k-1}\left(\frac{l}{k}\nu - \frac{j}{k}\eta\frac{d\nu}{d\eta}\right) = 0.\end{aligned} \tag{3.8}$$

Since $y^{l/k-1} \neq 0$, it follows that:

$$l\nu - j\eta\frac{d\nu}{d\eta} = 0, \tag{3.9}$$

is the sought equation. The solution is:

$$\nu = \text{const.}\,\eta^{l/j}; \qquad uy^{-l/k} = \text{const.}\,(xy^{-j/k})^{l/j}; \qquad u = \text{const.}\,x^{l/j}.$$

4. Application and Results

The author [*6, 7*] generalized the above technique to n-type reduction. This enables one to reduce a partial differential equation in n-independent variables to an ordinary differential equation. Some results are given below: Laplace equation in two- and three-dimensions; Laplace and Poisson's equations in n-dimensions; Schroedinger's time independent wave equation in m-dimensions; the unsteady three-dimensional equations of motion of a viscous, compressible, perfect gas; the most recent application of this technique is to prove the ergodic theorem in a particular case of a vibrating infinite membrane. The equation of such a vibrating membrane in Cartesian (x, y)-coordinates or in polar (r, θ)-coordinates is:

$$\frac{\partial^2 u}{\partial x^2} + \frac{\partial^2 u}{\partial y^2} = v^{-2}u_{,tt}; \qquad r^{-1}(ru_{,r})_{,r} + r^2 u_{,\theta\theta} = v^{-2}u_{,tt}, \tag{4.1}$$

where v is the constant velocity of wave propagation and u is the displacement at right angles to the (x, y)-plane. With $\eta = x^2 + y^2$, $2vt = T$, this equation is represented in the form:

$$\eta u_{,\eta\eta} + u_{,\eta} = u_{,TT}, \qquad u = u(\eta, T). \tag{4.2}$$

By choosing a unique function $u(\eta, T)$ which possesses the property that $u(\eta, \tau) = u(\eta)$, in the form:

$$u(\eta, T) = f_2(T - \tau)\left[\int_0^\eta df_1(s) + f_1(0)\right], \qquad f_2(0) = 1, \tag{4.3}$$

one deals with an ordinary differential equation in $u(\eta)$ to which one can apply the technique developed by the author based primarily upon the philosophy proposed by Kampé de Fériet.

Let us choose the function space, Ω, as the set of all real valued functions $f_1(s)$ of a real variable, s, $-\infty < s < +\infty$, satisfying certain conditions, superimposed by the original problem in question. One may show that Ω is a separable Banach space. Let us introduce a one parameter Abelian group of one to one transformations of Ω into itself, and let us establish a one to one correspondence between the point of a trajectory $\Gamma(\omega) \in \Omega$ and the points of a simple geometrical element. Next, by means of introducing linear functionals and operating in a product space one shows that in some case there exists a metric transitivity for the transformation of the phase-space, Ω, into itself. With a finite measure $m(\Omega)$ the ergodicity is equivalent to the metric transitivity which closes the proof.

As another example, consider a system describing the unsteady, two-dimensional flow of an incompressible fluid:

$$\frac{\partial u}{\partial t} + u\frac{\partial u}{\partial x} + v\frac{\partial u}{\partial y} = \nu\frac{\partial^2 u}{\partial y^2}; \qquad \frac{\partial u}{\partial x} + \frac{\partial v}{\partial y} = 0. \tag{4.4}$$

The transformation:

$$\bar{u} = \nu^{-1}u; \qquad \bar{v} = \nu^{-1}v; \qquad \bar{t} = \nu t; \qquad \bar{x} = x; \qquad \bar{y} = y,$$

removes ν from the above system of equations. Therefore, without any loss of generality, it is assumed that $\nu = 1$. Introducing the notion of the streamfunction, satisfying indentically the equation of continuity, $\psi = \psi(x, y, t)$ furnishes the following partial differential equation:

$$\frac{\partial^2\psi}{\partial y\partial t} + \frac{\partial\psi}{\partial y}\frac{\partial^2\psi}{\partial x\partial y} - \frac{\partial\psi}{\partial x}\frac{\partial^2\psi}{\partial y^2} = \frac{\partial^3\psi}{\partial y^3}. \tag{4.5}$$

This is conformally invariant under the group P:

$$P\colon\ \bar{x} = x + \gamma_1 a; \qquad \bar{y} = y + \gamma_2 a; \qquad \bar{t} = t + \gamma_3 a; \qquad \bar{\psi} = \psi + \gamma_4 a,$$

where γ_i, $i = 1$ to 4 are arbitrary constants. The invariants of P may be chosen as follows:

$$\eta_1 = \gamma_3 x - \gamma_1 t; \qquad \eta_2 = \gamma_3 y - \gamma_3 t; \qquad F(\eta_1, \eta_2) = \gamma_3\psi - \gamma_4 t. \tag{4.6}$$

The resultant equation is partial differential in two independent variables (η_1, η_2):

$$-\gamma_1\frac{\partial^2 F}{\partial\eta_1\partial\eta_2} - \gamma_2\frac{\partial^2 F}{\partial\eta_2^2} + \gamma_3\frac{\partial F}{\partial\eta_2}\frac{\partial^2 F}{\partial\eta_1\partial\eta_2} - \gamma_3\frac{\partial F}{\partial\eta_1}\frac{\partial^2 F}{\partial\eta_2^2} = \gamma_3^2\frac{\partial^3 F}{\partial\eta_2^3}. \tag{4.7}$$

Consider the case of $\gamma_1 \neq 0$, $\gamma_2 = 0$, $\gamma_3 \neq 0$, i.e., the variables x, y and ψ are all multiplied by an arbitrary "scale factor", γ_3; the constants γ_1 and γ_4 are a measure of the departure from steady flow conditions. The above equation with $\gamma_2 = 0$, is conformally invariant under the group

$$\Gamma_1'\colon\ \bar{\eta}_1 = b^n\eta_1; \qquad \bar{\eta}_2 = b^p\eta_2; \qquad \bar{F} = b^q F,$$

if and only if $n = 2p = 2q$. The invariants of Γ_1' may then be chosen as follows:

$$\begin{aligned} \alpha &= \eta_2\eta_1^{-1/2} = \gamma_3 y(\gamma_3 x - \gamma_1 t)^{-1/2}; \\ G(\alpha) &= F\eta_1^{-1/2} = (\gamma_3\psi - \gamma_4 t)(\gamma_3 x - \gamma_1 t)^{-1/2}. \end{aligned} \tag{4.8}$$

For $t = 0$, the above variables are in fact equal to those chosen by Blasius.

Inserting α, $G(\alpha)$ into the equation for F with $\gamma_2 = 0$, yields the following ordinary differential equation

$$2\gamma_3 \frac{d^3G}{d\alpha^3} + G \frac{d^2G}{d\alpha^2} - \gamma_1\gamma_3^{-1}\alpha \frac{d^2G}{d\alpha^2} = 0. \tag{4.9}$$

Setting $\gamma_1 = 0$, reduces this equation to Blasius' equation. The velocity components are given by the formulae:

$$u = \frac{\partial\psi}{\partial y} = \frac{dG}{d\alpha}; \qquad v = -\frac{\partial\psi}{\partial x} = \frac{1}{2}\left[\frac{\gamma_3 y}{(\gamma_3 x - \gamma_1 t)} \frac{dG}{d\alpha} - \frac{G}{(\gamma_3 x - \gamma_1 t)}\right]. \tag{4.10}$$

Therefore if the condition $u = v = 0$ when $y = 0$ is to be satisfied, one must have:

$$G = \frac{dG}{d\alpha} = 0 \qquad \text{for} \qquad \alpha = 0,$$

since $\alpha = 0$ when $y = 0$. These boundary conditions, dictated by the last expressions are exactly those found by Blasius. It is thus seen that solutions of equation for G, which satisfy condition $G = G' = 0$ for $\alpha = 0$, will yield velocity components which satisfy the usual boundary conditions in the physical plane, i.e., $u = v = 0$ for $y = 0$.

A few remarks may be made concerning the initial condition for $t = 0$. It seems farly obvious that not all initial conditions superposed upon the system in question can be satisfied by the obtained forms for u and v. Only some initial conditions can be satisfied by these expressions. The choice of another group and invariants will yield another form for u and v, which, of course, will satisfy again different initial conditions, and so on. It may be very difficult to forsee a priori, which group and which invariants may satisfy the given boundary and initial conditions. Rather, an inverse procedure may be useful: After the group and the invariants are chosen, one may investigate which boundary and initial conditions can be satisfied by the obtained forms for u and v. Probably, this procedure is more straight forward and even more logical.

The algebraic methods of solving differential equations, do not admit, as yet, any possibility of a direct attack on the boundary value and initial value problems for the simple reason that the notion of the boundary value and initial value problem is unknown in algebra. Hence, as the matter stands now, these notions must be introduced as a by-product of the entire formalism. It may be that future mathematics will solve this problem of finding invariants of groups and subgroups under the restrictive conditions that the values prescribed on the boundary curve (or curves) form some sort of "invariant conditions" superposed upon the system in question, which must be satisfied by the chosen groups, subgroups and invariants. For the time being, we have to be satisfied either by an inverse (indirect) procedure, like the one proposed above, or by a proper guess or choice of the groups and invariants so that either boundary conditions or initial conditions or both are satisfied by the group and invariants themselves. In the example, given above, the boundary conditions are satisfied by the group and invariants (i.e., $\alpha = 0$ for $y = 0$), but concerning the initial value problem, the question is open and requires some further discussion. In general, this question is too difficult to be solved in one paper or in one example, due to the facts, explained above. The matter requires a fundamental approach and introduction of the notion of the boundary and initial values in algebraic methods.

It seems that the following points may be considered in the future as those which require an additional improvement: (a) derive a technique of reduction of the number of independent variables starting from the general Michal's proof, so that it can be applied to vector functions (the above technique applies to point functions); (b) find more and more effective groups; (c) find a technique of finding invariants of groups; (d) attack the boundary value problem.

REFERENCES

[*1*] Baer, R., "Linear Algebra and Projective Geometry." Academic Press, New York, 1952.

[*2*] Birkhoff, G., Dimensional analysis of partial differential equations. *Elec. Eng.* **67**, 1185-1188 (1948).

[*3*] Birkhoff, G., "Hydrodynamics," Chapter IV. Princeton Univ. Press, Princeton, New Jersey, 1950; rev. ed., 1960.

[*4*] Dickson, L. E., Differential Equations from the Group Standpoint. *Ann. Math. (Princeton)* **25**, No. 4, 287-378 (1924).

[*5*] Eisenhart, L. P., "Continuous Groups of Transformations." Princeton Univ. Press, Princeton, New Jersey, 1933.

[*6*] Krzywoblocki, M. Z. v., "Ergodic Problem in the Theory of Vibrations and Wave Propagation." In press.

[*7*] Krzywoblocki, M. Z. v., and Roth, H., "On the Reduction of the Number of Independent Variables in Systems of Partial Differential Equations." To be published.

[*8*] Michal, A. D., General differential geometries and related topics. *Bull. AMS* **45**, 529-563 (1939).

[*9*] Michal, A. D., Differential invariants and invariant partial differential equations in normed linear spaces. *Proc. natl. Acad. Sci. U.S.* **37**, No. 9, 623-627 (1952).

[*10*] Morgan, A. J. A., The reduction by one of the number of independent variables in some systems of partial differential equations. *Quart. J. Math.* [N.S.] **3**, No. 12, 250-259 (1952).

Stochastic Approximation and Its Application to Problems of Prediction and Control Synthesis[1]

L. A. GARDNER, JR.
Lincoln Laboratory
Massachusetts Institute of Technology, Lexington, Massachusetts

1. The Stochastic Approximation Technique

The technique of Stochastic Approximation (abbr. SA) involves the familiar process of successive substitutions with a superimposed random term. Mathematical results thus constitute a certain class of random fixed point theorems, with the existence of the fixed point usually assumed a-priori. The initial SA schemes were motivated by two rather specific statistical problems, and the general approach was not taken until somewhat later.[2] These are now well-known as the Robbins-Monro (abbr. RM) and Kiefer-Wolfowitz (abbr. KW) processes for approximating the zero and minimum of an unknown regression function, i.e., the locus of expected values of an observable random variable indexed by a continuously adjustable parameter. They provide a good introduction to the subject matter.

Let $Y(x)$ be a family of observable random variables, with the "control variable" x varying over the real line, and put

$$f(x) = \mathscr{E} Y(x) \qquad \sigma^2(x) = \operatorname{Var} Y(x). \tag{1.1}$$

The problem is to estimate the point $x = \theta$ which solves the equation $f = L$, where L is a prescribed number. More specifically, assume

$$|f(x)| < A|x| + B \qquad \sigma^2(x) < C \qquad (-\infty < x < \infty) \tag{1.2}$$

where A, B, and C are some finite constants, and

$$\sup_{-k<x-\theta<-1/k} f(x) < L \qquad \inf_{1/k<x-\theta<k} f(x) > L \tag{1.3}$$

[1] Operated with support from the U.S. Army, Navy, and Air Force. MS-170A.

[2] All of the research has been done during the past decade. As the references indicate, the majority of published material appears in the mathematical statistics literature of this country and Czechoslovakia.

for every fixed positive integer k. Choose a positive null-sequence a_n such that

$$\sum a_n = \infty \qquad \sum a_n^2 < \infty. \tag{1.4}$$

If x_1 denotes an initial guess at θ, then the RM procedure generates an infinite sequence of approximations by means of the first order recursion

$$x_{n+1} = x_n - a_n(y_n - L) \qquad (n = 1, 2, \ldots). \tag{1.5}$$

Here y_n is a random variable whose conditional distribution given x_1, ..., x_n is assumed to be the distribution of $Y(x_n)$. It then follows from (1.2) and (1.3) that $x_n \to \theta$ both with probability one and in mean square.[1] The similarity between (1.5) and variations on Newton's method is clear. There is, however, an important distinction in that the latter generally presuppose knowledge of functional form. We consequently say that the RM scheme is a nonparametric SA procedure. From the application point of view this is a property we wish to maintain whenever possible. (See remarks at the end of this section). It might also be noted that the conditions of the theorem do not assume f actually attains the value L at θ, nor do they exclude f tending to L as $|x| \to \infty$. The first condition in (1.2) is necessary if the conclusion is to hold for all sequences a_n satisfying (1.4). It is automatically satisfied, of course, whenever the iteration is confined to a finite interval over which f is bounded.[2]

The KW process is obtained when we replace $y_n - L$ in (1.5) by the slope estimate

$$\frac{y_{2n} - y_{2n-1}}{2c_n}, \tag{1.6}$$

where y_{2n-1} and y_{2n} are conditionally independent random variables distributed as $Y(x_n - c_n)$ and $Y(x_n + c_n)$. The positive null-sequence c_n is to be chosen so that

$$\sum a_n^2/c_n^2 < \infty. \tag{1.7}$$

In analogy to (1.2) and (1.3), sufficient conditions for probability one

[1] The original paper by Robbins and Monro proved mean square convergence under stronger assumptions. The theorem as quoted is the best attained to date, and we refer the reader to Derman [*3*] for a comparative discussion of the several papers dealing with the RM and KW processes. (Ref. [*3*] also contains a bibliography of work done up through 1956.)

[2] Throughout this section we avoid unnecessary details by assuming an unbounded iteration space. In practice the domain of the functions definition is often a finite interval. A "truncation" logic is then added which is appropriate to the particular type of situation.

and mean square convergence to the unique point θ at which f assumes its minimum value are

$$|f(x+1)-f(x)| < A|x| + B \qquad \sigma^2(x) < C \qquad (-\infty < x < \infty) \tag{1.8}$$

and

$$\sup_{-k<x-\theta<-1/k} \bar{D}f(x) < 0 \qquad \inf_{1/k<x-\theta<k} \underline{D}f(x) > 0, \tag{1.9}$$

where $\underline{D}f(x)$ and $\bar{D}f(x)$ respectively denote the lower and upper derivative of the regression curve at x. We see that the function need not be differentiable, and can have negative left or right curvature at the solution θ. The obvious questions concerning convergence rate and how best to pick the process sequences within the restrictions have been considered by several authors, and have led to the development of some generally useful mathematical tools. (1.9) is of course too wide to permit specific results, and it is necessary to restrict attention to a class of regression curves defined by common functional properties. In this fashion Dupač [5], for example, uses Chung's Lemmas to derive power sequences a_n and c_n which yield the most rapid decrease to zero of process mean square error. Using the Central Limit Theorem for dependent random variables and evaluation of divergent sums, Sacks [*11*] gives conditions under which the properly scaled deviation $x_n - \theta$ possesses an asymptotic normal distribution. For smooth regression curves the limiting statistical behavior of the iterates is governed mainly by the curvature and a measure of asymmetry at θ. To illustrate this dependence, suppose f has a continuous second derivative throughout some neighborhood of the solution with value $0 < K < \infty$ at θ. Let $x = \theta(h)$ denote the solution of the equation $f(x-h) = f(x+h)$, and assume $\theta(h) = \theta + O(h^{1+\rho})$ as $h \to 0$ for some positive number ρ. Let us use the KW procedure with defining sequences

$$a_n = a/n^\alpha \qquad c_n = c/n^\gamma$$
$$(a, c > 0;\ \tfrac{1}{2} < \alpha \leq 1; \qquad 0 < \gamma < \alpha - \tfrac{1}{2}). \tag{1.10}$$

Then, if

$$a > 1/2K \qquad \alpha = 1 \qquad \gamma = 1/2(2+\rho) \tag{1.11}$$

we obtain

$$\mathscr{E}(x_n - \theta)^2 \sim 1/n^{1-2\gamma}, \tag{1.12}$$

and otherwise the n-dependence is worse. One can delineate choices for a and c whenever the limiting constant in (1.12) can be evaluated. It turns out that we should take $a = 1/K$. Assuming $\sigma^2(x)$ continuous with value $\sigma^2 > 0$ at θ, the best choice for c depends not only on σ^2 but explicit constants appearing in $\theta(h)$. Burkholder [2] applies his results to a modification of the KW process and shows how to construct asymptotic confidence intervals on θ which are free of unknowns (e.g. K and σ^2).

A very general SA scheme was introduced by Dvoretzky [6] of which the RM and KW processes are very special cases. The starting point is a sequence of transformations

$$T_n: R^n \to R^1 \qquad (n = 1, 2, ...) \tag{1.13}$$

where R^n is n-dimensional reals. Suppose the transformations satisfy

$$| T_n(r_1, ..., r_n) - \theta | \leq \max [\alpha_n, (1 + \beta_n) | r_n - \theta | - \gamma_n] \tag{1.14}$$

for every $(r_1, ..., r_n)$ in R^n and some real number θ. Here α_n, β_n, γ_n are real-valued nonnegative functions for which

$$\alpha_n \to 0 \qquad \sum \beta_n < \infty \qquad \sum \gamma_n = \infty \tag{1.15}$$

uniformly in all sequences $r_1, ..., r_n, ...$. The SA procedure is the process of noisy successive substitutions with T_n starting from a given random variable x_1:

$$x_{n+1} = T_n(x_1, ..., x_n) + w_n. \tag{1.16}$$

If (1.14)-(1.15) holds (at least for all n beyond some n_i) and the sequence of random variables w_n satisfies

$$\sum \text{a.s.} \sup_{x_1, \ldots, x_n} | \mathscr{E}\{w_n \mid x_1, ..., x_n\} | < \infty \qquad \sum \mathscr{E} \mid w_n \mid^2 < \infty, \tag{1.17}$$

then

$$P\{\lim_n | x_n - \theta | = 0\} = 1 \qquad \lim_n \mathscr{E} \mid x_n - \theta \mid^2 = 0. \tag{1.18}$$

Dvoretzky proves various generalizations. For example, he gives conditions under which the transformations can be random, so that the value of T_n is no longer determined by $x_1, ..., x_n$. The conclusion remains valid when x_n, T_n, etc. are vector-valued random variables, i.e., with R^1 replaced by Euclidean p-space. It is necessary only to interpret any absolute sign in (1.14)-(1.18) as length. In the present formulation we have for the RM process

$$T_n(x_n) = x_n - a_n[f(x_n) - L] \qquad w_n = - a_n[y_n - f(x_n)]. \tag{1.19}$$

The first statement in (1.17) is true because $\mathscr{E}\{w_n \mid x_1, ..., x_n\}$ is, by assumption, zero with probability one. The sum appearing in the second statement is majorized by $C \sum a_n^2 < \infty$. The divergence of Σa_n together with (1.2) and (1.3) are enough to guarantee (1.14)-(1.15). Similar arguments establish convergence of the KW process under (1.7)-(1.9). Another general (but somewhat different) approach using generalized random variables is taken by Hanš [9]. By applying his random fixed point theorems for separable Banach spaces, he gives conditions for probability one convergence of certain stochastic successive substitutions (based on averaging).

The first condition in (1.17) ensures convergence of "biased" SA schemes, i.e., those in which the random disturbances are not (conditionally) centered on zero. This significantly widens the range of application beyond regression problems, but at the same time introduces some new difficulties. Indeed, $\mathscr{E}\{w_n \mid x_1, ..., x_n\}$ will not only depend in a significant way on the various sequences defining the iteration (which appear in α_n, β_n, γ_n) but also on one or more unknown parameters of the model. One would like to select a-priori the process sequences so that (1.17) holds no matter what the values of the model parameters. In many cases there exists such a specification independent of "nature", but the result is generally poor convergence properties. One then looks for a way to calculate the process sequences from the available iterates, as well as from additional observations, in order to retain the desired nonparametric property and an acceptable decrease in process mean square as a function of the total number of observations. This is a fruitful, but relatively untouched, area for research.[1]

To give a concrete instance leading to a biased KW process, consider estimating the frequency θ at which the spectral density function $f(x)$ of an observable time series U_t takes on its minimum value. This application is interesting on several counts. First, our ability to estimate the value of $f(x)$ depends upon (sometimes unknown) differentiability properties of the function. As a result, the deterministic aspects of the transformations T_n and the statistical aspects of the noise sequence w_n are highly related (whereas in most applications they are not). Second, if only one realization of U_t is available, there are two sources of bias: the statistical bias at fixed frequency arising from finite sample size; and the conditional dependence of successive slope estimates along the iteration resulting from the use of correlated data. The latter is not present when an arbitrarily large number of statistically independent realizations are available, but even then there remains the problem of

[1] A procedure with the properties indicated above has been worked out by Gardner [7] for minimizing an extreme value function.

designing an efficient iterative scheme for making use of existing smoothness.

A related problem area to which the SA point of view can be successfully utilized is the design of "adaptive predictors". Given a single realization of a weakly stationary time series (say, again, U_t) we wish to sequentially synthesize the linear least squares predictor of a given order p. Although there are several approaches to this problem, one which can yield desirable convergence properties makes use of the idea of solving, by successive substitution, the linear equations $\mathbf{\Phi x} = \boldsymbol{\varphi}$ for the p-vector $\mathbf{x} = \boldsymbol{\theta}$ defining the optimum coefficients. Here $\mathbf{\Phi}$ is the matrix with ijth entry $\varphi(i-j)$ and $\boldsymbol{\varphi}$ the column vector with ith entry $\varphi(i)$ $(i, j = 1, 2, ..., p)$, where $\varphi(k) = \mathscr{E} U_t U_{t+|k|}$ for all integers k and t. We assume the unknown covariance sequence is summable, and that U_t has a summable fourth order moment function. From the first n observations $u_1, u_2, ..., u_n$ we compute an estimate of $\varphi(k)$ over $|k| \leq n-1$ by

$$\varphi_n(k) = \frac{1}{n - |k|} \sum_{t=1}^{n-|k|} u_t u_{t+|k|}. \tag{1.20}$$

Let $\mathbf{\Phi}_n$ and $\boldsymbol{\varphi}_n$ be the matrix $\mathbf{\Phi}$ and vector $\boldsymbol{\varphi}$ when the entry $\varphi(k)$ is replaced by $\varphi_n(k)$ $(k = 0, 1, ..., p)$. For any vector $\mathbf{x}$ and positive integer n, define the random transformation from Euclidean p-space E_p into itself

$$\mathbf{T}_n(\mathbf{x}) = \mathbf{x} - a_n(\mathbf{\Phi}_n \mathbf{x} - \boldsymbol{\varphi}_n), \tag{1.21}$$

where a_n abbreviates A/n $(A > 0)$. Corresponding to an initial guess $\mathbf{x}_1$, we then generate successive random approximations to $\boldsymbol{\theta}$ by

$$\mathbf{x}_{n+1} = \mathbf{T}_n(\mathbf{x}_n) \qquad (n = 1, 2, ...).^1 \tag{1.22}$$

If we assume that the (real) spectral density function

$$f(\omega) = \sum_{k=-\infty}^{\infty} \varphi(k) e^{ik\omega} \tag{1.23}$$

has a minimum at $\omega_0 \epsilon [0, \pi]$ with $f(\omega_0) > 0$, then it can be shown that $\mathscr{E} \| \mathbf{x}_n - \boldsymbol{\theta} \|^2$ goes to zero like $1/n$ provided $A > 1/2f(\omega_0)$. Thus, if we know a nonzero lower bound on the values of f, we can ensure regular estimation properties of (1.22) by fixing the constant in the gain sequence sufficiently large. In the absence of such knowledge there are several alternatives. An interesting one (chiefly from a theretical point of view)

[1] Hanš and Špaček [*10*] consider the continuous time version with $p = 2$, i.e., a pair of random differential equations.

makes further use of the covariance estimates in a concurrent KW process designed to locate the point ω_0 which minimizes f (assuming its uniqueness). If ω_n is the nth such approximation and f_n the appropriate estimate of the function f, then the sequence a_n in (1.21) is replaced by one of random variables $1/nf_n(\omega_n)$. Finally, it should be pointed out that the present approach can be used to obtain the optimum infinite memory predictor. To do this we select a sequence of integers $0 = n_0 < n_1 < n_2 < \ldots$ such that $n_p \gg p$ and $n_p - n_{p-1} \to \infty$ as $p \to \infty$. We perform $n_p - n_{p-1} - 1$ iterations in E_p with

$$\mathbf{x}_{n+1} = \mathbf{x}_n - a_{n-n_{p-1}}(\mathbf{\Phi}_n \mathbf{x}_n - \boldsymbol{\varphi}_n)$$
$$(n = n_{p-1} + 1, n_{p-1} + 2, \ldots, n_p - 1), \tag{1.24}$$

where the initial vector in E_p is formed by increasing the dimension of $\mathbf{x}_{n_{p-1}} \epsilon E_{p-1}$ with a zero. The gain sequence is reset when p is stepped but the calculation of the estimates in $\mathbf{\Phi}$ and $\boldsymbol{\varphi}$ continues. The main theoretical reason that (1.24) can be successfully used rests in the fact that the p eigenvalues of $\mathbf{\Phi}$ are bounded away from zero by $f(\omega_0)$ uniformly in $p = 1, 2, \ldots$. A formula for the sequence n_p can be derived in terms of the tail of the infinite dimensional solution $\theta_1, \theta_2, \ldots$ (in l_2). Thus, setting $\Theta_p^2 = \theta_{p+1}^2 + \theta_{p+2}^2 + \ldots$, we have for p tending to infinity

$$n_p \sim n_{p-1} + B^2 \frac{p^2}{\Theta_p^2} \tag{1.25}$$

where B^2 does not contain p. If only the first q components of the solution are nonzero then $n_q = \infty$, i.e., we continue to iterate in E_q ad. inf. The rather obvious drawback to practical use of (1.22) or (1.24) is the computation of covariance estimates for increasing sample sizes. This can be obviated, generally at the sacrifice of convergence rate, with a KW process which iterates in the space of the predictor coefficients to minimize the mean square prediction error. There is an advantage in this approach from the standpoint of making use of a-priori knowledge. Indeed, in many practical problems the covariance sequence is known up to the values of a small number of parameters. One then iterates in this space using given coefficient functions. A detailed treatment of this application, including a comparison with other methods, is to be published [8].

2. A Problem in Control

Consider a system governed by a first order differential equation of the form $\dot{x}(t) = K(t)\, g(t)$ $(0 \leq t < \infty)$ where $g(t)$ is the control function. The objective is to drive the solution, starting from an arbitrary initial value $x(0)$, to a prescribed point θ. However, the function $K(t)$ is unknown and the only observable output is $y(t) = x(t) + u(t)$, where $u(t)$ represents noise (i.e., a realization of a random function). We are therefore faced with a (rather trivial but, as will be seen, instructive) problem in stochastic control. Suppose first that

$$K(t) \text{ is continuous, } 0 < K' \leq K(t) \leq K'' < \infty$$
$$\lim_{T\to\infty} \frac{1}{T} \int_0^T K(t)\, dt = K_0 \tag{2.1}$$

for all t and some constants K', K'' and K_0; and furthermore, with probability one,

$$u(t) \text{ is continuous}$$
$$\lim_{T\to\infty} \frac{1}{T} \int_0^T u(t)\, dt = 0. \tag{2.2}$$

Define the function

$$a(t) = \begin{cases} A & \text{for} \quad 0 \leq t < 1 \\ A/t & \text{for} \quad 1 \leq t < \infty. \end{cases} \tag{2.3}$$

Then, if we take (without feedback delay)

$$g(t) = -\, a(t)[y(t) - \theta], \tag{2.4}$$

our controller is parameterized by the single constant $0 < A < \infty$. For any $g = g_A$ the output of the controlled system satisfies

$$\dot{x}(t) = -\, a(t)\, Y(x(t), t) \tag{2.5}$$

where $Y(x, t) = K(t)\,[x - \theta + u(t)]$ is a random function of $-\infty < x < \infty$ and $0 \leq t < \infty$. Under (2.1) and (2.2) it follows immediately, from a result of Driml and Nedoma [*4*], that the solution of the random differential equation (2.5) exists with probability one, and satisfies

$$P\{\lim_{t\to\infty} | x(t) - \theta | = 0\} = 1 \tag{2.6}$$

independent of $x(0)$. It is not difficult to see that we are merely solving (at time equal to infinity) the equation $K_0(x - \theta) = 0$. The conclusion is unaltered if $K(t)$ is a random function satisfying (2.1) with probability one.[1]

[1] An interesting nontrivial problem arises when $K(t)$ is bounded in absolute value with probability one, but has positive probability of infinitely many sign changes on $(0, \infty)$.

The question arises as to how rapidly the solution of the controlled system converges to the desired point as a function of our choice of $0 < A < \infty$. This is easily answered when we work with "sampled-data". Continuity no longer enters, and the only statement in (2.1) we need actually retain is the boundedness. However, more explicit results are attainable when the function $K(t)$ does posses a time average. We therefore introduce sampling instants $0 < t_1 < t_2 < \dots \to \infty$, and for the present leave them unspecified. On the other hand, since we will measure convergence rate by

$$e_n^2 = \mathscr{E}[x(t_n) - \theta]^2 \tag{2.7}$$

as n tends to infinity, we require certain properties of the covariance function

$$\varphi(t, s) = \mathscr{E}u(t)\, u(s) \qquad (-\infty < t, s < \infty) \tag{2.8}$$

on the sampling grid. Let $\mathbf{\Phi}_n$ be the $n \times n$ symmetric matrix with ijth entry $\varphi(t_i, t_j)$, and let w_n be an indeterminate vector with kth entry w_{nk} $(k = 1, 2, \dots, n)$.[1] For $n = 1, 2, \dots$ introduce the sequence of (necessarily) nonnegative functions

$$\rho_n^2(\mathbf{w}_n) = \frac{\mathbf{w}_n' \mathbf{\Phi}_n \mathbf{w}_n}{\| \mathbf{w}_n \|^2}, \tag{2.9}$$

where the dependence on the sequence t_n is to be remembered. We will always suppose that the largest eigenvalue of $\mathbf{\Phi}_n$ is uniformly bounded in $n = 1, 2, \dots$ by a finite constant M'', and consequently the same number serves as a uniform bound on (2.9) for any choice of $\mathbf{w}_1, \mathbf{w}_2, \dots$. In particular, for $\mathbf{w}_n = \mathbf{1}_n$ (i.e., $w_{nk} = 1$ for each $k = 1, 2, \dots, n$), we have

$$\rho_n^2(\mathbf{1}_n) = \frac{1}{n} \sum_{i=1}^{n} \sum_{j=1}^{n} \varphi(t_i, t_j) \leq M'' \tag{2.10}$$

for every n.

If the control function $g(t)$ is constantly equal to $g_n = g(t_n)$ for all $t_n \leq t < t_{n+1}$, then the differential equation reduces to the difference equation $x_{n+1} = x_n + \psi_n \cdot g_n \Delta t_n$. Here $x_n = x(t_n)$, $\Delta t_n = t_{n+1} - t_n$, and

$$\psi_n = \frac{1}{\Delta t_n} \int_{t_n}^{t_{n+1}} K(t)\, dt \tag{2.11}$$

[1] All vectors are considered to be columns, and prime denotes transposition. Note that $\mathbf{\Phi}_n$ here has meaning different from that used in Section 1 under (1.20).

by hypothesis belongs to $[K', K'']$ for every n. A discrete time version of $g = g_A$ in (2.3)-(2.4) is given by

$$g_n = -\frac{a_n}{\Delta t_n}(y_n - \theta) \qquad a_n = \frac{A}{n+1} \tag{2.12}$$

where the division by $n + 1$ rather than n is for notational convenience. There results the simple first order stochastic difference equation with correlated disturbances,

$$x_{n+1} - \theta = (1 - \psi_n a_n)(x_n - \theta) + \psi_n a_n u_n \qquad (n = 1, 2, \ldots), \tag{2.13}$$

where we have written $-u_n$ in place of $u(t_n)$.[1] The iteration is to be initialized by taking for $g(t)$, $0 \leq t < t_1$, any random function independent of $u(t)$ over $t \geq t_1$. For example, if $g(t) \equiv 0$ on $[0, t_1)$, x_1 is just $x(0)$. Introduce the products

$$\begin{aligned} B_n &= \prod_{j=1}^{n} (1 - \psi_j a_j) \\ b_{nk} &= \psi_k a_k \prod_{j=k+1}^{n} (1 - \psi_j a_j) = \psi_k a_k B_n / B_k \qquad (k = 1, 2, \ldots, n) \end{aligned} \tag{2.14}$$

where, as the notation implies, the void product is to be read as unity. We assume, until stated otherwise, that $1 - \psi_n a_n$ vanishes at no positive integer n. Since $\Sigma \psi_n a_n = \infty$, we have $B_n \to 0$ as $n \to \infty$. Moreover, it is easily shown that

$$\sum_{k=1}^{n} b_{nk} = 1 - B_n \tag{2.15}$$

identically in n. If we now iterate (2.13) we obtain

$$x_{n+1} - \theta = B_n(x_1 - \theta) + \sum_{k=1}^{n} b_{nk} u_k. \tag{2.16}$$

Squaring, and taking expected values, there results

$$e_{n+1}^2 = B_n^2 e_1^2 + \sum_{i=1}^{n} \sum_{j=1}^{n} b_{ni} \varphi(t_i, t_j)\, b_{nj}. \tag{2.17}$$

[1] The arguments which follow can be extended without difficulty to handle the vector version of (2.13). It should be pointed out that Bertram [1] has discussed the present problem from the control engineer's point of view, and has given conditions for mean square convergence of (2.13). (His conditions for the vector case are not generally consistent). Our calculations will yield convergence rates and, when they exist, limiting constants.

There is no cross term since $x_1 - \theta$ is uncorrelated with $u_1, u_2, \ldots, u_n$ and, by hypothesis, the mean of each is zero. Letting $\mathbf{b}_n$ be the vector with kth entry b_{nk} $(k = 1, 2, \ldots, n)$ we obtain

$$\frac{e_{n+1}^2}{\|\mathbf{b}_n\|^2} = \frac{B_n^2 e_1^2}{\|\mathbf{b}_n\|^2} + \rho_n^2(\mathbf{b}_n) \tag{2.18}$$

in the notation of (2.9).

Let us investigate the simplest case first when the function $K(t)$, and hence ψ_n, is identically equal to K_0. Setting

$$z = K_0 A > 0 \tag{2.19}$$

the products in (2.14) can be explicitly evaluated. Indeed, as $n \to \infty$,

$$(n+1)^z B_n = \frac{1}{-z(1-z)\,\Gamma_{n+1}(-z)} \to \frac{1}{\Gamma(2-z)} \qquad (z \neq 1, 2, \ldots) \tag{2.20}$$

where $\Gamma_{n+1}(-z)$ is Euler's approximation to the value of the Gamma Function $\Gamma(-z)$. When z is a positive integer

$$(n+1)^z B_n = \frac{1}{\Gamma(2-z)} = \begin{cases} 1 & (z = 1) \\ 0 & (z = 2, 3, \ldots) \end{cases} \tag{2.21}$$

for every n. The initial assumption $1 - \psi_n a_n \neq 0$ for all n means we are excluding the points $z = 2, 3, \ldots$ from $0 < z < \infty$ (see next paragraph). The kth entry of the vector $\mathbf{b}_n$ is simply

$$b_{nk} = \frac{z}{k+1} \prod_{j=k+1}^{n} \left(1 - \frac{z}{j+1}\right) = \begin{cases} \dfrac{1}{n+1} & (z = 1) \\[2ex] z\,\dfrac{\Gamma_{k+1}(-z)}{\Gamma_{n+1}(-z)}\,\dfrac{(k+1)^{z-1}}{(n+1)^z} & (z \neq 1). \end{cases} \tag{2.22}$$

Introduce

$$c_{nk} = \frac{b_{nk}}{\|\mathbf{b}_n\|} \qquad (k = 1, 2, \ldots, n) \tag{2.23}$$

so that $\|\mathbf{c}_n\| \equiv 1$ in n. From (2.22) we see that $c_{nk}^2 \to 0$ for each fixed k, provided z exceeds $\frac{1}{2}$. Consequently, by the Toeplitz Lemma,

$$\lim_n \xi_n = \xi \Rightarrow \lim_n \sum_{k=1}^{n} c_{nk}^2 \xi_k = \xi \qquad (z > \tfrac{1}{2}) \tag{2.24}$$

for any ξ (finite or not). In virtue of (2.14) and (2.15) this is true, without restriction on z, when c_{nk}^2 is replaced by b_{nk}. With the aid of (2.24) and the fact

$$\| \mathbf{b}_n \|^2 \cong \frac{z^2}{2z-1} \frac{1}{n} \qquad (z > \tfrac{1}{2}), \tag{2.25}$$

we can asymptotically evaluate the mean square error as $n \to \infty$.[1] Since $B_n^2 \sim 1/n^{2z}$, the first term in (2.18) tends to zero as $n \to \infty$. From (2.9), (2.23), and (2.25) it therefore follows that

$$n\mathscr{E}[x(t_n) - \theta]^2 \cong \frac{z^2}{2z-1} \sum_{i=1}^{n} \sum_{j=1}^{n} c_{ni} \varphi(t_i, t_j)\, c_{nj} \qquad (z > \tfrac{1}{2}). \tag{2.26}$$

Although the square summable sequence $c_{n1}, c_{n2}, \ldots, c_{nn}$ depends on z, there are important special cases in which the limiting value of the double sum in (2.26), i.e., of $\rho_n^2(\mathbf{b}_n)$, exists and is independent of z. Thus, suppose the random function $u(t)$ is weakly stationary, so that $\varphi(t, s) = \varphi(t - s)$ for all real numbers t and s. Since, for $z = 1$, $c_{nk} = 1/\sqrt{n}$ for all $k = 1, 2, \ldots, n$ we have $\rho_n^2(\mathbf{b}_n)\,|_{z=1} = \rho_n^2(\mathbf{1}_n)$ of (2.10). In the stationary case the latter is

$$\frac{1}{n} \sum_{i=1}^{n} \sum_{j=1}^{n} \varphi(t_i - t_j) = \sum_{|\nu| \leq n-1} \left(1 - \frac{|\nu|}{n}\right) \varphi(\nu h) \tag{2.27}$$

provided we use a constant sampling interval, i.e., $t_n = nh$ for all $n \geq 1$ and some $0 < h < \infty$. The statement that $\rho_n^2(\mathbf{1}_n)$ has a limit as $n \to \infty$ is now equivalent to the summability of $\varphi(0), \varphi(h), \varphi(2h), \ldots$; in other words, the existence of a (real) spectral density function

$$f(\omega; B) = \sum_{\nu=-\infty}^{\infty} \varphi\left(\frac{\nu}{B}\right) e^{i\nu\omega} \qquad (|\,\omega\,| \leq \pi). \tag{2.28}$$

Here we have introduced the sampling frequency parameter $B = 1/h$. From (2.27) it then follows that the limiting value of $\rho_n^2(\mathbf{1}_n)$ is $f(0; B)$. It remains to compute

$$\rho_n^2(\mathbf{b}_n) = \varphi(0) + 2 \sum\sum_{1 \leq i < j \leq n} c_{ni} \varphi((i - j)\, h)\, c_{nj} \tag{2.29}$$

when z is different from one. If we call this double sum P_n we have

$$P_n = \sum_{k=1}^{n-1} c_{nk}^2 \sum_{\nu=1}^{n-k} \xi_{k,k+\nu}\, \varphi(\nu h) \tag{2.30}$$

[1] (2.25) is a special case of Lemma 5 in Sacks [*11*].

wherein, with $z \neq 1$,

$$\xi_{k,k+\nu} = \frac{b_{n,k+\nu}}{b_{nk}} = \frac{\Gamma_{k+\nu+1}(-z)}{\Gamma_{k+1}(-z)} \left(\frac{k+1}{k+\nu+1}\right)^{1-z} \tag{2.31}$$

is independent of n. Let m be a large fixed integer and consider the truncated sequence $\varphi(h), ..., \varphi(mh), 0, 0, ...$ We find, after some rearranging, that

$$\begin{aligned} P_n^{(m)} &= \sum_{k=1}^{n-m} c_{nk}^2 \sum_{\nu=1}^{m} \xi_{k,k+\nu}\, \varphi(\nu h) + \sum_{k=n-m+1}^{n-1} c_{nk}^2 \sum_{\nu=1}^{n-k} \xi_{k,k+\nu}\, \varphi(\nu h) \\ &= \sum_{k=1}^{n-1} c_{nk}^2 \sum_{\nu=1}^{m} \xi_{k,k+\nu}\, \varphi(\nu h) - \sum_{k=n-m+1}^{n-1} c_{nk}^2 \sum_{\nu=n-k+1}^{m} \xi_{k,k+\nu}\, \varphi(\nu h). \end{aligned} \tag{2.32}$$

If we interchange the order of summation in the first term, we get

$$P_n^{(m)} = \sum_{\nu=1}^{m} \varphi(\nu h) \left[\sum_{k=1}^{n-1} c_{nk}^2\, \xi_{k,k+\nu}\right] - \sum_{k=n-m+1}^{n-1} c_{nk}^2 \left[\sum_{\nu=n-k+1}^{m} \xi_{k,k+\nu}\, \varphi(\nu h)\right]. \tag{2.33}$$

Since

$$\lim_{k\to\infty} \xi_{k,k+\nu} = 1$$

for every fixed ν the sum within the first square bracket, in virtue of (2.24), tends to unity as $n \to \infty$. The sum within the second square bracket is uniformly bounded, and hence the entire second term goes to zero as $n \to \infty$ because it is dominated by the tail of a convergent series. Furthermore, it is clear that we can allow m to increase with n provided $m/n \to 0$. It follows that

$$\lim_n P_n = \sum_{\nu=1}^{\infty} \varphi(\nu h) < \infty,$$

which combines with (2.29) to give

$$\lim_n t_n \mathscr{E}[x(t_n) - \theta]^2 = \frac{z^2}{2z-1} \cdot \frac{f(0; B)}{B} \qquad (z > \tfrac{1}{2}) \tag{2.34}$$

as the stationary version of (2.26). If we assume the discrete time process $u(nh)$ is nondeterministic (i.e., that $-\log f(\omega; B)$ is integrable over $|\omega| \leq \pi$) then we also get asymptotic normality. Specifically, if we reintroduce the gain constant A and set

$$\begin{gathered} \sigma^2(A, B) = \frac{K_0 A}{2K_0 A - 1} \cdot \frac{f(0; B)}{B} \\ (1/2K_0 < A < \infty,\ 0 < B < \infty), \end{gathered} \tag{2.35}$$

then for any $y \geqslant 0$

$$\lim_{n} P\{t_n^{1/2} \mid x(t_n) - \theta \mid > y\sigma(A, B)\} = \sqrt{\frac{2}{\pi}} \int_y^{\infty} e^{-\frac{1}{2}\xi^2} d\xi \tag{2.36}$$

where $\Delta t_n = 1/B$ is the constant time between observations. Since the function $z^2/(2z - 1)$ has a single minimum on $(\frac{1}{2}, \infty)$ at $z = 1$, the best setting for A is $1/K_0$ (which coincides with the convential time-independent gain used for a static noiseless system). In cases of practical interest, the function $f(0; B)/B$ has a minimum with respect to B.

Before turning to the general case, we should take care of the situations $z = 2, 3, \ldots$ and $0 < z \leq \frac{1}{2}$. For the first, the above results are unaltered. Indeed, suppose $z = N + 1$ for some finite integer $N \geq 1$. Then $\psi_n a_n = (N + 1)/(n + 1)$ for all $n \geq 1$, and from (2.13) $x_{N+1} - \theta = u_N$. Iterating back to $N + 1$ we get

$$x_{n+1} - \theta = b_{nN} u_N + \sum_{k=N+1}^{n} b_{nk} u_k = \sum_{k=N}^{n} b_{nk} u_k \tag{2.37}$$

where b_{nk} $(k = N, N + 1, \ldots, n)$ is still defined by (2.14). Consequently, for all n large enough,

$$\frac{e_{n+1}^2}{\| \mathbf{b}_{n-N+1} \|^2} = \rho_{n-N+1}^2(\mathbf{b}_{n-N+1}) \tag{2.38}$$

providing we interpret the subscript as the number of components when the first is now labelled with the fixed index N. The asymptotic behavior, of (2.38) is thus the same as (2.18) with $z = N + 1 > \frac{1}{2}$. The fact that the first $N + 1$ iterations are wasted is reflected, in the limit, by the steadily increasing function $z^2/(2z - 1)$. The behavior of (2.18) over $0 < z \leq \frac{1}{2}$ for large n is different from that discussed previously. In place of (2.25) we now have

$$\| \mathbf{b}_n \|^2 \sim 1/n^{2z} \qquad (0 < z < \tfrac{1}{2}), \tag{2.39}$$

and consequently, in view of (2.20), the limit is not independent of e_1^2.[11] Moreover, c_{nk} in (2.23) is essentially independent of n, and the inference in (2.24) is false. The asymptotic behavior of $\rho_n^2(\mathbf{b}_n)$ now depends on z. For the stationary case, the first square bracket in (2.33) has a limit depending on the index ν as well as z. It might be pointed

[1] We ignore $z = \frac{1}{2}$, in which case the left side of (2.39) is asymptotically equal to $\log n/4n$.

out that the "K_0A problem" can be avoided if we are willing to accept the behavior $e_n^2 = o(1/n^{1-\epsilon})$ for any $\epsilon > 0$. Indeed, for

$$a_n = A' \log(n+1)/(n+1)$$

this is true without restriction on $A' > 0$.

Suppose, now, we drop the assumption that $K(t)$ is constant for all $t \geq 0$. When $K(t)$ is arbitrary within the restriction $0 < K' \leq K(t) K' < \infty$ explicit results are of course impossible. By methods quite similar to the preceeding ones it is not difficult to show, however, that

$$e_n^2 = O(1/n) \qquad (A > 1/2K'). \tag{2.40}$$

The upper bound K'' ensures the existence of a finite index beyond which the sequence $1 - \psi_n a_n$ is positive. Finally, it is possible to calculate a nonconstant sampling sequence t_n and a gain constant A in certain cases when $K(t)$ possesses a time average. To do this we would require an estimate of how rapidly the limit K_0 in (2.1) is approached as a function of T. Moreover, since Δt_n would increase to infinity with n, more specific information about the covariance function $\varphi(t, s)$ would be needed.

It is interesting to compare the mean square error estimates derived above with that attainable when we drop the restriction on a_n imposed in (2.12). On the contrary, then, suppose the first n terms of the gain sequence are free to be a priori chosen in such a way as to minimize

$$e_{n+1}^2(\mathbf{b}_n) = (1 - \mathbf{1}_n'\mathbf{b}_n)^2 e_1^2 + \mathbf{b}_n'\mathbf{\Phi}_n\mathbf{b}_n \tag{2.41}$$

for an arbitrary fixed n. (See (2.15) and (2.17)). If this is done (2.41) must be minimum with respect to the n components of $\mathbf{b}_n$ which are related to $a_{n1}, a_{n2}, \ldots, a_{nn}$ by

$$b_{nk} = \psi_k a_{nk} \prod_{j=k+1}^{n} (1 - \psi_j a_{nj}) \qquad (k = 1, 2, \ldots, n). \tag{2.42}$$

For simplicity we consider only the stationary case and sampling instants $t_n = nh$ $(n = 1, 2, \ldots)$. In order to guarantee a unique minimum, we assume that $f(\omega; B)$ of (2.28) has the property

$$0 < M' \leq f(\omega; B) \leq M'' < \infty \qquad \text{for all} \qquad |\omega| \leq \pi,$$

where $B = 1/h$ is fixed. (Only the upper bound was heretofore assumed.)[1] The set of n eigenvalues of $\mathbf{\Phi}_n$ is then contained in $[M', M'']$

[1] It is to be understood that M' and M'' respectively stand for the greatest lower and least upper bounds on f.

for every n. The same holds true for every member of the sequence $\tau_n^2(\mathbf{1}_n)$ defined by

$$\frac{1}{\tau_n^2(\mathbf{1}_n)} = \frac{\mathbf{1}_n'\mathbf{\Phi}_n^{-1}\mathbf{1}_n}{\|\mathbf{1}_n\|^2}. \tag{2.43}$$

The gradient of $e_{n+1}^2(\mathbf{b}_n)$ has a unique zero at the point

$$\mathbf{b}_n^* = e_1^2 B_n^* \mathbf{\Phi}_n^{-1}\mathbf{1}_n, \tag{2.44}$$

But from the identity (2.15) we have $\mathbf{1}_n'\mathbf{b}_n = 1 - B_n$ in every case, and this combines with (2.44) to give

$$B_n^* = \frac{\tau_n^2}{e_1^2 n + \tau_n^2} \tag{2.45}$$

wherein τ_n^2 abbreviates $\tau_n^2(\mathbf{1}_n)$. Substituting back into (2.41) we find that the value of the minimum mean square error after the nth step is

$$e_{n+1}^{*2} = B_n^{*2} e_1^2 + \left[\frac{1}{1 + \tau_n^2/ne_1^2}\right]^2 \frac{\tau_n^2}{n}.^{1} \tag{2.46}$$

Note that this does not depend on $\psi_1, \psi_2, \ldots, \psi_n$, i.e. on the function $K(t)$ over $t \geq t_1$. Moreover, independent of the initial error

$$e_n^{*2} \cong \frac{\tau_n^2}{n} \qquad (n \to \infty) \tag{2.47}$$

which is a function of $\varphi(0)$, $\varphi(t_1)$, $\varphi(t_2)$, ... in accordance with (2.43). In order to synthesize the ideal controller for given n we must, of course, know the $2n$ numbers $\varphi(0)$, $\varphi(t_1)$, ..., $\varphi(t_{n-1})$, ψ_1, ..., ψ_n as well as e_1^2. The control applied at time $t_k = kh$ is

$$g_{nk}^* = -\frac{a_{nk}^*}{h}(y_k - \theta) \tag{2.48}$$

where the optimum gain sequence is a-priori computed from

$$a_{nk}^* = \frac{\tau_k^2}{\psi_k[k + \tau_k^2/e_1^2]} \sum_{i=1}^{n} \varphi^{ki}(n) \qquad (k = 1, 2, \ldots, n). \tag{2.49}$$

[1] This is the best that can be done with a control law which depends only on the present observation. (2.46) can be reduced by a control $g_{nk}(y_1, y_2, \ldots, y_k)$ at time t_k which depends on the past.

Here $\varphi^{ij}(n)$ stands for the ijth entry of the matrix $\mathbf{\Phi}_n^{-1}$. (2.49) follows immediately upon substitution of (2.44) and (2.45) into (2.42). There are obvious difficulties associated with the computation of

$$\frac{1}{\tau_k^2} = \frac{1}{k} \sum_{i=1}^{k} \sum_{j=1}^{k} \varphi^{ij}(k) \tag{2.50}$$

for each $k = 1, 2, \ldots, n$. The formulae can be expressed in terms of f if we introduce the spectral representation

$$\mathbf{\Phi}_n^{-1} = \sum_{i=1}^{n} \frac{1}{\lambda_{ni}} \mathbf{r}_{ni}\mathbf{r}'_{ni}, \tag{2.51}$$

where $\mathbf{r}_{n1}, \mathbf{r}_{n2}, \ldots, \mathbf{r}_{nn}$ denote the orthonormal eigenvectors of $\mathbf{\Phi}_n$ corresponding to the (not necessarily distinct) eigenvalues $\lambda_{n1}, \lambda_{n2}, \ldots, \lambda_{nn}$. To analytically evaluate these is difficult except in special cases. In fact, only when u_n is uncorrelated does the computation of the optimum control really become feasible. For then, $\tau_n^2 \equiv \varphi(0)$, and (2.49) reduces to

$$a_{nk}^* = \frac{1}{\psi_k[k + \varphi(0)/e_1^2]} = a_k^* \text{ (say)} \tag{2.52}$$

which is independent of the terminal time t_n. The optimum gain at time t_k when k is large is $1/(k+1)\,\psi_k + O(1/k^2)$. Thus, when we specialize the model to $K(t) \equiv K_0$, we get $a_k^* \simeq 1/(k+1)\,K_0$. This provides an independent proof of the fact that the best member of the one parameter family of control laws in (2.12) is defined by $A = 1/K_0$. Under this specialization, the asymptotic efficiency of the latter relative to (2.47) is

$$\frac{e_n^{*^2}}{e_n^2} \simeq \frac{\tau_n^2}{\rho_n^2} \tag{2.53}$$

where ρ_n^2 abbreviates $\rho_n^2(\mathbf{1}_n)$. We have previously shown, however, that $\lim_n \rho_n^2 = f(0; B)$. Consequently, since $e_n^{*^2} \leq e_n^2$ must hold for every n,

$$\overline{\lim_n}\, \tau_n^2 \leq f(0; B). \tag{2.54}$$

When we combine this with

$$M' \leq \underline{\lim_n}\, \tau_n^2 \leq \overline{\lim_n}\, \tau_n^2 \leq M'' \tag{2.55}$$

we see that $f(0; B) = M'$ is a sufficient condition for

$$\lim_n \frac{e_n^{*^2}}{e_n^2} = 1. \tag{2.56}$$

The class of stationary noise situations for which the calculations (2.49), when $\psi_k \equiv K_0$, are in the limit ineffectual is thus extensively wider than "white".

REFERENCES

[*1*] Bertram, J. E., Control by Stochastic Adjustment, AIEE Applications and Industry Report (1960).

[*2*] Burkholder, D. L., On a class of stochastic approximation processes. *Ann. Math. Statist.* **27**, 1044-1059 (1956).

[*3*] Derman, C., Stochastic approximation. *Ann. Math. Statist.* **27**, 879-886 (1956).

[*4*] Driml, M., and Nedoma, J., Stochastic approximations for continuous random processes. *Trans. 2nd Prague Conf. on Information Theory, Statistical Decision Functions, and Random Processes* pp. 145-158. Publ. House Czech. Acad. of Sci., Prague, 1960.

[*5*] Dupač, V., O Kiefer-Wolfowitzove aproximačni methode. *Casopis pro Pestovani Mat.* **82**, 47-75 (1957). English transl. MIT Lincoln Lab. Report No. 22G-0008 (1960).

[*6*] Dvoretzky, A., On stochastic approximation. *Proc. 3rd Berkeley Symposium on Math. Statistics and Probability* **1**, 39-55. Univ. of California Press, Berkeley, California, 1956.

[*7*] Gardner, L. A., Jr., Stochastic Approximation and 'Minimax' Problems, MIT Lincoln Lab. Tech. Report 219, April 1960.

[*8*] Gardner, L. A., Jr., Adaptivi predictors. To be published.

[*9*] Hanš, O., Random fixed point theorems. *Trans. 1st Prague Conf. on Information Theory, Statistical Decision Functions, and Random Processes* pp. 105-125. Pub. House Czech. Acad. Sci., Prague, 1956.

[*10*] Hanš, O., and Špaček, A., Random fixed point approximation by differentiable trajectories. *Trans. 2nd Prague Conf. on Information Theory, Statistical Decision Functions, and Random Processes* pp. 203-213. Publ. House Czech. Acad. Sci., Prague, 1960.

[*11*] Sacks, J., Asymptotic distribution of stochastic approximation procedures. *Ann. Math. Statist.* **29**, 373-405 (1958).

Boundedness and Stability[1]

H. A. ANTOSIEWICZ
Department of Mathematics
University of Southern California, Los Angeles, California

It is well known that the concepts of boundedness and stability are, in general, independent. Yet, many methods that are used in studying one of these concepts can often be so modified as to be of use also in problems involving the other. In this lecture I would like to give a brief account of two such methods which seem particularly suited to questions of conditional stability and of the boundedness of some (but not necessarily every) solution of a differential equation

$$x' = f(t, x). \tag{1}$$

My aim is to outline in each case the general approach and to indicate only in passing various old and new results that can be obtained this way in a unified manner. Throughout I make no attempt at completeness of presentation.

For the sake of simplicity in exposition, I will assume in the sequel that f in (1) is a continuous mapping of $R_+ \times R^n$ into R^n. This guarantees that for each $t_0 \in R_+$ and each $x_0 \in R^n$ there exists at least one solution u of (1) which is defined in some open neighborhood (in R_+) of t_0 and satisfies $u(t_0) = x_0$.

1. Let me begin with some remarks concerning equations (1) than can be written in the form

$$x' = A(t)x + h(t, x) \tag{2}$$

where $h(t, 0) = 0$ for every $t \in R_+$ so that the constant mapping $u_0 : t \to 0$ is a solution in R_+. As is to be expected, the behavior of the solutions of (2) depends very strongly upon the properties of the solutions of the linear equation

$$x' = A(t)x, \tag{3}$$

particularly when h satisfies appropriate smallness conditions. There are many results which relate the stability of the trivial solution of (2)

[1] This work was done with partial support by the Office of Naval Research under Contract N-onr 228[09].

to that of the trivial solution of (3). A typical example is the following classical theorem which is essentially due to Lyapunov [*21*].

Theorem. If the trivial solution of (3) is exponential-asymptotically stable so is the trivial solution of (2) provided h satisfies in $R_+ \times R^n$

$$\| h(t, x) \| = o(\| x \|) \tag{4}$$

as $\| x \| \to 0$ uniformly with respect to t in R_+.

This terminology implies that there exist positive constants N, ν such that, given any $t_0 \in R_+$ and any $x_0 \in R^n$ with $\| x_0 \|$ sufficiently small, every solution u of (2) with $u(t_0) = x_0$ is defined in $[t_0, \infty)$ and satisfies there

$$\| u(t) \| \leq N \| u(t_0) \| \exp [- \nu(t - t_0)].$$

Lyapunov's theorem as well as its many variants (see for example Cesari [*7*], III, Sections 6.1 and 6.2 or Coddington and Levinson [*8*], Chapter 13, Sections 1 and 3) all involve assertions as to the behavior of *all* solutions u of (2) (with $\| u(t_0) \|$ small enough), and these assertions are consequences of properties of *all* solutions of (3). Thus, in most cases, they can be proved in a straight-forward manner from considerations of the integral representation for the solutions of (2),

$$u(t) = U(t) \left[U^{-1}(t_0)x_0 + \int_{t_0}^{t} U^{-1}(s)\, h(s, u(s))\, ds \right], \tag{5}$$

where U is a fundamental matrix of solutions of (3).

More interesting, and correspondingly more difficult to treat, are questions such as of the existence of some bounded solution of (2) when not all solutions of (3) are assumed to be bounded in R_+ and h, as above, has property (4) in $R_+ \times R^n$.

The special case when the coefficient matrix in (3) is constant and has no eigenvalues with zero real part can still be discussed in a fairly simple way, particularly under stronger assumptions on h guaranteeing the uniqueness of a solution of (3). However, the general case in which no considerations of eigenvalues are available requires an entirely different approach.

It was Perron [*30*] who first observed in 1930 that considerable information about the properties of the solutions of (3) can be gained by investigating the behavior of the solutions of the non-homogeneous equation

$$x' = A(t)x + b(t) \tag{6}$$

relative to various classes of forcing functions b. This idea was subsequently taken up by various authors (e.g. [*3*, *18*, *22*]) and developed systematically by Massera and Schäffer in the series of papers [*25*, *26*, *27*, *31*] where they introduce the following notion of a dichotomy of the solutions of (3).

A linear subspace X_0 of R^n is said to induce an exponential dichotomy of the solutions u of (3) if there exist positive constants N, N', ν, ν' and γ such that:

(i) $u(0) \in X_0$ implies $\| u(t) \| \leq N \| u(t_0) \| e^{-\nu(t-t_0)}$ for $t \geq t_0 \geq 0$;

(ii) $u(0) \in' X_0$ implies $\| u(t) \| \geq N' \| u(t_0) \| e^{\nu'(t-t_0)}$ for $t \geq t_0 \geq 0$;

(iii) if u_1, u_2 are any non-trivial solutions with $u_1(0) \in X_0$, $u_2(0) \in' X_0$ then $\| u_1(t)/\| u_1(t) \| - u_2(t)/\| u_2(t) \| \| \geq \gamma$ in R_+.

This clearly generalizes the situation that occurs if the coefficient matrix in (3) is a constant matrix A none of whose eigenvalues has zero real part. In this case the dimension of the linear subspace which induces an exponential dichotomy equals the number of eigenvalues of A having negative real part.

Using this concept Massera [*24*] proved by means of functional-analytic arguments the following result.

Theorem. If there exists a linear subspace X_0 of R^n which induces an exponential dichotomy of the solutions of (3) and if h satisfies (4) in $R_+ \times R^n$ then every solution u of (2) with $P_0 u(0) \in X_0$ is defined and bounded in R_+. (Here P_0 denotes the projection associated with X_0.)

Actually, Massera obtained this theorem in a considerably more general setting with the underlying space being a general (real or complex) reflexive Banach space X and h assumed to be weakly continuous in $R_+ \times X$. In the finite-dimensional case an analogous result was proved slightly earlier by Corduneanu [*11*] using the methods of Massera and Schäffer in [*25*].

The question as to when a linear subspace exists that induces an exponential dichotomy has been answered completely and in great generality in [*27*, *31*]. A simple sufficient condition is, for example, the following: if $t \to A(t)$ in (3) is continuous in R_+ and such that

$$\sup_{R_+} \int_t^{t+1} \| A(s) \| \, ds < \infty \tag{7}$$

then (6) should have at least one bounded solution in R_+ for every continuous mapping b of R_+ into R^n which has the same property (7). The condition (7) is equivalent to the requirement that, for any constant

$\alpha > 0$, either $t \to e^{\alpha t} \int_t^\infty e^{-\alpha s} \| A(s) \| ds$ exist and be bounded in R_+ or that $t \to e^{-\alpha t} \int_0^t e^{\alpha s} \| A(s) \| ds$ be bounded in R_+. It holds evidently when $t \to \| A(t) \|$ is bounded in R_+.

Thus, Massera's theorem includes as special cases the classical result of Lyapunov [*21*] (see for example Coddington and Levinson [*8*], Chapter 13, Section 4) and the more recent results of Perron [*30*], Maizel [*22*] and Demidovich [*13*], all of which impose more stringent conditions on h (such as that h be Lipschitzian in $R_+ \times R^n$ for a sufficiently small constant).

2. Let me now turn to the general situation when (1) need not contain terms that are linear in x. In this case, too, most qualitative results rely on some kind of comparison technique. However, there generally exists no representation, such as (5), for the solutions of (1) in terms of the solutions of some suitable equation of "first approximation". Thus, all such comparison techniques are forced to use indirect arguments of entirely different nature. A great number of them is based on considerations which are closely linked to the fundamental ideas that underly the well-known Second Method of Lyapunov.

As in that method, a continuous real valued (though not necessarily positive definite) function V in $R_+ \times R^n$ is introduced with which there are associated two functions V_1', V_2' that are defined by setting for each $t \in R_+$ and each $x \in R^n$

$$V_i'(t, x) = (-1)^i \liminf_{h \to 0} \frac{(-1)^i}{h} [V(t + h, x + hf(t, x)) - V(t, x)], \qquad i = 1, 2.$$

These functions are then related to the differential equation (1) in such a way that a qualitative description of a solution u of (1) in R_+ can be obtained from the properties of the function $\bar{V}$ defined in R_+ by $\bar{V}(t) = V(t, u(t))$. The basis for any such description is furnished by the following Lemma which can be traced, in its earliest version, to Kamke [*17*] who used it for questions concerning the uniqueness of a solution of (1) (see also Antosiewicz [*2*], Brauer [*5*], and Conti [*9*]).

Lemma. Let V be a continuous locally Lipschitzian real valued function in $R_+ \times R^n$ and suppose there exist two continuous real valued functions ω_1, ω_2 in $R_+ \times R$ such that for each $t \in R_+$ and each $x \in R^n$

$$(-1)^i V_i'(t, x) \leq (-1)^i \omega_i[t, V(t, x)], \qquad i = 1, 2.$$

Let u be a solution of (1) in some interval $I = [0, T) \subset R_+$, let $(-1)^i W_i$

be the maximal solution, corresponding to $(0, (-1)^i V(0, u(0)))$, of the equation

$$w_i' = (-1)^i \omega_i(t, (-1)^i w_i), \qquad i = 1, 2, \tag{8}$$

and suppose that each W_i is defined in I.

Then for each $t \in I$

$$W_1(t) \leq V[t, u(t)] \leq W_2(t).$$

This Lemma yields immediately the following general boundedness theorem.

Theorem. Suppose that the hypotheses of the Lemma are satisfied and that, in addition,

$$\lim_{||x|| \to \infty} V(t, x) = \infty$$

for each $t \in R_+$. If there exists a constant $V_0 \in \{V(t, x) : t \in R_+, x \in R^n\}$ such that the maximal solution $(-1)^i W_i$ of (8) corresponding to $(0, (-1)^i V_0)$ is defined and bounded in R_+ then every solution u of (1) for which $V(0, u(0)) = V_0$ is also defined and bounded in R_+.

Observe that not all solutions of (8) are required to be bounded in R_+.

Another direct consequence of the Lemma is the following general stability theorem. Here it is assumed that f in (1) is such that $f(t, 0) = 0$ in R_+ and that, given any $t_0 \in R_+$, every solution u of (1) with $|| u(t_0) ||$ sufficiently small is defined in $[t_0, \infty)$. This latter condition is automatically satisfied if the requirement that

$$\lim_{||x|| \to \infty} V(t, x) = \infty$$

for each $t \in R_+$ is added to the hypotheses of the Theorem (cf. ref. [9].)

Theorem. Let V be a continuous real valued function in $R_+ \times R^n$ which is locally Lipschitzian and positive definite, and suppose that there exists a continuous real valued function ω in $R_+ \times R_+$ such that for each $t \in R_+$ and each $x \in R^n$

$$V_2'(t, x) \leq \omega[t, V(t, x)].$$

Suppose, further, that $\omega(t, 0) = 0$ in R_+ and that for each $t_0 \in R_+$ and every sufficiently small $| w_0 |$ there exists a unique solution of

$$w' = \omega(t, w) \tag{9}$$

which is defined in $[t_0, \infty)$ and satisfies $w(t_0) = w_0$. Then the stability properties of the trivial solution of (1) are the same as those of the trivial solution of (9).

Clearly, in any particular case, the usefulness of these Theorems depends to a large extent on our ability to choose the functions V, ω_1, ω_2 appropriately to fit the specific properties of f in (1). That this can be done is shown by the large number of results which these Theorems contain as special cases.

For example, in boundedness questions, the simplest, although often not the most natural choice for V is to take $V(t, x) = \| x \|$ in $R_+ \times R^n$ so that $| V_i'(t, x) | \leq \| f(t, x) \|$. Thus, if f satisfies in $R_+ \times R^n$ a condition of the form $\| f(t, x) \| \leq \omega(t, \| x \|)$ where ω is continuous in $R_+ \times R_+$, then the existence of bounded solutions of (1) hinges on the existence of at least one bounded solution of the equation $w' = \omega(t, w)$. For this, many sufficient conditions are readily available (see for example Wintner [*33*]). Therefore, this approach leads in general to the same type of results, under essentially the same conditions, as does the method that has been employed recently by Stokes [*32*] and Hale and Stokes [*14*] which is based on a systematic use of Tychonoff's fixed point theorem.

Frequently a more judicious choice of V will give better results, as is shown by the very simple system of equations

$$x' = y, \qquad y' = -a(t, x, y)y - b(x).$$

Under the assumptions that a, b are continuous functions such that $a(t, x, y) \geq 0$ in $R_+ \times R \times R$ and $xb(x) > 0$ for $x \neq 0$ and

$$\lim_{|x| \to \infty} \int_0^x b(r)\, dr = \infty,$$

we can choose here

$$V(t, x, y) = y^2/2 + \int_0^x b(r)\, dr$$

which yields

$$V_1'(t, x, y) \equiv V_2'(t, x, y) \leq 0.$$

Hence all solutions are defined and bounded in R_+, and the trivial solution is (uniformly asymptotically) stable.

Of the numerous results which depend either directly or indirectly on considerations of the same or similar type I mention here only some of the most recent ones, such as in refs. [*2*, *5*, *19*, *20*, *28*, *29*], the latter of which are extensions of previous results of my own [*1*]. For a general investigation of boundedness questions which is based on a very similar method of approach I refer to the work of Yoshizawa [*34-36*] and of Hayashi [*15*, *16*].

As to the second Theorem above, it is clear that it yields a simple unified approach to the basic theorems of Lyapunov's Second Method. As special cases, it includes, for example, the results in Bihari [*4*] and Conti [*10*] and theorems 15, 17, 18 in Massera [*23*]. It is also the basis for many perturbation type results, such as in Corduneanu [*12*], which involve questions of total stability.

REFERENCES

[*1*] Antosiewicz, H. A., On non-linear differential equations of the second order with integrable forcing term. *J. London math. Soc.* **30**, 64-67 (1955).

[*2*] Antosiewicz, H. A., An inequality for approximate solutions of ordinary differential equations. *Math. Z.* **78**, 44-52 (1962).

[*3*] Bellman, R., On an application of a Banach-Steinhaus theorem to the study of the boundedness of solutions of non-linear differential and difference equations. *Ann. Math. (Princeton)* **49**, 515-522 (1948).

[*4*] Bihari, I., Researches of the boundedness and stability of the solutions of non-linear differential equations. *Acta Math. Acad. Sci. Hung.* **8**, 261-278 (1957).

[*5*] Brauer, F., Global behavior of solutions of ordinary differential equations. *J. math. Anal. and Appl.* **2**, 145-158 (1961).

[*6*] Brauer, F., and Sternberg, S., Local uniqueness, existence in the large and the convergence of successive approximations. *Am. J. Math.* **80**, 421-430 (1958); **81**, 797 (1959).

[*7*] Cesari, L., "Asymptotic Behavior and Stability Problems in Ordinary Differential Equations." Springer, Berlin, 1959.

[*8*] Coddington, E. A., and Levinson, N., "Theory of Ordinary Differential Equations." McGraw-Hill, New York, 1955.

[*9*] Conti, R., Sulla prolungabilità delle soluzioni di un sistema di equazioni differenziali ordinarie. *Boll. Unione mat. ital.* **11**, 510-514 (1956).

[*10*] Conti, R., Limitazioni "in ampiezza" delle soluzioni di un sistema di equazioni differenziali e applicazioni. *Boll. Unione mat. ital.* **11**, 344-350 (1956).

[*11*] Corduneanu, C., Sur un théorème de Perron. *Analele sti. Univ. "Al. I. Cuza" Iasi, Sect. I* [N.S.] **5**, 33-36 (1959).

[*12*] Corduneanu, C., Sur la stabilité asymptotique, I. *Analele sti. Univ. "Al. I. Cuza" Iasi Sect. I* [N.S.] **5**, 37-40 (1959); II. *Acad. rep. pop. Romîne, Fil. Iasi, Stud. cerc. sti., Ser. I Mat.* **10**, 209-213 (1959).

[*13*] Demidovich, B. P., On bounded solutions of some systems of ordinary differential equations. *Mat. Sbornik* [N.S.] **40**, No. 82, 73-94 (1956) (in Russian).

[*14*] Hale, J. K., and Stokes, A. P., Behavior of solutions near integral manifolds. *Arch. rational Mech. Anal.* **6**, 133-170 (1960).

[*15*] Hayashi, K., On the strong stability and boundedness of solutions of ordinary differential equations. *Mem. Coll. Sci., Univ. Kyoto* **A32**, 281-295 (1959).

[*16*] Hayashi, K., Strong stability and boundedness with respect to the weight $\omega(t)$. *Mem. Coll. Sci., Univ. Kyoto* **A32**, 413-444 (1960).

[*17*] Kamke, E., Über die eindeutige Bestimmtheit der Integrale von Differentialgleichungen, II. *Sitzber. Heidelb. Akad. Wiss., Math.-naturwiss. Kl., Abhandl.* **17**, 15 pp. (1930).

[*18*] Kucer, D. L., On some criteria for the boundedness of the solutions of a system of differential equations. *Doklady Akad. Nauk S.S.S.R.* **69**, 603-606 (1949) (in Russian).

[*19*] Levin, J. J., On the global asymptotic behavior of non-linear systems of differential equations. *Arch. rational Mech. Anal.* **6**, 65-74 (1960).

[*20*] Levin, J. J., and Nohel, J. A., Global asymptotic stability for nonlinear systems of differential equations and applications to reactor dynamics. *Arch. rational mech. Anal.*, **5**, 194-211 (1960).

[*21*] Liapounoff, A. M., "Problème Général de la Stabilité du Mouvement." Princeton Univ. Press, Princeton, New Jersey, 1949.

[*22*] Maizel, A. D., On the stability of solutions of systems of differential equations. *Trudy Ural. Politekh. Inst.* **51**, 20-50 (1954) (in Russian).

[*23*] Massera, J. L., Contributions to stability theory. *Ann. Math. (Princeton)* **64**, 182-206 (1956).

[*24*] Massera, J. L., Sur l'existence de solutions bornées et periodiques des systèmes quasi-linéaires d'équations différentielles. *Ann. mat. pura appl.* [4] **51**, 95-105 (1960).

[*25*] Massera, J. L., and Schäffer, J. J., Linear differential equations and functional analysis, I. *Ann. Math. (Princeton)* **67**, 517-573 (1958).

[*26*] Massera, J. L., and Schäffer, J. J., Linear differential equations and functional analysis, III. Lyapunov's second method in the case of conditional stability. *Ann. Math. (Princeton)* **69**, 535-574 (1959).

[*27*] Massera, J. L., and Schäffer, J. J., Linear differential equations and functional analysis, IV. *Math. Ann. (Leipzig)* **139**, 287-342 (1960).

[*28*] Opial, Z. Sur la stabilité asymptotique des solutions d'un système d'équations différentielles. *Ann. polon. Math.* **7**, 259-267 (1960).

[*29*] Opial, Z., Sur une équation différentielle non-linéaire du second ordre. *Ann. polon. Math.* **8**, 65-69 (1960).

[*30*] Perron, O., Die Stabilitätsfrage bei Differentialgleichungen. *Math. Z.* **32**, 703-728 (1930).

[*31*] Schäffer, J. J., Linear differential equations and functional analysis, V. *Math. Ann. (Leipzig)* **140**, 308-321 (1960).

[*32*] Stokes, A. P., The applications of a fixed point theorem to a variety of nonlinear stability problems. *In* "Contributions to the Theory of Nonlinear Oscillations" (S. Lefschetz, ed.), Vol. 5. Princeton Univ. Press, Princeton, New Jersey, 1960.

[*33*] Wintner, A., An Abelian lemma concerning asymptotic equilibria. *Am. J. Math.* **68**, 451-454 (1946).

[*34*] T. Yoshizawa, Liapunov's function and boundedness of solutions. *Funk. Ekv.* **2**, 95-142 (1959).

[*35*] Yoshizawa, T., Existence of a bounded solution and existence of a periodic solution of the differential equations of the second order. *Mem. Coll. Sci., Univ. Kyoto* **A33**, 301-308 (1960).

[*36*] T. Yoshizawa, Stability and boundedness of systems. *Arch. rational Mech. Anal.* **6**, 409-421 (1960).

Discrete Time Optimal Control Systems[1]

LUCIEN W. NEUSTADT
The Aerospace Corporation
El Segundo, California

1. Statement of the Problem

The well known time optimal control problem may be formulated as follows. We consider a control system whose state at any time t is described by an n-dimensional vector function $\mathbf{x}(t)$. The coordinates of $\mathbf{x}$ are sometimes referred to as phase coordinates. This vector satisfies the differential equation

$$\dot{\mathbf{x}}(t) = \mathbf{f}(\mathbf{x}(t), \mathbf{u}(t), t), \tag{1}$$

where $\mathbf{u}(t)$ is the *control function*. We shall assume that $\mathbf{u}(t)$ is an r-dimensional vector. In general we are also given a region U in r-space. Functions $\mathbf{u}(t)$ whose values belong to U, and for which equation (1) has a solution, will be called *admissible*. Clearly, the solutions of (1) depend on the choice of $\mathbf{u}(t)$. If $\mathbf{x}(t)$ is a solution of equation (1) for a particular function $\mathbf{u}(t)$, with initial value $\mathbf{x}(0) = \mathbf{x}^0$, and if $\mathbf{x}(t_1) = \mathbf{x}'$ (for $t_1 > 0$), we shall say that $\mathbf{u}(t)$ transfers $\mathbf{x}^0$ to $\mathbf{x}'$ in t_1 seconds. If two points $\mathbf{x}^0$ and $\mathbf{x}^F$, equation (1), and U are given, the time optimal control problem consists of finding an admissible control which transfers $\mathbf{x}^0$ to $\mathbf{x}^F$ in minimal time.

In this paper we shall restrict ourselves to control functions which arise in so-called sampled-data control systems. Thus, we shall suppose that a "sampling period" T is given beforehand, and that $\mathbf{u}(t)$ is constant for $(i-1)\,T \leqq t < iT$ $(i = 1, 2, \ldots)$. A control function $\mathbf{u}(t)$ of this type:

$$\mathbf{u}(t) = \mathbf{u}^i \qquad \text{for} \qquad (i-1)T \leqslant t < iT, \qquad i = 1, 2, \ldots \tag{2}$$

whose values are in U ($\mathbf{u}^i \in U$, $i = 1, 2, \ldots$) will be called a *discrete admissible control function*. If $\mathbf{x}(t)$ is the solution of equation (1) with a particular discrete control function $\mathbf{u}(t)$, with $\mathbf{x}(0) = \mathbf{x}^0$; and if $\mathbf{x}(mT) = \mathbf{x}'$ for some positive integer m, we shall say that $\mathbf{u}(t)$ transfers $\mathbf{x}^0$ to $\mathbf{x}'$ in m steps. If $\mathbf{u}(t)$ is given by (2) we shall also say that the sequence $\{\mathbf{u}^1, \ldots, \mathbf{u}^m\}$ transfers $\mathbf{x}^0$ to $\mathbf{x}'$.

[1] This work was carried out while the author was with Space Technology Laboratories, Inc. as a part of the company sponsored research program.

The discrete time optimal problem consists of finding the discrete admissible controls $\mathbf{u}(t)$ which transfer $\mathbf{x}^0$ to $\mathbf{x}^F$ in a minimal number of steps. Such controls will be called time optimal discrete controls. Unless otherwise specified, we shall assume that $\mathbf{x}^0$, $\mathbf{x}^F$, and T are fixed.

In this paper we shall further restrict ourselves to control systems where equation (1) is linear, i.e., of the form of equation (3). We shall derive the form of particular time optimal controls, and since these controls are not, in general, unique, we shall discuss some side optimizations. We shall also describe methods for computing these optimal discrete controls. Finally we shall show that under appropriate restrictions, time optimal discrete controls approach the continuous time optimal control if the sampling period is allowed to go to zero.

The discrete optimal control problem was first considered by Krasovskii [*6*]. Our approach is similar to his, and our results generalize many of Krasovskii's. Desoer and Wing [*2, 3*] have also treated certain aspects of the same problem, and a comparison between some of their results and those of this paper are described in section 7. The continuous time optimal control problem has been extensively described in the literature (see for example refs. [*1, 4, 5, 7*]).

2. Introduction

We consider the discrete time optimal control problem where equation (1) has the form

$$\dot{\mathbf{x}} = A\mathbf{x} + B\mathbf{u}, \tag{3}$$

where A is an $n \times n$ matrix, and B is an $n \times r$ matrix. Both may depend on time. We shall also assume that U is the unit cube, so that $\mathbf{u} \in U$ if and only if[1] $| u_j | \leqslant 1$ for $j = 1, ..., r$. The solution of (1) may be written in the form

$$\mathbf{x}(t) = X(t) \left[\mathbf{x}^0 + \int_0^t X^{-1}(s)\, B(s)\, \mathbf{u}(s)\, ds\right], \tag{4}$$

where $X(t)$ is the matrix solution of $\dot{X}(t) = AX(t)$, $X(0) = I$, the identity matrix.

In particular, if $\mathbf{u}(t)$ is given by (2), and m is a positive integer,

$$\mathbf{x}(mT) = X(mT) \left[\mathbf{x}^0 + \int_0^{mT} X^{-1}(s)\, B(s)\, \mathbf{u}(s)\, ds\right]$$

$$= X(mT) \left\{\mathbf{x}^0 + \sum_{i=1}^{m} \int_{(i-1)T}^{iT} X^{-1}(s)\, B(s)\, ds\, \mathbf{u}^i\right\}. \tag{5}$$

[1] If $\mathbf{v}$ is a vector, we shall denote the components of $\mathbf{v}$ by v_x.

If we set $\mathbf{x}(mT) = \mathbf{x}^m$, and $X(mT) = X_m$, it is easily seen that

$$\mathbf{x}^m - \mathbf{x}^{m-1} = K_m \mathbf{x}^{m-1} + L_m \mathbf{u}^m, \tag{6}$$

where

$$K_m = X_m X_{m-1}^{-1} - I, \tag{7}$$

and

$$L_m = \int_{(m-1)T}^{mT} X_m X^{-1}(s)\, B(s)\, ds. \tag{8}$$

Also, from (5),

$$\mathbf{x}^m = X_m \left[\mathbf{x}^0 + \sum_{i=1}^{m} X_i^{-1} L_i \mathbf{u}^i\right]. \tag{9}$$

If we wish, we may use equation (6) as our point of departure, and consider that our discrete control system is described by this difference equation. Equation (9) may be derived directly from (6) if we define X_m as the solution of

$$X_m - X_{m-1} = K_m X_{m-1}, \qquad X_0 = I.$$

When we use this point of view, we shall refer to *admissible sequences* $\{\mathbf{u}^1, \ldots, \mathbf{u}^m\}$, defined by the property that $\mathbf{u}^i \in U$ for $i = 1, \ldots, m$.

3. The Sets Ω_m and C_m

We define the sets Ω_m and C_m by

$$\Omega_m = \left\{X_m \left[\mathbf{x}^0 + \sum_{i=1}^{m} X_i^{-1} L_i \boldsymbol{\xi}^i\right]; \qquad |\xi_j^i| \leqslant 1 \qquad i = 1, \ldots, m,\ j = 1, \ldots, r\right\},$$

$$C_m = \left\{- \sum_{i=1}^{m} X_i^{-1} L_i \boldsymbol{\xi}^i; \qquad |\xi_j^i| \leqslant 1 \qquad \text{all } i \text{ and } j\right\}.$$

If there is an admissible sequence $\{\mathbf{u}^1, \ldots, \mathbf{u}^m\}$ which transfers $\mathbf{x}^0$ to $\mathbf{x}'$, equation (9) shows that $\mathbf{x}' \in \Omega_m$ (let $\boldsymbol{\xi}^i = \mathbf{u}^i$). Conversely, if $\mathbf{x}' \in \Omega_m$, there is an admissible sequence, $\{\boldsymbol{\xi}^1, \ldots, \boldsymbol{\xi}^m\}$, which transfers $\mathbf{x}^0$ to $\mathbf{x}'$ in m steps. Thus, Ω_m is the set of all points to which $\mathbf{x}^0$ can be transferred in m steps, using an admissible discrete control. Similarly, it is easy to see that C_m is the set of all points which can be transferred to the origin in m steps, using an admissible discrete control. Note that,

$$\Omega_m = X_m[\mathbf{x}^0 - C_m]. \tag{10}$$

The sets Ω_m and C_m are discrete analogs of the sets $\Omega(t)$ in ref. [4] and $K(t)$ in ref. [7].

The discrete time optimal control problem consists of finding the smallest positive integer m such that $\mathbf{x}^F \in \Omega_m$.

Let

$$\mathbf{b}^{ij} = X_i^{-1} \boldsymbol{l}^{ij},$$

where $\boldsymbol{l}^{ij}$ is the j-th column vector of L_i. Then

$$C_m = \left\{ \sum_{i=1}^{m} \sum_{j=1}^{r} - \mathbf{b}^{ij} \xi_j^i ;\ |\xi_j^i| \leqslant 1, \text{ all } i \text{ and } j \right\}. \tag{11}$$

Let us denote the set of the $\mathbf{b}^{ij}$, for $i = 1, ..., m$ and $j = 1, ..., r$ by $\mathscr{B}_m$. It is easy to see that C_m is the convex hull of the finite point set

$$\left\{ \sum_{i=1}^{m} \sum_{j=1}^{r} \epsilon_{ij} \mathbf{b}^{ij} ;\ \epsilon_{ij} = \pm 1, \text{ all } i \text{ and } j \right\},$$

so that C_m is a convex polyhedron in the space $\mathscr{E}^l$ (linear space of dimension $l \leqslant n$) generated by $\mathscr{B}_m$.

Let us characterize the $(l - 1)$-dimensional faces of C_m (assuming that $l > 1$). Every vector $\boldsymbol{\psi}$ in $\mathscr{E}^l$ is an external normal of some support plane to C_m, defined by

$$\{\mathbf{x} : \boldsymbol{\psi} \cdot \mathbf{x} = \max_{\mathbf{y} \in C_m} \boldsymbol{\psi} \cdot \mathbf{y}\}.$$

But if $\mathbf{y} \in C_m$ is defined as in (11),

$$\boldsymbol{\psi} \cdot \mathbf{y} = \sum_{i=1}^{m} \sum_{j=1}^{r} - (\boldsymbol{\psi} \cdot \mathbf{b}^{ij})\, \xi_j^i, \qquad |\xi_j^i| \leqslant 1.$$

Considered as a function of $\mathbf{y}$, this expression takes on its maximum for $\mathbf{y}$ in C_m when $\xi_j^i = -\operatorname{sign}(\boldsymbol{\psi} \cdot \mathbf{b}^{ij})$ if $\boldsymbol{\psi} \cdot \mathbf{b}^{ij} \neq 0$, and for any ξ_j^i, $|\xi_j^i| \leqslant 1$, if $\boldsymbol{\psi} \cdot \mathbf{b}^{ij} = 0$. Thus,

$$\max_{\mathbf{y} \in C_m} \boldsymbol{\psi} \cdot \mathbf{y} = \sum_{i=1}^{m} \sum_{j=1}^{r} |\boldsymbol{\psi} \cdot \mathbf{b}^{ij}|. \tag{12}$$

If $\boldsymbol{\psi}$ is the normal to an $(l - 1)$-dimensional face of C_m, relation (12) must hold for an $(l - 1)$-dimensional set of vectors $\mathbf{y} \in C_m$. By what was said above, this will be true if and only if $\boldsymbol{\psi}$ is orthogonal to $(l - 1)$ linearly independent vectors in $\mathscr{B}_m$. Let $\mathscr{S}_1, \mathscr{S}_2, ..., \mathscr{S}_k$ be all the $(l - 1)$-dimensional subspaces spanned by subsets of $\mathscr{B}_m$, and let

$\pm \boldsymbol{\psi}^1, ..., \pm \boldsymbol{\psi}^k$ be the unit normals to $\mathscr{S}_1, ..., \mathscr{S}_k$, in $\mathscr{E}^l$. Then the $(l-1)$-dimensional faces of C_m (which are $2k$ in number) are defined by

$$\left\{\mathbf{x}: \boldsymbol{\psi}^s \cdot \mathbf{x} = \epsilon \sum_{i=1}^{m} \sum_{j=1}^{r} | \boldsymbol{\psi}^s \cdot \mathbf{b}^{ij} |, \mathbf{x} \in C_m \right\}, \qquad s = 1, ..., k, \epsilon = \pm 1.$$

Since a convex polyhedron is the intersection of all the half spaces determined by its faces, $\mathbf{x} \in C_m$ if and only if 1) $\mathbf{x} \in \mathscr{E}^l$, and 2)

$$|\boldsymbol{\psi}^s \cdot \mathbf{x}| \leq \sum_{i=1}^{m} \sum_{j=1}^{r} |\boldsymbol{\psi}^s \cdot \mathbf{b}^{ij}| \qquad \text{for } s = 1, ..., k. \tag{13}$$

We shall say that the control system is *normal in the discrete sense* if, for every $j = 1, ..., r$, every set of n vectors $\mathbf{b}^{i_1 j}, ..., \mathbf{b}^{i_n j}$ with $1 \leqslant i_1 < i_2 < ... < i_n$ is linearly independent. This definition is analogous to that of a normal continuous control system [7].

4. The Characterization of Optimal Controls

Consider a reformulation of the given problem.[1] For m fixed, consider the set $\mathscr{I}_m$ of all sequences (not necessarily admissible) $\{\mathbf{u}^1, ..., \mathbf{u}^m\} = \sigma$ which transfer $\mathbf{x}^0$ to $\mathbf{x}^F$ in m steps. Define the function

$$g(\sigma) = \max_{\substack{1 \leq i \leq m \\ 1 \leq j \leq r}} | u_j^i |, \tag{14}$$

and let

$$\alpha(m) = \inf_{\sigma \in \mathscr{I}_m} g(\sigma).$$

It is easily seen that this lower bound is achieved for some $\sigma \in \mathscr{I}_m$. Define

$$\mathbf{y}^m = \mathbf{x}^0 - X_m^{-1} \mathbf{x}^F. \tag{15}$$

Note that $\mathbf{x}^F \in \Omega_m$ if and only if $\mathbf{y}^m \in C_m$. It follows from the definition of C_m that $\alpha(m)$ is also the g.l.b. of all positive numbers μ for which $(1/\mu)\, \mathbf{y}^m$ belongs to C_m. (Note that $\mathscr{I}_m$ may be empty, in which case no positive multiple of $\mathbf{y}^m$ belongs to C_m.)

Our original problem is equivalent to finding the smallest m for which $\alpha(m) \leqslant 1$.

[1] This formulation is due to Krasovskii [5, 6].

Theorem 1. If $\alpha(m)$ is defined (i.e., if $\mathscr{I}_m$ is not empty) and positive, then

$$\frac{1}{\alpha(m)} = \min_s \frac{\sum_{i=1}^{m} \sum_{j=1}^{r} |\boldsymbol{\psi}^s \cdot \mathbf{b}^{ij}|}{|\boldsymbol{\psi}^s \cdot \mathbf{y}^m|}, \tag{16}$$

where the $\boldsymbol{\psi}^s$ are the vectors defined in section 3.

Proof: Since C_m is closed, $(1/\alpha(m))\, \mathbf{y}^m$ belongs to C_m. Since no larger multiple of $\mathbf{y}^m$ does so, $(1/\alpha(m))\, \mathbf{y}^m$ is on the boundary of C_m; i.e., $(1/\alpha(m))\, \mathbf{y}^m$ belongs to one of the $(l-1)$-dimensional faces of C_m. Let the external normal, in $\mathscr{E}^l$, of this face be $\boldsymbol{\psi}^a$. Then, by (12),

$$0 < \frac{1}{\alpha(m)} \boldsymbol{\psi}^a \cdot \mathbf{y}^m = \sum_{i=1}^{m} \sum_{j=1}^{r} |\boldsymbol{\psi}^a \cdot \mathbf{b}^{ij}|.$$

On the other hand, by (13),

$$\frac{1}{\alpha(m)} |\boldsymbol{\psi}^s \cdot \mathbf{y}^m| \leqslant \sum_{i} \sum_{j} |\boldsymbol{\psi}^s \cdot \mathbf{b}^{ij}| \qquad \text{for all } s,$$

which proves our assertion.

Let us denote the vector $\boldsymbol{\psi}^a$ of the theorem by $\boldsymbol{\phi}^m$.

Theorem 2. If there is a (not necessarily admissible) sequence which transfers $\mathbf{x}^0$ to $\mathbf{x}^F$ in m steps, there is another sequence $\hat{\sigma} = \{\hat{\mathbf{u}}^1, ..., \hat{\mathbf{u}}^m\}$ such that

(a) $\hat{\sigma}$ transfers $\mathbf{x}^0$ to $\mathbf{x}^F$,

(b) $g(\hat{\sigma}) = \alpha(m)$,

(c) if σ is any sequence which transfers $\mathbf{x}^0$ to $\mathbf{x}^F$ in m steps then $g(\sigma) \geqslant g(\hat{\sigma})$, and

(d) $\hat{u}_j^i = -\alpha(m) \operatorname{sign}(\boldsymbol{\phi}^m \cdot \mathbf{b}^{ij})$ if $\boldsymbol{\phi}^m \cdot \mathbf{b}^{ij} \neq 0$. (17)

Proof: By hypothesis $\mathscr{I}_m$ is not empty. Assume $\alpha(m) > 0$ (the results are immediate if $\alpha(m) = 0$). Since $(1/\alpha(m))\, \mathbf{y}^m$ belongs to a face of C_m (with external normal $\boldsymbol{\phi}^m$),

$$\frac{1}{\alpha(m)} \mathbf{y}^m = -\sum_{i=1}^{m} \sum_{j=1}^{r} \mathbf{b}^{ij} \xi_j^i$$

with $\xi_j^i = -\operatorname{sign}(\boldsymbol{\phi}^m \cdot \mathbf{b}^{ij})$ if $\boldsymbol{\phi}^m \cdot \mathbf{b}^{ij} \neq 0$ and $|\xi_j^i| \leqslant 1$ for all i and j. Using (15) and the definition of the $\mathbf{b}^{ij}$, it follows that

$$\mathbf{x}^F = X_m \left\{ \mathbf{x}^0 + \sum_{i,j} X_i^{-1} \boldsymbol{l}^{ij} [\alpha(m)\, \xi_j^i] \right\}.$$

Thus, if we set $\hat{u}_j^i = \alpha(m)\, \xi_j^i$, $\hat{\sigma}$ has properties (a), (b) and (d). If σ is any sequence which transfers $\mathbf{x}^0$ to $\mathbf{x}^F$ in m steps, $(1/g(\sigma))\, \mathbf{y}^m \in C_m$, so that (c) follows from the alternate characterization of $\alpha(m)$. Q.E.D.

The following theorem characterizes certain time optimal discrete controls.

Theorem 3. If there is an admissible discrete control which transfers $\mathbf{x}^0$ to $\mathbf{x}^F$, there is a time optimal discrete control $\hat{u}(t)$ (which also transfers $\mathbf{x}^0$ to $\mathbf{x}^F$) given by

$$\hat{\mathbf{u}}(t) = \hat{\mathbf{u}}^i \qquad (i-1)\,T \leqslant t < iT \qquad i = 1, \ldots, M, \tag{18}$$

$$\hat{u}_j^i = -A \operatorname{sign}(\boldsymbol{\phi}^M \cdot \mathbf{b}^{ij}) \qquad \text{if} \qquad \boldsymbol{\phi}^M \cdot \mathbf{b}^{ij} \neq 0, \tag{19}$$

$$|\hat{u}_j^i| \leqslant A \leqslant 1 \text{ all } i \text{ and } j.$$

If σ is any other admissible sequence which transfers $\mathbf{x}^0$ to $\mathbf{x}^F$ in M steps, $g(\sigma) \geqslant A$. Furthermore, $A = \alpha(M)$ is given by formula (16), where the minimizing vector is $\boldsymbol{\phi}^M$; and $\alpha(M-1) > 1$.

Proof: This is an immediate consequence of theorems 1 and 2.

Note the lack of uniqueness for time optimal discrete controls. This is due to the fact that the vector $\mathbf{y}^M$ is in general not on the boundary of C_M (or equivalently, $\mathbf{x}^F$ is not on the boundary of Ω_M). Then the representation of $\mathbf{y}^M$ in the form (11) is unique only if the vectors $\mathbf{b}^{ij}$ for $i = 1, \ldots, M$, $j = 1, \ldots, r$ are linearly independent, which is only true if $M \leqslant n/r$. Even the control described by theorem 3 may be ambiguous, since (18) does not determine $\hat{u}_j^i$ if $\mathbf{b}^{ij}$ is orthogonal to $\boldsymbol{\phi}^M$.

Theorem 3 in one sense defines the "best" time optimal discrete control, in that it minimizes the maximum amount of control required: $g(\hat{\sigma}) \leq g(\sigma)$. Also, note that

$$\boldsymbol{\phi}^M \cdot \mathbf{b}^{ij} = \boldsymbol{\phi}^M \cdot X_i^{-1} \boldsymbol{l}^{ij} = \boldsymbol{\theta}^i \cdot \boldsymbol{l}^{ij},$$

where $\boldsymbol{\theta}^i = \boldsymbol{\phi}^M \cdot X_i^{-1}$ (both $\boldsymbol{\phi}^M$ and $\boldsymbol{\theta}^i$ are row vectors). It follows from (7) that

$$\boldsymbol{\theta}^{i-1} - \boldsymbol{\theta}^i = \boldsymbol{\theta}^i K_i,$$

so that (19) is analogous to the maximum principle for linear systems [*4*].

If $\mathbf{x}^F = 0$, there is an optimal control which comes near to being "bang-bang" as is shown in theorem 4.

Theorem 4. Suppose that there is an admissible discrete control which transfers $\mathbf{x}^0$ to the origin. Then there is a time optimal discrete control $\hat{\mathbf{u}}(t)$ which transfers $\mathbf{x}^0$ to the origin (say in M steps) and is given by (18) where, for $j = 1, ..., r$,

$$\hat{u}^i_j = \text{sign}\,(\boldsymbol{\psi} \cdot \mathbf{b}^{ij}) \quad \text{if} \quad (\boldsymbol{\psi} \cdot \mathbf{b}^{ij}) \neq 0 \quad \text{and} \quad i = 1, ..., M-1,$$

$$\hat{u}^M_j = \beta\,\text{sign}\,(\boldsymbol{\psi} \cdot \mathbf{b}^{Mj}) \quad \text{if} \quad (\boldsymbol{\psi} \cdot \mathbf{b}^{Mj}) \neq 0.$$

The vector $\boldsymbol{\psi}$ is orthogonal to a face of C_M, and $0 < \beta \leqslant 1$. If the control system is normal in the discrete sense, then for every j, $u^i_j = \pm 1$ except for at most n values of i, $1 \leqslant i \leqslant M$.

Proof: By hypothesis, M is the least number of steps in which $\mathbf{x}^0$ can be transferred to 0 with an admissible sequence. Then $\mathbf{x}^0 \in C_M$ and $\mathbf{x}^0 \notin C_{M-1}$. Define

$$J_M = \left\{ -\sum_{j=1}^{r} \mathbf{b}^{Mj} \xi^M_j;\ |\, \xi^M_j \,| \leqslant 1,\ j = 1, ..., r \right\}.$$

Clearly $C_M = C_{M-1} + J_M$. Let β be the g.l.b. of the positive numbers η for which $\mathbf{x}^0 \in C_{M-1} + \eta J_M$. Then it is easy to see that $0 < \beta \leqslant 1$, and that $\mathbf{x}^0$ is a boundary point of $C_{M-1} + \beta J_M$. But

$$C_{M-1} + \beta J_M = \left\{ -\sum_{i=1}^{M} \sum_{j=1}^{r} \mathbf{b}^{ij} \xi^i_j;\ |\, \xi^i_j \,| \leqslant 1,\ i = 1, ..., M-1,\ j = 1, ..., r; \right.$$
$$\left. |\, \xi^M_j \,| \leqslant \beta,\ j = 1, ..., r \right\},$$

so that this set is a convex polyhedron whose faces are parallel to those of C_M. Since $\mathbf{x}^0$ lies on one of these faces, we can show as before (letting $\boldsymbol{\psi}$ be the internal normal to this face) that

$$\mathbf{x}^0 = -\sum_{i=1}^{M} \sum_{j=1}^{r} \mathbf{b}^{ij} \zeta^i_j,$$

where $\zeta^i_j = \text{sign}\,(\boldsymbol{\psi} \cdot \mathbf{b}^{ij})$ for $i = 1, ..., M-1$, $j = 1, ..., r$, and $\zeta^M_j = \beta\,\text{sign}\,(\boldsymbol{\psi} \cdot \mathbf{b}^{Mj})$, $j = 1, ..., r$. Thus, we need only set $\hat{u}^i_j = \zeta^i_j$. If the system is normal in the discrete sense, $\boldsymbol{\psi}$ can be orthogonal to at most $n - 1$ of the vectors $\mathbf{b}^{1j}, ..., \mathbf{b}^{M-1,j}$, which completes the proof.

Since the discrete time optimal control is not, in general, unique, it may be of interest to find the "best" time optimal control. One such meaning for "best" has been described above (the best control is the one that minimizes max $| u_j^i |$). In particular applications, other side optimizations may be of interest. For example, it may be desirable to minimize a measure of the total amount of control:

$$L = \sum_{i,j} \gamma_{ij} \, | u_j^i |, \qquad \gamma_{ij} > 0.$$

To derive the form of the time optimal discrete control which minimizes L, we consider the sets (for $\beta \geqslant 0$),

$$C_{M\beta} = \left\{ \mathbf{x} : \mathbf{x} = - \sum_{i=1}^{M} \sum_{j=1}^{r} \mathbf{b}^{ij} \xi_j^i ; \, | \xi_j^i | \leqslant 1, \sum_{i,j} \gamma_{ij} \, | \xi_j^i | \leqslant \beta \right\},$$

where M is the number of steps in a time optimal discrete control. It is clear that $C_{M\beta}$ is a convex subset of C_M, that $C_{M\beta} \subset C_{M\alpha}$ for $\beta < \alpha$, that $C_M = C_{M\beta}$ for β large enough, and that

$$\bigcap_{\beta > 0} C_{M\beta} = \{0\}.$$

If

$$\beta' = \inf_{\mathbf{y}^M \in C_{M\beta}} \beta,$$

it is easily seen that $\mathbf{y}^M$ is a boundary point of $C_{M\beta'}$. Thus,

$$\mathbf{y}^M = - \sum_{i=1}^{M} \sum_{j=1}^{r} \mathbf{b}^{ij} \zeta_j^i,$$

where

$$| \zeta_j^i | \leqslant 1 \qquad \text{and} \qquad \sum_{i,j} \gamma_{ij} \, | \zeta_j^i | = \beta',$$

and, for some $\boldsymbol{\psi}$,

$$\boldsymbol{\psi} \cdot \mathbf{y}^M \geqslant \boldsymbol{\psi} \cdot \mathbf{x}, \qquad \text{for all} \qquad \mathbf{x} \in C_{M\beta'}.$$

This is only possible if

$$\zeta_j^i = \begin{cases} - \operatorname{sign} (\boldsymbol{\psi} \cdot \mathbf{b}^{ij}) & \text{if} \quad | \boldsymbol{\psi} \cdot \mathbf{b}^{ij} / \gamma_{ij} | > \lambda \\ 0 & \text{if} \quad | \boldsymbol{\psi} \cdot \mathbf{b}^{ij} / \gamma_{ij} | < \lambda, \end{cases}$$

where λ is a non-negative number. By letting $u_j^i = \zeta_j^i$, we have a time optimal discrete control which minimizes L.

Instead of the time optimal control it may be of interest to find that admissible discrete control which transfers $\mathbf{x}^0$ to $\mathbf{x}^F$ while minimizing L (without necessarily being time optimal). This problem can be studied in a similar fashion. Different definitions of L also give rise to new formulations for the optimal control.

The results of theorems 1, 2, and 3 were essentially derived by Krasovskii [6], although the difference equations he considered arose from finite difference approximations to equation (1), rather than from piecewise constant control functions. Krasovskii also allowed an extra term $\mathbf{f}(t)$ in the right side of equation (3), and let $\mathbf{x}^F$ be time varying. By appropriately redefining $\mathbf{y}^m$ (equation (15)), our results carry over to this case. Finally, Krasovskii restricted himself to the case where A and B are constant matrices; but his proofs are easily generalized to the time-varying case.

Theorem 1 above is somewhat stronger than the corresponding result in ref. [6]. It is easily seen that relation (16) can also be given the form

$$\frac{1}{\alpha(m)} = \min \frac{\sum_{i=1}^{m}\sum_{j=1}^{r} \boldsymbol{\psi}\cdot\mathbf{b}^{ij}}{|\boldsymbol{\psi}\cdot\mathbf{y}^m|},$$

where the minimum is taken over *all* vectors $\boldsymbol{\psi}$ in $\mathscr{E}^l$; or equivalently

$$\frac{1}{\alpha(m)} = \min_{|\psi\cdot\mathbf{y}^m|=1} \sum_{i=1}^{m}\sum_{j=1}^{r} |\boldsymbol{\psi}\cdot\mathbf{b}^{ij}|.$$

This is essentially the same as equation (13) in ref. [6]. Equation (16) is more useful for computational purposes since it requires finding the minimum with respect to a finite number of vectors $\boldsymbol{\psi}$.

Krasovskii's approach differs from the one in this paper. He represented the sequence $\{\mathbf{u}^1, ..., \mathbf{u}^m\}$ as a linear functional on a finite $(m \times r)$-dimensional vector space, and reduced the problem to minimizing the norm of this functional in the presence of n constraints on its values. By using results on the so-called "L-problem", he derived equation (17) and the modified form of equation (16).

Krasovskii treated the continuous time optimal problem in the same way [5, 6], by replacing the finite dimensional space by the function space L^1. It is interesting to note that the continuous analogs of relations (16) and (17) can be derived geometrically as was done above for the discrete problem.

5. Computing the Time Optimal Control

In order to compute the discrete optimal control it is first necessary to determine the smallest m for which $\mathbf{x}^F \in \Omega_m$ (or equivalently, the smallest m for which $\mathbf{y}^m \in C_m$), and then to express $\mathbf{y}^m$ in the form (11), thereby determining an optimal control.

To accomplish the first objective, it suffices to compute the function $\alpha(m)$ from formula (16) for $m = 1, 2, \ldots$, and then to note the first value of m for which $\alpha(m) \leqslant 1$. If $\alpha(m)$ is computed from (16), a discrete (not necessarily admissible) control, which transfers $\mathbf{x}^0$ to $\mathbf{x}^F$ in m steps, is defined simultaneously through formula (17).

Thus, finding an optimal control is equivalent to determining the vectors $\boldsymbol{\phi}^m$ (beginning with the first m for which $\alpha(m)$ is defined). By direct application of (16), this involves a finite number of trials for each m; but this number may be as large as $\binom{rm}{n-1}$. However, if $\boldsymbol{\phi}^{m-1}$ is known, the number of trials required to determine $\boldsymbol{\phi}^m$ should be much smaller. Determining $\boldsymbol{\phi}^m$ is equivalent to finding that face of C_m through which passes the ray joining the origin with $\mathbf{y}^m$. By continuously building up C_m from C_{m-1}, and letting $\mathbf{y}^{m-1}$ go continuously to $\mathbf{y}^m$, the corresponding ray can be thought of as moving continuously, passing through one face, and then into an adjoining one. Thus instead of trying all the faces of C_m to determine $\boldsymbol{\phi}^m$, one need only try a finite sequence of adjoining faces.

6. Continuous Time Optimal Controls as the Limit of Discrete Controls

It appears plausible that time optimal discrete controls (for fixed $\mathbf{x}^0$ and $\mathbf{x}^F$), considered for different sampling periods, should in some sense approach the continuous time optimal control, as the sampling period goes to zero. This statement is made precise in theorem 5.

Let us state some of the well known results on the continuous time optimal control problem for systems described by equation (3) (see refs. [*1*, *4*, *7*]). The sets $\Omega(t)$ and $C(t)$ are defined by (using the notation of section 2):

$$\Omega(t) = \left\{ X(t) \left[\mathbf{x}^0 + \int_0^t X^{-1}(s)\, B(s)\, \mathbf{u}(s)\, ds \right];\ |\, u_j(s)\,| \leqslant 1\ j = 1, \ldots, r, 0 \leqslant s \leqslant t \right\},$$

$$C(t) = \left\{ - \int_0^t X^{-1}(s)\, B(s)\, \mathbf{u}(s)\, ds;\ |\, u_j(s)\,| \leqslant 1,\ j = 1, \ldots r, 0 \leqslant s \leqslant t \right\};$$

$\Omega(t)$ is the set of all points to which $\mathbf{x}^0$ can be transferred in t seconds, using an admissible control, and

$$C(t) = \mathbf{x}^0 - X^{-1}(t)\,\Omega(t).$$

Both sets are convex for all t. We define

$$\mathbf{y}(t) = \mathbf{x}^0 - X^{-1}(t)\,\mathbf{x}^F,$$

so that $\mathbf{x}^F \in \Omega(t)$ if an only if $\mathbf{y}(t) \in C(t)$.

Let t^* be the g.l.b. of all positive t such that $\mathbf{x}^F \in \Omega(t)$. Then $\mathbf{x}^F \in \Omega(t^*)$. Let a corresponding control (which is time optimal) be $\mathbf{u}^*(t)$. Then there is a non-trivial (row) vector solution $\boldsymbol{\psi}(t)$ of the equation

$$\dot{\boldsymbol{\psi}} = -\,\boldsymbol{\psi} A \tag{20}$$

such that

$$u_j^*(t) = \operatorname{sign}[\boldsymbol{\psi}(t)\cdot\mathbf{b}^j(t)] \qquad j = 1, ..., r, \qquad 0 \leqslant t \leqslant t^*, \tag{21}$$

where $\mathbf{b}^j$ is the j-th column of B.

The pair $(\mathbf{x}^0, \mathbf{x}^F)$ will be called *non-singular* if there exists a positive number ω such that $\mathbf{x}^F$ belongs to the *interior* of $\Omega(t)$ for all t, $t^* < t < t + \omega$; or equivalently, if $\mathbf{y}(t)$ belongs to the interior of $C(t)$ for $t^* < t < t + \omega$.

The control system will be called *normal* if for every non-trivial solution $\boldsymbol{\psi}(t)$ of equation (20), and for every $j = 1, ..., r$, the set $\{t : \boldsymbol{\psi}(t)\cdot\mathbf{b}^j(t) = 0,\ t > 0\}$ has measure zero. The system will be called *proper* if the set of positive t for which the vector $\boldsymbol{\psi}(t)\cdot B(t)$ vanishes has measure zero (for every non-trivial solution $\boldsymbol{\psi}(t)$ of equation (20)).[1]

In order to distinguish among discrete admissible controls with different sampling periods, we shall say that a function given by equation (2) is a discrete admissible control function with period T. We shall denote by $\mathcal{O}(T)$ the set of all discrete admissible controls with period T which transfer $\mathbf{x}^0$ to $\mathbf{x}^F$ in a minimal number of steps. If $\mathcal{O}(T)$ is not empty, and the minimal number of steps is $m(T)$, we shall denote the minimal time $m(t)\cdot T$ by $\tau(T)$. We then have the following theorem.

Theorem 5. Let $\mathbf{u}^*(t)$ be a time optimal control which transfers $\mathbf{x}^0$ to $\mathbf{x}^F$ as a solution of equation (3), so that $\mathbf{u}^*(t)$ is given by equations (21) and (20), where t^* is the minimal transfer time. Let T_i be a sequence of positive numbers such that $T_i \xrightarrow[i\to\infty]{} 0$. Let $\mathbf{v}^i(t)$ be any member of

[1] These definitions differ slightly from those given by La Salle [7].

$\mathcal{O}(T_i)$, and let $\tau_i = \tau(T_i)$. Then, if the pair $(\mathbf{x}^0, \mathbf{x}^F)$ is non-singular, τ_i is defined for i large enough, and $\tau_i \xrightarrow[i\to\infty]{} t^*$. Furthermore, for every j, the sequence $v_j^i(t)$ converges in measure to $u_j^*(t)$ on the set

$$\{t : \boldsymbol{\psi}(t) \cdot \mathbf{b}^j(t) \neq 0, \qquad 0 \leq t \leq t^*\}.$$

In particular, if the control system is normal, the $v_j^i(t)$ converge in measure to $u_j^*(t)$ on $(0, t^*)$.

Proof. Since the pair $(\mathbf{x}^0, \mathbf{x}^F)$ is non-singular, $\mathbf{y}(t^* + \epsilon)$ belongs to the interior of $C(t^* + \epsilon)$ for all $\epsilon > 0$ small enough. Let us fix such an ϵ. There is a cube with center at $\mathbf{y}(t^* + \epsilon)$ which is contained in the interior of $C(t^* + \epsilon)$. Let this cube have sides of length γ, and denote its vertices by $\mathbf{P}^1, \mathbf{P}^2, \ldots$. Thus there exist admissible functions $\mathbf{w}^1(s), \mathbf{w}^2(s), \ldots$ such that for $k = 1, 2, \ldots,$

$$\mathbf{P}^k = -\int_0^{t^*+\epsilon} X^{-1}(s)\, B(s)\, \mathbf{w}^k(s)\, ds.$$

Since $\mathbf{y}(t)$ is continuous, there is a $\delta > 0$ such that

$$\| \mathbf{y}(t) - \mathbf{y}(t^* + \epsilon) \| < \gamma/3$$

whenever $| t - t^* - \epsilon | < \delta$. In addition, δ may be chosen in such a way that for every T, $0 < T < \delta$, there exist discrete admissible function $\tilde{\mathbf{w}}^k(t)$, with period T, such that the points $\mathbf{Q}^k$, given by

$$\mathbf{Q}^k = -\int_0^{[(t^*+\epsilon)/T]T} X^{-1}(s)\, B(s)\, \tilde{\mathbf{w}}^k(s)\, ds, \tag{22}$$

are at a distance less than $\gamma/3$ from $\mathbf{P}^k$, for each k. (The symbol $[x]$ in the above integral denotes the greatest integer not exceeding x.) This is a consequence of a well known theorem on approximating measurable function by step functions.

Let $m_i = [(t^* + \epsilon)/T_i]$. Since $T_i \to 0$, there is an I such that $T_i < \delta$ for all $i > I$. Also, $| m_i T_i - (t^* + \epsilon) | < \delta$ for $i > I$. Let the corresponding functions in (22) be $\mathbf{w}^{ki}(t)$. Fix an $i > I$, and let

$$\mathbf{Q}^{ki} = -\int_0^{m_i T_i} X^{-1}(s)\, B(s)\, \tilde{\mathbf{w}}^{ki}(s)\, ds.$$

Then, by definition of δ, $\mathbf{y}(m_i T_i)$ belongs to the convex hull of the points $\mathbf{Q}^{ki}$. The sets C_{m_i} (for period T_i) defined in section 3 are convex, and contain the $\mathbf{Q}^{ki}$, so that $\mathbf{y}(m_i T_i) \in C_{m_i}$ for $i > I$. Thus,

$$\mathbf{y}(m_i T_i) = -\int_0^{m_i T_i} X^{-1}(s)\, B(s)\, \hat{\mathbf{w}}^i(s)\, ds$$

for some discrete admissible function $\hat{\mathbf{w}}^i(t)$ with period T_i. From the definition of $\mathbf{y}(t)$ and $\Omega(t)$, this implies that $\hat{\mathbf{w}}^i(t)$ transfers $\mathbf{x}^0$ to $\mathbf{x}^F$ in m_i steps, or $m_i T_i$ seconds. Hence, $\tau_i \leq m_i T_i \leq t^* + \epsilon$. But $t^* \leq \tau_i$ by definition of t^*. Therefore, $t^* \leq \tau_i \leq t^* + \epsilon$ for $i > I$. Since ϵ is arbitrary, we have proved that $\tau_i \underset{i\to\infty}{\longrightarrow} t^*$.

Since $\mathbf{v}^i(t)$ transfers $\mathbf{x}^0$ to $\mathbf{x}^F$ in time τ_i,

$$\mathbf{y}(\tau_i) = -\int_0^{\tau_i} X^{-1}(s)\, B(s)\, \mathbf{v}^i(s)\, ds.$$

Let

$$\mathbf{z}^i = -\int_0^{t^*} X^{-1}(s)\, B(s)\, \mathbf{v}^i(s)\, ds.$$

Since $\tau_i \to t^*$ and the integrands above are uniformly bounded, $\mathbf{z}^i \underset{i\to\infty}{\longrightarrow} \mathbf{y}(t^*)$. Also,

$$\mathbf{y}(t^*) = -\int_0^{t^*} X^{-1}(s)\, B(s)\, \mathbf{u}^*(s)\, ds.$$

By hypothesis, $u_j^*(s) = \operatorname{sign} \boldsymbol{\psi}(s) \cdot \mathbf{b}^j(s)$, where $\boldsymbol{\psi}(s)$ is a solution of equation (20) and is therefore of the form $\boldsymbol{\eta} \cdot X^{-1}(s)$ where $\boldsymbol{\eta}$ is a non-zero constant (row) vector. Now $\boldsymbol{\eta} \cdot \mathbf{z}^i \underset{i\to\infty}{\longrightarrow} \boldsymbol{\eta} \cdot \mathbf{y}(t^*)$, so that

$$\sum_{j=1}^{r} \int_0^{t^*} \{|\boldsymbol{\psi}(s) \cdot \mathbf{b}^j(s)| - \boldsymbol{\psi}(s) \cdot \mathbf{b}^j(s)\, v_j^i(s)\}\, ds \underset{i\to\infty}{\longrightarrow} 0.$$

Since $|v_j^i(s)| \leqslant 1$, the integrands above are non-negative. Consequently, if we set $\phi_j(s) = \boldsymbol{\psi}(s) \cdot \mathbf{b}^j(s)$,

$$\int_0^{t^*} \{|\phi_j(s)| - \phi_j(s)\, v_j^i(s)\}\, ds \underset{i\to\infty}{\longrightarrow} 0 \qquad \text{for} \qquad j = 1, ..., r.$$

For fixed $\epsilon > 0$, let

$$E_{ij} = \{t: |u_j^*(t) - v_j^i(t)| > \epsilon\}.$$

Then,

$$\epsilon \int_{E_{ij}} |\phi_j(s)|\, ds \leqslant \int_{E_{ij}} \{|\phi_j(s)| - \phi_j(s)\, v_j^i(s)\}\, ds$$

$$\leqslant \int_0^{t^*} \{|\phi_j(s)| - \phi_j(s)\, v_j^i(s)\}\, ds,$$

and

$$\int_{E_{ij}} |\phi_j(s)|\, ds \underset{i\to\infty}{\longrightarrow} 0, \qquad j = 1, ..., r. \tag{23}$$

If

$$E'_{ij} = E_{ij} \cap \{t : \phi_j(t) \neq 0\},$$

equation (23) implies that the measure of $E'_{ij} \underset{i\to\infty}{\longrightarrow} 0$, which is the desired result.

Corollary. If $\mathbf{x}^F = 0$, the condition that $(\mathbf{x}^0, \mathbf{x}^F)$ be non-singular in theorem 5 may be replaced by the condition that the control system is proper.

Proof. If $\mathbf{x}^F = 0$, $\mathbf{y}(t) \equiv \mathbf{x}^0$, so that it is sufficient to prove that $\mathbf{x}^0$ is an interior point of $C(t)$ for $t > t^*$. Suppose the contrary, so that $\mathbf{x}^0$ is on the boundary of $C(t)$ for some $t > t^*$. It is easily seen that $C(t) \supset C(t^*)$. Let $\boldsymbol{\eta}$ be the external normal to the support plane to $C(t)$ at $\mathbf{x}^0$. This plane is also a support plane to $C(t^*)$. Since $\mathbf{x}^0 \in C(t) \cap C(t^*)$,

$$\mathbf{x}^0 = -\int_0^{t^*} X^{-1}(s)\, B(s)\, \mathbf{u}'(s)\, ds = -\int_0^{t} X^{-1}(s)\, B(s)\, \mathbf{u}''(s)\, ds$$

for some admissible functions $\mathbf{u}'(s)$ and $\mathbf{u}''(s)$. Since the function $\boldsymbol{\eta} \cdot \mathbf{x}$, for $\mathbf{x} \in C(t)$, takes on its maximum at $\mathbf{x} = \mathbf{x}^0$,

$$u'_j(s) = u''_j(s) = -\operatorname{sign} \boldsymbol{\eta} \cdot X^{-1}(s)\, \mathbf{b}^j(s),$$

and

$$\boldsymbol{\eta} \cdot \mathbf{x}^0 = \sum_{j=1}^{r} \int_0^{t^*} |\boldsymbol{\eta} \cdot X^{-1}(s)\, \mathbf{b}^j(s)|\, ds = \sum_{j=1}^{r} \int_0^{t} |\boldsymbol{\eta} \cdot X^{-1}(s)\, \mathbf{b}^j(s)|\, ds.$$

Consequently, if we let $\boldsymbol{\psi}(s) = \boldsymbol{\eta} \cdot X^{-1}(s)$,

$$\sum_{j=1}^{r} \int_{t^*}^{t} |\boldsymbol{\psi}(s) \cdot \mathbf{b}^j(s)|\, ds = 0,$$

which implies that the vector $\boldsymbol{\psi}(s) \cdot B(s)$ vanishes for $t^* \leqslant s \leqslant t$. This contradicts the definition of proper control systems. Q.E.D.

A weaker form of this theorem was proved by Krasovskii [*6*], for the case of finite difference approximations to equation (1). Basing his proof on properties of $\alpha(m)$, his result was restricted to the case where $\mathbf{x}^F = 0$, and only for the particular time optimal discrete controls given by theorem 3.

7. Linear Autonomous Systems

Let us consider the special case where the matrices A and B in equation (3) do not depend on time. Then, $X(t) = e^{tA}$, so that if we define $X_m = X(mT)$ for any integer (positive, negative, or zero) m, we have that $X_m = (X_1)^m$, $X_{-m} = (X_m)^{-1} = (X_{-1})^m$, and $X_{m'}X_{m''} = X_{m'+m''}$. Therefore, equations (6)—(9) have the form

$$\mathbf{x}^m - \mathbf{x}^{m-1} = K\mathbf{x}^{m-1} + L\mathbf{u}^m,$$

$$K = X_1 - I = e^{TA} - I,$$

$$L = \int_0^T X(\tau)\, d\tau\, B,$$

$$\mathbf{x}^m = X_m \mathbf{x}^0 + \sum_{i=1}^{m} X_{m-i} L \mathbf{u}^i.$$

Thus, the difference equation which describes the control system is also time independent. The vectors $\mathbf{b}^{ij}$ are given by

$$\mathbf{b}^{ij} = (X_{-1})^i \, \mathbf{l}^j,$$

where $\mathbf{l}^j$ is the j-th column vector of L.

It is easily seen that an autonomous system is normal in the discrete sense only if it is normal. The converse is true if the eigenvalues of A are real and distinct. To prove this, suppose that A has eigenvalues $\lambda_1, \ldots, \lambda_n$. If $\boldsymbol{\psi}$ is any vector, define the functions

$$f_j(t; \boldsymbol{\psi}) = \boldsymbol{\psi} \cdot X(-t)\, \mathbf{l}^j = \sum_{i=1}^{n} c_{ij} e^{-\lambda_i t}. \tag{24}$$

The c_{ij} are constants which depend on $\boldsymbol{\psi}$. If the system is not normal in the discrete sense, there are vectors $\mathbf{b}^{i_1 j}, \ldots, \mathbf{b}^{i_n j}$ $(1 \leqslant i_1 < i_2 \ldots < i_n)$ which are linearly dependent. Let $\boldsymbol{\eta}$ be a non-zero vector orthogonal to all these $\mathbf{b}^{i_k j}$. Then,

$$0 = \boldsymbol{\eta} \cdot \mathbf{b}^{i_k j} = \boldsymbol{\eta} \cdot X(-i_k T)\, \mathbf{l}^j = f_j(-i_k T; \boldsymbol{\eta}), \qquad k = 1, \ldots, n.$$

But functions of the form (24), if they do not vanish identically, have at most $(n - 1)$ zeros. Therefore, $\boldsymbol{\eta} \cdot X(-t)\, \mathbf{l}^j \equiv 0$. But

$$\boldsymbol{\eta} \cdot X(-t)\, \mathbf{l}^j = \int_0^T \boldsymbol{\eta} \cdot X(\tau - t)\, \mathbf{b}^j d\tau. \tag{25}$$

The integrand in (25) is of the form

$$\sum_{i=1}^{n} c_i e^{\lambda_i(\tau - t)},$$

so that $f_j(t; \eta) \equiv 0$ only if this integrand vanishes identically. Since $\eta \cdot X(-s)$ is a solution of equation (20), the system is not normal.

Theorem 4 has a particularly simple interpretation for normal autonomous systems with real distinct eigenvalues. The values of $\hat{u}_j^i$ are determined by the sign of the value of the function $f_j(t; \psi)$ at the points iT. Since the functions f_j have no more than $(n - 1)$ zeros, the sequences $\{\hat{u}_j^1, \hat{u}_j^2, ..., \hat{u}_j^M\}$ consist of n, or fewer, alternating "blocks" of $+ 1$'s and $- 1$'s, corresponding to regions in which f_j is positive or negative. Between two successive blocks there may be a $\hat{u}_j^i$ whose absolute value is less than one, if f_j has a zero at a sampling instant. Some of these blocks will be empty if f_j has more than one (or a multiple) zero during an interval $iT \leqslant t \leqslant (i + 1) T$. The last element of these sequences may have absolute value less than one, but even including this element, a sequence can have at most $(n - 1)$ sign changes. Essentially the same result has been derived by Desoer and Wing [*3*], for the special case $r = 1$ and the eigenvalues of A negative, using a somewhat different approach.

If $\mathbf{x}^F \neq 0$, the optimal control described in theorem 3 may be similarly characterized.

REFERENCES

[*1*] Boltyanskii, V. G., Gamkrelidze, R. V., and Pontryagin, L. S., The theory of optimal processes. *Izvest. Akad. Nauk S.S.S.R., Ser. Mat.* **24**, 3-42 (1960); English transl., *AMS Translations* [2] **18**, 341-382 (1961).

[*2*] Desoer, C. A., and Wing, J., An optimal strategy for a saturating sampled-data system. *IRE Trans. on Automatic Control* **AC-6**, No. 1, 5-15 (1961).

[*3*] Desoer, C. A., and Wing, J., A minimal-time discrete system. *IRE Trans. on Automatic Control* **AC-6**, No. 2, 111-125 (1961).

[*4*] Gamkrelidze, R. V., The theory of time-optimal processes for linear systems. *Izvest. Akad. Nauk S.S.S.R., Ser. Mat.* **22**, 449-474 (1958); English transl., Univ. of California Dept. Engineering Report No. 61-7 (1961).

[*5*] Krasovskii, N. N., On the theory of optimal controls. *Avtomat. i Telemekh.* **18**, No. 11, 960-970 (1957); English transl. of journal, *Automation and Remote Control.*

[*6*] Krasovskii, N. N., On an optimal control problem. *Priklad. Mat. i Mekh.* **21**, No. 5, 670-677 (1957).

[*7*] La Salle, J. P., Time optimal control systems. *Proc. nat. Acad. Sci. U.S.* **45**, 573-577 (1959).

The Principle of Optimal Evolution[1]

HUBERT HALKIN
RIAS, Baltimore, Maryland, and
Université de Liége, Belgium

1. Introduction

Most of the problems arising in the theory of optimal control of servomechanisms or of optimal performances in astrodynamics can be considered as Mayer problems of the calculus of variations subjected to unilateral constraints as side conditions.

Without these unilateral constraints the problem of Mayer—or its equivalents, the problems of Lagrange and Bolza—is relatively simple and has been extensively studied. (For Mayer, see Bolza [*7*], Bliss [*5*], and Carathéodory [*11*].)

As soon as unilateral constraints are introduced, however, the problem becomes much more difficult. Two different approaches to this enlarged problem may be noticed. The first approach, an attempt to generalize the classical results of Bolza and Bliss, was mainly considered in connection with problems of optimal performances in Aero- and Astronautics (Breakwell [*8*], Cicala [*14*], Fraeys de Veubeke [*19*], Halkin [*21*], Leitmann [*26*], Miele [*28*]). Another class of research related to automatic control and optimal performances of servomechanisms gave rise to the study of direct methods to solve this problem (Bellman [*4*], Bushaw [*9*], Flügge-Lotz [*17*], La Salle [*25*], Pontryagin [*6*]). We will mention also a third class of research devoted to the application of the steepest descent or gradient method to this generalized problem (Bryson [*10*], Kelley [*24*]).

Our method, on the other hand, is an extension of the Hamiltonian point of view in calculus of variations (Huygens, Hamilton, Carathéodory [*11*], and more recently Bellman [*3*] and Kalman [*23*]). Our starting point is what we have called "the Principle of Optimal Evolution" by which, "Every event of an optimal process belongs to the boundary of the set of possible events" (these terms will be defined below).[2]

[1] This research was supported in part by the Office of Naval Research Contract Nonr. 225 (11).

[2] This principle has been also developed independently and in a slightly different direction by Dr. E. Roxin.

It should be noted that this principle even with its great simplicity carries, as we shall show, very important local as well as global implications. From this particularly simple principle and with a method directly inspired by the Huygens' Principle of geometrical optics we have transformed our variational problem into a problem of differential geometry. In the classical case, i.e. without unilateral constraints in the variational problem, the resulting problem of differential geometry can be easily solved by the theory of contact transformations, which leads to all the classical results of the calculus of variations. In the solution of the problem of differential geometry arising from a variational problem with unilateral constraints, the theory of contact transformations is no longer sufficient and we have supplemented it by new direct arguments.

2. Statement of the Problem

The simplest problem of the calculus of variations in n dimensional space can be stated as follows:

Given: (a) two points ξ_1 and ξ_2 in E^n, the Euclidian n dimensional space
(b) two points $t_1 < t_2$ in E'
(c) a function $L(x, \dot{x}, t)$ continuous and sufficiently differentiable in all its arguments x and $\dot{x}$ in E^n, t in E'
(d) the class F_1 of piecewise continuous vector valued functions $u(t)$ in E^n

find a vector valued function $x(t)$ such that

$$\text{(i) } \dot{x}(t) \in F_1 \qquad \text{for} \qquad t \in [t_1, t_2] \tag{1}$$

$$\text{(ii) } x(t_1) = \xi_1 \tag{2}$$

$$\text{(iii) } x(t_2) = \xi_2 \tag{3}$$

$$\text{(iv) } J = \int_{t_1}^{t_2} L(x, \dot{x}, t)\, dt \text{ is minimum.} \tag{4}$$

We introduce a new coordinate x^{n+1} by the following relations:

$$\dot{x}^{n+1} = L(x, \dot{x}, t) \tag{5}$$

$$x^{n+1}(t_1) = 0. \tag{6}$$

$$\text{(iv) can now be written : } x^{n+1}(t_2) \text{ is minimum.} \tag{7}$$

The vector $\bar{x} = (x, x^{n+1}) \in E^{n+1}$ is called the state vector.

The problem just defined will be called P_1. We will now define the problem P_2 by adding a set of unilateral constraints on the vector valued functions $\dot{x}(t)$ of the problem P_1. Instead of choosing $\dot{x}$ in E^n we will restrict this choice to a subset of E^n defined as the mapping in E^n of a subset Ω of E^r through the system of equations

$$\dot{x} = f(x, t, u) \qquad \text{with} \qquad u \in \Omega \subset E^r. \tag{8}$$

We will assume that the vector valued function f is continuous and sufficiently differentiable in all its arguments. We shall consider curves $x(t)$ such that $\dot{x}(t)$ satisfies the requirements (8). This will define corresponding curves $u(t)$ in the parameter space E^r. We shall suppose that $u(t)$ is also piecewise continuous and we will call F_2 the class of admissible $u(t)$ so defined.

The vector u is called the control vector.

The condition (1) should now read

$$(1') \qquad u(t) \in F_2 \qquad \text{such that} \qquad \dot{x} = f(x, t, u) \quad \text{for} \quad t \in [t_1, t_2]. \tag{9}$$

We have now $\dot{x}^{n+1} = L(x, f(x, t, u), t) = f^{n+1}(x, t, u)$.

For the sake of symmetry we will admit a function $f^{n+1}(x, t, u)$ under the same restrictions as the vector valued function $f(x, t, u)$ without requiring the existence of a function $L(x, \dot{x}, t)$ such that

$$f^{n+1}(x, t, u) = L(x, f(x, t, u), t).$$

We will even generalize it to the case

$$\dot{\bar{x}} = \bar{f}(\bar{x}, t, u) = (f(\bar{x}, t, u), f^{n+1}(\bar{x}, t, u)). \tag{10}$$

In the following we will always refer to F without subscript as the class of admissible controls. F is to be understood as F_1 or F_2 according to the problem P_1 or P_2 then considered.

3. The Set of Possible Events

The vector $y = (t, \bar{x}) \in E^{n+2}$ is called an event.

We will denote by $A = (t_1, \xi_1, 0)$ the initial event and by $B = (t_2, \xi_2, E')$ the set of virtual final events.

We define now an operator valued functional which takes an event y_a into an event y_e as follows:

$R(u, \theta)\, y_a$ where $\theta \in E'$ and $\theta \geq 0$

$$y_a = (t_a, \bar{x}_a) \in E^{n+2}$$

$$u \in F$$

is the element $y_e = (t_e, \bar{x}_e)$ of E^{n+2} defined by

$$t_e = t_a + \theta \tag{11}$$

$$\bar{x}_e = \bar{x}_a + \int_{t_a}^{t_e} \bar{\bar{f}}(\bar{x}, u, t)\, dt. \tag{12}$$

We will now introduce the notion of set of possible events. If

$$y \in E^{n+2} \qquad \text{let} \qquad H(y) = \{R(u, \theta)y : u \in F, \theta \geq 0\} \tag{13}$$

$H(y)$ is the set of events which can be reached from y by an admissible control in a non-negative time. $H(y)$ is called the set of possible events from y.

We have then the following trivial properties:

$$\text{(i) } b \in H(a) \quad \text{implies} \quad R(u, \theta)b \in H(a) \quad \text{for any} \quad u \in F, \theta \geq 0 \tag{14}$$

$$\text{(ii) } b \in H(a) \quad \text{and} \quad c \in H(b) \quad \text{imply} \quad c \in H(a) \tag{15}$$

$$\text{(iii) } b \in H(a) \quad \text{implies} \quad H(b) \subset H(a). \tag{16}$$

When we will mention a possible event (or a set of possible events) without giving explicitly its (or their) origin, the later one will always be assumed to be A. Hence $H(A)$ the "set of possible events from A" will usually be referred to as the "set of possible events".

With these definitions both problems P_1 and P_2 can be formulated in the following way: Given A and B find the element $u \in F$ such that:

$$\text{(i) } R(u, t_2 - t_1)\, A \in B \tag{17}$$

$$\text{(ii) } \langle K \mid R(u, t_2 - t_1)A \rangle = \min_{y \in H(A) \cap B.} \langle K \mid y \rangle \tag{18}$$

where $K = (0, \ldots, 0, 1) \in E^{n+2}$ and where $\langle a \mid b \rangle$ means the scalar product of a and b.

4. The Principle of Optimal Evolution

A pair (u, τ) where u is an element of F and τ is a non-negative scalar is called a possible strategy. It is called an acceptable strategy if $\tau = t_2 - t_1$ and if u satisfies the condition (17). An acceptable strategy satisfying the condition (18) is called an optimal strategy.

The trajectory corresponding to the strategy (u, τ) and the initial conditions (2) and (6) will be denoted by $\phi_{u,\tau}$

$$\phi_{u,\tau} = \{R(u, t)\, A : t \in [0, \tau]\} \tag{20}$$

We will apply to the trajectory $\phi_{u,\tau}$ the same names as to the strategy (u, τ).

Theorem I. Every event of an optimal trajectory belongs to the boundary of the set of possible events.

We call Principle of Optimal Evolution the property of the optimal trajectory given in theorem I [*18*].

Indeed the existence of the final event[1]

$$y_e = \arg\min_{y \in H(A) \cap B} \langle K \mid y \rangle \tag{21}$$

is trivially connected to the existence of a solution of our variational problem, see (18). Let us assume that such a solution exists and that $(u, t_2 - t_1)$ is a strategy such that

$$y_e = R(u, t_2 - t_1)A. \tag{22}$$

According to our definition $(u, t_2 - t_1)$ is then an optimal strategy of our problem.

Let y_θ be an arbitrary event of the optimal trajectory ϕ_{u,t_2-t_1}, i.e.

$$y_\theta = R(u, \theta)A \qquad \text{with} \qquad \theta \in [0, t_2 - t_1] \tag{23}$$

$$y_e = R(u, t_2 - t_1 - \theta) y_\theta. \tag{24}$$

We have to prove that[2]

$$y_\theta \in \dot{H}(A). \tag{25}$$

If y_θ is interior to $H(A)$ there is a closed ϵ neighborhood of y_θ belonging to $H(A)$

$$N_\epsilon(y_\theta) \subset H(A). \tag{26}$$

By application of (14) we have

$$R(u, t_2 - t_1 - \theta)\, N_\epsilon(y_\theta) \subset H(A). \tag{27}$$

But, since the vector function $\bar{f}(x, u, t)$ is sufficiently differentiable with respect to $\bar{x}$ there is an $\eta > 0$ such that

$$N_\eta(R(u, t_2 - t_1 - \theta)\, y_\theta) \subset R(u, t_2 - t_1 - \theta)\, N_\epsilon(y_\theta). \tag{28}$$

[1] By convention when: $f(a) = \min_{x \in X} f(x)$ we will write $a = \arg\min_{x \in X} f(x)$.

[2] By convention $\dot{X}$ is the boundary of the set X.

From (24), (27) and (28) we conclude

$$N_\eta(y_e) \subset H(A). \tag{29}$$

We have also

$$(y_e - \eta K) \in N_\eta(y_e) \cap B \tag{30}$$

hence

$$(y_e - \eta K) \in H(A) \cap B \tag{31}$$

and

$$\langle K \mid y_e - \eta K \rangle = \langle K \mid y_e \rangle - \eta < \langle K \mid y_e \rangle \tag{32}$$

The relations (21), (31) and (32) are incompatible. This contradiction leads to the rejection of the assumption (27), hence

$$y_0 = R(u, \theta)\, A \in \dot{H}(A) \tag{33}$$

which proves theorem I.

Definitions

b is called an extremal event of a if $b \in \dot{H}(a) \cap H(a)$.

b is called a conjugate event of a if b is an extremal event of a such that $\dot{H}(a) \cap \dot{H}(b)$ contains no element other than b.

A possible trajectory whose elements are all extremal events of a is called an extremal of a.

Corollary I. An optimal trajectory is an extremal.

Corollary II. If b is an element of an extremal of a any later element c of this extremal belongs to $\dot{H}(a) \cap \dot{H}(b)$.

By definition $c \in \dot{H}(a)$. We have also $c \in \dot{H}(b)$ because if c is interior to $H(b)$ it is also interior to $H(a)$ since $H(b) \subset H(a)$. Hence the proof.

Corollary III. If b is a conjugate event of a there does not exist a c different from b such that abc is an extremal.

The corollary III is just a transformation of corollary II with the help of the definition of a conjugate event.

5. Generalized Huygens' Principle

Let $W_A(\bar{x}, t) = 0$ be the equation of $\dot{H}(A)$.

For the problem P_1 the hypersurface $W_A(\bar{x}, t) = 0$ can always be written in the form $V_A(x, t) - x^{n+1} = 0$ where $V_A(x, t)$ is continuous and differentiable on both arguments. In such a case $V_A(x, t)$ is called the principal function of Hamilton (see refs. [*13*, *15*].)

Let $W_{A,\tau}(\bar{x}) = 0$ be the intersection of $W_A(\bar{x}, t) = 0$ by the hyperplane $t = \tau$. By the Principle of Optimal Evolution we know that $W_{A,\tau+h}(\bar{x}) = 0$, where $h > 0$, can only be reached from $W_{A,\tau}(\bar{x}) = 0$. Hence $W_{A,\tau}(\bar{x}) = 0$ may be visualized as the wavefront in an abstract medium E^{n+1} of the propagation of an initial perturbation corresponding to the event A.

Let us call "wavelet" associated with the event $(\bar{x}, t)$ the set of points which can be reached from $\bar{x}$ after a time dt with an admissible control. The wavelet associated with the event $(\bar{x}, t)$ is then the set of points $\bar{x} + d\bar{x}$ of E^{n+1} defined in the case of problem P_1 by

$$dx^{n+1} = L\left(x, \frac{dx}{dt}, t\right) dt \text{ with an arbitrary } dx \tag{34}$$

and in the case of problem P_2 by

$$d\bar{x} = \bar{f}(\bar{x}, u, t)\, dt \qquad \text{with} \qquad u \in \Omega. \tag{35}$$

For the problem P_2 the wavelet associated with the event $(\bar{x}, t)$ is then, up to a factor dt and a translation $\bar{x}$, the mapping in E^{n+1} of the subset $\Omega \subset E^r$ through the relations

$$\dot{\bar{x}} = \bar{f}(\bar{x}, u, t).$$

In the case of the problem P_1 the equation (34) is sometimes called the equation of Monge (see Cartan [*13*]).

The hypersurface $W_{A,t+dt}(\bar{x}) = 0$ is the forward boundary of the union of the wavelets associated with the events of $W_{A,t}(\bar{x}) = 0$. More simply we will say that $W_{A,t+dt}(\bar{x}) = 0$ is the "envelope" of the wavelets from $W_{A,t}(\bar{x}) = 0$.

When the hypersurface $W_{A,t}(\bar{x}) = 0$ is differentiable let $\bar{p}(\bar{x}, t)$ be its outward normal.[1]

i.e.
$$\bar{p}(\bar{x}, t) = \frac{\partial W_{A,t}(\bar{x})}{\partial \bar{x}} \tag{36}$$

[1] We use here the classical notation

$$\frac{\partial f(x)}{\partial x} \qquad \text{for} \qquad \left(\frac{\partial f(x)}{\partial x'}, \ldots, \frac{\partial f(x)}{\partial x^n}\right)$$

when $f(x)$ is a scalar function of

$$x = (x', \ldots, x^n) \in E^n.$$

Similarly we will use later

In this case the condition for an element of $W_{A,t}(\bar{x}) = 0$ to be sent into an element of $W_{A,t+dt}(\bar{x}) = 0$ is:

$$\dot{x} = \arg\max_{\dot{x}\in E^n} \langle \bar{p} \mid (\dot{x}, L(x, \dot{x}, t)\rangle \qquad \text{for} \qquad P_1 \tag{37}$$

$$u = \arg\max_{u\in\Omega} \langle \bar{p} \mid \bar{\bar{f}}(\bar{x}, u, t)\rangle \qquad \text{for} \qquad P_2. \tag{38}$$

The wavefronts $W_{A,t}(\bar{x}) = 0$ will sometimes be referred to as the isochrones.

6. Hamiltonian Formulation of the Principle of Optimal Evolution

We will introduce

$$H(\bar{x}, \dot{x}, \bar{p}, t) = \langle \bar{p} \mid (\dot{x}, L(x, \dot{x}, t))\rangle \qquad \text{for} \qquad P_1 \tag{39}$$

$$H(\bar{x}, u, \bar{p}, t) = \langle \bar{p} \mid \bar{\bar{f}}(\bar{x}, u, t)\rangle \qquad \text{for} \qquad P_2 \tag{40}$$

and

$$H^0(\bar{x}, \bar{p}, t) = \max_{\dot{x}\in E^n} H(\bar{x}, \dot{x}, \bar{p}, t) \qquad \text{for} \qquad P_1 \tag{41}$$

$$H^0(\bar{x}, \bar{p}, t) = \max_{u\in\Omega} H(\bar{x}, u, \bar{p}, t) \qquad \text{for} \qquad P_2. \tag{42}$$

In the case of the problem P_1 the condition (37) for a trajectory to be an extremal can now be written

$$\dot{x} = \arg\max_{\dot{x}\in E^n} H(\bar{x}, \dot{x}, \bar{p}, t). \tag{43}$$

We will assume that the relation (43) defines a single valued function

$$\dot{x} = c(\bar{x}, \bar{p}, t). \tag{44}$$

From the definition (41) we may now write

$$H^0(\bar{x}, \bar{p}, t) = H(\bar{x}, c(\bar{x}, \bar{p}, t), \bar{p}, \bar{t}) \tag{45}$$

$$\frac{\partial g(x)}{\partial x} \quad \text{for} \quad \begin{bmatrix} \dfrac{\partial g'(x)}{\partial x} & \cdots & \dfrac{\partial g^m(x)}{\partial x'} \\ \vdots & & \vdots \\ \dfrac{\partial g'(x)}{\partial x^n} & \cdots & \dfrac{\partial g^m(x)}{\partial x^n} \end{bmatrix}$$

when $g(x)$ is a vector valued function $(g'(x), \ldots, g^m(x))$ of

$$x = (x', \ldots, x^n) \in E^n.$$

In the case of the problem P_2 similar considerations will be made: the condition (38) may be written

$$u = \arg\max_{u \in \Omega} H(\bar{x}, u, \bar{p}, t). \tag{46}$$

We will again assume that (46) defines a single valued function

$$u = c(\bar{x}, \bar{p}, t). \tag{47}$$

The relation (42) is then

$$H^0(\bar{x}, \bar{p}, t) = H(\bar{x}, c(\bar{x}, \bar{p}, t), \bar{p}, t). \tag{48}$$

Let us first consider the problem P_1.

We will now derive the partial differential equation of Hamilton-Jacobi from the Principle of Optimal Evolution.

We have seen that $H(\bar{x}, \dot{x}, \bar{p}, t)$ must take its maximum value with respect to $\dot{x} \in E^n$ at any point of an extremal. We have assumed $L(x, \dot{x}, t)$ to be sufficiently differentiable in all its arguments hence $H(\bar{x}, \dot{x}, \bar{p}, t)$ will also be sufficiently differentiable for all $\dot{x}$ on E^n. The maximization of the differentiable function $H(\bar{x}, \dot{x}, \bar{p}, t)$ over the unbounded domain E^n of the $\dot{x}$ implies

$$\frac{\partial H}{\partial \dot{x}} = 0 \tag{49}$$

i.e.

$$\frac{\partial H}{\partial \dot{x}^i} = 0 \qquad \text{for} \qquad i = 1, \ldots, n. \tag{50}$$

This leads to

$$p_i(\bar{x}, t) + p_{n+1}(\bar{x}, t) \frac{\partial L(x, \dot{x}, t)}{\partial \dot{x}^i} = 0 \qquad i = 1, \ldots, n \tag{51}$$

i.e.

$$p_i(\bar{x}, t) = - \frac{\partial L(x, \dot{x}, t)}{\partial \dot{x}^i} p_{n+1}(\bar{x}, t) \qquad i = 1, \ldots, n. \tag{52}$$

The relations (52) express the fact that $\bar{p}(\bar{x}, t)$ is normal to the wavelet

$$\bar{x} + \bar{d}\bar{x} = \bar{x} + \left(\frac{dx}{dt}, L\left(x, \frac{dx}{dt}, t\right)\right) dt \tag{53}$$

But $\bar{p}(x, t)$ was defined as the normal to the wavefront

$$W_{A,t}(\bar{x}) = 0 \tag{54}$$

hence we may conclude that for any point of the hypersurface $W_{A,t+dt}(\bar{x}) = 0$ there is a wavelet $\bar{x} + \bar{d}\bar{x}$ associated with a point of $W_{A,t}(\bar{x}) = 0$ and tangent to $W_{A,t+dt}(\bar{x}) = 0$.

Since we have assumed that there is a unique single valued function

$$\dot{x} = c(\bar{x}, \bar{p}, t) \tag{55}$$

determined by the maximization of $H(\bar{x}, \dot{x}, \bar{p}, t)$ we must have

$$\left| \frac{\partial^2 H(\bar{x}, \dot{x}, \bar{p}, t)}{\partial x^i \, \partial x^j} \right| < 0 \tag{56}$$

or

$$p_{n+1}(\bar{x}, t) \left| \frac{\partial^2 L(x, \dot{x}, t)}{\partial x^i \, \partial x^j} \right| < 0. \tag{57}$$

But for the problem P_1 we have

$$p_{n+1}(\bar{x}, t) < 0 \tag{58}$$

hence

$$\left| \frac{\partial^2 L(x, \dot{x}, t)}{\partial \dot{x}^i \, \partial \dot{x}^j} \right| > 0. \tag{59}$$

The inequality (57) is the Clebsch-Legendre Condition [5, *20*].

We will now establish a useful lemma.

Lemma I

(a)
$$\frac{\partial H(\bar{x}, c(\bar{x}, \bar{p}, t), \bar{p}, t)}{\partial \bar{p}} = \frac{\partial H(\bar{x}, \dot{x}, \bar{p}, t)}{\partial \bar{p}} \bigg|_{\dot{x} = c(\bar{x}, \bar{p}, t)} \tag{60}$$

(b)
$$\frac{\partial H(\bar{x}, c(\bar{x}, \bar{p}, t), \bar{p}, t)}{\partial \bar{x}} = \frac{\partial H(\bar{x}, \dot{x}, \bar{p}, t)}{\partial \bar{x}} \bigg|_{\dot{x} = c(\bar{x}, \bar{p}, t)}. \tag{61}$$

Indeed

$$\frac{\partial H(\bar{x}, c(\bar{x}, \bar{p}, t), \bar{p}, t)}{\partial \bar{p}}$$

$$= \frac{\partial H(\bar{x}, \dot{x}, \bar{p}, t)}{\partial \bar{p}} \bigg|_{\dot{x} = c(\bar{x}, \bar{p}, t)} + \frac{\partial H(\bar{x}, c(\bar{x}, \bar{p}, t), \bar{\pi}, t)}{\partial \bar{p}} \bigg|_{\bar{\pi} = \bar{p}} \tag{62}$$

but

$$\frac{\partial H(\bar{x}, c(\bar{x}, \bar{p}, t), \bar{\pi}, t)}{\partial \bar{p}} \bigg|_{\bar{\pi} = \bar{p}} = \frac{\partial H}{\partial c} \frac{\partial c}{\partial p} = \frac{\partial H}{\partial \dot{x}} \frac{\partial c}{\partial \bar{p}} = 0 \tag{63}$$

$$\text{since} \quad \frac{\partial H}{\partial \dot{x}} = 0 \quad \text{from} \quad (49).$$

Hence the proof of (a) and by analogy the proof of (b).

The relations (60) and (61) can also be given under the form

$$\frac{\partial H^0}{\partial \bar{p}} = \frac{\partial H}{\partial \bar{p}} \quad \text{along an extremal} \tag{64}$$

$$\frac{\partial H^0}{\partial \bar{x}} = \frac{\partial H}{\partial \bar{x}} \quad \text{along an extremal.} \tag{65}$$

By the Principle of Optimal Evolution we know that

$$W_A(\bar{x}, t) = 0 \tag{66}$$

is identically satisfied along an extremal, hence

$$\frac{d}{dt} W_A(\bar{x}, t) = 0 \tag{67}$$

which gives

$$\frac{\partial W_A(\bar{x}, t)}{\partial t} + \sum_{i=1}^{n} \frac{\partial W_A(\bar{x}, t)}{\partial x^i} \dot{x}^i + \frac{\partial W_A(\bar{x}, t)}{\partial x^{n+1}} L(x, \dot{x}, t) = 0. \tag{68}$$

With the definition (39) the equation (68) becomes

$$\frac{\partial W_A(\bar{x}, t)}{\partial t} + H\left(\bar{x}, \dot{x}, \frac{\partial W_A(\bar{x}, t)}{\partial x}, t\right) = 0. \tag{69}$$

Since we are on an extremal we have

$$\dot{x} = c(\bar{x}, \bar{p}, t). \tag{70}$$

From the definition of $W_{A,t}(\bar{x}) = 0$ follows that

$$\frac{\partial W_{A,t}(\bar{x})}{\partial \bar{x}} = \frac{\partial W_A(\bar{x}, t)}{\partial \bar{x}}. \tag{71}$$

With the help of the relation (71) and the definition (36) the equation (70) becomes

$$\dot{x} = c\left(\bar{x}, \frac{\partial W_A(\bar{x}, t)}{\partial \bar{x}}, t\right) \tag{72}$$

which by introduction in (69) gives

$$\frac{\partial W_A(\bar{x}, t)}{\partial t} + H\left(\bar{x}, c\left(\bar{x}, \frac{\partial W_A(\bar{x}, t)}{\partial \bar{x}}, t\right), \frac{\partial W_A(\bar{x}, t)}{\partial \bar{x}}, t\right) = 0 \tag{73}$$

or

$$\frac{\partial W_A(\bar{x}, t)}{\partial t} + H^0\left(\bar{x}, \frac{\partial W_A(\bar{x}, t)}{\partial \bar{x}}, t\right) = 0 \tag{74}$$

by the relation (45).

The partial differential equation (74) is called the equation of Hamilton-Jacobi and $H^0(\bar{x}, \bar{p}, t)$ is the classical Hamiltonian of the problem P_1 [*15*].

For the sake of completeness we will now derive the canonical system of Hamilton associated with this problem.

From the definition (39) we have

$$\dot{\bar{x}} = \frac{\partial H(\bar{x}, \dot{x}, \bar{p}, t)}{\partial \bar{p}}, \tag{75}$$

but we have seen in Lemma I that

$$\frac{\partial H(\bar{x}, \dot{x}, \bar{p}, t)}{\partial \bar{p}} = \frac{\partial H^0(\bar{x}, \bar{p}, t)}{\partial \bar{p}} \quad \text{on an extremal} \tag{76}$$

hence

$$\dot{\bar{x}} = \frac{\partial H^0(\bar{x}, \bar{p}, t)}{\partial \bar{p}} \quad \text{on an extremal.} \tag{77}$$

The relation (74) is an identity on $W_A(\bar{x}, t) = 0$ hence by partial differentiation with respect to $\bar{x}$ we obtain:

$$\frac{\partial^2 W_A(\bar{x}, t)}{\partial t\, \partial x} + \frac{\partial H^0(\bar{x}, \bar{p}, t)}{\partial \bar{x}} + \frac{\partial H^0(\bar{x}, \bar{p}, t)}{\partial \bar{p}} \frac{\partial^2 W_A(\bar{x}, t)}{\partial^2 \bar{x}} = 0 \tag{78}$$

i.e. by introduction of (77)

$$\frac{\partial^2 W_A(\bar{x}, t)}{\partial t\, \partial x} + \frac{\partial H^0(\bar{x}, \bar{p}, t)}{\partial \bar{x}} + \dot{\bar{x}} \frac{\partial^2 W_A(\bar{x}, t)}{\partial^2 \bar{x}} = 0. \tag{79}$$

By total differentiation with respect to time of

$$\bar{p} = \frac{\partial W_A(\bar{x}, t)}{\partial \bar{x}} \tag{80}$$

we obtain

$$\dot{\bar{p}} = \frac{\partial^2 \mathrm{W}_A(\bar{x}, t)}{\partial t\, \partial \bar{x}} + \frac{\partial^2 W_A(\bar{x}, t)}{\partial^2 \bar{x}} \dot{\bar{x}}. \tag{81}$$

By comparison of (79) and (81) we have then

$$\dot{\bar{p}} = -\frac{\partial H^0(\bar{x}, \bar{p}, t)}{\partial \bar{x}}. \tag{82}$$

The relations (77) and (82) are the canonical system of Hamilton.

It should be noticed that these results have been obtained without using either the fundamental lemma of the Calculus of Variations or any of its consequences such as the equation of Euler-Lagrange. This equation can be derived, a posteriori, from the equation of Hamilton-Jacobi and the canonical system of Hamilton.

7. Study of the Problem P_2

For the problem P_2 we do not have any relation similar to (51). Nevertheless when the hypersurface $W_A(\bar{x}, t) = 0$ is continuous and differentiable, which implies the existence of a unique normal $\bar{p}(\bar{x}, t)$, we will show that for the problem P_2 also we can derive a canonical system of the form (77) and (82).

Let $u^*(t)$ be an arbitrary member of the class F.

Let $\bar{x} = \bar{\phi}(t)$ be a solution of

$$\dot{\bar{x}} = \bar{f}(\bar{x}, u^*(t), t) \tag{83}$$

with the boundary condition $\bar{x} = \bar{x}_\alpha$ for $t = \tau$.

If $\bar{k}$ is an arbitrary small vector in E^{n+1} the solution of (83) satisfying the condition $\bar{x} = \bar{x}_\alpha + \bar{k}$ for $t = \tau$ may be written under the form $\bar{x} = \bar{\phi}(t) + \bar{\xi}(t)$ up to a term of higher order in k.
where

$$\bar{\xi}(\tau) = \bar{k}$$

and

$$\dot{\bar{\xi}}(t) = D\bar{\xi}(t)$$

with

$$D = \frac{\partial \bar{f}(\bar{x}, u^*(t), t)}{\partial \bar{x}}. \tag{84}$$

Let u^* be the control given by the relation (46) for the element $(\bar{x}_\alpha, \tau)$ of $W_{A,\tau}(\bar{x}) = 0$. Let $(\bar{x}_\beta, \tau + dt)$ be the element of $W_{A,\tau+dt}(\bar{x}) = 0$ corresponding to the integration of the system (83) with the control u^* from the element $(\bar{x}_\alpha, \tau)$. Let $\tilde{W}_{A,\tau+dt}(\bar{x}) = 0$ be the locus of the transformation of the elements of $W_{A,\tau}(\bar{x}) = 0$ with the fixed control u^*. It should be noticed that $\tilde{W}_{A,\tau+dt}(\bar{x}) = 0$ does not generally correspond to a wavefront in our propagation analogy. By construction the surfaces $\tilde{W}_{A,\tau+dt}(\bar{x}) = 0$ and $W_{A,\tau+dt}(\bar{x}) = 0$ have the element $(\bar{x}_\beta, t + dt)$ in common.

Let us consider n arbitrary small independent vectors $\bar{k}_{(i)}$, $i = 1, \ldots, n$, tangent to $W_{A,\tau}(\bar{x}) = 0$ at the point $\bar{x} = \bar{x}_\alpha$. Let $\bar{\xi}_{(i)}(t)$ be the solution of the system (84) corresponding to the initial conditions $\bar{\xi}_{(i)}(\tau) = k_{(i)}$.

By construction the vectors $\xi_{(i)}(\tau + dt)$ will be tangent to $\tilde{W}_{A,\tau+dt}(\bar{x}) = 0$ at the point $\bar{x} = \bar{x}_\beta$. If we call $\bar{p}(\tau)$ the normal to $W_{A,\tau}(\bar{x}) = 0$ at $\bar{x} = \bar{x}_\alpha$,

$$\tilde{\bar{p}}(\tau + dt) \quad \text{the normal to} \quad \tilde{W}_{A,\tau+dt}(\bar{x}) = 0 \quad \text{at} \quad \bar{x} = \bar{x}_f,$$

we have then the following relations

$$\langle \bar{\xi}_{(i)}(\tau) \mid \bar{p}(\tau) \rangle = 0 \qquad i = 1, \ldots, n \tag{85}$$

$$\langle \bar{\xi}_{(i)}(\tau + dt) \mid \tilde{\bar{p}}(\tau + dt) \rangle = 0 \qquad i = 1, \ldots, n \tag{86}$$

Theorem II. The outward normal of $W_{A,\tau+dt}(\bar{x}) = 0$ and $\tilde{W}_{A,\tau+dt}(\bar{x}) = 0$ at $(\bar{x}_\beta, \tau + dt)$ are identical.

We have then to prove that

$$\bar{p}(\tau + dt) = \tilde{\bar{p}}(\tau + dt). \tag{87}$$

Indeed the relation (87) is a direct conclusion of the two following statements:

(i) The hypersurface $\tilde{W}_{A,\tau+dt}(\bar{x}) = 0$ is continuous and differentiable since it is the transformation of the continuous and differentiable hypersurface $W_{A,\tau}(\bar{x}) = 0$ by the sufficiently differentiable system (83).

(ii) The hypersurface $\tilde{W}_{A\tau,+dt}(\bar{x}) = 0$ is entirely located on one side of the hypersurface $W_{A,\tau+dt}(\bar{x}) = 0$ by the Principle of Optimal Evolution. From (86) and (87) we obtain

$$\langle \bar{\xi}_{(i)}(\tau + dt) \mid \bar{p}(\tau + dt) \rangle = 0 \qquad i = 1, \ldots, n. \tag{88}$$

From (85) and (88) we obtain at the limit for $dt = 0$

$$\frac{d}{dt} \langle \bar{\xi}_{(i)}(t) \mid \bar{p}(t) \rangle = 0 \qquad i = 1, \ldots, n \tag{89}$$

i.e.

$$\langle \dot{\bar{\xi}}_{(i)}(t) \mid \bar{p}(t) \rangle + \langle \bar{\xi}_{(i)}(t) \mid \dot{\bar{p}}(t) \rangle = 0. \tag{90}$$

But

$$\dot{\bar{\xi}}_{(i)}(t) = D\bar{\xi}_{(i)}(t) \tag{84}$$

hence

$$\langle D\bar{\xi}_{(i)}(t) \mid \bar{p}(t) \rangle + \langle \bar{\xi}_{(i)}(t) \mid \dot{\bar{p}}(t) \rangle = 0 \tag{91}$$

or

$$\langle \bar{\xi}_{(i)}(t) \mid \tilde{D}\bar{p}(t) \rangle + \langle \bar{\xi}_{(i)}(t) \mid \dot{\bar{p}}(t) \rangle = 0 \tag{92}$$

where $\tilde{D}$ is the transpose of D, a real matrix by construction.

From (92) we have now

$$\langle \bar{\xi}_{(i)}(t) \mid \tilde{D}\bar{p}(t) + \dot{\bar{p}}(t) \rangle = 0. \tag{93}$$

Since (93) is valid for n independent vectors $\bar{\xi}_{(i)}$ in E^{n+1} and since $\bar{p}(t)$ is orthogonal to them we may write

$$\tilde{D}\bar{p}(t) + \dot{\bar{p}}(t) = g(\bar{x}, t)\,\bar{p} \tag{94}$$

where $g(\bar{x}, t)$ is an undetermined scalar function.

We will define

$$\bar{p}^*(t) = \bar{p}(t) \exp\left[-\int_{t_0}^{t} g(\bar{x}, t)\, dt\right]. \tag{95}$$

The vector $\bar{p}(t)$ is by definition normal to the wavefront $W_{A,t}(\bar{x}) = 0$ but its length is not yet determined. The vector $\bar{p}^*(t)$ will also be normal to $W_{A,t}(\bar{x}) = 0$ and we take advantage of the liberty of determination of the length of the normal by choosing it in order to simplify the system (94). From (94) and (95) we obtain the differential system

$$\dot{\bar{p}}^*(t) = -\tilde{D}\bar{p}^*(t). \tag{96}$$

Since $\bar{p}^*(t)$ has all the properties required from $\bar{p}(t)$ we will use it instead of $\bar{p}(t)$ and we will drop the superscripts * in the expression (96) and use the differential system:

$$\dot{\bar{p}}(t) = -\tilde{D}\bar{p}(t). \tag{97}$$

The fundamental system

$$\dot{\bar{x}} = f(\bar{x}, k(\bar{x}, \bar{p}, t), t)$$

is identically equivalent to

$$\dot{\bar{x}} = \left.\frac{\partial H(\bar{x}, u, \bar{p}, t)}{\partial \bar{p}}\right|_{u = k(\bar{x}, \bar{p}, t)} \tag{98}$$

by the definition (40).

Similarly the differential system (97) may be written

$$\dot{\bar{p}}(t) = -\left.\frac{\partial H(\bar{x}, u, \bar{p}, t)}{\partial \bar{x}}\right|_{u = k(\bar{x}, \bar{p}, t)} \tag{99}$$

The lemma I introduced earlier for the problem P_1 is still valid for the problem P_2 in which case the proof is slightly different [*23*].

We have then

$$\dot{\bar{x}} = \frac{\partial H^0(\bar{x}, \bar{p}, t)}{\partial \bar{p}} \tag{100}$$

$$\dot{\bar{p}} = -\frac{\partial H^0(\bar{x}, \bar{p}, t)}{\partial \bar{x}}. \tag{101}$$

The equations (100) and (101) constitute the canonical system we were looking for.

8. The Condition of Weierstrass and the Maximum Principle of Pontryagin

For the problem P_1 we have found

$$\dot{x} = \arg\max_{x \in E^n} H(\bar{x}, \dot{x}, \bar{p}, t). \tag{102}$$

In other words, see (39), the control $\dot{x}$ is always choosen in order to maximize the velocity of the wavefront; this means that the wavelet is entirely located on one side of its tangent plane. If we call $E(\bar{x}, t, \dot{x}, \dot{x}^*)$ the distance measured parallely to the $n + 1$st axis between a point $(\dot{x}^*, L(x, \dot{x}^*, t))$ of the wavelet and the tangent plane to the wavelet at the point of contact $(\dot{x}, L(x, \dot{x}, t))$, we must have

$$E(\bar{x}, t, \dot{x}, \dot{x}^*) \geq 0 \qquad \text{for all} \qquad \dot{x}^* \in E^n. \tag{103}$$

Simple algebraic manipulations lead immediately to

$$E(\bar{x}, t, \dot{x}, \dot{x}^*) = L(x, \dot{x}, t) - L(x, \dot{x}^*, t) + \left\langle \dot{x}^* - \dot{x} \,\middle|\, \frac{\delta L(x, \dot{x}, t)}{\delta \dot{x}} \right\rangle. \tag{104}$$

The expressions (103) and (104) constitute the well known Weierstrass E-test [*11-20*].

For the problem P_2, as we have seen, there is no relation similar to (51) and we must write:

$$E(\bar{x}, t, \dot{x}, \dot{x}^*) = L(x, \dot{x}, t) - L(x, \dot{x}^*, t) + \langle \dot{x}^* - \dot{x} \mid p(\bar{x}, t) \rangle. \tag{105}$$

The relation (105) is neither elegant nor useful since $p(\bar{x}, t)$ cannot be expressed in terms of $L(x, \dot{x}, t)$ and its derivatives. We will generally use the relation

$$u = \arg\max_{u \in \Omega} H(\bar{x}, u, \bar{p}, t)$$

which with the canonical system introduced earlier, (100) and (101), constitutes the Maximum Principle of L. S. Pontryagin.

9. Non-Differentiable Isochrones

When the hypersurface $W_{A,t}(\bar{x}) = 0$ does not have a unique normal at a point of an extremal the application of the generalized Huygens' Principle leads to the possiblity of branching for this extremal. The structure of the extremals in such a case depends on the particular problem considered. We have, nevertheless, proved two general theorems to simplify this situation.

Theorem III. The regularity of the Wavefront along an extremal is non-decreasing up to but excluding a conjugate point.

By the word "non-decreasing regularity" we mean the conservation of the differentiability properties of the wavefront along an extremal.

Theorem IV. The structure of the Wavefront is given entirely by a set of normals which can be classified into two categories.

(1) The past-dependent normals which are solutions of a canonical system (100) and (101).

(2) The past-independent normals which are determined only by the local structure of the fundamental system $\dot{\bar{x}} = \bar{f}(\bar{x}, t, u)$ i.e. by the local wavelets.

A past-independent normal is always the boundary value of a past-dependent normal of branching extremal in accordance to theorem III.

10. Acknowledgments

This research owes much to Professor M. Schiffer and the author wishes to express his gratitude for his invaluable assistance and support.

He thanks also the Belgian American Educational Foundation who made possible his stay in the United States through a C. R. B. Graduate Fellowship (Perrin C. Galpin Fellow 1960-61).

REFERENCES

[*1*] André, J., and Seibert, P., After end-point motions of general discontinuous control systems and their stability properties. *Proc. 1st intern. Congr. on Automatic Control, Moscow, (1961)*.

[*2*] Bass, R. W., Optimal Feedback Control System Design by the Adjoint System. Aeronca-Aerospace Tech. Report No. GO-22 (1960).

[*3*] Bellman, R., "Adaptive Control Processes, a Guided Tour." Princeton Univ. Press, Princeton, New Jersey, 1961.

[4] Bellman, R., Glickberg, I., and Gross, O., On the 'bang-bang' control problems. *Quart. appl. Math.* **14**, 11-18 (1956).
[5] Bliss, G. A., "Lectures on the Calculus of Variations." Univ. of Chicago Press, Chicago, Illinois, 1946.
[6] Boltyanskii, V. G., Gamkrelidze, R. V., and Pontryagin, L. S., Theory of optimal processes. *Izvest. Akad. Nauk S.S.S.R., Ser. Mat.* **24**, No. 1 (1960) (in Russian).
[7] Bolza, O., "Vorlesungen über Variationsrechnung." Teubner, Leipzig, 1909.
[8] Breakwell, J. V., The optimization of trajectories. *J. Soc. ind. appl. Math.* **7**, No. 2, 215-247 (1959).
[9] Bushaw, D., Optimal discontinuous forcing terms. *In* "Contributions to the Theory of Nonlinear Oscillations" (S. Lefschetz, ed.), Vol. 4, pp. 29-52. Princeton Univ. Press, Princeton, New Jersey, 1958.
[10] Bryson, A. E., and Denham, W. F., A Steepest-Ascent Method for Solving Optimum Programming Problems. Raytheon Company Missile and Space Div. Tech. Report No. BR-1303 (1961).
[11] Carathéodory, C., "Variationsrechnung und partielle Differentialgleichungen erster Ordnung." Teubner, Leipzig, 1935.
[12] Carathéodory, C., "Geometrische Optik." Springer, Berlin, 1937.
[13] Cartan, E., "Leçons sur les Invariants Integraux." Hermann, Paris, 1932.
[14] Cicala, P., Le evoluzioni ottime di un aero. *Atti acad. sci. Torino* **89**, 1-9 (1954-1955).
[15] Courant, R., and Hilbert, D., "Methoden der mathematischen Physik." Springer, Berlin, 1937.
[16] Desoer, C. A., Pontryagin's maximum principle and the principle of optimality. *J. Franklin Inst.* **271**, No. 5 (1961).
[17] Flügge-Lotz, I., "Discontinuous Automatic Control." Princeton Univ. Press, Princeton, New Jersey, 1953.
[18] Flügge-Lotz, I., and Halkin, H., Pontryagin's Maximum Principle and Optimal Control. Stanford Univ. Dept. Engineering Mech. Techn. Report No. 130 (1961).
[19] Fraeys de Veubeke, B., Méthodes variationnelles et performances optimales en aéronautique. *Bull. Soc. Math. belg.* **8**, No. 2, 136-157 (1956).
[20] Hadamard, J., "Leçons sur le Calcul des Variations." Hermann, Paris, 1910.
[21] Halkin, H., Trajectoires Optimales. Engineering Dissertation, Université de Liège (1960).
[22] Hamel, G., Über eine mit dem Problem der Rakete zusammenhängende Aufgabe der Variationsrechnung. *Z. angew. Math. u. Mech.* **7**, No. 6 (1927).
[23] Kalman, R., The Theory of Optimal Control and the Calculus of Variations. RIAS Tech. Report No. 61-3 (1961).
[24] Kelley, H. J., Method of gradient. *In* "Optimization Techniques" (G. Leitmann, ed.), Chapter 6. Academic Press, New York, in press.
[25] La Salle, J. P., Time optimal control. *Bol. soc. mat. mex.* **5**, No. 1, 102-119 (1960).
[26] Leitmann, G., On a class of variational problems in rocket flight. *J. Aero/Space Sci.* **26**, No. 9, 586-591 (1959).
[27] Lie, S., and Scheffers, G., "Geometrie der Berührungs-Transformationen." Teubner, Leipzig, 1896.
[28] Miele, A., A Survey of the Problem of Optimizing Flight Paths of Aircrafts and Missiles. Boeing Sci. Research Lab. Report (1960).

[*29*] Morse, M., "The Calculus of Variations in the large." Am. Math. Soc., New York, 1934.

[*30*] Neustadt, L. W., Time Optimal Systems with Position and Integral Limits. Space Technol. Lab. Report (1961).

[*31*] Paiewonsky, B. H., A Study of Time Optimal Control. ARAP (Princeton) Report No. 33 (1961).

[*32*] Roxin, E., and Spinadel, V. W. de, Reachable zones in autonomous differential systems. *Contribs. Differential Equations* **1** (1961).

[*33*] Zermelo, E., Über das Navigationsproblem bei ruhender oder veränderlicher Windverteilung. *Z. angew. Math. u. Mech.* **11**, 114-124 (1931).

A Geometric Interpretation of Pontryagin's Maximum Principle[1]

EMILIO ROXIN
RIAS, Baltimore, Maryland

1. Introduction

In connection with problems of control mechanisms, there have been studied systems of differential equations in which some more or less arbitrary "control functions" appear. An important question is how to choose those control functions in order to minimize (or maximize) a certain functional of the solution. Physically speaking, this may mean to operate the system in the "best" possible way, minimizing a "cost" function or simply the time of operation [*5*].

Pontryagin and his collaborators [*1*] have pointed out a general principle (the "maximum principle"), which applies to this problem.

The problem of what can be accomplished by adequate election of the control function, disregarding the optimization problem, has been studied independently by Kalman [*4*] and the author [*8*, *9*].

In this paper we want to look at these problems from a geometrical point of view; our aim is to show some interesting relations which remain hidden in a more analytical treatment, like that of Pontryagin, Boltyanskii and Gamkrelidze.

We shall study the system

$$\dot{x}^i = f^i(t, x^1, ..., x^n, u^1, ..., u^m); \qquad i = 1, 2, ..., n, \tag{1.1}$$

where, as usual $\dot{x} = dx/dt$. Let us write

$$x^0 = t, \qquad f^0 = 1 \tag{1.2}$$

[1] This research was partially supported by the United States Air Force through the Air Force Office of Scientific Research of the Air Research and Development Command, under Contract Number AF 49(638)-382. Reproduction in whole or in part is permitted for any purpose of the United States Government.

and introduce the following vectors

$$
\begin{aligned}
\hat{x} &= (x^1, \ldots, x^n) \\
x &= (x^0, x^1, \ldots, x^n) = (x^0, \hat{x}) \\
\hat{f} &= (f^1, \ldots, f^n) \\
f &= (f^0, f^1, \ldots, f^n) = (f^0, \hat{f}) \\
u &= (u^1, \ldots, u^m).
\end{aligned}
\tag{1.3}
$$

These vectors are considered in euclidean vector spaces, with the usual norm: $\| x \|^2 = \Sigma(x^i)^2$. Equation (1.1) becomes, then,

$$\dot{\hat{x}} = \hat{f}(t, \hat{x}, u) \tag{1.4}$$

or

$$\dot{x} = f(x, u). \tag{1.5}$$

We make the following assumptions:

(i) $f(x, u)$ is defined in $X \times U$, where X is the x-space (the half-space corresponding to $x^0 = t \geqq 0$ could also be chosen) and U is a compact set of the u-space.

(ii) $f(x, u)$ is continuous in (x, u).

(iii) The Lipschitz condition holds:

$$\| f(x_2, u) - f(x_1, u) \| < K \cdot \| x_2 - x_1 \|.$$

(iv) $\| f(x, u) \| = 0(\| x \|)$ for $\| x \| \to \infty$. This assumption is made in order to avoid a finite escape-time for the solutions.

We shall consider the parameter u as a function of t. $u(t)$ will be called an "admissible control function" if

(v) $u(t)$ is measurable,

(vi) for each t, $u(t) \in U$.

(vii) "Convexity condition": $f(x, U) = \{ f(x, u) \mid u \in U \}$ is a convex set for each x.

We note that being U compact and $f(x, u)$ continuous, the set $f(x, U)$ is compact too.

According to well known theorems (see Carathéodory [*2*], Sansone and Conti [*11*], these conditions guarantee the existence and uniqueness of the solution of (1.5) for any admissible control function $u(t)$. The importance of the convexity condition (vii) will appear in the theorem stating that the set of reachable points is a closed set.

2. Reachable Set

Given the equation (1.5), we shall say that the point x_1 is reachable from x_0, if there exists an admissible control function $u(t)$ defined in the time interval $(t_0, t_1) = (x_0^0, x_1^0)$ where $t_1 \geqq t_0$ such that equation (1.5), where $u = u(t)$, together with the initial condition

$$x(t_0) = x_0$$

has the solution $x(t)$ satisfying

$$x(t_1) = x_1.$$

The set of all points x which are reachable from x_0, will be called "*reachable set from* x_0", and denoted by R_{x_0}. When there is no possibility of misunderstanding, the subindex x_0 will be dropped.

In many simple cases it is easy to see which points are reachable and which are not; the linear case has been treated in detail by La Salle [*5*] and Gamkrelidze [*3*]. Here we are concerned with rather general properties of the reachable set.

Theorem 2.1. If the equation (1.5) fulfills the conditions (i) to (vii), then for any given initial x_0, the reachable set R_{x_0} is a closed set.

The proof of this theorem is given in ref. [*7*]. A similar theorem under less general conditions is proved by Lee and Markus [*6*]. The importance of this theorem lies in the fact that it may be used as an existence theorem for solutions of optimality problems like those which are the concern of Pontriagin's Maximum Principle.

3. Some Properties of the Boundary of the Reachable Set

It is important to study the properties of the boundary ∂R in the space $X = \{(x^0, x^1, ..., x^n)\}$, because there we might expect to find the solution of problems of the variational type. Let us take three examples.

(a) Minimum time for reaching a given set $\hat{G}$ of the space $X = \{(x\ , ..., x^n)\}$.

Let us start at $x_0 = (x_0^0, x_0^1, ..., x_0^n)$, where $x_0^0 = t_0$; we want to reach any point $x_1 = (x_1^0, x_1^1, ..., x_1^n)$, such that $\hat{x}_1 = (x_1^1, ..., x_1^n) \in \hat{G}$ and $t_1 = x_1^0$ has the least possible value. Let G be the cylindrical set $I^+ \times \hat{G}$, where I^+ is that part of the x^0 axis, for which $t \geqq t_0$. Then, what we are looking for is a point $x_1 \in R \cap G$ which minimizes $x_1^0 = t_1$. This point will belong to the boundary of R, because in every neighborhood of it there are points of G with smaller x^0 coordinates, which cannot belong to R because they do not belong to $G \cap R$.

(b) If, in the preceding example, instead of minimizing the time $t_1 = x_1^0$, we want to minimize some function of the coordinates $\psi(x_1^0, x_1^1, ..., x_1^n)$ then some additional assumptions have to be made. If on the set G, the function $\psi(x)$ has some minimum, attained at the set $G^* \subseteq G$, and if $G^* \cap R \neq \phi$ (some part of G^* is reachable), then any point of $G^* \cap R$ gives us a solution of our problem. These points do not need to belong to the boundary of R, and this case is called "degenerate" by Rozonoer [*10*], because Pontryagin's principle does not apply to it. In fact (as we shall see), this principle applies if $x_1 \in G^* \cap R$ belongs to the boundary ∂R.

(c) If we want to minimize some functional of the type

$$\int_{t_0}^{t_1} \varphi(x(t))\, dt, \tag{3.1}$$

the classical procedure consists in adding one coordinate to the X-space, putting

$$\dot{x}^{n+1} = \varphi(x); \qquad x^{n+1}(0) = 0,$$

and minimizing $\psi(x_1) = x_1^{n+1}$.

In the example (b) it is also possible to write $x^{n+1} = \psi(x^0, x^1, ..., x^n)$ and minimize x^{n+1}, but in that case R is contained in the hypersurface $x^{n+1} = \psi(x^0, ..., x^n)$, so that R coincides with its boundary and those properties which depend on the fact that the solution point belongs to ∂R, become trivial.

In this case, the value of the integral (3.1) depends in general on the trajectory $x(t)$, so that the set R does not need to be contained in some hypersurface. If we want to reach the set G we have, therefore, to look for the points on $G \cap R$ with minimum x^{n+1}. These are, of course, on the boundary of R because the set G is cylindrical with respect to the coordinate x^{n+1}.

We now pass to establish a basic property of the trajectories belonging to the boundary of R.

Theorem 3.1. If for equation (1.5), R is the reachable set from the point ξ_0, if the point $\xi_1 \in \partial R$ and if $x_1(t)$ $(t_0 \leqq t \leqq t_1)$ is a trajectory going from $x_1(t_0) = \xi_0$ to $x_1(t_1) = \xi_1$, which corresponds to an admissible control $u_1(t)$ $(t_0 \leqq t \leqq t_1)$, then the whole trajectory $x_1(t)$ $(t_0 \leqq t \leqq t_1)$ belongs to the boundary of R.

The fact that $\xi_1 \in R$ guarantees the existence of the functions $u_1(t)$ and $x_1(t)$ mentioned. Let us assume that for $t = t_2$, $(t_0 \leqq t_2 \leqq t_1)$, the point $\xi_2 = x_1(t_2)$ is interior to R. There exists, therefore, a neighborhood A_2 of ξ_2, contained in R. Consider the equation

$$\dot{\xi} = f(\xi, u_1(t)) \tag{3.2}$$

in the interval (t_2, t_1), where now the control function $u_1(t)$ is fixed. For the initial condition $\xi(t_2) = \xi_2$ we obtain $\xi(t_1) = \xi_1$. Thus there is a certain neighborhood A_1 of ξ_1 such that $\xi(t_1) \in A_1$ implies $\xi(t_2) \in A_2$. We shall now prove that $A_1 \subseteq R$, which contradicts the fact that $\xi_1 \in \partial R$.

Let ξ_1' be any point of A_1. Equation (3.2) together with $\xi(t_1) = \xi_1'$ defines a trajectory such that $\xi(t_2) = \xi_2' \in A_2$. Since $\xi_2' \in R$, there exists an admissible control function $u_2(t)$ defined in the interval (t_0, t_2') (where $t_2' = \xi_2'^0$), such that

$$\dot{x}_2(t) = f(x_2(t), u_2(t));\ x_2(t_0) = \xi_0 \tag{3.3}$$

gives $x_2(t_2') = \xi_2'$.

We now extend this trajectory, coupling it with the path given by (3.2), in the following manner:

$$\left.\begin{aligned} &u_2(t) = u_1(t_2 + t - t_2'); \\ &\dot{x}_2(t) = f(x_2(t), u_2(t));\ x_2(t_2') = \xi_2' \end{aligned}\right\}\ t_2' \leqq t \leqq t_1' \tag{3.4}$$

As this trajectory coincides with the former $\xi(t)$ (we have shifted the t but not the x-values), we obtain $x_2(t_1) = \xi_1'$ (see Figure 1). Hence, $\xi_1' \in R$ as we wanted to prove.

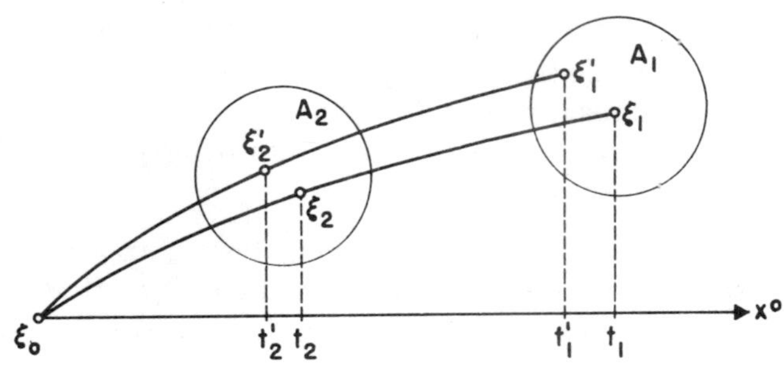

FIG. 1

The Variational and the Adjoint Equation

It is well known that, for a fixed control function $u^*(t)$ and the corresponding solution $x^*(t)$ of equation (1.5), which we suppose given in the time interval (t_1, t_2), the neighboring solutions of (1.5) can be represented by the variational equation

$$\dot{y} = \left[\frac{\partial f}{\partial x}\right]_{\substack{u=u^*(t)\\ x=x^*(t)}} \cdot y. \tag{3.5}$$

Here

$$y(t) = x(t) - x^*(t) + o(\| x(t_1) - x^*(t_1) \|)^1 \tag{3.6}$$

is the first variation of $x^*(t)$ and $\partial f/\partial x$ is the Jacobian matrix.

Since the equation (3.5) is linear, its solution is of the form

$$y(t) = A(t, t_0) \cdot y(t_0), \tag{3.7}$$

where $A(t, t_0)$ is a linear operator (i.e., a matrix).

The y is a contravariant (column) vector, of the tangent space, and if $p = p(t)$ is a covariant (row) vector such that

$$p(t) \cdot y(t) = \text{const.}, \tag{3.8}$$

it is known that $p(t)$ satisfies the adjoint equation

$$\dot{p} = -p \cdot \left[\frac{\partial f}{\partial x}\right]_{\substack{u=u^*(t)\\ x=x^*(t)}}, \tag{3.9}$$

the solution of which is

$$p(t) = p(t_0) \cdot A(t_0, t). \tag{3.10}$$

This vector p is very useful for defining hyperplanes in the tangent space, as we shall see.

Theorem 3.2. If, for equation (1.5), R is the reachable set from a certain point ξ_0, if $x^*(t)$, $(t_1 \leqq t \leqq t_2)$ is an admissible trajectory of (1.5) belonging to the boundary of R and if at the points of $x^*(t)$ boundary of R is a smooth surface with a tangent plane given by $p(t) \cdot (x - x^*(t)) = 0$, then $p(t)$ satisfies equation (3.9).

Proof. Let $x(t)$ $(t_1 \leqq t \leqq t_2)$ be a neighboring trajectory of $x^*(t)$, which corresponds to the same control function $u^*(t)$. If $x(t_1) \in R$ then evidently $x(t) \in R$ for any $t \in (t_1, t_2)$. The fact that $p \cdot (x - x^*)$ is the tangent plane means that

$$p \cdot (x - x^*) + o(\| x - x^* \|) \geqq 0 \qquad \text{for} \qquad x \in R$$

and

$$p \cdot (x - x^*) + o(\| x - x^* \|) \leqq 0 \qquad \text{for} \qquad x \notin R$$

[1] $y = o(x)$ means that $\lim y/x = 0$.

where we have supposed, for example, that the set R is on that side of the tangent plane where the scalar product is positive. As $o(\| x - x^* \|)$ refers to $x \to x^*$, we are interested in a first order approximation and may substitute $x - x^*$ by y as given in (3.6). So, if $x(t_1) \in R$ and $y(t_1) = x(t_1) - x^*(t_1)$, we have

$$p(t) \cdot y(t) + o(\| y(t_1) \|) \geqq 0 \qquad (t_1 \leqq t \leqq t_2) \tag{3.11}$$

and this determines completely the position of the tangent plane given by $p(t)$, since the mapping $y(t_1) \to y(t)$ is linear and transforms a half space into a half space. Besides, the vector $p^*(t)$ defined by equation (3.9) and the initial value $p^*(t_1) = p(t_1)$ also satisfies the relation (3.11) because $p^*(t) \cdot y(t) =$ const., so that identically $p^*(t) = p(t)$.

4. Tangent Cone to a Point Set

In order to simplify the statement of the properties of a reachable set, we introduce the concept of tangent cone to a given point set and start proving some elementary relations.

Definition 4.1 Given a point set A of the space X and a point $x_0 \in X$, we shall call tangent cone of the set A at the point x_0, the set of vectors y such that there exist a sequence of positive numbers c_i $(i = 1, 2, 3, \ldots)$ and a sequence of points $x_i \in A$ such that

$$\lim_{i\to\infty} x_i = x_0,$$

$$\lim_{i\to\infty} c_i(x_i - x_0) = y. \tag{4.1}$$

We shall denote the tangent cone of the set A at the point x_0 by $T_{x_0}[A]$, or simply by $T[A]$ when there is no possibility of misunderstanding.

According to the definition, if x_0 is an exterior point of the set A, the cone $T_{x_0}[A]$ will be empty. If x_0 is an interior point of A, it will be the whole tangent space (all vectors y). If the set A is a differentiable r-dimensional hypersurface and $x_0 \in A$, then the cone $T_{x_0}[A]$ is the r-dimensional tangent linear space.

Lemma 4.1. If $A = A_1 \cup A_2$, then

$$T[A] = T[A_1] \cup T[A_2].$$

Proof: Obviously, if $A_1 \subseteq A$, also $T[A_1] \subseteq T[A]$. Therefore

$$T[A] \supseteq T[A_1] \cup T[A_2].$$

For proving the opposite relation, it is sufficient to see that if $y \in T[A_1 \cup A_2]$, there exist c_i, x_i such that

$$y = \lim c_i(x_i - x_0)$$

with $x_i \in A_1 \cup A_2$, then some infinite subsequence x_{i_j} belongs to one or both sets A_1 or A_2.

Lemma 4.2. The tangent cone of any set is closed.

Proof: We have to show that if

$$y = \lim_{i \to \infty} y_i \tag{4.2}$$

and

$$y_i = \lim_{j \to \infty} c_{ij}(x_{ij} - x_0) \qquad i = 1, 2, 3, \ldots \tag{4.3}$$

where

$$c_{ij} > 0; \qquad x_{ij} \in A; \qquad \lim_{j \to \infty} x_{ij} = x_0, \tag{4.4}$$

then there exist some sequences

$$\gamma_\alpha > 0, \qquad x_\alpha \in A, \qquad \alpha = 1, 2, 3, \ldots$$

such that

$$\lim_{\alpha \to \infty} x_\alpha = x_0; \qquad \lim_{\alpha \to \infty} \gamma_\alpha(x_\alpha - x_0) = y. \tag{4.5}$$

We shall construct the sequence x_α in the following way. By virtue of (4.3) and (4.4), for each positive integer α we can select a value of j, which we call j_α, such that both inequalities

$$\| x_{\alpha j_\alpha} - x_0 \| < \frac{1}{\alpha}$$

and

$$\| c_{\alpha j_\alpha}(x_{\alpha j_\alpha} - x_0) - y_\alpha \| < \frac{1}{\alpha}$$

hold. Now we put

$$x_\alpha = x_{\alpha j_\alpha}; \qquad \gamma_\alpha = c_{\alpha j_\alpha}.$$

Obviously, the first of the inequalities (4.5) holds. For showing that the second holds too, we write

$$\| \gamma_\alpha(x_\alpha - x_0) - y \| \leqq \| \gamma_\alpha(x_\alpha - x_0) - y_\alpha \| + \| y_\alpha - y \|$$

and see that for any $\epsilon > 0$, we have

$$\| y_\alpha - y \| < \frac{\epsilon}{2} \quad \text{and} \quad \| \gamma_\alpha(x_\alpha - x_0) - y_\alpha \| < \frac{\epsilon}{2}$$

for all α greater than some $N(\epsilon)$.

Lemma 4.3. For any set A (and any point x_0)

$$T[\partial A] \subseteq T[A].$$

The proof is similar to the preceding one. Suppose that $y \in T[\partial A]$. There exist sequences $c_i > 0$ and $x_i \to x_0$ such that $c_i(x_i - x_0) \to y$. Furthermore, as $x_i \in \partial A$, we may write

$$x_i = \lim_{j\to\infty} x_{ij}, \qquad x_{ij} \in A.$$

For each positive integer α we can select a value of $j = j_\alpha$ such that

$$\| x_{\alpha j_\alpha} - x_\alpha \| = o(\| x_\alpha - x_0 \|).$$

Hence

$$\lim_{\alpha\to\infty} (x_{\alpha j_\alpha} - x_\alpha) = \lim_{\alpha\to\infty} (x_\alpha - x_0) = 0$$

and

$$\lim_{\alpha\to\infty} c_\alpha(x_{\alpha j_\alpha} - x_0) = \lim_{\alpha\to\infty} c_\alpha(x_\alpha - x_0) = y,$$

this last relation being valid because

$$x_{\alpha j_\alpha} - x_0 = (x_{\alpha j_\alpha} - x_\alpha) + (x_\alpha - x_0)$$

is an infinitesimal which is equivalent to $x_\alpha - x_0$. So, since in the last equality, $x_{\alpha j_\alpha} \in A$, we conclude that $y \in T[A]$.

Corollary 4.1. $T[\bar{A}] = T[A]$, where $\bar{A}$ denotes the closure of A.

Proof.

$$T[\bar{A}] = T[A \cup \partial A] = T[A] \cup T[\partial A] = T[A].$$

Theorem 4.1. For any set A

$$\partial T[A] \subseteq T[\partial A].$$

We start by proving an auxiliary lemma.

Lemma 4.4. If Y_0 is an open cone (a cone projecting an open set), if A is a given set and if y_0 is a non-zero vector fulfilling the following conditions:

$$\text{(i) } y_0 \in Y_0$$

$$\text{(ii) } y_0 \in T_{x_0}[A]$$

then, for any $\epsilon > 0$ there exists a point $x \in A$ such that

$$0 < \| x - x_0 \| \leqq \epsilon \qquad \text{and} \qquad x - x_0 \in Y_0.$$

Indeed, since $y_0 \in T_{x_0}[A]$, there exists a sequence $x_i \to x_0$, $x_i \in A$, $c_i(x_i - x_0) \to y_0$, where $c_i > 0$. Therefore, $\| x_i - x_0 \| \leqq \epsilon$ for i greater than some N_1. Since Y_0 is open, $c_i(x_i - x_0) \in Y_0$ for i greater than some N_2. Taking $i > \max(N_1, N_2)$, x_i fulfills the desired requirements.

Now we return to the proof of Theorem 4.1. It is easy to see that $\partial T[A]$ is a cone, i.e., that if $y_0 \in \partial T[A]$ then the whole ray $\{cy_0 \mid c > 0\}$ belongs to $\partial T[A]$. Let us suppose that

$$y_0 \in \partial T_{x_0}[A]; \tag{4.6}$$

if we show that y_0 necessarily belongs to $T_{x_0}[\partial A]$, the theorem will be proved.

We take a sequence of open connected cones Y_i such that

$$Y_{i+1} \subseteq Y_i \qquad \text{and} \qquad \bigcap_{i+1}^{\infty} Y_i$$

is the single ray cy_0, $(x > 0)$. We may take, for example, on the unit sphere the sets

$$\left\{ y \mid \left\| y - \frac{y_0}{\| y_0 \|} \right\| < \epsilon_i \right\}$$

with $\epsilon_i < 0$, $\epsilon_i \to 0$, and project them from the center of the sphere.

Let us call now Y_i^* the intersection of the cone Y_i with the ball of radius ϵ_i, i.e.,

$$Y_i^* = \{ y \mid y \in Y_i ; \| y \| < \epsilon_i \},$$

where always $\epsilon_i \to 0$ (see Figure 2).

Since $y_0 \in \partial T_{x_0}[A]$, in any open cone Y_i there exist two rays, one belonging to $T[A]$ and the other to its complement $CT[A]$. From the preceding lemma we deduce, therefore, that in any neighborhood of the first ray there is a point $x' \neq x_0$, such that $x_i' \in A$, $x_i' - x_0 \in Y_i^*$.

On the second ray there exists some point $x_i'' \neq x_0$ such that $x_i'' \notin A$ and $x_i'' - x_0 \in Y_i^*$ (indeed, a whole segment of this ray, starting at x_0, does not belong to A, otherwise this ray would belong to $T[A]$.) Therefore, the set $Y_i^* - \{x_0\}$ being connected, there exists a point $x_i \neq x_0$ such that $x_i \in \partial A$, $x_i - x_0 \in Y_i^*$ and $\| x_i - x_0 \| < \epsilon_i$. From here we conclude that

(i) $x_i \to x_0$ as $i \to \infty$;

(ii) $c_i(x_i - x_0) \to y_0$, if we take $c_i = \dfrac{\| y_0 \|}{\| x_i - x_0 \|} > 0.$

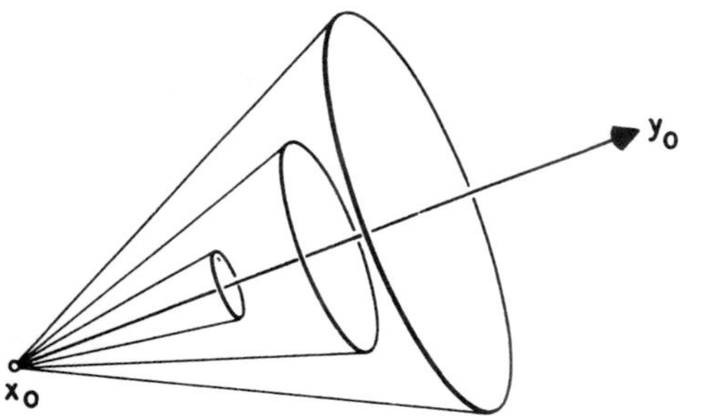

FIG. 2

Remark. It is important to note that the converse is not necessarily true, that is,

$$T[\partial A] \text{ not necessarily } \subseteq \partial T[A]$$

as shown in an example, in Figure 3.

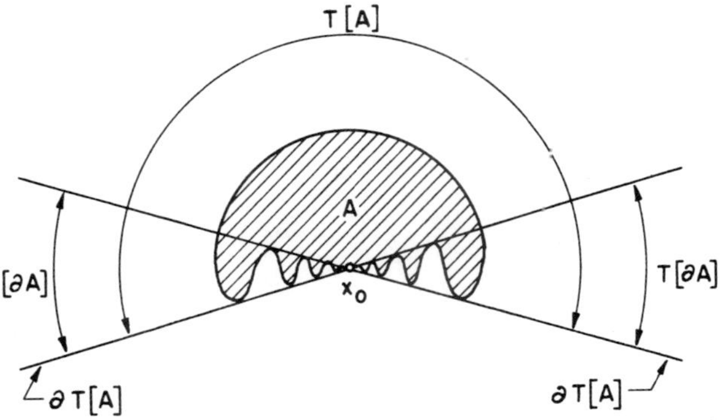

FIG. 3

5. The Reachable Cone

Definition: Given equation (1.5) and a certain point x_0, let R_{x_0} be the reachable set from x_0; we shall call "reachable cone from the point x_0 at the point x", and denote by $C_{x_0}[x]$ or simply $C[x]$, the tangent cone of R_{x_0} at the point x:

$$C_{x_0}[x] = T_x[R_{x_0}]. \tag{5.1}$$

Taking into account that R_{x_0} is closed and connected, it is easily seen that a necessary and sufficient condition for $C_{x_0}[x]$ to be not empty is that $x \in R_{x_0}$.

Theorem 5.1. Let $u(t)$ be an admissible control function of equation (1.5), defined in the interval (t_1, t_2) where $t_1 < t_2$, $x(t)$ the corresponding trajectory, let $C[x(t)] = C_{x_0}[x(t)]$ defined as above for some point x_0, and let $A(t, t_1)$ be the matrix defined in (3.7) as solution of the variational equation along $x(t)$; then

$$C[x(t_2)] \supset A(t_2, t_1)\, C[x(t_1)]. \tag{5.2}$$

Proof: Suppose $y \in C[x(t_1)]$, so that there exists a sequence $x_i \to x(t_1)$, $x_i \in R$, $c_i(x_i - x(t_1)) = y_i \to y$, $c_i > 0$. If we show that $A(t_2, t_1) \cdot y \in C[x(t_2)]$, we will have proved the theorem.

Let x_i' be the point obtained by translation of x_i along the trajectory corresponding to the same control function $u(t)$ during the time interval (t_1, t_2) (similarly as in the proof of Theorem 3.1). Let y_i' be the value for $t = t_2$ of the solution of the variational equation starting at y_i for $t = t_1$, so that (see Figure 4)

$$y_i' = A(t_2, t_1)\, y_i. \tag{5.3}$$

Between x_i' and y_i' exists the relation

$$x_i' - x(t_2) = \frac{y_i'}{c_i} + o(\| x_i - x(t_1) \|).$$

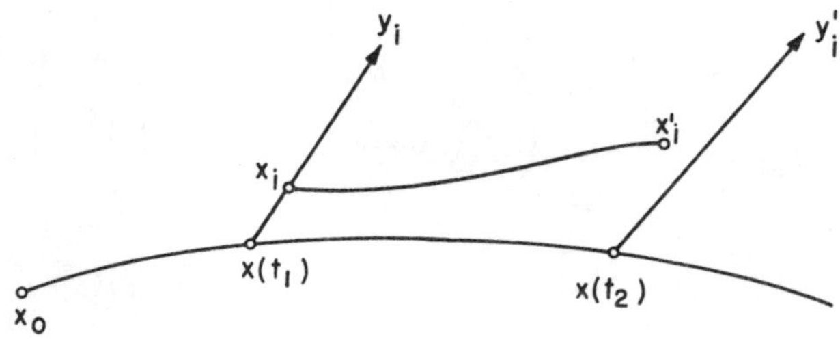

FIG. 4

Therefore, supposing the existence of these limits,

$$y' = \lim y_i' = \lim c_i(x_i' - x(t_2)). \tag{5.4}$$

Since the linear operator $A(t_2, t_1)$ is continuous, we obtain from (5.3)

$$y' = \lim y_i' = A(t_2, t_1) \lim y_i = A(t_2, t_1)y$$

which, by the way, proves the existence of the limit. Moreover, from the way they are obtained, it follows that $x_i' \in R$, hence (5.4) proves that $A(t_2, t_1)\, y \in C[x(t_2)]$, which was our task.

Theorem 5.2. If $x(t)$ is the trajectory of (1.5) corresponding to the admissible control frunction $u(t)$, defined in the interval (t_0, t_1), if $x(t_0) = x_0$, $x(t_1) = x_1$; and if at the point x_1 the tangent cone $c_{x_0}[x_1]$ is contained in the half-space defined by $p_1 \cdot y \geqq 0$, where p_1 is a covariant (row-)vector, then, at any point $x(t)$ with $t_0 \leqq t \leqq t_1$, the cone $C_{x_0}[x(t)]$ is contained in the half-space $p(t) \cdot y \geqq 0$ where $p(t) = p_1 \cdot A(t_1, t)$ is is the solution of the adjoint variational equation.

Proof: If for some $y_t \in C[x(t)]$ it would be

$$p(t) \cdot y_t = p_1 \cdot A(t_1, t)\, y_t < 0,$$

applying the preceding theorem we see that

$$y_1 = A(t_1, t)\, y_t \in C[x_1]$$

and $p_1 y_1 < 0$, contradicting our hypothesis.

This theorem generalizes the theorem (3.2), and here the advantage is that we are not assuming any kind of smoothness of the boundary of R_{x_0}.

Corollary 5.1. If, under the conditions of the last theorem, the reachable set R_{x_0} is "flat" in the neighborhood of the point x_1, that is if $C_{x_0}[x_1]$ is contained in an r-dimensional linear subspace, the same happens along the trajectory $x(t)$ $(t_0 \leqq t \leqq t_1)$.

Corollary 5.2. If, under the same conditions $C_{x_0}[x_1]$ has a convex edge or, more generally, a convex vertex:

$$p_1 \cdot y \geqq 0, \ldots, p_r' y \geqq 0; \qquad y \in C_{x_0}[x_1],$$

the same happens along the whole trajectory $x(t)$ $(t_0 \leqq t \leqq t_1)$.

When $C_{x_0}[x]$ is convex for all x, as in the linear case, we obtain the result that any trajectory on the boundary of R_{x_0} cannot move from a smooth to a "rougher" part of ∂R (it can move from a vertex to an edge and from there to a region where ∂R has a tangent plane, but never in the opposite direction).

6. The Admissible Cone

Definition: Given equation (1.5) and the set U of admissible values of the variable u, we shall call "admissible cone at point x_0", and denote by $F[x_0]$, the cone which projects the set $f(x_0, U)$; this is

$$y \in F[x_0]$$

if and only if there exists a value $u_0 \in U$ and a positive number c such that

$$y = cf(x_0, u_0). \tag{6.1}$$

$F[x]$ is closed and convex, since the set $f(x, U)$ has those properties.

Note that $f(x, U)$ is the section of the cone $F[x]$ with the plane $y^0 = 1$.

Our task now is to study the relation between the admissible cone $F[x]$ and the reachable set R, and especially its tangent cone $T_x[R]$.

Lemma 6.1. If, under the conditions stated in Section 1, $u(t)$ is an admissible control function and $x(t)$ the corresponding trajectory of (1.5), then

$$\frac{\Delta x}{\epsilon} = \frac{1}{\epsilon}\int_{t_0}^{t_0+\epsilon} f(x(t), u(t))\, dt \tag{6.2}$$

is a vector belonging to the convex hull of the set $\{f(x, U) \mid x = x(t);\ t_0 \leqq t \leqq t_0 + \epsilon\}$.

This is a vectorial generalization of the classical mean value theorem. For proving it we multiply (6.2) by any vector p, obtaining

$$p \cdot \frac{\Delta x}{\epsilon} = \frac{1}{\epsilon}\int_{t_0}^{t_0+\epsilon} p \cdot f(x(t), u(t))\, dt.$$

Now the integrand is a scalar function, so that we have

$$\text{l.u.b. } \{p \cdot f(x(t), U)\} \geqq p \cdot \frac{\Delta x}{\epsilon} \geqq \text{g.l.b. } \{p \cdot f(x(t), U)\} \tag{6.3}$$

for $0 \leqq t - t_0 \leqq \epsilon$. Since these inequalities are valid for any vector p, they define precisely the convex hull of the set $f(x(t), U)$, for $t_0 \leqq t \leqq t_0 + \epsilon$.

Lemma 6.2. Let

$$\delta = \delta(x_0, \epsilon) = \text{l.u.b.}\ \{\| f(x, u) - f(x_0, u) \|\} \tag{6.4}$$

subject to the conditions

$$\| x - x_0 \| \leqq \epsilon; \qquad u \in U. \tag{6.4'}$$

Then, under the conditions of § 1,

$$\delta(x_0, \epsilon) \to 0 \qquad \text{for} \qquad \epsilon \to 0. \tag{6.5}$$

Proof. As $\delta(x_0, \epsilon)$ is a nondecreasing function of ϵ for $\epsilon > 0$, there exists the limit

$$l = \lim_{\epsilon \to 0} \delta(x_0, \epsilon).$$

Let us take the sequences of positive numbers $\epsilon_i \to 0$, $\rho_i \to 0$ for $i = 1, 2, 3, \ldots$. For any i we can select a value $u_i \in U$ and a value x_i such that $\| x_0 - x_i \| \leqq \epsilon_i$ and

$$\delta(x_0, \epsilon_i) \geqq \| f(x_i, u_i) - f(x_0, u_i) \| \geqq \delta(x_0, \epsilon_i) - \rho_i.$$

As $i \to \infty$, $\rho_i \to 0$ and therefore

$$\| f(x_i, u_i) - f(x_0, u_i) \| \to l.$$

According to (6.4'), $x_i \to x_0$. Since U is compact, we can select a subsequence such that the lim $u_i = u_0$ exists; we shall suppose that $i = 1, 2, 3, \ldots$, refers to that subsequence. Hence, the continuity of $f(x, u)$ for $x = x_0$ assumes that

$$l = \lim_{i \to \infty} \| f(x_i, u_i) - f(x_0, u_i) \| = 0.$$

With these elements we can prove for rather general conditions, what in the simplest cases is quite obvious, namely that "in first approximation" the reachable points from a given x_0, are those which are inside the cone $F[x_0]$.

Theorem 6.1. If the differential equation (1.5) satisfies the conditions of § 1, then

$$T_{x_0}[R_{x_0}] = F[x_0]. \tag{6.6}$$

Proof. First we shall prove that if $y \in F[x_0]$, then $y \in T_{x_0}[R_{x_0}]$. This is easy, because in this case there exists a positive number c and a value $u_0 \in U$ such that $c \cdot f(x_0, u_0) = y$, and if we integrate the equation

(1.5) with $u(t) = u_0 =$ const. and the initial condition $x(t_0) = x_0$, we obtain a trajectory $x(t)$ belonging evidently to R_{x_0}, for which the vector y is tangent at the point x_0, and therefore $y \in T_{x_0}[R_{x_0}]$.

For proving the converse, we suppose that $y \in T_{x_0}[R_{x_0}]$. Hence, there exist sequences $c_i > 0$, $x_i \in R_{x_0}$ $(i = 1, 2, 3, \dots)$ such that

$$x_0 = \lim x; \qquad y = \lim c_i(x_i - x_0).$$

The relation $x_i \in R_{x_0}$ implies the existence of control functions $u_i{}^*(t)$ and corresponding trajectories of the differential equation $x_i^*(t)$, leading from $x_i^*(t_0) = x_0$ to $x_i{}^*(t_0 + \epsilon_i) = x_i$, $\epsilon_i > 0$. We now apply Lemma 6.1 to the equality

$$x_i - x_0 = \int_{t_0}^{t_0+\epsilon_i} f(x_i{}^*(t), u_i{}^*(t))\, dt,$$

concluding that $(x_i - x_0)/\epsilon_i$ belongs to the convex hull of

$$\{f(x, U) \mid \| x - x_0 \| < \eta_i\},$$

where $\eta_i = \eta_i(\epsilon_i)$ might be taken of the form $\eta_i = K \cdot \epsilon_i$, taking for K an upper bound of $\| f(x, u) \|$ in a suitable neighborhood of x_0. The fact that the component $f^0(x, u) \equiv 1$ assures that $\| (x_i - x_0)/\epsilon_i \| \geqq 1$ and therefore $\epsilon_i \to 0$ for $x_i \to x_0$.

It is easy to see that the sequence $(x_i - x_0)/\epsilon_i$ contains a convergent subsequence; in fact, it is contained in the convex hull of

$$\{f(x, u) \mid \| x - x_0 \| < \max \eta_i\},$$

which is a bounded set. Let us suppose in the following that the indices $i = 1, 2, 3, \dots$ refer to that convergent subsequence.

From the relation of Lemma 6.1 we deduce that for any vector p

$$\begin{aligned} \text{g.l.b. } \{p \cdot f(x, U) \mid \| x - x_0 \| < \eta_i\} &\leqq p \cdot \frac{x_i - x_0}{\epsilon_i} \\ &\leqq \text{l.u.b. } \{p \cdot f(x, U) \mid \| x - x_0 \| < \eta_i\} \end{aligned} \tag{6.7}$$

this following from the fact that

$$\begin{aligned} &\text{g.l.b. } [p \cdot \text{convex hull } \{f(x, U) \mid \| x - x_0 \| < \eta_i\}] \\ &\quad = \text{g.l.b. } [p \cdot \{f(x, U) \mid \| x - x_0 \| < \eta_i\}] \end{aligned}$$

and similarly for the l.u.b.

We may write

$$\text{l.u.b. } \{p \cdot f(x, U) \mid \| x - x_0 \| < \eta_i\} \leqq \text{l.u.b. } \{p \cdot f(x_0, U)\}$$
$$+ \text{ l.u.b. } \{p[f(x, u) - f(x_0, u)] \mid u \in U, \| x - x_0 \| < \eta_i\}$$

and

$$\text{g.l.b. } \{p \cdot f(x, U) \mid \| x - x_0 \| < \eta_i\} \geqq \text{g.l.b. } \{p \cdot f(x_0, U)\}$$
$$- \text{ l.u.b. } \{p[f(x, u) - f(x_0, u)] \mid u \in U, \| x - x_0 \| < \eta_i\}.$$

By Lemma 6.2, the last term in both inequalities tends to zero for $i \to \infty$, therefore taking limits in (6.7) we get

$$\text{g.l.b. } \{p \cdot f(x_0, U)\} \leqq p \cdot \lim \frac{x_i - x_0}{\epsilon_i} \leqq \text{l.u.b. } \{p \cdot f(x_0, U)\}. \tag{6.8}$$

Since $\{f(x_0, U)\}$ is compact and convex by hypothesis, and (6.8) holds for any vector p,

$$\lim \frac{x_i - x_0}{\epsilon_i} \in f(x_0, U). \tag{6.9}$$

Besides, since $\| (x_i - x_0)/\epsilon_i \| \geqq 1$ and bounded, its limit is neither zero nor infinity, therefore

$$y = \lim c_i(x_i - x_0) = (\lim c_i\epsilon_i) \left(\lim \frac{x_i - x_0}{\epsilon_i}\right) = k \lim \frac{x_i - x_0}{\epsilon_i}$$

which, together with (6.9) implies that $y \in F[x_0]$ by definition of the admissible cone.

This theorem can be applied to compare the relative position of the admissible cone $F[x_1]$ at any other point than x_0. Indeed, $x_1 \in R_{x_0}$ implies $R_{x_1} \subseteq R_{x_0}$ and $T_{x_1}[R_{x_1}] \subseteq T_{x_1}[R_{x_0}]$. Applying the last theorem we obtain immediately the following theorem.

Theorem 6.2. If the differential equation (1.5) satisfies the conditions of § 1, and the point x_1 is reachable from x_0, then

$$F[x_1] \subseteq T_{x_1}[R_{x_0}]. \tag{6.10}$$

These theorems do not assert anything about the actual reachability of the points located in the admissible cone. This will be the concern of the following ones.

Theorem 6.3. If, under the assumptions of Theorem 6.1, Y is a closed cone belonging to the interior of the cone $F[x_0]$, then there exists an $\eta > 0$ such that all points of the form

$$\{x_0 + y \mid y \in Y, \|y\| < \eta\}$$

are reachable from x_0.

Proof. If $x(t)$ is the solution of the differential equation corresponding to the control function $u(t) = u =$ const., starting at x_0, we may write

$$\left\| \int_{t_0}^{t_0+} [f(x(t), u) - f(x_0, u)]\, dt \right\| \leqq K \int_{t_0}^{t_0+\epsilon} \| x(t) - x_0 \|\, dt$$

$$\leqq K \int_{t_0}^{t_0+\epsilon} dt \int_{t_0}^{t} \| f(x(t), u) \|\, dt \leqq KM \frac{\epsilon^2}{2},$$

where K and M are suitable constants which can be taken independently of u for all $u \in U$ (since the set U is compact), provided ϵ is restricted to some neighborhood of zero. Therefore

$$\lim_{\epsilon \to 0} \frac{1}{\epsilon} \int_{t_0}^{t_0+\epsilon} f(x(t), u)\, dt = f(x_0, u)$$

uniformly in $u \in U$.

The set $f(x_0, U)$ is really n-dimensional (not $n + 1$ dimensional) because its zero component is always 1. We shall consider in what follows the n-dimensional interior of this set, this is the interior according to the topology of the n-dimensional hyperplane $f^0 \equiv 1$ (see fig. 5).

For any fixed ϵ, the mapping of n-space into n-space, which maps $f(x_0, u)$ in

$$\frac{1}{\epsilon} \int_{t_0}^{t_0+\epsilon} f(x(t), u)\, dt$$

is continuous in its argument $f(x_0, u)$, therefore it maps the compact set $f(x_0, U)$ onto a compact set, and according to what was shown before, this mapping tends uniformly to the identity for $\epsilon \to 0$. Taking any closed set Y' belonging ot the n-dimensional interior of $f(x_0, U)$, there exists an $\eta > 0$ such that for every positive $\epsilon < \eta$ the image of $f(x_0, U)$ in the above defined mapping, covers Y'; for seeing this it is sufficient to take η so small that for every pair (ϵ, u), $0 < \epsilon < \eta$, $u \in U$ the following inequality holds:

$$\left\| f(x_0, u) - \frac{1}{\epsilon} \int_{t_0}^{t_0+\epsilon} f(x(t), u)\, dt \right\| < \begin{array}{r}\text{distance from set } Y' \text{ to the } n\text{–dimensional} \\ \text{boundary of } f(x_0, U).\end{array}$$

Now we take as set Y' the intersection of the closed cone Y of our hypothesis with the plane $(x - x_0)^0 = 1$. Since Y is closed and contained in the $(n + 1$ dimensional) interior of $F[x_0]$, also Y' will be closed and contained in the (n-dimensional) interior of $f(x_0, U)$. We may, therefore, apply the preceding result stating that there exists an $\eta > 0$ such that for any $y' \in Y'$ and any positive $\epsilon < \eta$, there exists an $u \in U$ such that

$$\epsilon \cdot y' = \int_{t_0}^{t_0+\epsilon} f(x(t), u)\, dt.$$

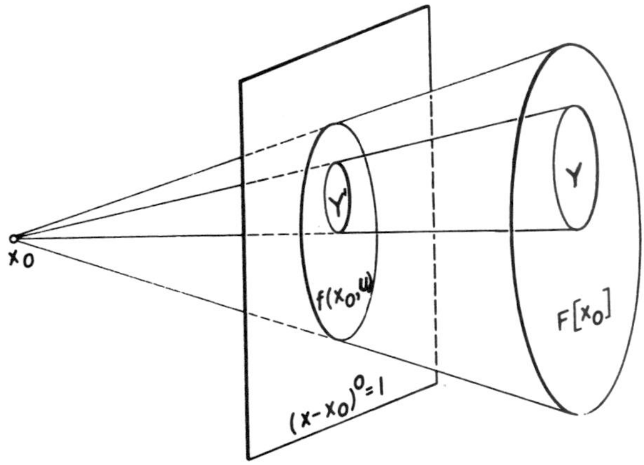

FIG. 5

We shall prove that this η fulfills the requirement of the theorem. Indeed, if $y \in Y$ and $\| y \| < \eta$, we take $y' = y/y^0$ and $\epsilon = y^0 \leqq \| y \| < \eta$ (y^0 is the zero component of the vector y, which is positive because $y \in F[x_0]$). Then $y' \in f(x_0, U)$ and we may apply the last result, obtaining that

$$x_0 + y = x_0 + \epsilon y' = x_0 + \int_{t_0}^{t_0+\epsilon} f(x(t), u)\, dt$$

for some value of $u \in U$, i.e., $x_0 + y$ is actually reachable from x_0.

Corollary 6.1. If R is the reachable set from the point x_0, Q is the complement of R (i.e., the set of not reachable points) and if x_1 is reachable from x_0, then, under the assumptions of the preceding theorems, the intersection $F[x_1] \cap T_{x_1}[Q]$ is void or belongs to the boundary of $F[x_1]$.

Indeed, if we suppose that y belongs to the interior of $F[x_1]$, there exists a closed cone Y which belongs to the interior of $F[x_1]$ and contains y in its interior. Consider a sequence of points $x_i \to x_1$ $(i = 2, 3, 4, \ldots)$, such that $c_i(x_i - x_1) \to y$, $(c_i > 0)$. We shall prove that $x_i \in Q$ is impossible (and therefore $y \in T_{x_1}[Q]$ is impossible). In fact, $c_i(x_i - x_1) \to y$ and y is interior to the cone Y, therefore $x_i - x_0 \in Y$ for sufficiently large i, and by Theorem 6.3 x_i is reachable from x_1 for i larger than some i_0.

7. Relations with the Boundary of *R*

Theorem 7.1. Given the differential equations (1.5) satisfying the conditions stated there, denoting by $R = R_{x_0}$ the reachable set from the point x_0, ∂R its boundary, and x_1 any point, the intersection of the cones $F[x_1]$ and $T_{x_1}[\partial R]$ is void or belongs to the boundary of $F[x_1]$.

The proof is immediate if we apply the corollary 6.1, taking into account that by Theorem 5.3,

$$T_{x_1}[\partial R] = T_{x_1}[\partial Q] \subseteq T_{x_1}[Q]$$

where Q is the complement of R.

Theorem 7.2. If, under the same assumptions, the trajectory $x(t)$ belongs to the boundary of R for $t_0 \leqq t \leqq t_2$, and $t_0 \leqq t_1 < t_2$, then denoting $x(t_1) = x_1$, the cones $F[x_1]$ and $T_{x_1}[\partial R]$ have a non-void intersection, which belongs to the boundary of $F[x_1]$.

Taking into account the preceding theorem, it remains only to prove the existence of at least one common ray of $T_{x_1}[\partial R]$ and $F[x_1]$. This ray is easily obtained by taking, on the trajectory $x(t)$, a decreasing sequence $t_i \to t_1$ $(i = 3, 4, 5, \ldots)$ and the corresponding points $x_i = x(t_i) \in \partial R$. From the sequence

$$c_i(x_i - x_1), \qquad c_i = \frac{1}{\| x_i - x_1 \|}$$

we take a converging subsequence; its limit belongs to $F[x_1]$ by Lemma 6.1 and a reasoning similar to that of Theorem 6.1. Therefore $y \in F[x_1]$, $y \in T_{x_1}[\partial R]$.

From here we get the relation with Pontryagin's "Maximum Principle". Indeed, this principle applies to optimization problems characterized precisely by the fact that any optimal trajectory (i.e., solution of the optimization problem) lies on the boundary of the reachable set (from the initial point). Any such trajectory has a definite tangent vector at almost every point and by Theorem (7.2) this tangent vector, given by $f(x(t), u(t))$, belongs to the boundary of $F[x(t)]$. The corresponding value of $u(t)$ may or may not belong to the boundary of U, this fact depending on the mapping of U into $f(x, U)$. In any case, since $F[x]$ is convex, there exists a vector $p(t)$ such that $u(t)$ maximizes the scalar product $p(t) \cdot f(x(t), u(t))$ with respect to all possible values of $u \in U$, and this is the maximum principle.

If along the trajectory the boundary of the reachable set R has a tangent plane, then Theorem 3.2 show that $p(t)$ satisfies the adjoint variational equation (3.10).

If at some point of the trajectory the boundary of R has a tangent plane, Theorem 5.2 shows that the vector $p(t)$ satisfying the maximum principle may be taken to satisfy the adjoint variational equation even in the case when for previous values of the time the trajectory goes along an (r-dimensional) edge of the boundary of R.

The main difference between the approach of Pontryagin and this paper, is that here we tried to use global properties of the reachable set, while Pontryagin uses small variations of the optimizing trajectory.

A nice picture of the reachable set R is obtained by interpreting the previous results in terms of wave propagation. We imagine $x_0 = (t_0, \hat{x}_0)$ as the origin of perturbations propagated along all admissible trajectories. Then, for any fixed value of $t > t_0$, the section of R is the pertrubed region. The section of the boundary of R is the wave front. The boundary of R, imagined as a surface given by an equation

$$t_{\min} = S(x)$$

is an envelope of the cones $F[x]$ and may be interpreted as a (generalized) solution of a partial differential equation of the first order, with $F[x]$ the characteristic cone at each point. This clearly points out the relation with the Hamilton-Jacobi theory.

REFERENCES

[1] Boltyanskii, R. V., Gamkrelidze, R. V., and Pontryagin, L. S., The theory of optimal processes and the maximum principle. *Izvest. Akad. Nauk S.S.S.R., Ser. Mat.* **24**, 3-24 (1960) (in Russian).

[2] Carathéodory, C., "Vorlesungen über reelle Funktionen" 2nd ed. Teubner, Leipzig, 1927 (reprinted by Chelsea Publ. Co., New York, 1948).

[3] Gamkrelidze, R. V., Theory of time-optimal processes for linear systems. *Izvest. Akad. Nauk S.S.S.R., Ser. Mat.* **22**, 449-474 (1958) (in Russian).

[4] Kalman, R. E., Contributions to the theory of control. *Bol. soc. mat. mex.* pp. 102-119 (1960).

[5] La Salle, J. P., The time optimal control problem. *In* "Contributions to the theory of Nonlinear Oscillations" (S. Lefschetz, ed.), Vol. 5, pp. 1-24. Princeton Univ. Press, Princeton, New Jersey, 1960.

[6] Lee, E. B., and Markus, L., Optimal control for nonlinear processes. *Arch. rational Mech. Anal.* 8, No. 1, 36-58 (1961).

[7] Roxin, E., On the existence of optimal controls. *Michigan J. Math.* to be published.

[8] Roxin, E., Reachable zones in autonomous differential systems. *Bol. soc. mat. mex.* pp. 125-135 (1960).

[9] Roxin, E., and de Spinadel, V. W., Reachable zones in autonomous differential systems. *Contribs. Differential Equations* 1 (1960).

[10] Rozonoer, L. I., Prontryagin's "maximum principle" in the theory of optimum systems. *Avtomat. i Telemekh.* **20**, No. 1, 36-58 (1961).

[11] Sansone, G., and Conti, R., "Equazioni differenziali non lineari." Edizione Cremonese, Rome, 1956.

On the Accessibility Problem in Control Theory[1]

ROBERT HERMANN
Lincoln Laboratory
Massachusetts Institute of Technology, Lexington, Massachusetts, and
University of California, Berkeley, California

1. Introduction

The problem can be stated as follows: Given a system of ordinary differential equations

$$g_a(x_i, \dot{x}_i, t) = 0, \qquad \dot{x}_i = \frac{dx_i}{dt} \tag{1.1}$$

$$(1 \leq a, b, ... \leq m; 1 \leq i, j, ... \leq n; \text{summation convention})$$

which points of R^n can be reached along solutions starting from a given point x_i^0? Because of its evident importance for control theory, we present a short survey here of what is known about the differential-geometric side of this problem. (Thus, we will consider only piecewise smooth solutions of 1.1). The basic result we will use is due to Chow [*3*], but does not seem to be well-known.[2] In turn, Chow's theorem is a generalization of a famous theorem of Carathéodory in thermodynamics giving a geometric condition that a Pfaffian form be completely integrable. Some of Carathéodory's work on the Lagrange variational problem [*1*] is also applicable to the accessibility problem.

The proofs of most of the new results mentioned here will be published [*4*, *5*].

2. Pfaffian Systems and Monge Systems

All manifolds, tensor-fields, curves, maps, etc. will be of differentiability class C^∞ unless mentioned otherwise. (Of course, this is only a technical device to avoid counting the precise order of differentiability necessary in each problem.) Manifolds will be connected and paracompact, and a

[1] This work was done at Lincoln Laboratory, Massachusetts Institute of Technology, but the paper was written with support from the National Science Foundation, 6-19136.

[2] I am indebted to N. Lichtblau for pointing this out.

path will be a continuous, piecewise C^∞ curve. Let M be a manifold. For $x \in M$, M_x denotes the tangent space to M at x. If $\alpha: M \to M'$ is a map of manifolds, $\alpha_* : M_x \to M'_{\phi(x)}$ denotes the linear map α induces on tangent vectors. If w is a differential form on M', $\alpha^*(w)$ is the form on M induced by α. If $\sigma : [a, b] \to M$ is a curve, $\sigma'(t) \in M_{\sigma(t)}$ is, for $a \leq t \leq b$, the tangent vector to σ at t.

Let $V(M)$ denote the set of vector fields on M, considered as a module over the ring $C(M)$ of C^∞ real valued functions on M and as a real (infinite dimensional) Lie algebra under the Jacobi bracket operation: $X, Y \to [X, Y]$, for $X, Y \in V(M)$. A real subspace $\mathbf{H} \subset V(M)$ is said to define a *Pfaffian system* on M. For $x \in M$, let

$$\mathbf{H}_x = \{X(x) : X \in \mathbf{H}\} \subset M_x.$$

$\mathbf{H}$ is said to be *non-singular* if dim $\mathbf{H}_x$ is constant for $x \in M$. We will mainly consider non-singular systems in this paper.

Define: $D^1(\mathbf{H}) = \mathbf{H} + [\mathbf{H}, \mathbf{H}]$, the sum of linear combinations of elements of $\mathbf{H}$ and their Jacobi brackets with coefficients from $C(M)$. $D^2(\mathbf{H}) = D^1(D^1(\mathbf{H}))$, etc. $D(\mathbf{H}) = \cup_j D^j(\mathbf{H})$, the *derived system* of $\mathbf{H}$. Note that $[D(\mathbf{H}), D(\mathbf{H})] \subset D(\mathbf{H})$. If $D(\mathbf{H})$ is non-singular, it defines a foliation ("involutive distribution," in the language of Chevalley [*2*]) on M, and through every $x \in M$ there is a leaf, i.e. a maximal connected integral manifold L^x with $L^x_y = D(\mathbf{H})_y$, i.e. the tangent space to L^x at each point is precisely the value of $D(\mathbf{H})$ at that point. This result on the existence of integral manifolds (the Frobenius theorem) can be extended to the singular case if one of the following conditions is satisfied [*4*]:

(a) dim $D(\mathbf{H})_{\sigma(t)}$ is constant along each curve σ that is an integral curve of $D(\mathbf{H})$ i.e. satisfies $\sigma'(t) \in D(\mathbf{H})_{\sigma(t)}$ for $a \leq t \leq b$.

(b) $D(\mathbf{H})$ is *locally finitely generated* in the sense that each point of M has a neighborhood U and a finite number of elements of $D(\mathbf{H})$ such that every element of $D(\mathbf{H})$ can, in U, be written as a linear combination of these elements with coefficients from $C(U)$. (2.1)

(c) M and $\mathbf{H}$ are real analytic.

(It is not quite proven in ref. [*4*] that 2.1c) implies the existence of leaves, but the proof involves similar ideas.)

Let $P(\mathbf{H})$ be the set of integral paths of $\mathbf{H}$. Let $P(\mathbf{H}; x_0, x_1)$ (resp. $P(\mathbf{H}; x_0)$) denote those integral paths beginning and ending (resp. beginning) at points x_0 and x_1 of M (resp. at x_0). Let $\phi : P(\mathbf{H}; x_0) \to M$

be the map such that $\phi(\sigma) = \sigma(b)$ if $\sigma : [a, b] \to M$ is an integral path of $\mathbf{H}$.

Chow's Theorem [3]. If $\mathbf{H}$ and $D(\mathbf{H})$ are non-singular, then $\phi(P(\mathbf{H}, x_0))$ is the maximal connected integral submanifold of $D(\mathbf{H})$ that passes through x_0.

We proceed to restate this result in forms suggesting generalizations to more general path systems. A real valued function f in an open set U is said to be an *integral* of $\mathbf{H}$ if $f(\sigma(t))$ is constant in t, for each integral curve σ of $\mathbf{H}$. Such f's satisfy the system of first-order linear partial differential equations:

$$X(f) = 0, \qquad \text{for all} \qquad X \in \mathbf{H}, \tag{2.2}$$

where X is interpreted as a first-order partial differential operator. Let $I(\mathbf{H}, U)$ be the ring of integrals defined in an open set U. Notice that 2.2. implies that $X(f) = 0$ for all $X \in D(\mathbf{H})$. But, the submanifolds obtained locally by setting solutions of $D(\mathbf{H})\,(f) = 0$ equal to constants are precisely the integral manifolds of maximal dimension of $D(\mathbf{H})$ (if $D(\mathbf{H})$ is non-singular). Hence:

Reformulation A. If $D(\mathbf{H})$ is non-singular, then $\phi(P(\mathbf{H}, x_0))$ is the set of points of M that can be reached on all paths σ in M such that:

(a) x_0 is the initial point of σ.

(b) For each open set U, each $f \in I(\mathbf{H}, U)$, $f(\sigma) =$ constant for the portion of σ in U.

Intuitively, every point is accessible from x_0 that is not "obviously" inaccessible.

Reformulation B. If $D(\mathbf{H})$ satisfies 2.1. (a), (b), or (c), then $\phi(P(\mathbf{H}, x_0))$ is a submanifold L of M such that:

(a) $\mathbf{H}$ is tangent to L.

(b) $\mathbf{H}$ restricted to L has, even locally, no integrals.

The point of stating these corollaries in this form is that they seem to be true for more general systems of paths on manifolds, or at least suggest conjectures. For example, suppose P is a set of paths on M which locally are solutions of a system of differential equations 1.1, i.e., the paths in P form what is called classically a *Monge system*. The path system can then be "lifted" to one on a manifold M' sitting over M which is linear on M', i.e., is defined as the set of integral paths of a Pfaffian system on M'.

For 1.1, locally M' is the subset of $(x_i, \dot{x}_i, t)$-space defined by $g_a = 0$. The Pfaffian system is that defined by the Pfaffian forms $w_i = dx_i - \dot{x}_i dt$ restricted to this subset. Suppose that the classical regularity conditions are satisfied:

$$\text{The matrix } \left(\frac{\partial g_a}{\partial \dot{x}_i}\right) \text{ has rank } m. \tag{2.3}$$

The coordinate system can be chosen so that 1.1. takes the form:

$$\frac{dx_a}{dt} = h_a\left(x_i, \frac{dx_u}{dt}, t\right), \qquad (m + 1 \leq u, v, ... \leq n.) \tag{1.1'}$$

Hence, M' can be taken as $(x_i, \dot{x}_u, t)$-space, and the Pfaffian system is that defined by:

$$dx_a - h_a dt = 0, \qquad dx_u - \dot{x}_u dt = 0. \tag{2.4}$$

Notice that the system of vector fields dual to 2.4 contains the operators $\partial/\partial \dot{x}_u$, i.e., a function on $(x_i, \dot{x}_u, t)$ space which is an integral of 2.4 is independent of $\dot{x}_u$, hence arises from a function on (x_i, t)-space which is an integral of the path system defined by 1.1′. This means, with suitable local regularity hypotheses, that A and/or B are true for the path system in (x_i, t)-space defined by 1.1′.

Example.

$$\frac{dx_a}{dt} = A_{ab}(t)\, x_b + V_{au}(t)\, y_u(t). \tag{2.5}$$

The corresponding Pfaffian system in (x_a, y_u, t)-space is defined by:

$$dx_a - (A_{ab}(t)\, x_b + V_{au}(t)\, y_u)\, dt = 0.$$

The dual space of vector fields is spanned by:

$$Y_u = \frac{\partial}{\partial y_u}, \qquad Z = \frac{\partial}{\partial t} + (A_{ab}x_b + V_{au}y_u)\frac{\partial}{\partial x_a}.$$

Introduce a matrix notation:

$$A = (A_{ab}), \qquad \text{an } m \times m \text{ matrix.}$$

$V_u = (V_{au})$, an $m \times 1$ matrix. (u is then merely a counting index for $m \times 1$-vectors.)

$\partial/\partial x = (\partial/\partial x_a)$, an $1 \times m$ matrix of first order differential operators. Then,

$$Z = \frac{\partial}{\partial t} + \frac{\partial}{\partial x}(Ax + V_u y_u)$$

$$[Z, Y_u] = -\frac{\partial}{\partial x} V_u.$$

If V is an $m \times 1$-matrix function of t, define ΔV, also an $m \times 1$-matrix function of t, as follows:

$$\Delta V = \frac{d}{dt} V + AV.$$

Then,

$$[Z, [Z, V_u]] = -\frac{\partial}{\partial x}(\Delta V_u)$$

$$[Z, [Z, [Z, V_u]]] = -\frac{\partial}{\partial x}(\Delta^2 V_u), \text{ etc.}$$

Hence, using Chow's theorem, we have:

Theorem 1. If the dimension of the space of $m \times 1$-vectors spanned by $V_u(t)$, $\Delta V_u(t)$, $\Delta^2 V_u(t)$, ... is, for each t, equal to m, then every point of (x_a)-space is accessible from the origin on paths that are solutions of 2.5, for some choice of the "forcing" terms $y_u(t)$.

3. Local and Infinitesimal Accessibility

Suppose $\mathbf{H} \subset V(M)$ defines a non-singular Pfaffian system on M, with $D(\mathbf{H}) = V(M)$. Then, for $x_0 \in M$, $\varphi(P(\mathbf{H}, x_0)) = M$, i.e., every point is accessible from x_0. However, this leaves open the following "stability" problem:

Problem. If $\sigma : [0, 1] \to M$ is an integral path of $\mathbf{H}$ and x is a point of M close to $\sigma(1)$, can x be reached from x_0 along an element of $P(\mathbf{H}, x_0)$ that is everywhere close to σ?

We are not yet prepared to completely answer this question as stated, but instead shall deal with the analogous infinitesimal question. First, we must introduce some more notation.

It is well-known that $P(x_0)$, the space of paths of M beginning at x_0, may be considered as an infinite dimensional manifold. If $\sigma \in P(x_0)$, the "tangent space" $P(x_0)_\sigma$ to σ may be considered as the space of vector-fields $v : t \to v(t) \in M_{\sigma(t)}$ along σ. *A* "curve" in $P(x_0)$ is a deformation σ_s, $0 \leq s \leq 1$. If $\sigma_0 = \sigma$, the corresponding infinitesimal deformation $v \in P(x_0)_\sigma$ is the vector field on σ such that $v(t)$ is the tangent vector to the curve $s \to \sigma_s(t)$ at $s = 0$. If $\sigma \in P(\mathbf{H}, x_0)$, those $v \in P(x_0)_\sigma$ that appear as the infinitesimal deformation of σ in $P(\mathbf{H}, x_0)$ will be thought of as the "tangent space" of $P(\mathbf{H}, x_0)$ at σ, denoted by: $P(\mathbf{H}, x_0)_\sigma$.

It is found that these v are solutions of a certain system of linear differential equations, the *linear variational equations* of the Pfaffian system **H** at σ. To describe these equations explicitly, suppose that (w_i), $1 \leq i, j, \ldots \leq n$, is a basis for 1-forms on M, and that $w_a = 0$ defines **H**, $1 \leq a, b, \ldots \leq m$.

$$v : t \to v(t) \in M_{\sigma(t)} \text{ is in } P(\mathbf{H}, x_0)_\sigma \text{ iff.}$$

$$\text{(a) } v(0) = 0 \tag{3.1}$$

$$\text{(b) } \frac{d}{dt} w_a(v(t)) = dw_a(v(t), \sigma, (t)).$$

For $X \in \mathbf{H}$, define $D(\mathbf{H}, X)$ as the smallest $C(M)$-submodule of $V(M)$ containing $\mathbf{H}$, $[X, \mathbf{H}]$, $[X, [X, \mathbf{H}]]$, etc. For $x \in M$, $v \in \mathbf{H}_x$, define $D(\mathbf{H}, v)$ as $D(\mathbf{H}, X)_x$, where X is any vector field in **H** with $X(x) = v$. (It is easily seen that this definition of $D(\mathbf{H}, v)$ is independent of the X chosen).

Theorem 2. With the above notations, the "differential" $\phi_* : P(\mathbf{H}, x_0)_\sigma \to M_{\sigma(1)}$ of the end-point map $\phi : P(\mathbf{H}, x_0) \to M$ at σ is the map which assigns $v(1)$ to each vector field v along σ satisfying 3.1. ϕ_* is onto $M_{\sigma(1)}$ i.e., ϕ_* is a "maximal rank" mapping about σ, if:

$$D(\mathbf{H}, \sigma'(t)) = M_{\sigma(t)} \qquad \text{for} \qquad 0 \leq t \leq 1. \tag{3.2}$$

This result then gives a reasonable criterion for "infinitesimal accessibility" of points near $\sigma(1)$ by curves near σ. Since generalizing the implicit function theorem for such infinite dimensional spaces as $P(\mathbf{H}, x_0)$ is not a routine matter, we cannot yet infer that curves σ satisfying 3.2. also satisfy "local accessibility," i.e., Problem. However, this certainly seems to be a reasonable conjecture, especially since, as we show in Section 4, it can be verified in case σ is a sufficiently small extremal of a Lagrange variational problem whose "constraints" are defined by **H**.

The proof of Theorem 2 uses Theorem 1, since the differential equations 3.1b) are of type 2.5, with the terms $w_u(v(t))$ identified with the "forcing terms" $V_u(t)$ in 2.5.

4. Infinitesimal and Local Accessibility by Extremals of Lagrange Variational Problems

Our work in this section may be considered as a complement to some of Carathéodory's on the Lagrange variational problem [*1*]. We will only work locally; M can be taken as the space of the variables (x_i).

Consider a Hamiltonian function $H(x_i, y_i, t)$ of $2n$-variables, and the corresponding Hamilton equations:

$$\frac{d}{dt} x_i(t) = H_{n+i}(x, y, t)$$

$$\frac{d}{dt} y_i(t) = - H_i(x, y, t), \tag{4.1}$$

$$H_{n+i} = \frac{\partial H}{\partial y_i}, \qquad H_{i,j} = \frac{\partial H}{\partial x_i \, \partial x_j}, \qquad \text{etc.}$$

Suppose we normalize the initial conditions satisfied by $x_i(t)$ to: $x_i(0) = 0$. There is then an "exponential mapping" from the space of variables (y_i) to the space of x_i : (y_i^0) is mapped into $(x_i(1))$, where $x_i(t)$, $0 \leq t \leq 1$ is a part of a solution of 4.1. corresponding to the initial conditions $x_i(0) = 0$, $y_i(0) = y_i^0$. Carathéodory proves the following results [*1*], using the basic existence theorems of the calculus of variations.

Let $(x_i(t), y_i(t))$, $0 \leq t \leq 1$, be a "sufficiently small" solution of 4.1. If there are *no* solutions of the following equations for functions $Y_i(t)$:

$$\begin{aligned} \frac{d}{dt} Y_i(t) &= - H_{i,n+j}(x(t), y(t), t)\, Y_j(t) \\ H_{n+i,n+j}(x(t), y(t), t)\, Y_j(t) &= 0 \end{aligned} \tag{4.2}$$

and if the following condition is satisfied:

The symmetric matrix $H_{n+i,n+j}$ is everywhere positive semi-definite and of constant rank (4.3)

then the exponential mapping is of maximal rank about y_i^0. We may say in this case that all points near $x_i(1)$ in x_i-space are *infinitesimally accessible by extremals*, i.e., by projections in (x_i)-space of solutions of 4.1. Since the exponential mapping involves only finite dimensional manifolds, we may say that infinitesimal extremal accessibility *implies* local extremal accessibility. Suppose now that $w_a = A_{ai} dx_i = 0$ define a Pfaffian system $\mathbf{H}$ and that:

(a) $A_{ai} H_{n+i} = 0$, i.e., the curves $x_i(t)$ that occur as solutions of 4.1. are integral curves of $\mathbf{H}$. (4.4)

(b) n–(rank $(H_{n+i,n+j})$) = rank (A_{ai}).

Theorem 3. With these assumptions, suppose in addition that the curve σ in (x_i) space determined by the solution $x_i(t)$ of 4.1. satisfies

3.2., i.e., $\phi : P(H, x_0)$ is of "maximal rank" about σ. Then, points close to $x_i(1)$ are infinitesimally accessible by extremals from σ. In particular, points close to $x_i(1)$ are locally accessible from σ.

The next step should be to extend this result to more general Lagrange variational problems, i.e., those whose constraints are not necessarily linear, and to consider the analogous question for paths that are broken extremals. However, we will defer to a later work carrying this program further along.

REFERENCES

[*1*] Carathéodory, C., Über die Einteilung der Variationsprobleme von Lagrange nach Klassen. *Comm. Math. Helv.* **5**, 1-9 (1933); or see "Gesammelte mathematische Schriften," Vol. I, pp. 270-288. C. H. Beck, Munich, 1954.

[*2*] Chevalley, C., "Theory of Lie Groups." Princeton Univ. Press, Princeton, New Jersey, 1946 (new ed. 1960).

[*3*] Chow, W. L., Systeme von linearen partiellen Differentialgleichungen erster Ordnung. *Math. Ann. (Leipzig)* **117**, 98-105 (1939).

[*4*] Hermann, R., The differential geometry of foliations, II. *J. Math. and Mech.* **11**, 303-316 (1962).

[*5*] Hermann, R., Some differential geometric aspects of the Lagrange variational problem. Illinois, *J. Math.* to be published.

Time Optimal Control of Linear Systems with Bounded Controls

B. PAIEWONSKY
Aeronautical Research Associates of Princeton, Inc.
Princeton, New Jersey

I. Introduction

The performance requirements of many vehicles and dynamic systems require optimization of the overall response to commands in order to insure successful missions. One of the more important optimization criteria is minimum error time. A time optimal system strives to reduce the error and its derivatives to zero in the least possible time.

Although the main problem studied and reported on here was not directed towards immediate specific applications it seems appropriate to consider briefly some situations in which the use of such a controller would be advantageous. There are several different examples that can be used to illustrate the applicability of time optimal control or maximum effort systems.

Attitude control of space vehicle boosters is required during the ascent phase of a launch trajectory in order to insure that the desired flight path will be achieved in the presence of disturbances and to prevent the occurrence of large or destructive transient loads on the vehicle due to external inputs such as winds. The control system must act to prevent excessive flight loads from building up, at the same time striving to reduce drifting or other deviations from the flight path. The limited control forces available from the gimballed engines (or other means of thrust vector control) suggests the possibility of control saturation occurring in the event of a large disturbance. Analysis of the behavior of systems with saturated controls or other constraints on the system output variables, such as angle of attack limits, leads to the consideration of maximum effort or switching controllers [*23*, *43*, *49*, *57*]. The use of a switching controller in the case of a large rocket arises mainly from the limited control forces available. The time optimal nature of this form of control may be only a secondary consideration.

Another application of maximum effort system is in the depth control of a submarine [*52*].

In this case it is desired to go from one depth to another in the least time with limitations on the maximum rate of control deflection, maximum amount of control deflection and also limits on the maximum pitch angle of the deck. For large depth changes the limitation on deck pitch angle may cause the system to remain in a steady state dive determined by the maximum allowable pitch angle. The optimization then takes place with respect to the time required to establish the steady dive and to recover from the steady dive.

Many industrial processes can realize substantial gains in performance by following time optimal programs for startup, shut down, and changing operating set-point. Several examples of this type of operation can be found in the chemical industry [*5*, *24*].

In addition, many problems in dynamic performance of aircraft and other high speed vehicles can be formulated as time optimal problems. In general the equations governing the motion of the vehicles are nonlinear. This investigation is concerned with linear processes and optimization problems in flight mechanics will not be treated. Extensive bibliographies on flight mechanics and astrodynamics have been prepared by Miele [*46*] and Baker [*2*]. Noteworthy papers on dynamic performance can be found in the works of Garfinkel [*22*], Leitmann [*42*], Kelley [*30*, *31*], and Breakwell [*11*], among others [*6*, *55*].

II. Discussion

The problem under consideration is time optimal regulation of linear systems with bounded controls. The system is described by a state vector $\mathbf{x}(t)$, satisfying the differential equation

$$\dot{\mathbf{x}} = A\mathbf{x} + B\mathbf{u} \tag{1}$$

where

$\mathbf{x}$ is an n–vector

A is an $n \times n$ matrix

B is an $n \times r$ vector

$\mathbf{u}$ is an r–vector, $|u_k| \leq M_k$

and $u(t)$ is piecewise smooth.

It is desired to find a control law $u^0(t)$ such that the system is brought from an initial state $\mathbf{x}(0)$ to the desired terminal state (in this case $\mathbf{x}(T) = 0$) in the least time T^0.

It is already well known that the time optimal control law for a large class of linear systems with bounded control is of bang-bang or relay type. Relay or on-off controllers have been studied for a considerable time. The existing mathematical foundation had its start with the work of Hazen [*25*], Flügge-Lotz [*20*], McDonald [*44*], Bushaw [*12*], and Rose [*51*].

Bellman [*7*] proved that for a time invariant linear system of order n with bounded control, and real distinct eigenvalues the bang-bang control law was time optimal and furthermore at most $(n - 1)$ switches were required to bring the error and its derivatives to zero. La Salle [*39, 40*] extended these results and made precise the conditions under which they apply. A result due to La Salle [*40*] states that the optimal control is bang-bang and unique if the system (1) is normal. A system is normal if no component of the expression $\eta X^{-1}(t) B(t)$ vanishes identically for any vector $\eta \neq 0$. If the system is normal then the optimal control law is given by the expression $u^0(t) = \operatorname{sgn}(\eta X^{-1}(t) B(t))$.

In addition to this work considerable attention has been paid to time optimal control in Russia. Krassovskii [*34-36*] has investigated the optimal control problem for a range of restrictions on the controls. Pontryagin [*50*] (see also Ronzoner [*54*]) has obtained a result of great generality called the Maximum Principle that enables the optimum control to be determined for many problems. The maximum principle in effect defines a generalized distance and velocity and shows how the control law should be chosen in order to maximize this "velocity" and thereby minimize the value of the performance criteria evaluated at the terminal conditions. However in the case where the terminal time is not specified in advance (i.e. time optimal control) the maximum principle gives only the form of the control law and certain problems still remain. For example in the case of the time optimal control the maximum principle gives the bang-bang form for the control law but the switch times are still unknown.

Several important results arise from this mathematical foundation. One of these results shows the possibility of constructing switching hypersurfaces such that the optimal control can be found as a function of position in the state phase space. These switching surfaces however are often difficult to obtain for high order systems. Considerable effort has been spent by many investigators [*17, 47*] to obtain approximate representations and physical realizations of the optimum switching surfaces.

The solutions of the system equations of motion in the phase space for the unknown switching times is an alternative procedure. The switching surfaces and the switching times are in principal obtainable from the basic system equations and are equivalent representations of the optimal control law. The switching surfaces represent the control law as a function of position in phase space whereas the switch times represent the control as a function of time.

The switching surface method requires a computer to either store or compute values of a function of many variables over a large range of values of the independent variables. The switching time method requires a computer to solve a set of non-linear (transcendental) equations of order n. This method of solving for the switching times is limited to those systems having at most $(n - 1)$ switches to reach the origin of the phase space. All systems that have real distinct and negative eigenvalues have this property for all initial points in the phase space.

Other systems have this property only for limited neighborhoods of the origin. For example a second order system characterized by a lightly damped oscillation may require 2 or 3 or more switches to be brought to rest after a large disturbance, whereas a second order system with real distinct eigenvalues would require at most 1 switch irrespective of the initial condition. This means that the number of unknown switch times for the oscillatory system may exceed the number of equations available, and a solution is not obtainable in these cases by solving the trajectory equations directly. Additional relations between the switching times must be considered.

An equally important result is the idea of the adjoint system. Bass [*3*] and Desoer [*16*], among others [*21*, *54*], observed that if the system to be controlled were linear then the optimal times for switching the sign of the forcing function correspond to the zeros of linear combinations of solutions of the system: $\dot{Y} = -YA$ adjoint to the given one: $\dot{X} = AX$. That is, if a model of the adjoint system were constructed and if the proper initial values, η were known, then the times of the zero crossing of the resulting response of the adjoint (due to the initial conditions) are the times for the optimal switching.

The problem is to find the optimal initial state of the adjoint corresponding to the state of the physical system.

The adjoint approach seems particularly attractive for systems possessing oscillatory modes. In such cases only $(n - 1)$ components of the initial adjoint vector need be found whereas the number of switch times may exceed $(n - 1)$. This property of the adjoint system overcomes the difficulty previously noted regarding the trajectory equations and the number of unknown switch times.

A variational approach has been used by several investigators in connection with the general problem of optimal control as well as the more specific problems of optimal control of vehicles. Desoer [*16*] studies the time optimal control problem for nonautonomous linear systems whose eigenvalues are all real and negative. An isoperimetric problem is formulated and an iterative solution is proposed for the Lagrange multipliers. It was pointed out that there is a direct correspondence between the Lagrange multipliers of the variational calculus approach and the adjoint equations. Dreyfus [*18*] treated a somewhat more general problem. Given a set of initial conditions $x_i(0) = c_i$, $i = 1 \dots n$, a set of terminal conditions $x_i(T) = d_i$, $i = 1 \dots n$ where the terminal time T is not specified in advance, and a set of differential equations $\dot{x}_i = f_i(x_j, t, u)$ with $|u| \leq 1$.

The problem is to find the control law $u(t)$ bringing the system state point from the initial position to the terminal position in the least time. It was shown that the solution involves n Lagrange multipliers λ_i $i = 1, \dots, n$. Furthermore, if the system $\dot{x}_i = f_i(x_j, u)$ does not involve the time explicitly then a first integral of the Euler equations exists; namely $\Sigma_i \lambda_i f_i = -1$. This relation holds all along the optimal trajectory. The Euler equations for the Lagrange multipliers are

$$\frac{d\lambda_i}{dt} = -\frac{\partial f_j}{\partial x_i} \lambda_j, \qquad i = 1, \dots, n. \tag{2}$$

These equations are the same as the adjoint equations obtained previously by other considerations. The Euler equations require $(n - 1)$ constants to be found in order to satisfy the terminal conditions. These constants are the $(n - 1)$ independent components of the initial state of the adjoint vector.

Hestenes [*26*] using the results of Bliss [*8*] and Valentine [*56*] formulated a very general problem in the calculus of variations for the purpose of studying minimum time flight paths. These results can be applied to the optimal control problem.

It was shown by Hestenes that under certain conditions there exist unique multipliers $\lambda_i(t)$ such that if the function H is defined as below

$$H(t, x, \lambda, u) = \lambda_i f_i = -1 \tag{3}$$

then the optimal trajectory has the following properties:

I. Along the optimal trajectory the equations below hold.

$$\dot{x}_i = \frac{\partial H}{\partial \lambda_i} \qquad \dot{\lambda}_i = -\frac{\partial H}{\partial x_i} \tag{4}$$

also

$$\frac{d}{dt} H = H_t. \tag{5}$$

II. The necessary condition of Weierstrass along the optimal trajectory takes the form

$$H(t, x^0, \lambda^0, u^0) \leq H(t, x, \lambda, u). \tag{6}$$

A comparison shows that the Pontryagin Maximum Principle is equivalent to the requirement that the Weierstrass necessary condition be satisfied. The problem of finding the $n - 1$ independent initial components of the adjoint corresponds to the solution of the 2-pt. boundary problem associated with the Euler equations.

Alternative formulations based on functional analysis were made by Krassovskii [*35*], Kirillova [*33*], and Kulikowski [*37*]. More extensive discussions of the various formulations can be found in ref. [*32*] and [*48*]. Lee [*41*], Kurzweil [*38*], Ho [*27*], and Kalman [*28*] have also formulated problems in optimal control.

III. The Optimal Isochrones

The correspondence between the results of the previous investigations on time optimal control depends on the fact that the optimal process time T^0 satisfies the requirements for a Liapunov function. That is, T^0 is a positive definite function whose time derivative evaluated along the optimal trajectories is always negative definite. The surfaces $T^0 =$ constant are called optimal isochrones.

These surfaces fill the entire region of controllability in phase space. If the system is stable then the region of controllability consists of the entire space. Through each point in the phase space passes one and only one optimal isochrone.

The fact that the surfaces $T^0 =$ constant are closed and devoid of holes follows from Bellman's result by noting that the T^0 surfaces are generated by a mapping that preserves connectivity. In the case of linear systems $\dot{\mathbf{x}} = A\mathbf{x} + B\mathbf{u}$ this mapping generates an expanding convex set

$$c(t) = \left\{ \int_0^t X^{-1}(\tau)\, B(\tau)\, u(\tau)\, d\tau;\; u \in \Omega \right\}$$

whose boundary at any instant corresponds to an optimal isochrone. In addition a Hamilton Jacobi equation can be written and the optimal process can be interpreted in terms of a propagating wave front corresponding to $c(t)$.

Anderson [*1*] and Rose [*51*] have also shown by different arguments that these surfaces are topologically equivalent to a hypersphere. Their questions arose in connection with a different aspect of the optimal control problem.

The Liapunov functions define a metric or distance function with respect to the origin in the phase space. This can also be used to define velocities. With this in mind it is now possible to develop a very important property of the isochronal surfaces.

Along the optimal trajectory the following equation holds:

$$\frac{dT^0}{dt} = \frac{\partial T^0}{\partial t} + \nabla T^0 \cdot \mathbf{f} = -1. \tag{7}$$

The time derivative of ∇T^0 along a trajectory is given by:

$$\frac{d}{dt}(\nabla T^0)_j = \frac{\partial}{\partial x_i}(\nabla T^0)_j \frac{dx_i}{dt} = \frac{\partial}{\partial x_i}(\nabla T^0)_j f_i.$$

If the system is time independent then the derivative of dT^0/dt with respect to x_j is

$$\frac{\partial}{\partial x_j}[(\nabla T^0) \cdot f] = \left[\frac{\partial}{\partial x_j}(\nabla T^0)_i\right] \cdot f_i + \nabla T_i^0 \frac{\partial f_i}{\partial x_j} = 0.$$

The final result is obtained directly by substitution.

$$\frac{d}{dt}(\nabla T^0)_j = -\nabla T_i^0 \cdot \frac{\partial f_i}{\partial x_j}. \tag{8}$$

This equation states that along an optimal trajectory the gradient of the optimal isochrones satisfies the adjoint equation.

This result provides an immediate connection between the previous investigations.

Let $V(x)$ be a Liapunov function. Then the time derivative of V, i.e. $\dot{V}(x)$ is:

$$\dot{V} = -\frac{\partial V}{\partial t} + \nabla V \cdot \mathbf{f}.$$

For example, suppose there is only one control.

$$\mathbf{F} = \mathbf{f} + \mathbf{b}u$$

$$\dot{V} = \frac{\partial V}{\partial t} + \nabla V \cdot (\mathbf{f} + \mathbf{b}u). \tag{9}$$

In order to minimize the time to the equilibrium point the control law $u(t)$ is chosen so that V is made as negative as possible. To do this u must be set equal to $-\operatorname{sgn}(\nabla V \cdot b)$. The maximum velocity occurs if $V = T^0$. Then we get

$$\nabla T^0 \cdot (f + bu) + \frac{\partial T^0}{\partial t} = -1 \tag{10}$$

and

$$u = -\operatorname{sgn}(\nabla T^0 \cdot b). \tag{11}$$

This means that the state of the adjoint vector corresponds directly to the gradient of the optimal isochrones. The Pontryagin Maximum Principle can be given a physical interpretation. The essence of the Pontryagin Maximum Principle is that it requires the control law to be chosen so that the component of system state velocity in phase space along the gradient of the optimal isochrone is always a maximum. The equivalence of the Maximum Principle and the Weierstrass necessary condition has already been noted.

The optimal isochrones form a family of closed shells in the phase space. These shells however are not smooth surfaces. In fact they have sharp corners and pointed edges. This property of the isochrones can be shown by the following argument. Consider a point in the phase space close to the optimal switching surface, and suppose that this point lies on a trajectory that will soon intersect the switching surface. Suppose also that there is only one control acting and that the b-vector is of the form $(0, 0, 0, \ldots 1)$. In this situation the nth component of the adjoint vector must be close to zero since

$$u^0(t) = \operatorname{sgn} p_n(t)$$

and the time interval to the next switch is short. The initial state of the adjoint corresponds to the direction of the gradient of the optimal isochrone. Thus the gradient of the isochrone must be nearly horizontal. At the instant of the switching of the control the gradient of the optimal isochrone is horizontal, meaning that p_n vanishes. Consider now a point on the opposite side of the switching surface, perhaps on the continuation of the original trajectory. The time to the next switch is long compared to the conditions first discussed. Therefore the gradient of the optimal isochrone through that point must be large since p_n will not vanish for a while.

The magnitude of the discontinuity in the gradient of the isochrones across the switching curve is a measure of the difference in switching times between points on opposite sides of the switching surface.

The fact that the optimal isochrones yield solutions to the time optimal control problem suggests the possibility of making an approximation to these surfaces and using the gradient of these approximate functions to determine the switching of the control.

It is interesting to observe that if quadratic forms are chosen by the Liapunov corollary to be used as a performance criterion or as approximate optimal isochrones, and no further refinements are made, the result is linear switching about a hyperplane defined by the locus of horizontal gradients of the approximating surfaces. In a recent paper Kalman [*29*] uses quadratic Liapunov functions, $V(x)$, and selects the control law that maximizes V.

The suitability of linear switching might be determined by the extent to which the system optimal isochrones may be approximated by quadratic forms.

IV. Iteration Procedures

The previous discussion has indicated the relationships between the different approaches and has shown the significance of the adjoint system. The question of finding the optimal initial state of the adjoint for arbitrary systems under general conditions of initial state and input signals has never been completely answered. In this section the idea of searching for the optimal initial state of the adjoint in a systematic manner will be presented. If such a searching scheme is to be used on a real system then the necessary calculations must be performed very rapidly compared with the natural time scale of the process being controlled.

A block diagram of a computer controlled process is shown in figure 1. Although the diagram appears to be the same as a conventional feedback control system there is a difference. The computer contains a model of the system and adjoint that operates on a very fast time scale. An ideal system would work as follows: The computer would sample the system state variables intermittently. On the basis of its knowledge of the process dynamics it would compute the optimal initial state of the adjoint (i.e. the gradient of the optimal isochrone passing through the state point in phase space) and thereby arrive at the optimal control. The optimal control is the one that maximizes $(\nabla T^0 \cdot \dot{\mathbf{x}})$ at all points along the trajectory. This process is repeated at frequent intervals of time. It may happen that not all of the components of the system state vector are available for direct measurement. In these cases estimated values must be computed using whatever measured quantities are available. In non-ideal systems the presence of noise and component imperfections must be considered. (See Darlington [*15*]). Imperfect information about the location of the system state point in phase space will lead to errors in switching time and will produce limit cycles about the error null. This investigation is restricted to ideal systems without noise and with perfect information about the process dynamics and the state of the system.

It is proposed to search for the proper initial state of the adjoint vector by iteration on a high speed computer. Although the adjoint space is of dimension n, as is the state phase space, it is only necessary to obtain

$(n-1)$ components of the inital adjoint vector. A control law of the form

$$u = \operatorname{sgn} P_n(t)$$

defines switch times as the roots of the equation

$$P_n^0(t) = P_1^0(0)\,\eta_{1n}(t) + P_2^0(0)\,\eta_{2n}(t) + \ldots P_n^0(0)\,\eta_{nn}(t) = 0 \tag{12}$$

where $P_j(0)$ are the adjoint initial conditions and the $\eta_{jn}(t)$ are the

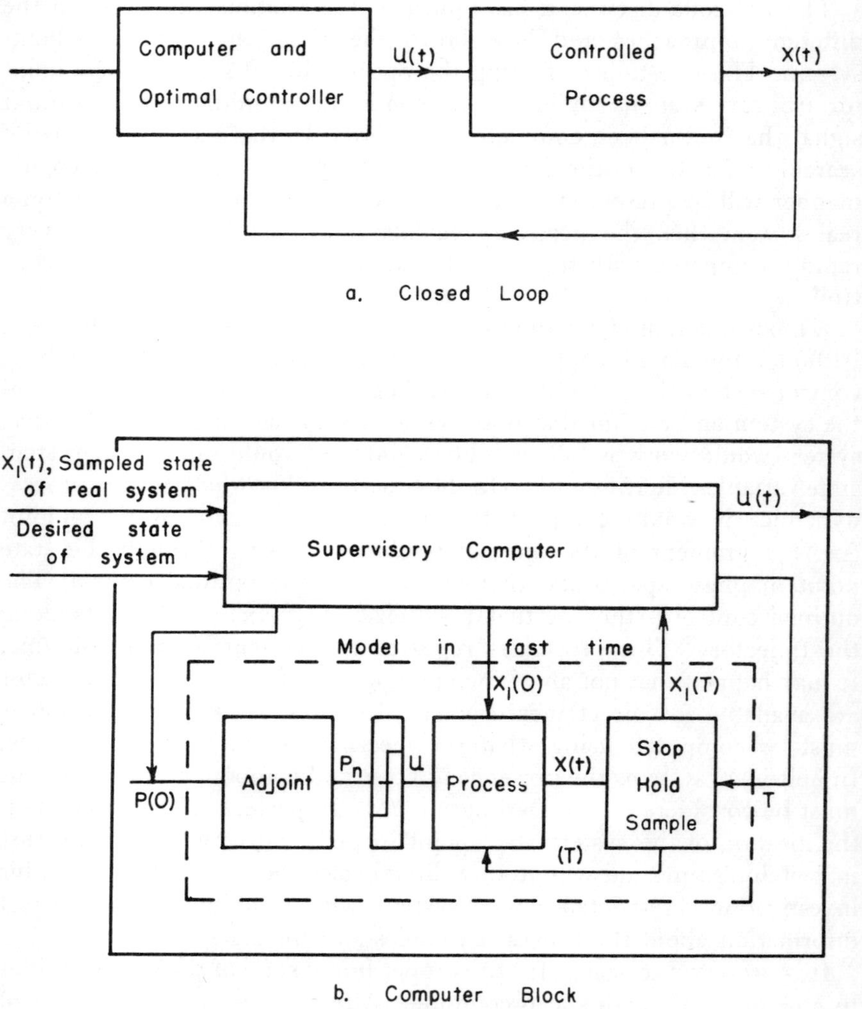

FIG. 1. Block diagram of computer controlled system

homogeneous solutions for the nth component of adjoint equations. Clearly only $(n - 1)$ of the $P_j^0(0)$ are required to define the switch times. It is important that only $(n - 1)$ numbers need be found irrespective of the number of switch times required for optimal control. This fact makes the use of the adjoint system seem attractive for the control of systems with oscillatory modes.

There are many paths open for the development of search techniques, but only one of these will be discussed here. The basic ideas underlying this iteration scheme can be stated briefly as follows. A first selection of an approximate initial state of the adjoint is made by some predetermined means according to the initial state of the physical system. The high speed computer determines the response of the basic system and the adjoint system for some finite time interval into the future. The computer predicts the system response and it also computes a certain function of an appropriate "distance" in the phase space. The object of the iteration procedure is to bring this distance function to zero. After the completion of an individual run and resampling the actual system, the computer must determine if a new choice for the adjoint vector is necessary and if so to compute it.

Questions regarding start up and convergence of the iterative schemes are treated in detail in ref. [*48*]. The computation schemes described here are based on the use of a high speed computer incorporating a "model" or mathematical description of the controlled process and its adjoint. The problem of optimum regulation can be solved if the correct initial state of the adjoint system can be found for all possible initial states of the physical system. Suppose the following experiment to be carried out on a computer. An initial choice of $P(0)$ is made and the future behavior of the system is predicted from the equations

$$\begin{aligned} \dot{x}_i &= A_{ij}x_j + b_i \operatorname{sgn} b_k p_k(t) \\ \dot{p}_j &= -A'_{ij}p_i : p_j(0) \end{aligned} \tag{13}$$

Select a positive definite function $V(x)$ as a norm in the phase space. The function $V(x(T))$ is considered an implicit function of $p_j(0)\, j = 1 \ldots n$ and T. Since V is positive definite we assume that the solution $V(x) = 0$ corresponds to the absolute minimum of V and also to the solution of the problem $X_i(T) = 0$. The problem is to find values of $p_j(0)$ and T that minimize V.

The optimal values $(p^0(0), T^0)$ are sought by searching for the minimum value of V, $V = 0$. It might be possible to solve this problem by an exploration of the surface following the methods of Box and Wilson [*10*] or Box [*9*]. If the exploration is thorough the solution will be found. Such an exploration would be a time consuming and costly process. An alternative is to seek the minimum of V by a more direct means such as the method of steepest descent.

Curry [*14*] (see also Courant [*13*] Rosenbloom [*53*] Fishbeck [*19*]) has investigated in detail the application of the steepest descent procedure to minimization problems. The method of steepest descent has the following geometrical interpretation. Starting at the point $(P(0), T)$ corresponding to the estimated initial state of the adjoint and the estimated process time, the direction of steepest descent of the surface $V(P, T)$ is determined from the gradient of V. The minimum value of V, representing the solution to the problem, is found by "running" downhill. Successive values of the independent variables $P(0)$ and T are selected according to the conditions $P^{\nu+1}(0) = P^{\nu} - \alpha \nabla_p V$ and $T^{\nu+1} = T^{\nu} - \alpha \nabla_T V$.

This in effect is a traversal of the V surface along the direction of gradient V with steps of size α. The process continues until either a valley floor is reached and the gradient direction is then changed by 90 degrees or a minimum point is obtained or the boundary of the domain is reached. The minimum occurs in a bowl shaped region and the closeness of the approach to the actual minimum value of V depends upon the size of the step α. If α is too large the iteration procedure may overshoot the minimum and oscillate about it. If α is too small it will take too long to reach the minimum point. In the cases studied the experiments have shown that the bowls are quite steep and it appears better to choose α to be small, rather than permitting overshoots of large amplitude. It should be noted that the step size can be varied during the iteration according to the particular problem. It is possible to take large steps at first and then refine the process and take smaller steps by reducing α as the minimum point is approached.

V. Start Up

The iteration scheme requires the selection of starting values of $P(0)$ and T. The rapidity of convergence of the iteration depends on the proper selection of initial values. The most difficult case occurs when no prior information is available. In practical cases the optimal controller will have the benefit of previous operation, and it is quite likely that a certain degree of adaptivity and memory will be incorporated into the system. Thus an optimal regulator operating in a repetitive manner in the presence of small disturbances could be expected to obtain trial values of T from previously computed results obtained during the computing cycle immediately past. The problem posed here assumes no *a priori* information of this nature. This corresponds to a start up or a large disturbance.

The basis of the selection of a starting value for the iteration depends on obtaining an approximation to the optimal isochrones and using the resulting approximate gradient as the trial initial state of the adjoint. An initial estimate of the process time must be made either by a trial run or by a specific approximation. Before proceeding further with the iteration scheme, some optimal isochrones will be examined in order to study their general nature and to assist in making initial estimates of the optimal adjoint vector.

To begin, consider a second order linear system with constant coefficients. These systems have been studied extensively and the optimum switching control law has been obtained by Bushaw [*12*] in the form of switching curves. The isochrones for several of these systems were generated by means of a combined analog digital computer.

Two examples of these isochrones are shown in figures 2 and 3. These figures show that the isochrones for the system $\ddot{x} = \pm 1$ are elongated and pointed while the isochrones for the system $\ddot{x} + \omega^2 x = \pm 1$ are rounded but with corners. In the vicinity of the origin the isochrones of both systems resemble each other. This is to be expected since the system $\ddot{x} + \omega^2 x = \pm 1$ behaves as a double integrator or inertial mass close to the origin.

In many cases of interest the principal disturbances to the system occur as step inputs or large displacement errors. Accordingly, attention will be focused on the case of optimal response to step inputs or initial displacement errors. The response to arbitrary initial errors can be obtained by the iteration method developed here. However, for the sake of minimizing the number of cases to be studied the restriction to step inputs or displacement disturbances will be imposed. It should be emphasized that this restriction on the cases studied on the computer is not a limitation on the applicability of the proposed method. In an actual operation the computer must be able to determine the optimal initial state of the adjoint at each sampling instant for arbitrary locations of the initial state point. The computer can accomplish this, in the cases where the initial error lies in the region of attainability, by utilizing the iteration procedure proposed here.

An examination of the optimal isochrones suggests several possibilities for the selection of starting values for the iteration. In regions away from the origin the rounded isochrones can be approximated by quadratic forms. For systems possessing elongated pointed isochrones the multiple integrator isochrones can be used as an approximation along the x_1 axis. The optimal control law can be found explicitly for the multiple integrator in terms of the zeros of the Tschebycheff polynomials (see Kulikowski [*37*]). Thus the initial state of the adjoint can be found for these cases (multiple integrators, step inputs).

A certain amount of difficulty is to be expected in attempting to fit a family of smooth functions (the quadratic form approximation) to a family of surfaces having corners, sharp edges and points. It can be seen from the figures presented that the isochrones for the second order systems are well behaved in the vicinity of the x_1 axis. The worst cases will occur in the vicinity of the switching curves where the real isochrones have corners.

It is possible to consider approximating the optimal isochrones by

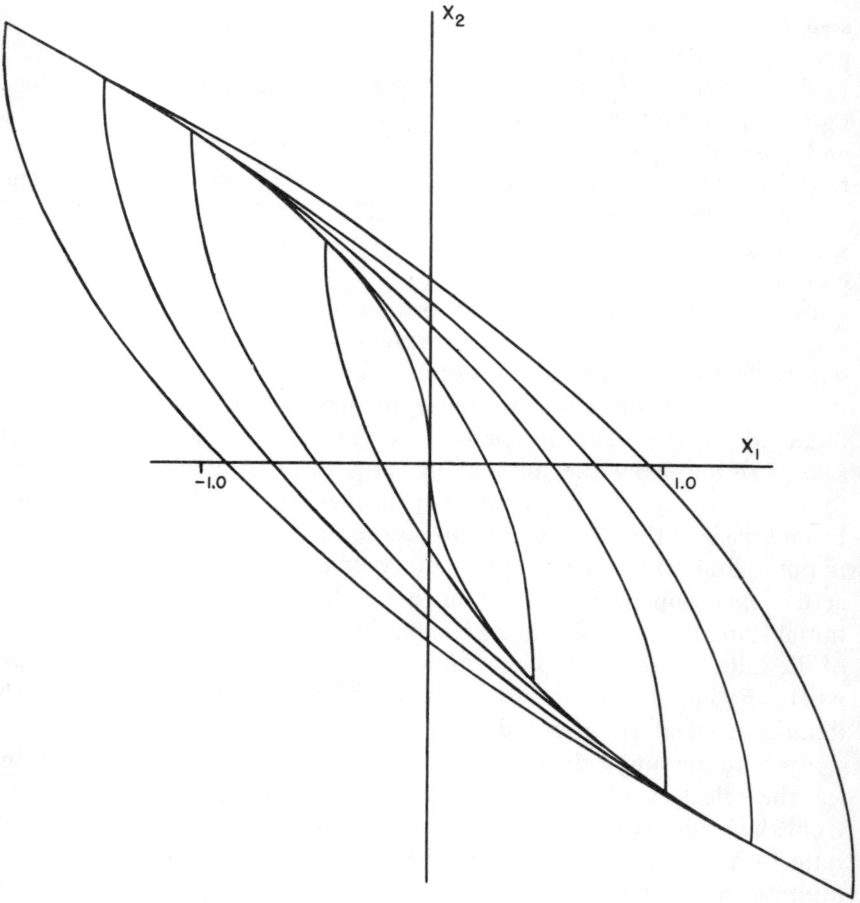

FIG. 2. Optimal isochrones, $\ddot{x} = \pm 1$

families of curves with corners or, in higher dimensions, by surfaces with sharp edges and points. However, there seems to be little advantage to be gained by such an attempt as the corners and edges of the optimal isochrones define the switching surfaces and these are in general unknown. If the switching surfaces were known the problem would in effect be solved.

The approximations to the optimal isochrones were developed in detail for several linear systems of 2nd and 3rd order and are presented in ref. [*48*].

The corollary of Liapunov can be applied to generate a family of quadratic forms to be used as approximations to the optimal isochronal Liapunov functions. The families of quadratic forms have the property that their gradient vectors have a constant direction along a ray from the origin. In particular there is a ray through the origin such that the gra-

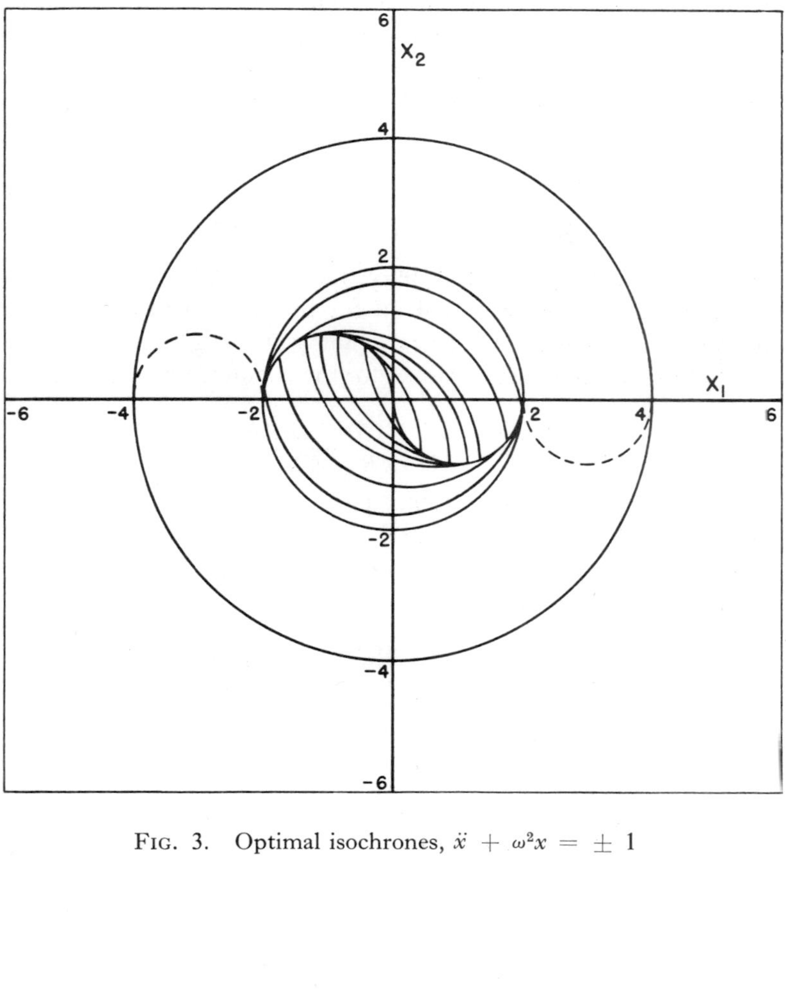

FIG. 3. Optimal isochrones, $\ddot{x} + \omega^2 x = \pm 1$

dient vectors $\nabla V(x)$ of the quadratic forms are horizontal everywhere along this ray. That ray corresponds to the switching line used in Kalman's linear switching method [*29*]. It is interesting to observe that if the adjoint system were forced to coincide with the gradient of the quadratic forms at all times, the result would be control by linear switching. Linear switching however is not optimum switching. Although the optimum switching and linear switching trajectories may nearly coincide over a large region of phase space, the chattering mode exhibited by the linear switching system causes a slow convergence to take place near the error null.

An examination of the optimal isochrones, figure 4, for the case $\zeta = 0$ shows that for large distance from the origin the curves resemble circles except near the corners. In particular for step inputs the optimal initial state of the adjoint approaches zero as $\omega x_1(0)$ increases.

The optimal initial direction of the adjoint vector for the double integrator approaches 90 degrees for large values of $x_1(0)$. For large disturbances the approximate gradients are close to the optimal gradients. As a specific example of the use of quadratic forms for approximating the optimal isochrones consider the following 3rd order system

$$\begin{aligned} \dot{x}_1 &= x_2 \\ \dot{x}_2 &= x_3 \\ \dot{x}_3 &= -Ax_3 + Bx_2 - Cx_1 + u(t). \end{aligned} \tag{14}$$

Application of the Liapunov Corollary, for $u = 0$ [*29*] yields six simultaneous linear equations to be solved for the coefficients of a positive definite symmetric matrix. After a laborious but straightforward calculation, the following values are obtained for the coefficients of the quadratic form

$$\begin{aligned} P_{11}(&= \frac{C(A^2 + AC + C^2) + B(AB - C)}{2C(AB - C)} \\ P_{12} &= \frac{A^2B + C^2(B + 1)}{2C(AB - C)} \\ P_{13} &= \frac{1}{2C} \\ P_{22} &= \frac{A^3 + C(A^2 + AC + B^2 + B + 1)}{2C(AB - C)} \\ P_{23} &= \frac{A^2 + AC + C^2}{2C(AB - C)} \\ P_{33} &= \frac{A + C(B + 1)}{2C(AB - C)} \end{aligned} \tag{15}$$

The approximation to the optimal isochrones are given by the quadratic forms

$$V(x) = P_{11}x_1^2 + 2P_{12}x_1x_2 + 2P_{13}x_1x_3 + 2P_{23}x_2x_3 + P_{22}x_2^2 + P_{33}x_3^2 = \text{const.} \tag{16}$$

The gradient of the approximate isochrones $\nabla V(x)$ is easily obtained.

The multiple integrator approximation for the isochrones are obtained for third order systems as another example:

$$\begin{aligned} \dot{x}_1 &= x_2 \\ \dot{x}_2 &= x_3 \\ \dot{x}_3 &= u(t) \end{aligned} \tag{17}$$

For the case of step inputs the optimal switching times are $\tau_1 = T^0/4$ and $\tau_2 = 3T^0/4$ where T^0 is the optimal process time. The optimal values of $P_2(0)/P_1(0)$ and $P_3(0)/P_1(0)$ can be shown to be:

$$\begin{aligned} \frac{P_2^0(0)}{P_1(0)} &= T^0/2 \\ \frac{P_3^0(0)}{P_1(0)} &= \tfrac{3}{8}\left(\frac{T^0}{2}\right)^2 \end{aligned} \tag{18}$$

The optimal process time T^0 is obtainable from the Tchebycheff Polynomial solution.

$$T^0 = (16x_1(0)^{1/3} \tag{19}$$

The locus of optimal initial state of the adjoint for step inputs in the $P_2(0)/P_1(0)$, $P_3(0)/P_1(0)$ subspace is described by the equation

$$\frac{P_3(0)}{P_1(0)} = \tfrac{3}{8}\left(\frac{P_2(0)}{P_1(0)}\right)^2 \tag{20}$$

The graph of this equation is shown in figure 7 as a heavy dashed line. Comparisons of actual and approximate isochrones are shown in figure 4 and 5.

VI. Convergence of the Iterations

In order to insure the convergence of the steepest descent procedure, Curry [*14*] has established that it is necessary and sufficient to show (1) that $V(P(0), T)$ has continuous first partial derivatives in the domain of interest (2) that V has a unique minimum $(P^0(0), T^0)$, (3) that the

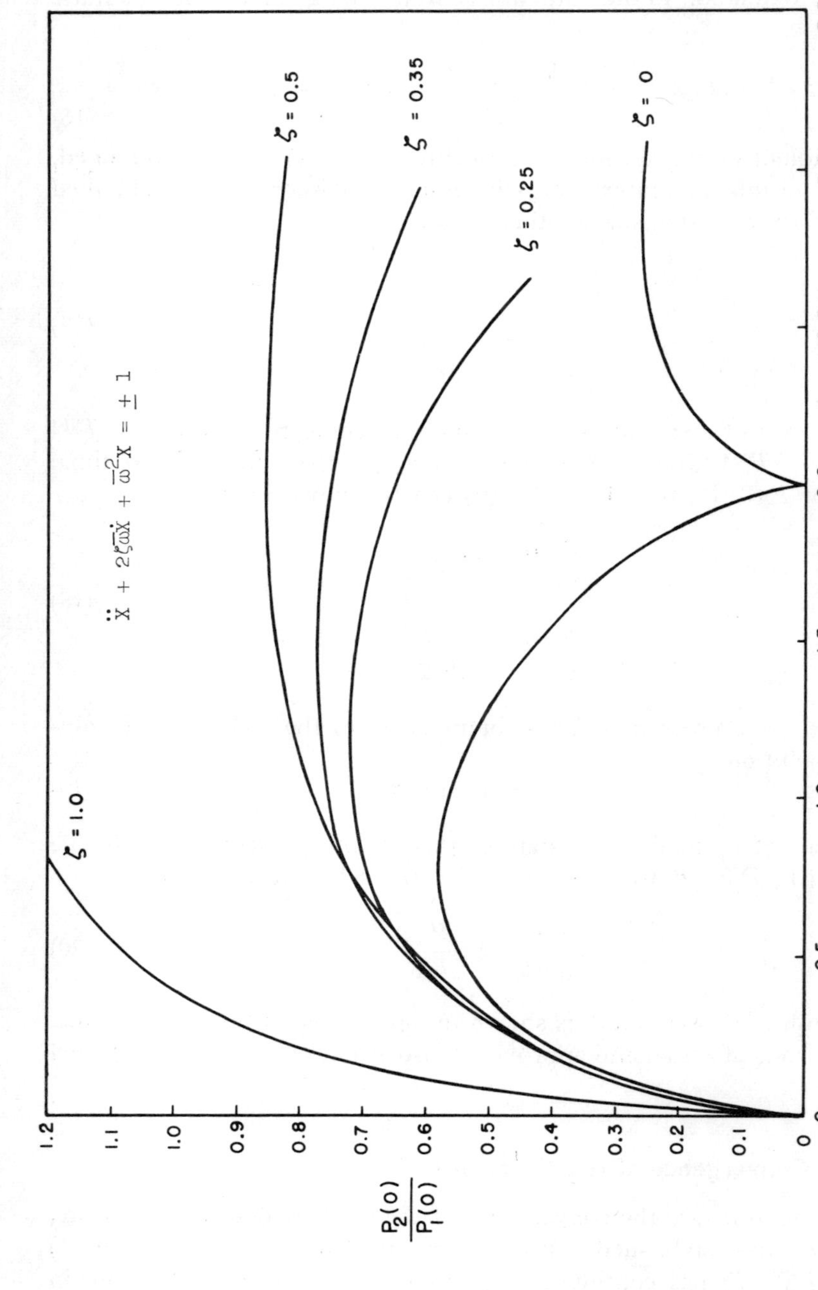

FIG. 4. Optimal initial state of the adjoint step inputs for various values of ζ

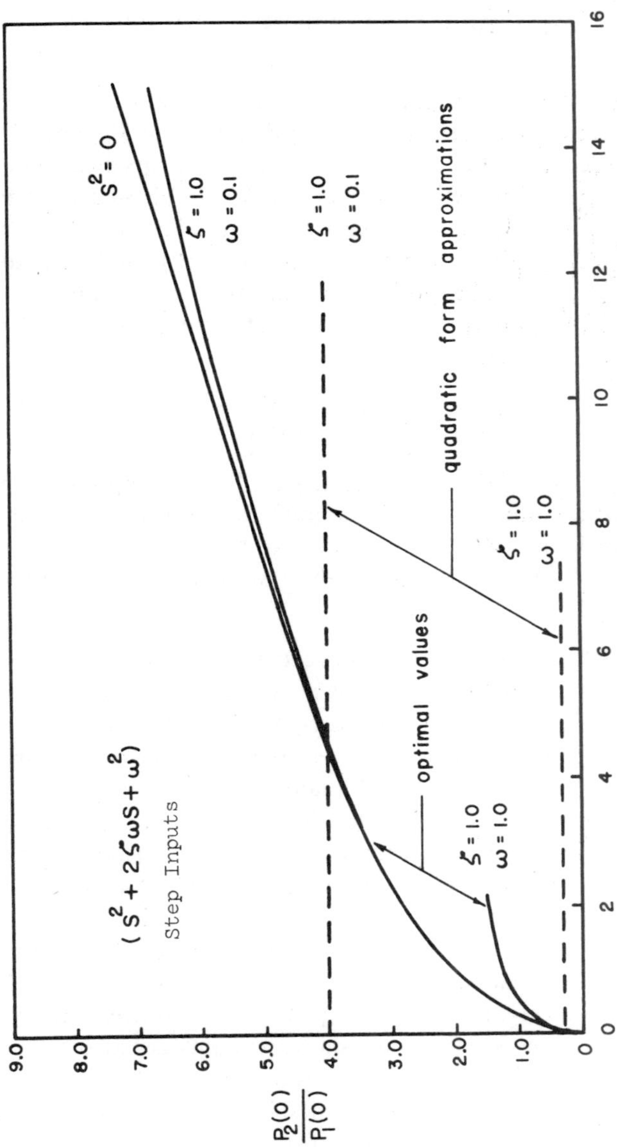

FIG. 5. Comparison of actual and approximate values of gradient of optimal isochrones

Jacobian of $V[\mathbf{x}(P, T)]$ does not vanish. These points will be taken up in order. It is not possible to present a complete discussion here of all aspects of these questions. Therefore reference is made to ARAP Report No. 33 [*48*] which contains a more extensive treatment of these questions.

1. A method for determining the partial derivatives of V has been found for the case of linear processes and quadratic V. This facilitates the actual numerical solution of the problem as the gradient of V can be calculated directly after each run. The result can also be used for error analysis. The basic idea is to obtain an expression for the effect of a change in switch time τ_k due to change in $P_j(0)$, and also to obtain change in $x_i(T)$ due to change in switching time $\Delta\tau_k$. The latter can be obtained directly from the use of the system impulse response. It is not difficult to show that

$$\frac{\partial \tau_k}{\partial P_j(0)} \cong -\frac{\eta_{jn}(\tau_k)}{P_n(\tau_k)} \tag{21}$$

where the zeros of $P_n(T)$ determine the τ_k. The η_{jn} represents the response of the nth component of the adjoint to a unit initial displacement of the jth coordinate $P_j(0)$. Combining and superposing the effects of the changes in $P_j(0)$ and summing over all switch times provides the expression for ∇V.

2. The possibility of local minima occurring in the domain of interest is the second aspect of the convergence problem to be considered. It is generally difficult to establish whether an arbitrary function $V(P, T)$ has a unique minimum and to exclude the possibility of undesired local minima. Proper selection of the domain of interest may eliminate the unwanted minima, however this may prove to be an impractical procedure. The domain of interest cannot be restricted to be a small region close to the true optimum point simply because the location of the true minimum point is not known in advance. In particular cases the nature of the surfaces $V(P, T)$ can be determined by actual exploration. Beale [*4*] has considered a breakout procedure for escaping from undesirable regions such as local minima. This procedure requires the system to take large exploratory steps from time to time as a protective measure against being hung up on a local minimum. In investigating a particular problem it may prove advantageous to select another norm, i.e., a different V that might be more suitable for the specific system. Although the possibility of local minima has not been definitely ruled out for all situations no such occurrences were observed during any of the experiments. The evidence obtained by direct exploration of V surfaces indicate the presence of a long valley with steep sides. These results can only serve as an indication of what can be expected and are by no means a conclusive demonstration of general non-existence of local minima.

3. The third condition for convergence of the iterative process hinges upon the behavior of the partial derivatives and the gradient of the function $V[x(P(0), T)]$. If the Jacobian vanishes at an isolated point such as a saddle point, an exploratory procedure can be used to get the new direction for the next descent. It will be shown however that regions of this space exist such that the Jacobian vanishes identically. In fact, table-like regions exist for which changes in P produce no changes in $V(P, T)$. If an initial choice of $P(0)$ selected for the iteration happened to lie in this region, the computer would be unable to decide what to do and the method may fail to work. There are procedures similar to those mentioned in connection with the local minima problem that could be devised to get the system out of such a situation. Clearly it would be much better to avoid these areas.

Consider a linear system with a single control $u(t)$. The optimal control law for this system is of the form

$$\begin{aligned} u^0(t) &= \operatorname{sgn} b_j P_j(t) \\ P_j(t) &= P_j(0)\, \eta_{ij}(t). \end{aligned} \tag{22}$$

It is possible to select values of $P(0)$ such that $b_j P_j(t)$ has no zeros in the interval $(0, T)$. According to the nature of the functions $\eta(t)$ it may be possible to find $P(0)$ such that there exists no zeros at all in any time interval. It is important to determine the regions in the P subspace where the Jacobian vanishes identically. This has been done for 2nd and 3rd order linear systems and a detailed account is contained in ref. [*48*].

The conditions under which the Jacobian vanishes can be seen by consideration of some examples. In the case of the second order system

$$\begin{aligned} \dot{x}_1 &= x_2 \\ \dot{x}_2 &= -\,2\zeta\omega x_2 - \omega^2 x_1 + u(t). \end{aligned} \tag{23}$$

The adjoint equations are

$$\begin{aligned} P_1' &= \omega^2 P_2 \\ P_2' &= -\,P_1 + 2\zeta\omega P_2, \end{aligned} \tag{24}$$

and the control law is

$$u(t) = \operatorname{sgn} P_2(t).$$

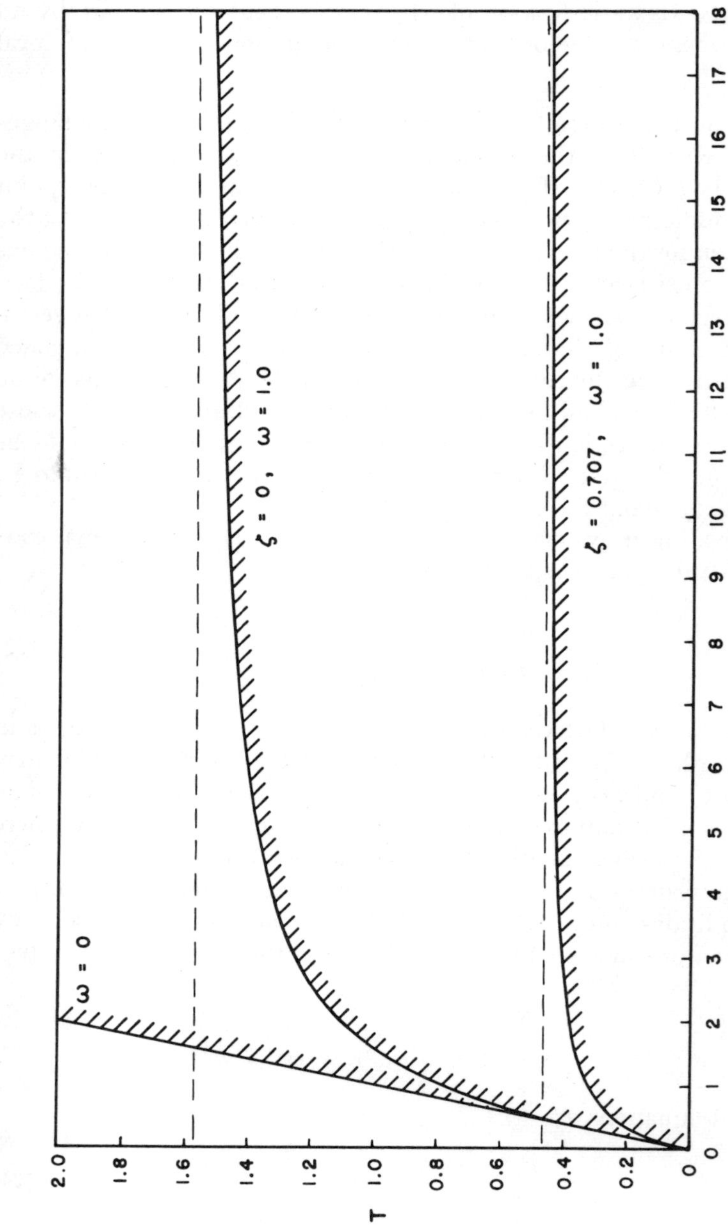

FIG. 6. Boundaries for no switches in the interval $0 \leq t \leq \infty$ $x + 2\zeta\omega x + \omega^2 x = \pm 1$

The question is to determine the initial conditions on the adjoint so that $P_2(t)$ has no zeros in the interval $[0, T]$. If there are no zeros in the interval $(0T)$ then small changes in $P(0)$ have no effect on $X(T)$ and the gradient of V vanishes. If the initial conditions $P_1(0)$, $P_2(0)$ are such that

$$\text{Tan}\, \omega_n T = \frac{P_2(0)/P_1(0)}{[P_2(0)/P_1(0)]\, 2\zeta\omega - 1} \tag{25}$$

then there will be no zeros, hence no switches in the interval $0 \leq t < T$. Figure 6 shows the boundary for $P_2(0)/P_1(0)$ *vs* $\omega_n T$. If the initial state of the adjoint corresponds to a point below the cross hatched boundary then there will be no zero crossings in the time interval of duration T.

The adjoints of second order systems with oscillatory modes always have zero crossings in long enough time intervals. This is not the case in higher order systems.

The boundaries for cases $\omega = 0$; and $\omega \neq 0$, $\zeta < 1$, are shown in figure 6. These boundaries delineate the region where the Jacobian of $V(P, \tau)$ vanishes identically. If a search procedure for the optimal initial state of the adjoint by the method of steepest descent is to succeed, the trial values of P and T must not lie in these regions.

Some examples are also presented for linear 3rd order systems. In these cases the iteration procedure seeks $P_2^0(0)/P_1^0(0)$, $P_3^0(0)/P_1^0(0)$ and T^0. For third order systems with an oscillatory mode it is possible to find values of initial conditions such that zero crossings of $P_3(t)$ never occur.

Severals sets of boundaries were investigated by means of computer studies. Prior to the start of the computer studies a brief root locus examination was made in order to determine the range of non-dimensional parameters to be used. These results are shown in figures 7 and 8. The coordinates used here are

$$ln \frac{P_2(0)}{P_1(0)}$$

and (26)

$$ln \frac{P_3(0)}{P_1(0)}\,.$$

In addition to the boundaries for no switches in the interval (i.e., the vanishing of the Jacobian) the locus of optimal initial states for all step inputs for triple integrators is shown as a heavy dashed line. The relative positions of the locus of approximating solution (only the case of step inputs is shown) and the boundaries for the vanishing of the gradient are shown in this figure. For example, figure 7 shows that the system

$\dddot{x} + \gamma\dot{x} = u(t)$ with $\gamma = 1.0$ cannot use the integrator approximation if the predicted $P(0)$ values are such that

$$\frac{P_2(0)}{P_1(0)} > 1.7 \tag{27}$$

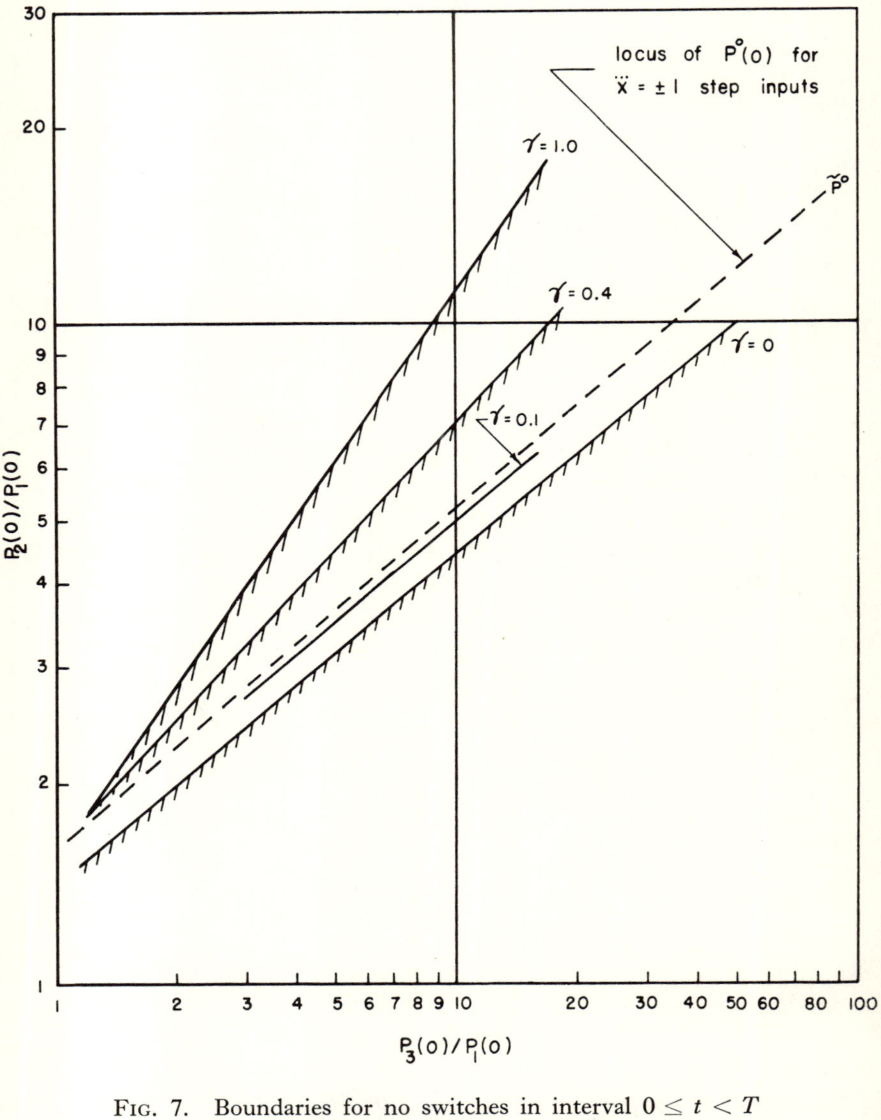

FIG. 7. Boundaries for no switches in interval $0 \leq t < T$
$S^2(S+\gamma)$, $\quad T = 14$ Seconds

and

$$\frac{P_2(0)}{P_1(0)} > 1.1 \tag{28}$$

corresponding to an estimated optimal process time of .85 secs. On the other hand, for $\gamma = 0.1$ the approximation is satisfactory for the range

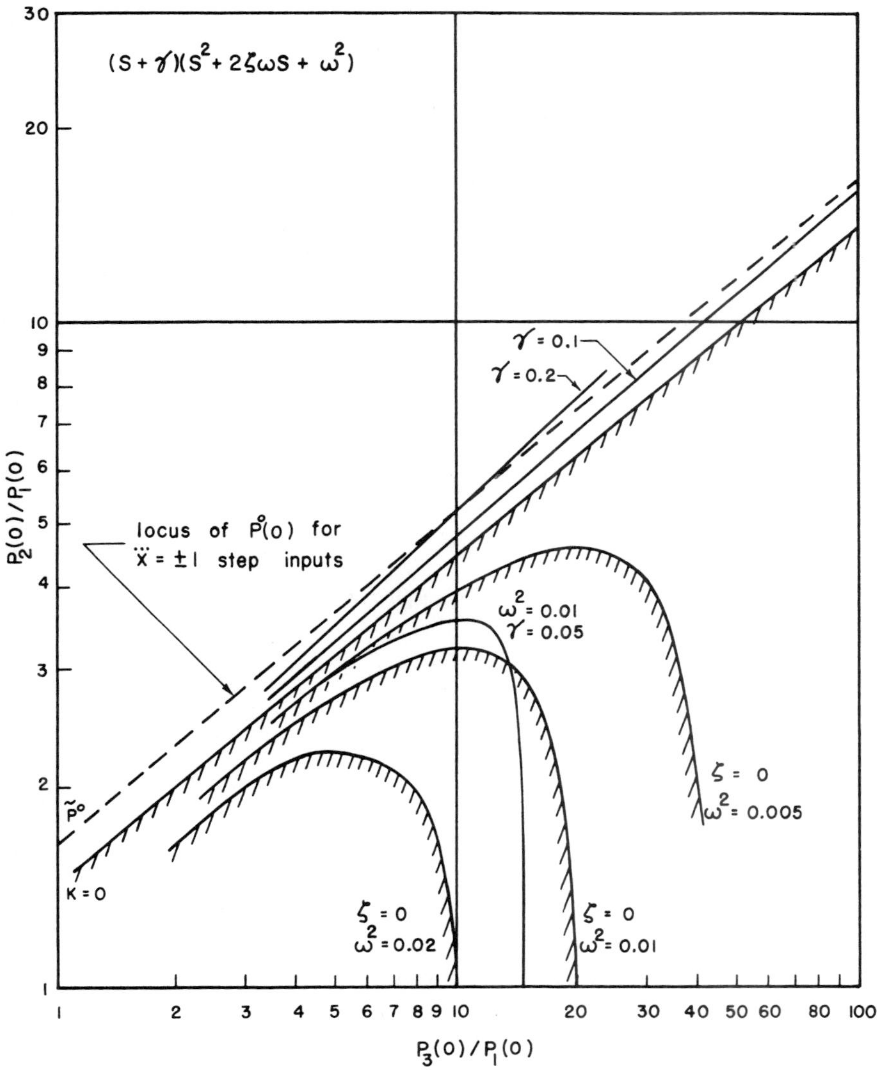

FIG. 8. Boundaries for no switches in interval $0 \leq t < \infty$.

of values shown on the figure. In the cases where the boundary $J = 0$ is known in advance it may prove to be possible to use this boundary to generate a locus of starting values. A curve similar to the boundary but shifted up into the safe region to provide a safety margin could replace the locus of optimal integrator values.

Figure 8 shows boundaries for the systems of the general operational form

$$(S + \gamma)(S^2 + 2\zeta\omega S + \omega^2) \tag{29}$$

for different locations of characteristic roots. Figure 7 shows the progression of the boundary for systems of the form $S^2(S + \gamma)$ with γ as the parameter. The configuration of the characteristic roots of the adjoint affects the boundaries mainly through the action of the most unstable mode of adjoint response. The values of the parameters associated with the curves are indicated on the figures.

Similar boundaries are shown in figure 8 for the case of $T = \infty$. This represents the most extreme case. Below this boundary the Jacobian vanishes for all time.

These results show that the general trend of the boundaries is to decrease the acceptable regions of the P subspace as the instability of the adjoint is increased. This agrees with the results obtained for the second order systems wherein an increase in ζ moved the boundaries into previously acceptable regions.

VII. Exploration of the Error Surfaces and Checkout of the Iteration Scheme

A series of exploratory runs was made on a combined analog digital computer. These runs consisted of traversals of the $P_2(0)/P_1(0)$, $P_3(0)/P_1(0)$ subspace for constant values of T. These runs enable profiles of the V surface to be studied with regard to the choice of the size of the steps, α, during the iteration. A typical cross section of the function $V(P(0), T)$ is shown in figure 9. An interesting practical aspect was uncovered during these tests. It was not possible to set the potentiometers by hand using the analog panel type voltmeter. The steepness of the surface in the $P_2(0)$, $P_3(0)$ subspace made it necessary to use the octal voltmeter on the digital computer for setting and reading the potentiometers. The apparent sensitivity of the system response to the initial conditions of the adjoint during analog simulation had been observed some time ago by Mengel [*45*] of the RAND Corp. in connection with another type of minimization problem. This is actually a question of resolution and can be overcome if the proper instrumentation is available.

The iteration procedure was tested for the case of step inputs for third order systems. In the case of a small real root and two zero roots the procedure using the triple integrator approximation converged rapidly, requiring two calculations of the gradient. The case of a complex pair of roots and a zero root caused some difficulty.

The method for obtaining initial values of P based on quadratic forms fails to provide a solution here. In the case of second order oscillatory systems it was observed that the initial adjoint vector was

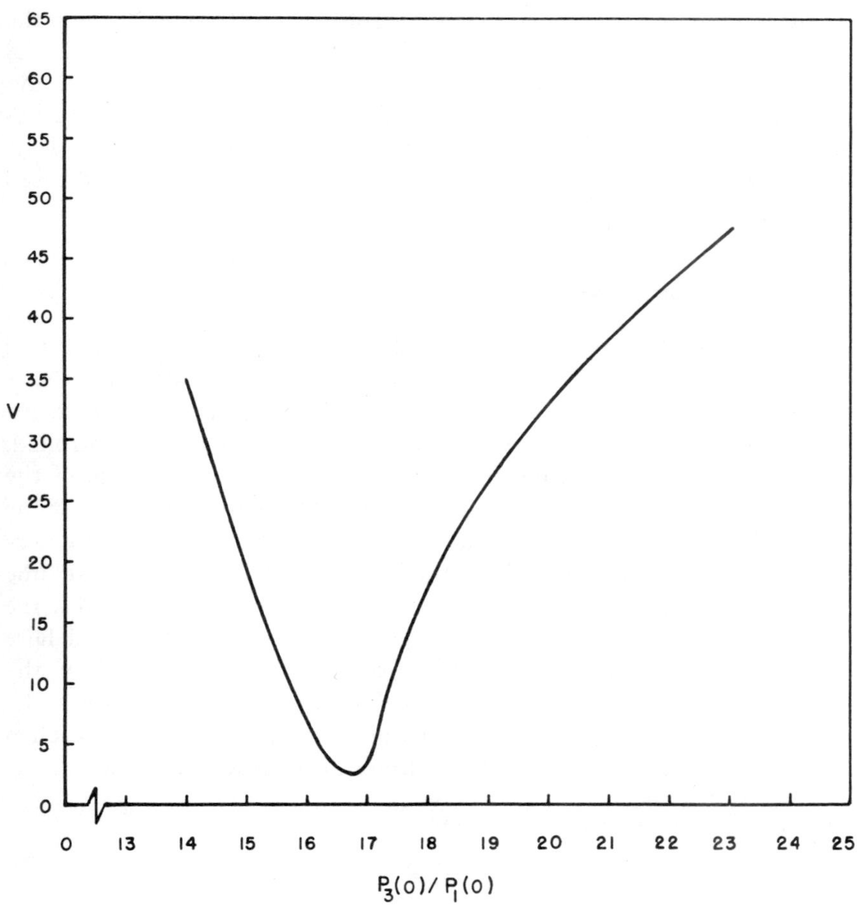

FIG. 9. V *vs.* $P_3(0)/P_1(0)$, $\gamma = 0.05$, $T = 13.5$ Seconds.

nearly opposite to the system state vector for large initial disturbances in position. The quadratic forms approximating the optimal isochrones were nearly circles for large values of $\omega^2 x_1(0)$. This idea was carried over to the third order system. As the zero switch boundary passes through the origin in the $P_2(0)/P_1(0)$, $P_3(0)/P_1(0)$ subspace it is not practical to set $P_2 = 0$ and $P_3 = 0$ as initial values for the iteration. Instead, values of $P_2(0)/P_1(0)$ and $P_3(0)/P_1(0)$ were selected that were close to the origin but in the region of the $P_2(0)/P_1(0)$, $P_3(0)/P_1(0)$ plane such that the Jacobian did not vanish.

The difficulty that was encountered was due to the problem of selecting a suitable step size for the steepest descent procedure. It was necessary to use the smallest step size obtainable (one octal digit on the voltmeter or 1/80 volt) to prevent overshooting the optimal point and arriving in the region of the vanishing Jacobian. The optimal values for the initial states of the adjoint $P_2^0(0)/P_1(0)$ and $P_3^0(0)/P_1(0)$ were in fact small and were located very close to the boundary where the Jacobian vanishes for all T. This accounts for the difficulty experienced by Flügge-Lotz and Yin in obtaining the switching surface for this case by a method based on the adjoint system.

VIII. Computer Requirements

In order to apply the iterative procedures of this report to an actual system it is necessary to obtain some preliminary estimates of the size and speed requirements of the computer. The time allowable for computation depends upon the time scale of the process being controlled. The most useful information for a digital system is an estimate of the number of elementary operations required. The required speed of the computer in terms of the elementary operation time can then be determined from the real time scale of the process and the ratio of fast time to real time. The outline of procedural steps shown below will serve as a review of the computational scheme. The method discussed here utilizes a modified steepest descent procedure. The gradient of the surface is not computed at each step. The gradient is found initially and steps of size α are taken in that direction until a valley floor is reached and a new gradient is calculated or a minimum point is attained.

1. Sample the state of the system.
2. Estimate $P(0)$, T by a preselected set of rules.
3. Find $V(P, T)$ (For example $V^2 = \Sigma_i x_i^2$.)
4. Obtain Gradient $V(P, T)$.

5. Select values for P, T by modified method of steepest descent.

$$P^{\nu+1}(0) = P^{\nu}(0) - \alpha \nabla_P V$$

$$T^{\nu+1} = T^{\nu} - \alpha \nabla_T V$$

6. Repeat steps 3 and 5 until the valley floor or a minimum point is reached.

7. Recompute the Gradient of V and repeat the process.

This outline shows that there are two major tasks facing the computer. In order to obtain $V(P, T)$ the computer must obtain the response of the controlled process and the adjoint for a time interval of duration T. There are several ways that this can be done. The process and the adjoint can be simulated in fast time on an analog computer. The time scale for the analog will depend upon the number of iterations anticipated in an interval of real time, the duration of the runs, and the time scale of the process being controlled.

Another possibility is the use of a digital computer to simulate the process and the adjoint. This requires an integration of the differential equations describing the process. This procedure may be costly in terms of time. An alternative use of a digital computer involves the use of stored tables of the process step response and adjoint fundamental solutions. This may be costly in terms of machine storage.

The second task for the computer is the evaluation of $\nabla V(P, T)$. There are also several possibilities for doing this. Although an explicit method for computing the gradient of $V(P, T)$ is available, the computation time for a 3rd order system was considered lengthy, and an experimental procedure was developed using the combined analog-digital computer. A series of runs was made on the analog with the digital computer controlling the duration and sampling the adjoint at the beginning of a run and the controlled system state at time T. In this way the gradients could be evaluated quite rapidly.

At the present time a promising procedure appears to be one based on the use of a high speed analog computer for simulating the process and the adjoint response and also for obtaining the gradients, coupled with a digital computer to act as an executive computer to make logical decisions and to supervise the operation of the analog.

IX. Summary and Conclusions

The previous work on the problem of time optimal control was briefly reviewed. It was shown that the surfaces of constant optimal time in the system phase space, i.e. the optimal isochrones, form a Liapunov function for the system. The optimal control law is obtainable from the gradient of the optimal isochrones. The gradient of the optimal isochrones satisfies the system adjoint equations.

It was shown that it is possible to obtain solutions to the two point boundary problem associated with the optimal control problem by an iterative method based on a modification of the method of steepest descent. This procedure will converge, in the absence of local minima if the Jacobian of the error surface does not vanish. This Jacobian does, in fact vanish identically over large regions of the adjoint subspace. Boundaries for these regions are shown for second and third order linear systems.

Several families of optimal isochrones were constructed for second order systems. An examination of these curves suggested generating approximations to the optimal isochrones and hence approximations to the optimal control law.

A series of computational tests were made on a combined analog-digital computer. An investigation of the error surfaces and checkout of the iterative procedure was performed for third order systems and step inputs. The convergence to the optimal initial state of the adjoint was rapid in most of the cases examined.

A brief discussion of the computational aspects of the iterative procedure was presented. It was concluded that a promising computational procedure appears to be the use of a high speed analog computer for simulating the system and the adjoint, coupled with a digital computer to make logical decisions and to act as an executive computer supervising the operation of the analog.

REFERENCES

[*1*] Anderson, C. A., Switching Criteria for Certain Contactor Servos with Arbitrary Inputs. Ph.D. Thesis, Univ. of Michigan (1958).

[*2*] Baker, R. M. L., Jr., Recent advances in astrodynamics 1960. *ARS Journal.* **30**, No. 12, 1127 (1960).

[*3*] Bass, R. W., Equivalent linearization—Nonlinear circuit synthesis, and the stabilization and optimization of control systems. *Proc. Symposium on Nonlinear Circuit Analysis, Brooklyn Polytech. Inst.* **16**, 163-198 (1956).

[*4*] Beale, E. M. L., On An Iterative Method for Finding a Local Minimum of a Function of More Than One Variable. Statist. Techniques Research Group (Dept. of Math.) Princeton Univ. Tech. Report No. 25 (1958).

[*5*] Beecher, A. E., and Gould, L. A., Considerations in the Design of a Dynamic Control System for the Generalized Chemical Processing Model Considered as a Nonlinear System. AIEE Paper No. CP-60-980 (1960).

[*6*] Behrbohm, H., Optimal Trajectories in the Vertical Plane. SAAB Report No. TN-34 (1955).

[7] Bellman, R., Glicksberg, I., and Gross, O., On the bang-bang control problem. *Quart. appl. Math.* **14**, 11-18 (1956).

[8] Bliss, G. A., "Lectures on the Calculus of Variations." Univ. of Chicago Press, Chicago, Illinois, 1949.

[9] Box, G. P., The exploration and exploitation of response surfaces: Some general considerations and examples. *Biometrics* **10**, 16 (1954).

[10] Box, G. P., and Wilson, K. B., Experimental attainment of optimum conditions. *J. Roy. Statist. Soc.* **B13**, No. 1, 1-45 (1951).

[11] Breakwell, J. V., The optimization of trajectories. *J. Soc. Ind. Appl. Math.* **7**, No. 2 (1959).

[12] Bushaw, D., Optimal discontinuous forcing terms. *In* "Contributions to the Theory of Nonlinear Oscillations" (S. Lefschetz, ed.), Vol. 4, pp. 29-52. Princeton Univ. Press, Princeton, New Jersey, 1958.

[13] Courant, R., Variational methods for the solution of problems of equilibrium and vibrations. *Bull. AMS* **49**, No. 1 (1943).

[14] Curry, H. B., The Method of Steepest Descent for Nonlinear Minimization Problems. *Quart. appl. Math.* **2**, No. 3 (1944).

[15] Darlington, S., Linear least-squares smoothing and prediction with applications. *Bell System Tech. J.* **37**, 1221 (1958).

[16] Desoer, C. A., The bang-bang servo problem treated by variational technique. *Information and Control.* **2**, No. 4, 333-348 (1959).

[17] Doll, H. G., and Stout, T. M., Design and analog computer analysis of an optimum third order nonlinear servomechanism. *Trans. ASME* **79**, 513 (1957).

[18] Dreyfus, S., Dynamic Programming and the Calculus of Variations. RAND Corp. Rept. No. P-1464 (1960).

[19] Fischbach, J. W., Some applications of gradient methods. *Proc. 7th AMS Symposium in appl. Math.* p. 59 (1956).

[20] Flügge-Lotz, I., "Discontinuous Automatic Control." Princeton Univ. Press, Princeton, New Jersey, 1953.

[21] Flügge-Lotz, I., and Yin, M., On the Optimum Response of Third-Order Contactor Control Systems. Stanford Univ. Tech. Report No. 125 (1960).

[22] Garfinkel, B., Minimal problems in airplane performance. *Quart. appl. Math.* **9**, No. 2, 149 (1951).

[23] Geissler, E., Problems in attitude stabilization of large guided missiles. *Aero/Space Eng.* **19**, No. 10 (1960).

[24] Gould, L., and Kipiniak, W., Dynamic Optimization and Control of a Stirred Tank Chemical Reactor. AIEE Paper No. CP-60-982 (1960).

[25] Hazen, H. L., Theory of servomechanisms. *J. Franklin Inst.* **218**, 279-330 (1934).

[26] Hestenes, M. R., A General Problem in the Calculus of Variations with Applications to Paths of Least Time. RAND Corp. Report No. RM-100 (1950).

[27] Ho, Yu-Chi, Solution Space Approach to Optimal Control Problems. ASME Paper No. 60-JAC-11 (1960).

[28] Kalman, R. E., The Theory of Optimal Control and the Calculus of Variations. RIAS Tech. Report No. 61-3 (1961).

[29] Kalman, R. E., and Bertram, J. E., Control system analysis and design via the "second method" of Liapunov, Part I. *ASME Trans., Ser. D: J. Basic Eng.* **82**, 371, June (1960).

[*30*] Kelley, H. J., An investigation of optimal zoom climb techniques. *J. Aero/Space Sci.* **26**, No. 12, 794 (1959).

[*31*] Kelley, H. J., Gradient Theory of Optimal Flight Paths. *ARS Journal.* **30**, No. 10, 947 (1960).

[*32*] Kipiniak, W., A Variational Approach to Dynamic Optimization and Control System Synthesis. Ph.D. Thesis, Mass. Inst. Technol. (1961).

[*33*] Kirillova, F. M., On a limiting process in the solution of a problem of optimum control. *Priklad. Mat. i Mekh.* **24**, No. 2 (1960).

[*34*] Krassovskii, N. N., On the theory of optimum regulation. *Automation and Remote Control.* **18**, No. 11, 1005 (1957) (English).

[*35*] Krassovskii, N. N., On a problem of optimal control of nonlinear systems. *Priklad. Mat. i Mekh.* **23**, No. 2, 303 (1959).

[*36*] Krassovskii, N. N., Choice of parameters of optimum stable systems. *Proc. Intern. Federation of Automatic Control Congr., Moscow* p. 677 (1960).

[*37*] Kulikowski, R., Optimizing processes and synthesis of optimizing automatic control systems with nonlinear invariable elements. *Proc. Intern. Federation of Automatic Control Congr., Moscow* p. 473 (1960).

[*38*] Kurzweil, F., Jr., Dynamic Synthesis of Higher Order Optimum Saturating Systems. ASME Paper No. 60-JAC-2 (1960).

[*39*] La Salle, J. P., The bang-bang principle. AFOSR TN-59-1142; also RIAS Tech. Report No. 59-5.

[*40*] La Salle, J. P., The time optimal control problem. *In* "Contributions to the Theory of Nonlinear Oscillations" (S. Lefschetz, ed.), Vol. 5, pp. 1-24. Princeton Univ. Press, Princeton, New Jersey, 1960.

[*41*] Lee, E. B., Design of optimum multivariable control system. ASME Paper No. 60-JAC-5 (1960).

[*42*] Leitmann, G., On a class of variational problems in rocket flight. *J. Aero/Space Sci.* **26**, No. 9, 586 (1959).

[*43*] Lukens, D. R., A method of analysis of a control system for a large flexible missile. *Proc. natl. Specialists Meeting on Guidance of Aerospace Vehicles, Boston*, p. 98 (1960).

[*44*] McDonald, D., Nonlinear techniques for improving servo performance. *Proc. Nat. Electronics Conf.* **6**, 400-421 (1950).

[*45*] Mengel, A. S., Optimum trajectories. *Proc. Project Cyclone Symposium I on REAC Techniques, New York 1951.*

[*46*] Miele, A., Some recent advances in the mechanics of terrestrial flight. *Jet Propulsion.* **28**, No. 9, 581-587 (1958).

[*47*] Oldenburger, R., Optimum nonlinear control. *Trans. ASME* **79**, 527 (1957).

[*48*] Paiewonsky, B. H., A Study of Time Optimal Control. ARAP (Princeton) Report No. 33 (1961).

[*49*] Pistiner, J. S., On-off control system for attitude stabilization of a space vehicle. *ARS Journal* **29**, No. 4, 283 (1959).

[*50*] Pontryagin, L. S., Boltyanskii, V. G., and Gamkrelidze, R. V., Optimal processes. *Proc. 1st IFAC Conf., Moscow, 1959.*

[*51*] Rose, N. J., Theoretical Aspects of Limit Control. Experimental Towing Tank. Stevens Inst. of Technol. Report No. 459 (1953).

[*52*] Rose, N. J., Optimum switching criteria for discontinuous automatic controls. *IRE Convention Record* Part 4, p. 61 (1956 Natl. Convention).

[*53*] Rosenbloom, P. C., The method of steepest descent. *Proc. 7th AMS Symposium in appl. math.* p. 127 (1956).

[54] Rozoner, L. I., L. S. Pontryagin's maximum principle in theory of optimal systems. *Automation and Remote Control* **20**, Oct., Nov., Dec. (1959) (English).

[55] Tsien, H. S., and Evans, R. C., Optimum thrust programming for a sounding rocket. *ARS Journal*, **21**, No. 5, 99 (1951).

[56] Valentine, F. A., The problem of Lagrange with differential inequalities as added side conditions. *In* "Contributions to Calculus of Variations, 1933-1937". Univ. of Chicago Press, Chicago, Illinois.

[57] Weiss, H. K., Analysis of relay servomechanisms. *J. Aeronaut. Sci.* **13**, No. 7, 364 (1946).

Controllable Stability and Equivalent Nonlinear Programming Problem

J. B. ROSEN
Shell Development Company, Emeryville, California

1. Introduction

A special type of stability for the differential equation system (2.1) will be considered in this paper. We consider a cylinder in the (t, x) space with a specified constant radius ρ and its axis coincident with the time axis. The cylinder may have a finite or infinite length. We will call the system (2.1) ρ-stable if every solution which starts in, or enters, the cylinder remains in the cylinder for its entire length (see definition in next section). This type of stability differs somewhat from that usually considered [*1*, *2*, *4*, *8*], but appears to have both practical and theoretical value for a system of the type (2.1) where the quantities u_i may be either fixed parameters (Section 2) or control variables (Sections 3 and 4). It is of practical interest in connection with a transient physical process, represented by a system of linear or nonlinear differential equations, where it is desired that certain physical quantities, represented by the solution, remain within a specified tolerance of specified values. This might be the situation, for example, in a chemical plant where the solution of the differential equation system gives the time dependent composition of the product stream. More general situations where it is desired that solutions remain within a specified tolerance (which may depend on time) of a specified trajectory can be reduced to the case treated here by an appropriate change of variables. This is shown at the end of the next section.

It should be remarked that a system may be unstable in the usual sense [*4*], not ρ-stable for sufficiently small ρ, but ρ-stable for larger ρ. This might be the case if the system is unstable for small perturbations but has a nonlinear restoring force. Solutions to such a system would not necessarily remain in a small cylinder but would always remain inside a larger cylinder. Conversely (and more commonly), a system might be ρ-stable for a specified ρ, but not for larger values of ρ.

Another practical reason for considering ρ-stability is that one can answer the question as to whether or not a particular system is ρ-stable by solving a single nonlinear mathematical programming problem [*7*, *11*].

This is described in the next section where the equivalence of a certain mathematical programming problem (2.2) and ρ-stability is demonstrated (Theorem 1). Efficient computational methods for the solution of such nonlinear programming problems are available for use with high speed computers [12]. It should be possible to study relatively large systems in this way.

In the proof of Theorem 1 we use the Euclidean norm $\| x(t) \|$ as a Liapunov function [1, 3] and apply a lemma due to Conti [5]. The proof also requires that the function $\Lambda(\alpha)$ defined by (2.2) satisfy a Lipschitz condition. It is shown that this follows directly from the assumption that the f_i in (2.1) satisfy such a condition. By means of similar arguments, comparable results for other norms can be obtained.

In Section 3 the control problem is investigated from the same point of view. We consider the system (2.1) where the u_i are now control variables which may be adjusted within specified limits. We want to know if any control adjustment within allowed limits will make the system (2.1) controllable ρ-stable. The interesting case is of course a system which is unstable without control. It is shown in Theorem 2 that the answer is given by a max min problem in the general case. However, if the system is linear in the control variables (not much of a restriction since the system may be nonlinear in the x_i) we show that the answer is given by solving the single nonlinear programming problem (3.5). These results are illustrated with two examples, a linear system and a nonlinear system. The ρ-stability of these systems is determined without control and with control.

We may interpret the relation $\psi(\rho) \leq 0$ in Theorem 2 as a test which determines whether there exist allowable controls which will make the system ρ-stable. If such controls exist we say that (2.1) is controllable ρ-stable. The existence of controls for the related optimal control problem [6, 10] has been investigated recently [9]. In Section 4, the completely linear system (4.1) is considered. We give a necessary and sufficient condition on the matrices A and B in order that the system be controllable ρ-stable (Theorem 3). This condition, given by (4.3), is related to the linear independence of rows of B where the choice of rows depends on A. This condition differs from that of Pontriagin [10] for optimum control which requires the linear independence of vectors Bw, ABw, ... $A^{n-1}Bw$. Theorem 3 also gives an explicit method for determining the largest value of ρ for which the controllable linear system (4.1) is ρ-stable with controls $| u_i(t) | \leq 1$. This is given by the solution of the programming problem (4.4). Another result of practical importance is given by the corollary and shows that for ρ-stability there must be at least as many controls $u_i(t)$ as there are positive eigenvalues of $(A + A^T)$.

2. Stability and Equivalent Programming Problem

We will consider the following differential equation system

$$\frac{dx_i}{dt} = f_i(t, x_1, \dots x_n, u_1, \dots u_m), \qquad i = 1, \dots n$$

or

$$\frac{dx}{dt} = f(t, x, u) \tag{2.1}$$

where the $f_i(t, x, u)$ are real continuous functions and satisfy a Lipschitz condition with respect to x. The vector u is to be selected from a compact convex subset Ω of the m-dimensional real number space R_m. The results in this section will be obtained with u held fixed for each solution of (2.1) (the quantities u_i may therefore be thought of as system parameters which do not depend on time.) It follows from the Carathéodory existence theorem (see ref. [*4*]) that a unique absolutely continuous solution $x(t)$ of (2.1) exists on a subinterval $t_0 \leq t \leq t_1$ for a prescribed initial condition $x(t_0) = x_0$.

We will use two vectors norms, the Euclidean norm

$$|| x || = (x^T x)^{1/2} = \left(\sum_{i=1}^{n} x_i^2\right)^{1/2},$$

and the norm

$$| x | = \sum_{i=1}^{n} | x_i |.$$

The hypersphere in R_n with radius ρ will be denoted by

$$R(\rho) = \{x \mid || x || \leq \rho\}.$$

We consider a time interval

$$I = \{t \mid t_0 \leq t < T\},$$

which may be finite or infinite ($T = \infty$). For any positive constant ρ, we define ρ-stability as follows:

Definition. The system (2.1) is ρ-stable, if for every $(t', u) \in I \times \Omega$, $x(t') \in R(\rho)$ implies $x(t) \in R(\rho)$ for $t' \leq t < T$.

That is, the system (2.1) is ρ-stable if every solution which starts in or enters the cylinder $I \times R(\rho)$, remains in the cylinder.

It will now be shown that ρ-stability of the system (2.1) can be determined by means of a certain nonlinear mathematical programming problem. In fact, a condition on the solution to the programming problem is shown to be equivalent to ρ-stability of the system. For a specified positive value of ρ, the mathematical programming problem pertinent to the system (2.1) is that of determining

$$\Lambda(\alpha) \equiv \max_{t,x,u} \{\alpha^{-1}x^T f(t, x, u) \mid \| x \| = \alpha, (t, u) \in I \times \Omega\} \tag{2.2}$$

for the value $\alpha = \rho$. In order to show this equivalence we must first establish the fact that the function $\Lambda(\alpha)$ satisfies a Lipschitz condition. We do this in the following.

Lemma. Let $\beta(t, x, u)$ be a scalar function of (vectors) x, u and (scalar) t which satisfies a Lipschitz condition with respect to x for $(t, x, u) \in I \times R(\rho) \times \Omega$. Let

$$\gamma(\alpha) = \max_{t,x,u} \{\beta(t, x, u) \mid \| x \| = \alpha, (t, u) \in I \times \Omega\} \tag{2.3}$$

Then $\gamma(\alpha)$ satisfies a Lipschitz condition for $0 \leq \alpha \leq \rho$.

Proof. Consider any two values of α, $0 \leq \alpha_1 < \alpha_2 \leq \rho$. Let $\gamma(\alpha_1)$ be attained at (t_1, x_1, u_1), so that

$$\gamma(\alpha_1) = \beta(t_1, x_1, u_1) \equiv \beta_1 \qquad \text{with} \qquad \| x_1 \| = \alpha_1$$

and similarly

$$\gamma(\alpha_2) = \beta(t_2, x_2, u_2) \equiv \beta_2 \qquad \text{with} \qquad \| x_2 \| = \alpha_2.$$

We first assume that $\gamma(\alpha_2) \geq \gamma(\alpha_1)$, and let x_1' be the point in R_r determined by the intersection of the radius through x_2 with the hypersphere $\| x \| = \alpha_1$. Since $\| x_1' \| = \alpha_1$, it follows from (2.3) that

$$\gamma(\alpha_1) = \beta(t_1, x_1, u_1) \geq \beta(t_2, x_1', u_2) \equiv \beta_1'.$$

Therefore

$$\gamma(\alpha_2) - \gamma(\alpha_1) = \beta_2 - \beta_1 \leq \beta_2 - \beta_1'.$$

Since β satisfies a Lipschitz condition there is a constant k such that

$$| \beta_2 - \beta_1' | \leq k \| x_2 - x_1' \| = k | \alpha_2 - \alpha_1 |,$$

so that

$$| \gamma(\alpha_2) - \gamma(\alpha_1) | \leq | \beta_2 - \beta_1' | \leq k | \alpha_2 - \alpha_1 |.$$

For $\gamma(\alpha_1) \geq \gamma(\alpha_2)$ we repeat the argument using the point x_2' on the radius through x_1 and the hypersphere $\| x \| = \alpha_2$. Q.E.D.

It follows from (2.2) and this lemma that if the functions $f_i(t, x, u)$, $i = 1, \dots n$, satisfy a Lipschitz condition for $(t, x, u) \in I \times R(\rho + \delta) \times \Omega$, then $\Lambda(\alpha)$ satisfies a Lipschitz condition for $\delta \leq \alpha \leq \rho + \delta$, where δ is a specified constant, $0 < \delta < \rho$.

Theorem 1. Let ρ and δ be specified constants $0 < \delta < \rho$, and let the $f_i(t, x, u)$ be real continuous functions satisfying a Lipschitz condition with respect to x for $(t, x, u) \in I \times R(\rho + \delta) \times \Omega$. Then a necessary and sufficient condition that the system (2.1) be ρ-stable is

$$\Lambda(\rho) \leq 0 \tag{2.4}$$

Proof. The sufficiency follows from a lemma due to Conti [5] (or see Lemma 1 of Brauer [3]). As applied here the lemma states that if a function $\omega(\alpha)$ exists such that

$$\frac{d \| x \|}{dt} \leq \omega(\| x \|), \qquad x \in R(\rho + \delta) \tag{2.5}$$

then the solution $x(t)$ of (2.1) with $\| x(\tau) \| \leq \rho$ satisfies the inequality $\| x(t) \| \leq \alpha(t)$ for all $t \geq \tau$ such that $\alpha(t) \leq \rho + \delta$. The bound $\alpha(t)$ is the maximum solution of

$$\frac{d\alpha}{dt} = \omega(\alpha), \qquad \alpha(\tau) = \| x(\tau) \| \tag{2.6}$$

We choose $\omega(\alpha) \equiv \Lambda(\alpha) - \Lambda(\rho) \geq \Lambda(\alpha)$, by (2.4). From (2.1)

$$\frac{d \| x \|}{dt} = \frac{x^T f(t, x, u)}{\| x \|} \tag{2.7}$$

and from (2.2)

$$\frac{x^T f(t, x, u)}{\| x \|} \leq \Lambda(\| x \|) \leq \omega(\| x \|), \qquad x \in R(\rho + \delta) \tag{2.8}$$

so that (2.5) is satisfied. Now consider any solution $x(t)$ of (2.1) such that $x(t') \in R(\rho)$. Either we will have $\| x(t) \| < \rho$ for $t' \leq t < T$, or since $x(t)$ is continuous there will be a smallest value of t, say τ, $t' \leq \tau < T$, such that $\| x(t) \| < \rho$ for $t' \leq t < \tau$ and $\| x(\tau) \| = \rho$. Since $\omega(\alpha)$ satisfies a Lipschitz condition for $\delta \leq \alpha \leq \rho + \delta$ and $\omega(\rho) = 0$, the unique solution to (2.6) is $\alpha(t) = \rho$ for $\tau \leq t < T$. It follows from the lemma that $\| x(t) \| \leq \rho$ for $\tau \leq t < T$ and the system is ρ-stable.

To show that (2.4) is necessary assume it is not satisfied, and $\Lambda(\rho) > 0$. Then there must exist $(t', u') \in I \times \Omega$ and $\| x' \| = \rho$ such that

$x'^T f(t', x', u') > 0$. Then from (2.7), $(d \| x' \|)/dt > 0$, and the system is not ρ-stable. Q.E.D.

Theorem 1 will be illustrated with two examples, a linear system and a nonlinear system. The linear system is $\dot{x} = A(t)\, x$, $t \in I$, where $A(t)$ is an $(n \times n)$ matrix whose coefficients may depend on t. We have

$$x^T f(t, x) = x^T A(t) x = \tfrac{1}{2} x^T [A(t) + A^T(t)] x.$$

From (2.2) it follows that $\Lambda(\alpha) = \alpha \lambda_m$, where λ_m is the maximum eigenvalue of the symmetric matrix $\frac{1}{2}[A(t) + A^T(t)]$ for $t \in I$. Therefore this linear system is ρ-stable for any $\rho > 0$, if, and only if, $\lambda_m \leq 0$.

The nonlinear system is $\dot{x} = -x + xy \sin^2 \nu t$, $\dot{y} = -y + xy$, where x and y are scalars and $0 \leq \nu t < \infty$. We have

$$x^T f(t, x, \nu) = -x^2(1 - y \sin^2 \nu t) - y^2(1 - x).$$

It is readily shown that $\Lambda(\alpha) \leq 0$ for $0 \leq \alpha \leq \sqrt{2}$, $\Lambda(\sqrt{2}) = 0$, and $\Lambda(\alpha) > 0$ for $\alpha > \sqrt{2}$, the maximum being attained for $x = y = \alpha/\sqrt{2}$ and $\nu t = \pi/2$. It follows from Theorem 1 that this system is ρ-stable for any positive $\rho \leq \sqrt{2}$, and not ρ-stable for any $\rho > \sqrt{2}$.

The concept of ρ-stability is more general than might first appear, since two more general cases can be included by an appropriate change of variables. The first such case is that of remaining within a tube of radius ρ around a specified trajectory $p(t) \in R_n$, where $p(t)$ is differentiable for $t \in I$. All solutions of (2.1) which start in or enter this tube will remain in it if, and only if, the system

$$\frac{dy}{dt} = f(t, y + p(t), u) - \frac{dp(t)}{dt} \equiv \bar{f}(t, y, u), \qquad (t, u) \in I \times \Omega \tag{2.9}$$

is ρ-stable. In the second case we require that the radius of the tube be a specified function $\rho(t)$ for $t \in I$. In particular we let $\rho(t) = \rho \xi(t)$ where $\xi(t_0) = 1$, $\xi(t)$ differentiable and $\xi(t) \neq 0$ for $t \in I$. Solutions of (2.1) will remain in this time-dependent tube if, and only if, the system

$$\frac{dy}{dt} = \xi^{-1}(t) \left[f(t, \xi(t)\, y, u) - y \frac{d\xi(t)}{dt} \right] \equiv \bar{\bar{f}}(t, y, u), \qquad (t, u) \in I \times \Omega, \tag{2.10}$$

is ρ-stable. This second case permits the investigation, for example, of systems for which $\| x(t) \|$ approaches zero at a specified rate.

A final remark on ρ-stability is that similar results can be obtained for the norm $| x |$ if we let

$$\Lambda(\alpha) = \max_{t,x,u} \left\{ \sum_{i=1}^{n} f_i(t, x, u) \operatorname{sgn} x_i \,\middle|\, | x | = \alpha, (t, u) \in I \times \Omega \right\} \tag{2.11}$$

3. Controllable Stability

We now consider the control problem for the system (2.1). The quantities u_i are now control variables which may be adjusted as functions of time to achieve the desired controlled response. In particular we now consider $u_i = u_i(t)$, where each $u_i(t)$ is a piecewise continuous function such that the graph of $u(t)$ lies in the convex subset Ω. The assumptions concerning the f_i stated at the beginning of the previous section are sufficient to insure that there is a unique continuous solution $x(t)$ to this control problem.

The concept of a ρ-stable system discussed previously can be extended to the control problem by defining a *controllable ρ-stable* system in the following way:

Definition: The system (2.1) is controllable ρ-stable if there exists a control function $u^*(t, x)$ with its graph in Ω such that the system

$$\frac{dx}{dt} = f^*(t, x) \equiv f(t, x, u^*(t, x)) \tag{3.1}$$

is ρ-stable.

On the basis of this definition and the equation (2.7) it is clear that we wish to select that value of $u \in \Omega$, say $u^*(t, x)$, which will *minimize* $x^T f(t, x, u)$ for each value of (t, x) considered. If $x^T f(t, x, u^*(t, x)) \leq 0$ for all (t, x) considered, then the system is controllable ρ-stable. The mathematical programming problem pertinent to controllable stability is therefore a max min problem. We may state this more precisely as

Theorem 2. Let

$$\psi(\alpha) \equiv \max_{t,x} \min_{u} \{\alpha^{-1} x^T f(t, x, u) \mid \| x \| = \alpha, (t, u) \in I \times \Omega\} \tag{3.2}$$

and assume that the $f_i(t, x, u)$ satisfy the same conditions as in Theorem 1. Then a necessary and sufficient condition that the system (2.1) be controllable ρ-stable is $\psi(\rho) \leq 0$.

Proof. Let $u^*(t, x)$ be a value of $u \in \Omega$ which minimizes $x^T f(t, x, u)$ for each value of $t \in I$ and $\| x \| \in R(\rho + \delta)$. Let

$$f^*(t, x) = f(t, x, u^*(t, x))$$

and

$$\Lambda^*(\alpha) = \max_{t,x} \{\alpha^{-1} x^T f^*(t, x) \mid \| x \| = \alpha, t \in I\} \tag{3.3}$$

It follows from (3.2) that $\psi(\alpha) = \Lambda^*(\alpha)$ for $\delta \leq \alpha \leq \rho + \delta$. By theorem 1, the system (3.1) is ρ-stable if, and only if, $\Lambda^*(\rho) \leq 0$. Q.E.D.

We will now specialize to functions $f(t, x, u)$ which are linear in the control variables, so that $f(t, x, u) = g(t, x) + B(t, x) u$, where $B(t, x)$ is an $(n \times m)$ matrix with columns $b_j(t, x)$, $j = 1, \ldots m$. We also specialize Ω to a rectangular region in R_m containing the origin, and for convenience take $\Omega^* = \{u \mid -1 \leq u_i \leq 1, i = 1, \ldots m\}$. For each fixed value of (t, x) the problem $\min_u \{x^T g(t, x) + x^T B(t, x) u \mid u \in \Omega^*\}$ is a linear programming problem. The solution $u^*(t, x)$ therefore is a vertex of Ω^*, $u_i = \pm 1$, $i = 1, \ldots m$. In particular

$$u_j^*(t, x) = - \operatorname{sgn} [x^T b_j(t, x)], \qquad j = 1, \ldots m \tag{3.4}$$

It follows that the desired function $\psi(\alpha)$ defined by (3.2) is given by the solution to the mathematical programming (maximization) problem

$$\psi(\alpha) = \max_{t,x} \{\alpha^{-1}[x^T g(t, x) - | x^T B(t, x) |] \mid \| x \| = \alpha, t \in I\} \tag{3.5}$$

By introducing two non-negative slack vectors y and z, we can eliminate the absolute value term in the objective function, and get the following equivalent maximization problem.

$$\psi(\alpha) = \max_{t,x,y,z} \left\{ \alpha^{-1}\Big[x^T g(t, x) - (y + z)\Big] \,\Big|\, \begin{matrix} \|x\| = \alpha, t \in I, y \geq 0, z \geq 0 \\ y \geq x^T B(t, x) \geq - z \end{matrix} \right\} \tag{3.6}$$

If $g(t, x)$ and $B(t, x)$ are differentiable, this problem can be solved by an available method [12].

As an example of a controllable ρ-stable system we again consider the nonlinear system discussed in Section 2, with a single control variable $-1 \leq u(t) \leq 1$. We now have

$$\dot{x} = -x + xy \sin^2 \nu t + u(t), \qquad \dot{y} = -y + xy + u(t),$$

so that

$$x^T g(t, x) = -x^2(1 - y \sin^2 \nu t) - y^2(1 - x), \qquad \text{and} \qquad x^T B u = (x + y)u.$$

From (3.5) we obtain

$$\alpha\psi(\alpha) = \max \{x^T g - | x + y | \mid \| x \| = \alpha, \nu t \geq 0\}.$$

As previously, it can be shown that a maximum is attained for $x = y = \alpha/\sqrt{2}$ and $\nu t = \pi/2$. However, with the control we now obtain $\psi(\alpha) \leq 0$ for $0 \leq \alpha \leq \frac{1}{2}(\sqrt{2} + \sqrt{10}) \cong 2.29$ and $\psi(\alpha) > 0$ for $\alpha > \frac{1}{2}(\sqrt{2} + \sqrt{10})$. The control has therefore increased the maximum permissible ρ for ρ-stability from $\rho = \sqrt{2}$ to $\rho = \frac{1}{2}(\sqrt{2} + \sqrt{10})$.

4. Linear Controllable System

In this section we consider the completely linear controllable system with allowable control vectors $u(t)$ in the hypercube Ω^*. The system will be written as

$$\frac{dx}{dt} = Ax + Bu, \qquad u \in \Omega^* \tag{4.1}$$

where A is an $(n \times n)$ and B is an $(n \times m)$ matrix. For simplicity we assume that A and B are constant matrices. Similar results can be obtained when A and B depend on time, by requiring that $A(t)$ and $B(t)$ have the required properties for all $t \in I$.

For this linear system we have $g(t, x) = Ax$ and $B(t, x) = B$, so that (3.5) gives

$$\begin{aligned}\psi(\alpha) &= \max_x \{\alpha^{-1}[x^T Ax - | B^T x |] \mid \| x \| = \alpha\} \\ &= \max_x \left\{\frac{\alpha}{2} x^T(A + A^T)x - | B^T x | \mid \| x \| = 1\right\}\end{aligned} \tag{4.2}$$

In the second expression we have used the identity

$$x^T Ax \equiv \tfrac{1}{2} x^T(A + A^T)x$$

to put the quadratic term in symmetric form. If the rows of B are linearly independent there will be no vector $x \neq 0$, such that $| B^T x | = 0$. This can only be the case if $m = n$ and the matrix B is nonsingular. We wish to make the following less restrictive requirement on the matrix B.

Definition. The matrix B is a control matrix for A if there exists no vector x, such that the conditions

$$x^T(A + A^T)x > 0 \qquad \text{and} \qquad | B^T x | = 0 \tag{4.3}$$

hold simultaneously.

We can now state the main result for the linear system (4.1) as

Theorem 3. There exists a $\rho > 0$, such that the system (4.1) is controllable ρ-stable if, and only if, B is a control matrix for A.

Proof. We let $\lambda(x) \equiv \frac{1}{2} x^T(A + A^T)\, x$ and $\varphi(\alpha, x) \equiv \alpha\lambda(x) - | B^T x |$, and first consider the case when $(A + A^T)$ is negative semidefinite. We then have $\lambda(x) \leq 0$ for all x, so that by definition every matrix B is a control matrix for A. For all positive α, $\varphi(\alpha, x) \leq 0$. It follows from (4.2) that $\psi(\rho) \leq 0$ and (4.1) is ρ-stable for all $\rho > 0$. This is of course the case which is ρ-stable without control (when B is the null matrix). We now consider the case when $(A + A^T)$ has at least one positive eigenvalue. For this case there exist vectors $x \in P \equiv \{x \mid \| x \| = 1, \lambda(x) > 0\}$, and we can define the quantity

$$\alpha_m \equiv \min_x \left\{ \frac{| B^T x |}{\lambda(x)} \,\middle|\, x \in P \right\} \tag{4.4}$$

If B is a control matrix for A, the conditions (4.3) cannot be satisfied and it follows that $\alpha_m > 0$. For $x \notin P$ we have $\varphi(\alpha, x) \leq 0$ for $\alpha \geq 0$. For $x \in P$ we have $\varphi(\alpha, x) = \lambda(x) [\alpha - \lambda^{-1}(x) \mid B^T x \mid]$. It follows from (4.4) that $\varphi(\alpha, x) \leq 0$ for $0 \leq \alpha \leq \alpha_m$ and all $\| x \| = 1$, so that $\psi(\alpha) \leq 0$ for $0 \leq \alpha \leq \alpha_m$, by (4.2). We can now apply Theorem 2 to the system (4.1) and conclude that the system (4.1) is controllable ρ-stable for every ρ, $0 < \rho \leq \alpha_m$, if B is a control matrix for A.

To prove necessity we assume that B is not a control matrix for A. Then there exists a vector $x = x'$ such that $\| x' \| = 1$, $\lambda(x') > 0$, and $| B^T x' | = 0$. Then $\psi(\alpha) = \alpha\lambda(x') > 0$ for every $\alpha \geq 0$, and the system (4.1) is not ρ-stable. Q.E.D.

For any specified matrices A and B the corresponding value of α_m is given by solving the nonlinear programming problem (4.4). If $\alpha_m > 0$, then B is a control matrix for A and the linear system is controllable ρ-stable for $0 < \rho \leq \alpha_m$. If $m = n$ and B is nonsingular, then B is a control matrix for every matrix A. This follows from the fact that for nonsingular B we cannot have $| B^T x | = 0$ except when $x = 0$. On the other hand as mentioned above, if $(A + A^T)$ is negative semidefinite, every matrix B (including the null matrix) is a control matrix for A.

Certain additional special results can be obtained by considering the (real) eigenvalues λ_i, and corresponding orthogonal eigenvectors q_i of the symmetric matrix $\frac{1}{2}(A + A^T)$, $i = 1, \dots n$. We assume that at least one eigenvalue is positive, and without loss of generality that $\lambda_i > 0$, $i = 1, \dots r$, and $\lambda_i \leq 0$, $i = r + 1, \dots n$. We have $\lambda_i = \lambda(q_i)$, $i = 1, \dots n$. Then an upper bound which follows directly from (4.4) is given by

$$\alpha_m \leq \min \left\{ \frac{| B^T q_i |}{\lambda_i} \,\middle|\, i = 1, \dots r \right\} \tag{4.5}$$

The minimum number of control variables $u_i(t)$, $i = 1, \dots m$, which will permit stable control is given by the following

Corollary. In order that the system (4.1) be controllable ρ-stable it is necessary that $m \geq r$, and that the r vectors $B^T q_i$, $i = 1, \dots r$, be linearly independent.

Proof: Assume the vectors $B^T q_i$, $i = 1, \ldots r$ are linearly dependent. Then there are coefficients β_i, not all zero, such that

$$\sum_{i=1}^{r} \beta_i B^T q_i = 0.$$

If we take

$$x = \sum_{i=1}^{r} \beta_i q_i,$$

we have

$$B^T x = 0 \qquad \text{and} \qquad \lambda(x) = \sum_{i=1}^{r} \lambda_i \beta_i^2 > 0.$$

It follows from (4.3) that B is not a control matrix for A and therefore the system is not controllable ρ-stable. It is also clear that since the vectors $B^T q_i$ have m components we must have $m \geq r$ in order that they be linearly independent.

REFERENCES

[1] Antosiewicz, H. A., A survey of Liapunov's second method. *In* "Contributions to the Theory of Nonlinear Oscillations" (S. Lefschetz, ed.), Vol. 4, pp. 141-166. Princeton Univ. Press, Princeton, New Jersey, 1958.

[2] Bellman, R., "Stability Theory of Differential Equations," Chapters 2, 3, and 4. McGraw-Hill, New York, 1953.

[3] Brauer, F., Global behavior of solutions of ordinary differential equations. *J. math. Anal. and Appl.* **2**, 145-158 (1961).

[4] Coddington, E. A., and Levinson, N., "Theory of Ordinary Differential Equations." Chapters 2 and 13. McGraw-Hill, New York, 1955.

[5] Conti, R., Sulla prolungabilita delle soluzioni di un sistema di equazioni differenziali ordinarie. *Boll. Unione mat. ital.* **11**, 510-514 (1956).

[6] Krasovskii, N. N., Theory of optimal control. *Priklad. Mat. i Mekh.* **23**, No. 4 (1959).

[7] Kuhn, H. W., and Tucker, A. W., Nonlinear programming, *Proc. 2nd Berkeley Symposium on math. Statistics and Probability*, pp. 481-492. Univ. California Press, Berkeley, California, 1950.

[8] La Salle, J. P., The extent of asymptotic stability. *Proc. natl. Acad. Sci. U.S.* **46**, 363-365 (1960).

[9] Markus, L., and Lee, E. B., On the Existence of Optimal Controls. ASME Paper No. 61-JAC2 (1961).

[10] Pontryagin, L. S., Optimal control processes. *Uspekhi mat. Nauk* **14**, No. 1, 13-20 (1959).

[11] Rosen, J. B., Stability of differential equations and equivalent nonlinear programming problem, Abstract 60T-13. *Notices AMS* **7**, 996 (1960).

[12] Rosen, J. B., The gradient projection method for nonlinear programming, Part I. Linear constraints. *J. Soc. ind. appl. Math* **8**, 181-217 (1960); Part II. Nonlinear constraints. *Ibid.* **9**, 514-532 (1961).

Periodic Solutions of Singular Perturbation Problems[1]

E. R. RANG
Minneapolis-Honeywell Regulator Company
Minneapolis, Minnesota

We establish conditions which insure the existence and uniqueness of a periodic solution of the ordinary differential equations

$$\frac{dx}{dt} = f(x, y, t, \epsilon)$$
$$\Omega \frac{dy}{dt} = g(x, y, t, \epsilon), \qquad \Omega = \operatorname{diag}\{\epsilon^{h_1}, ..., \epsilon^{h_N}\}, \tag{1}$$

when the reduced system

$$\frac{d\varphi}{dt} = f(\varphi, \psi, t, 0)$$
$$0 = g(\varphi, \psi, t, 0) \tag{2}$$

has a known periodic solution $\varphi(t)$, $\psi(t)$ and the real non-negative parameter ϵ is sufficiently small. We consider x, f as vectors with M-components, y, g are N-component vectors, and the exponents $h_1, ..., h_N$ in the diagonal matrix Ω are positive integers.

The functions f, g are assumed to be real analytic functions of the components of the dependent vectors x, y, and the parameter ϵ, and continuous in the independent variable t for all t, ϵ in an interval of the real ϵ-axis containing the origin, and x, y in a suitable neighborhood of the periodic solution of the reduced equation. We further assume that the reduced periodic solution $\varphi(t)$, $\psi(t)$ is continuous in t and has continuous first derivatives.

In this discussion we consider the non-autonomous case in which f, g depend explicitly on t and are periodic in t with a real period T^0 which is independent of ϵ and the other arguments. Then, assuming that the period of the periodic solution of the reduced system of equations is this T^0, we look for a periodic solution of period T^0 of the complete system (1) which tends to the given reduced solution as ϵ tends to zero through

[1] This discussion is an outline of the results contained in the thesis [*3*] done under the direction of Professor H. L. Turrittin.

positive values. The autonomous case can be studied after making the standard modifications. These details are given in ref. [*3*].

The periodic solution of system (1) is constructed as a convergent series using a formal procedure originally employed by I. M. Volk [*4-6*] and shown to be convergent in the case of a single *n*th order equation by W. R. Wasow [*7*]. Professor Wasow makes a skillful application of the theory of asymptotic expansions for linear differential equations to arrive at certain lemmas which are then used to show that the formal series solution is dominated by a series solution of a particular set of implicit equations. This solution exists and is convergent by virtue of the implicit function theorem. We adapt Wasow's convergence proof but establish the analogous lemmas by direct calculations. In this analysis we use several devices given by Flatto and Levinson [*1*] in their study of this problem. In that work all of the exponents h_j are taken equal and since an iterative type existence proof is used they require only continuity of the functions f and g and their derivatives.

A Preliminary Transformation

Following Flatto and Levinson we make the change to new variables ξ, η by the equations

$$\begin{aligned} x &= \xi + \varphi(t) \\ y &= \eta + \psi(t) - g_y^{-1}(t, \epsilon)\, g_x(t, \epsilon)\xi, \end{aligned} \tag{3}$$

where $g_x(t, \epsilon)$ and $g_y(t, \epsilon)$ are the matrices of partial derivatives with respect to components of the indicated vectors evaluated along the given reduced periodic solution. For example

$$g_y(t, \epsilon) = \left(\frac{\partial g_i(\varphi(t), \psi(t), t, \epsilon)}{\partial y_j}\right).$$

We therefore make the first of four basic hypotheses. It is:

Assumption A. Assume that the matrix $g_y(t, \epsilon)$ is non-singular for some interval on ϵ including the origin and all t. With this change of variables the system (1) becomes

$$\begin{aligned} \frac{d\xi}{dt} &= B(t, \epsilon)\xi + f_y(t, \epsilon)\eta + \epsilon\Delta f(t, \epsilon) + F(\xi, \eta, t, \epsilon) \\ \Omega \frac{d\eta}{dt} &= g_y(t, \epsilon)\eta + \epsilon\Delta g(t, \epsilon) + G(\xi, \eta, t, \epsilon) + \Omega H(\xi, \eta, t, \epsilon), \end{aligned} \tag{4}$$

where we have separated the linear terms from the higher powers in the components of ξ, η. In detail

$$B(t, \epsilon) = f_x(t, \epsilon) - f_y(t, \epsilon)\, g_y^{-1}(t, \epsilon)\, g_x(t, \epsilon)$$

and $f_x(t, \epsilon)$, $f_y(t, \epsilon)$ are abbreviations similar to $g_x(t, \epsilon)$;

$$\Delta f(t, \epsilon) = \frac{1}{\epsilon}\,[f(\varphi(t), \psi(t), t, \epsilon) - f(\varphi(t), \psi(t), t, 0)], \text{ etc.,}$$

and we presume the interpretation of the other notation is evident.

The Formal Procedure

We look for a series expansion of the solution of the form

$$\begin{aligned} \xi(t, \epsilon) &= \sum_{j=1}^{\infty} \epsilon^j p_j(t, \epsilon) \\ \eta(t, \epsilon) &= \sum_{j=1}^{\infty} \epsilon^j q_j(t, \epsilon). \end{aligned} \tag{5}$$

Following the procedure originated by Volk, we formally substitute these series into the equations (4), expand in powers of ϵ and then equate the coefficients of like powers of ϵ on both sides of the equation but we preserve the factor

$$\Omega = \text{diag}\,\{\epsilon^{h_1}, \ldots, \epsilon^{h_N}\}$$

on the left hand side of the second equation and also the matrix $g_y(t, \epsilon)$ on the right hand side.

This is equivalent to writing

$$\begin{aligned} \xi_k(t, \sigma, \epsilon) &= \sum_{j=0}^{\infty} \sigma^j p_j(t, \epsilon) \\ \eta_k(t, \sigma, \epsilon) &= \sum_{j=0}^{\infty} \sigma^j q_j(t, \epsilon), \end{aligned} \tag{6}$$

where $p_j(t, \epsilon)$, $q_j(t, \epsilon)$ are found from the iterations

$$p_0(t, \epsilon) = 0, \qquad q_0(t, \epsilon) = 0, \tag{7}$$

$$\frac{dp_r}{dt} = B(t, 0)\, p_r + f_y(t, 0)\, q_r$$

$$+ \frac{1}{r!} \frac{d^r}{d\sigma^r} [B(t, \sigma)\, \xi_{r-1} + f_y(t, \sigma)\, \eta_{r-1} + \sigma \Delta f(t, \sigma) + F(\xi_{r-1}, \eta_{r-1}, t, \sigma)] \,|_{\sigma=0}$$

$$\Omega \frac{dq_r}{dt} = g_y(t, \epsilon)\, q_r + \frac{1}{r!} \frac{d^r}{d\sigma^r}$$

$$\times [\sigma \Delta g(t, \sigma) + G(\xi_{r-1}, \eta_{r-1}, t, \sigma) + (\delta_{ij} \sigma^{h_j})\, H(\xi_{r-1}, \eta_{r-1}, t, \sigma)] \,|_{\sigma=0}, \tag{8}$$

for $r = 1, 2, 3, \ldots$, and we choose for p_r, q_r those solutions which are periodic of period T^0. We note that the iteration progresses, once p_0, p_1, $\ldots$, p_{r-1}; q_0, q_1, $\ldots$, q_{r-1} have been found, by calculating q_r from the second equation of (8) and substituting the result into the first so that p_r can be computed.

In order to facilitate this iteration, we make the following:

Assumption B. The equation

$$\frac{dz}{dt} = B(t, 0) z,$$

which is the "variational equation" with respect to the solution $\varphi(t)$, $\psi(t)$, has no nontrivial periodic solution of period T^0.

Then if q_r, ξ_{r-1}, η_{r-1} have been found, assumption B assures that the equation for p_r will have a unique real periodic solution of period T^0. That the second equation has such a solution turns, out to be a consequence of certain hypotheses on the matrix

$$(\delta_{ij} \epsilon^{h-h_j})\, g_y(t, \epsilon), \qquad h = \max h_j$$

which we will make.

Hence we expect that the solution of our problem will be given by

$$\xi(t, \epsilon) = \lim_{k \to \infty} \xi_k(t, \epsilon, \epsilon)$$

$$\eta(t, \epsilon) = \lim_{k \to \infty} \eta_k(t, \epsilon, \epsilon).$$

In fact, with assuptions C, D yet to be explained in the following paragraph, we state our conclusion as the

Theorem. If in addition to the analyticity and continuity requirements, assumptions A, B, C, D are satisfied, there is an $\epsilon_0 > 0$, such that if $0 \leqq \epsilon < \epsilon_0$, the equations (1) have a unique periodic solution of period T^0 with the representation

$$x(t, \epsilon) = \varphi(t) + \sum_{j=1}^{\infty} \epsilon^j p_j(t, \epsilon)$$
$$y(t, \epsilon) = \psi(t) + \sum_{j=1}^{\infty} \epsilon^j [q_j(t, \epsilon) - g_y^{-1}(t, \epsilon)\, g_x(t, \epsilon)\, p_j(t, \epsilon)]. \tag{9}$$

The series converge uniformly with respect to t and as ϵ tends to zero through positive values, this solution tends uniformly to $\varphi(t)$, $\psi(t)$, the solution of the reduced system (2). Further, each $q_j(t, \epsilon)$ will have the form

$$q_j(t, \epsilon) = -\, g_y^{-1}(t, \epsilon)\, \omega_j(t, \epsilon) + h_j(t, \epsilon),$$

where $\omega_j(t, \epsilon)$ is a certain nonhomogeneous term computed successively and $h_j(t, \epsilon)$ is such that norm $\{h_j(t, \epsilon)\} \to 0$ uniformly as $\epsilon \to 0^+$.

A Formula for Periodic Solutions

We included here a formula for periodic solutions of a nonhomogeneous linear system with periodic coefficients which was found helpful in the analysis. Consider the vector equation

$$\frac{dz}{dt} = P(t)z + r(t), \tag{10}$$

when $P(t)$, $r(t)$ are real and periodic with period T^0 and continuous for all real t. Let $\Phi(t)$ be a fundamental matrix solution of the homogeneous system

$$\frac{dz}{dt} = P(t)z. \tag{11}$$

Then by direct calculation, one finds that equation (10) has a unique periodic solution of period T^0 which can be written in the form

$$z(t) = [\Phi^{-1}(t + T^0) - \Phi^{-1}(t)]^{-1} \int_t^{t+T^0} \Phi^{-1}(\tau)\, r(\tau)\, d\tau, \tag{12}$$

providing that the matrix $[\Phi(t + T^0) - \Phi(t)]$ is nonsingular, that is, the homogeneous equation (11) has no nontrivial periodic solution of period T^0.

A Canonical Form

In order to analyze the periodic solutions of the second equation in system (8) which is of the form

$$\epsilon^h \frac{dq}{dt} = A(t, \epsilon)q + \bar{\Omega}w(t, \epsilon), \tag{13}$$

where

$$A(t, \epsilon) = (\delta_{ij}\epsilon^{h-h_j})\, g_y(t, \epsilon), \qquad h = \max h_j$$

$$\bar{\Omega} = (\delta_{ij}\epsilon^{h-h_j}),$$

we develope a canonical transformation $P(t, \epsilon)$ so that

$$P^{-1}(t, \epsilon)\, A(t, \epsilon)\, P(t, \epsilon) = D(t, \epsilon) + \epsilon^h R(t, \epsilon) \tag{14}$$

and the matrix $D(t, \epsilon)$ has a certain block diagonal form

$$\left\| \begin{array}{ccccc} \epsilon^{\nu_n} D_n^+(t, \epsilon) & 0 & \cdots & 0 & 0 \\ 0 & \epsilon^{\nu_n} D_n^-(t, \epsilon) & \cdots & 0 & 0 \\ & & \cdots & & \\ \cdots & \cdots & \cdots & \cdots & \cdots \\ & & \cdots & & \\ 0 & 0 & \cdots & \epsilon^{\nu_1} D_1^+(t, \epsilon) & 0 \\ & & \cdots & & \\ 0 & 0 & \cdots & 0 & \epsilon^{\nu_1} D_1^-(t, \epsilon) \end{array} \right\|,$$

where

$$D_j^{\pm}(t, \epsilon) = D_{0j}^{\pm}(t) + \epsilon D_{1j}^{\pm}(t) + \dots + \epsilon^{k_j} D_{k_j j}^{\pm}(t),$$

and the matrices $D_{0j}^{+}(t)$ have characteristic roots with positive real parts, $D_{0j}^{-}(t)$ have all characteristic roots with negative real parts. That such a canonical form exists and can be found by a sequence of transformations of the type $I + \epsilon^k Q_k(t)$ is a consequence of hypotheses C and D. The exact statements of the hypotheses and the canonical development are lengthy; for these details we refer to ref. [3].

We may utilize this transformation by applying an ingenious artifice of Flatto and Levinson. Evaluate matrix $A(t, \epsilon)$ at any fixed pont $t = \sigma$ and introduce it into equation (13) by adding and subtracting:

$$\epsilon^h \frac{dq}{dt} = A(\sigma, \epsilon)q + \{[A(t, \epsilon) - A(\sigma, \epsilon)]q + \bar{\Omega}\omega(t, \epsilon)\}. \tag{15}$$

Then with the substitution

$$q(t, \epsilon) = P(\sigma, \epsilon)\, r(t, \sigma, \epsilon),$$

where $P(\sigma, \epsilon)$ is the transformation matrix from (14), the resulting equation for $r(t, \sigma, \epsilon)$ has the form

$$\epsilon^h \frac{dr}{dt} = D(\sigma, \epsilon)r + p(t, \sigma, \epsilon) \tag{16}$$

and since the homogeneous equation has the matrix

$$\Phi(t) = \exp \{\epsilon^{-h} D(\sigma, \epsilon)t\}$$

as a fundamental solution, we may apply formula (12) and find the representation

$$q(t, \epsilon) = \epsilon^{-h} P(\sigma, \epsilon) \{\exp [- \epsilon^{-h} D(\sigma, \epsilon) T^0] - I\}^{-1}$$
$$\times \int_t^{t+T^0} \exp [- \epsilon^{-h} D(\sigma, \epsilon) (\tau - t)] \, p(\tau, \sigma, \epsilon) \, d\tau \tag{17}$$

for the periodic solution. But this formula is true for any σ and since equation (13) is independent of σ we may indentify σ with t. Since $D(t, \epsilon)$ is in the block diagonal form, formula (17) with $\sigma = t$ can be further analyzed and we arrive at lemmas analogous to those given by Wasow. In these estimates a lemma stated by Levin and Levinson [*2*] is used. Again, for details, we refer to ref. [*3*].

The convergence proof and validation of the solution are then completed by the construction of dominating series.

REFERENCES

[*1*] Flatto, L., and Levinson, N., Periodic solutions of singularly perturbed systems. *J. rational Mech. and Anal.* **4**, 943-950 (1955).

[*2*] Levin, J. J., and Levinson, N., Singular perturbations of nonlinear systems of differential equations and associated boundary layer equations. *J. rational Mech. and Anal.* **3**, 247-270 (1954).

[*3*] Rang, E. R., Ph.D. Thesis, University of Minnesota (1957).

[*4*] Volk, I. M., On the periodic solution of non-autonomous systems depending on a small parameter. *Priklad. Mat. i Mekh.* **10**, 559-574 (1946).

[*5*] Volk, I. M., Generalization of the method of a small parameter in the theory of periodic motions of non-autonomous systems. *Priklad. Mat. i Mekh.* **11**, 431-444 (1947).

[*6*] Volk, I. M., On periodic solutions of autonomous systems. *Priklad. Mat. i Mekh.* **12**, 29-38 (1948).

[*7*] Wasow, W. R., On the construction of periodic solutions of singular perturbation problem. *In* "Contributions to the Theory of Nonlinear Oscillations" (S. Lefschetz, ed.), Vol. 1, pp. 313-350. Princeton Univ. Press, Princeton, New Jersey, 1950.

Ultimately Periodic Behavior in a Class of Non-Linear Servomechanisms

V. E. BENEŠ
Bell Telephone Laboratories, Incorporated
Murray Hill, New Jersey

1. Introduction

Let $s(\cdot)$ be a given signal, $F(\cdot)$ a non-linear function, and $k(\cdot)$ the impulse response of a linear system. The equation

$$x(t) = s(t) - \int_0^\infty F(x(t-u))\, k(u)\, du \tag{1}$$

describes a general class of servomechanisms. This class can be studied by finding a fixed point (corresponding to a solution $x(\cdot)$ of the equation) in a specific set of a function space. In this way many properties of the class can be established without either specifying $s(\cdot)$, $F(\cdot)$, and $k(\cdot)$ in detail, or solving the equation explicitly. Since the function space and the set in which the fixed point is found can be chosen in many ways, the method described above can of course be used for a wide variety of problems. The choice to be made depends on the equation to be treated and on the properties of the solution that are of interest. For this reason the present paper is restricted to sample results on equation (1) that illustrate the argument for Tychonov's fixed point theorem. A similar study of solutions of (1) having finite energy appears in the author's paper [*1*].

The equation (1) corresponds to the system diagram of Fig. 1.

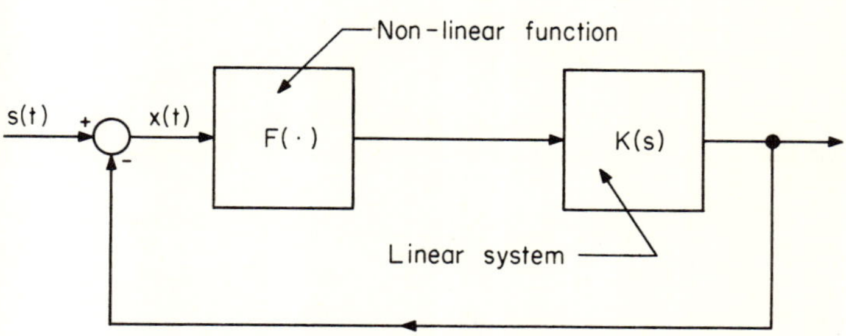

FIG. 1. System Diagram of Servomechanism

We shall assume that $F(0) = 0$, and that $F(\cdot)$ is bounded and satisfies the uniform Lipschitz condition

$$| F(x) - F(y) | \leq \beta | x - y |.$$

Also we assume that the function $k(\cdot)$ is the impulse response of a physically realizable network, vanishes for negative argument, and is absolutely integrable.

The equation (1) can be placed into the context of non-linear differential equations by assuming that $k(\cdot)$ has a rational transform $K(s)$,

$$K(s) = \frac{B(s)}{A(s)},$$

$$B(s) = \sum_{j=0}^{m} b_j s^j,$$

$$A(s) = \sum_{j=0}^{n} a_j s^j,$$

and proceeding quite formally with the aid of the Laplace transform $\mathscr{L}$. Let us introduce the values at 0,

$$x_j = x^{(j)}(0), \qquad j = 0, 1, ..., n$$

$$F_j = \{F(x)\}^{(j)}(0), \qquad j = 0, 1, ..., m$$

$$S_j = s^{(j)}(0), \qquad j = 0, 1, ..., n,$$

and set

$$p_j(s) = x_0 s^j + x_1 s^{j-1} + ... + x_j, \qquad j = 0, 1, ..., n-1$$

$$P(s) = a_n p_{n-1}(s) + a_{n-1} p_{n-2}(s) + ... + a_1 p_0(s)$$

$$q_j(s) = F_0 s^j + F_1 s^{j-1} + ... + F_j, \qquad j = 0, 1, ..., m-1$$

$$Q(s) = b_m q_{m-1}(s) + b_{m-1} q_{m-2}(s) + ... + b_1 q_0(s)$$

$$r_j(s) = S_0 s^j + S_1 s^{j-1} + ... + S_j, \qquad j = 0, 1, ..., n$$

$$R(s) = a_n r_{n-1}(s) + a_{n-1} r_{n-2}(s) + ... + a_1 r_0(s).$$

Then [*3*, p. 482]

$$\mathscr{L}\sum_{j=0}^{n} a_j x^{(j)}(t) = A(s)\cdot\mathscr{L}x(t) - P(s)$$

$$\mathscr{L}\sum_{j=0}^{m} b_j\{F(x)\}^{(j)}(t) = B(s)\cdot\mathscr{L}F(x(t)) - Q(s)$$

$$\mathscr{L}\sum_{j=0}^{n} a_j s^{(j)}(t) = A(s)\cdot\mathscr{L}s(t) - R(s)$$

The equation (1), restricted to $t \geq 0$, is then equivalent to the transform relationship

$$\mathscr{L}\left(\sum_{j=0}^{n} a_j x^{(j)}(t) + \sum_{j=0}^{m} b_j\{F(x)\}^{(j)}(t)\right) = \mathscr{L}\sum_{j=0}^{n} a_j s^{(j)}(t) + P(s) + Q(s) - R(s)$$

and so to the differential equation

$$\sum_{j=0}^{n} a_j(x - s)^{(j)}(t) + \sum_{j=0}^{m} b_j\{F(x)\}^{(j)}(t) = f(t),$$

where

$$f(t) = \mathscr{L}^{-1}\{P(s) + Q(s) - R(s)\}$$

is a linear combination of derivatives of delta-functions. It is thus apparent that the integral equation form of the problem is almost misleadingly simple, because it hides in the Green's function the complexity of the operator under study. In general, the integral equation form permits the deduction of general results more easily than the differential equation, while the latter is better suited to obtaining specific results.

2. Ultimately Periodic Functions

The (restricted, illustrative) topic we shall consider concerns the asymptotic behavior of a solution $x(\cdot)$ of (1). In many applications, the asymptotic behavior of the solution $x(\cdot)$ is of interest when that of $s(\cdot)$ has been specified. In particular, we shall concern ourselves with given signals $s(\cdot)$ and solutions $x(\cdot)$ that are ultimately periodic according to the following

Definition: A function $x(\cdot)$ is ultimately periodic with period $\tau > 0$ if there is a periodic function $p_x(\cdot)$ of period τ such that

$$x(n\tau + t) - p_x(t) \to 0 \qquad \text{as} \qquad n \to \infty \tag{2}$$

for $0 \leq t \leq \tau$. In this case we write that $x(\cdot)$ is u.p. $[\tau]$.

If $f(\cdot)$ is a continuous function, and $x(\cdot)$ is u.p. $[\tau]$, then so is $f(x(\cdot))$. If $y(\cdot)$ is an integrable function (i.e., L_1) and $x(\cdot)$ is bounded and u.p. $[\tau]$, then the convolution

$$\int_{-\infty}^{\infty} x(t-u)\, y(u)\, du$$

is also u.p. $[\tau]$, provided that the convergence in (2) is uniform on $[0, \tau]$. For then

$$\int_{-\infty}^{\infty} [x(n\tau + t - u) - p_x(t-u)]\, y(u)\, du \to 0 \tag{3}$$

as $n \to \infty$ for $0 \leq t \leq \tau$, by a simple Abelian argument.

The principal topic of this paper will be a sufficient condition under which if the input signal $s(\cdot)$ is u.p. $[\tau]$, then so is at least one solution $x(\cdot)$ of (1). The uniqueness of a solution of (1) will not concern us, since it can be investigated by standard successive approximation arguments.

3. Preliminaries

A convenient topology for studying ultimately periodic behavior is given by the following metric on the space $B \cap C$ of bounded continuous functions:

$$m(f, g) = \sum_{m=1}^{\infty} 2^{-m} \sup_{-m \leq u \leq m} |f(u) - g(u)| + \sup_{0 \leq t \leq \tau} \limsup_{n \to \infty} |f(n\tau + t) - g(n\tau + t)|$$

The sum term defines a metric for the topology of uniform convergence on compact sets, and the other term constrains "nearby" functions to be alike at infinity, if they are u.p. $[\tau]$. The number $\tau > 0$ is a parameter, the *period* of interest.

We shall assume that $s(\cdot)$ is uniformly continuous with modulus of continuity $\sigma(\cdot)$, and we define a new modulus by

$$\mu(\epsilon) = \sigma(\epsilon) + \sup_u |F(u)| \cdot \sup_{|\delta| \leq |\epsilon|} \int_{-\infty}^{\infty} |k(t+\delta) - k(t)|\, dt$$

$$\mu(\epsilon) = \mu(|\epsilon|)$$

We shall assume that the "forcing function" $s(\cdot)$ is u.p. $[\tau]$, approaching the continuous periodic function $p_s(\cdot)$ in such a way that the sequence

$$s_n = \sup_{0<t<\tau} |s(n\tau + t) - p_s(t)|, \qquad n \geq 0 \tag{4}$$

is absolutely summable. Then $s(\cdot)$ is also bounded, and we define

$$\sigma = \sup_u | s(u) | + \sup_u | F(u) | \cdot \int_{-\infty}^{\infty} | k(u) | \, du.$$

A condition on $k(\cdot)$, stronger than integrability, will also be used: this is that

$$k_n = \sup_{0<t<\tau} \int_{n\tau}^{(n+1)\tau} | k(t+u) | \, du, \qquad n \geq 0, \tag{5}$$

be an absolutely summable sequence. Finally we shall assume that there is a positive absolutely summable sequence $\{h_n, n \geq 0\}$ that satisfies the integral inequality

$$s_n + \beta \sum_{j=0}^{n} h_{n-j} k_j \leq h_n, \qquad n \geq 0. \tag{6}$$

A simple condition that ensures the existence of an absolutely summable, positive sequence $\{h_n, n \geq 0\}$ satisfying the integral inequality (6) is given by:

Lemma 1. If

$$\sum_{j=0}^{\infty} k_j < \beta^{-1},$$

then the function

$$\sum_{j=0}^{\infty} z^n s_n \Big/ \left(1 - \beta \sum_{j=0}^{\infty} z^n k_n\right), \qquad | z | \leq 1$$

is the generating function of a positive, absolutely summable sequence $\{h_n, n \geq 0\}$ satisfying (6) with equality.

Proof. This can be seen from a direct geometric series solution of (6) with equality.

Let S be the subset of functions $x(\cdot)$ from $B \cap C$ with the properties:

(i) There exists a continuous function $p_x(\cdot)$ of period ω such that

$$| x(n\tau + t) - p_x(t) | \leq h_n, \qquad \text{for} \qquad t \in [0, \tau], \qquad n \geq 0,$$

(ii) $| x(\cdot) | \leq \sigma$.

(iii) $x(\cdot)$ has modulus of continuity $\mu(\cdot)$.

Lemma 2. S is compact (in m-topology).

Proof. Let $x_m \in S$ be a sequence converging to x. Clearly $|x| \leq \sigma$. The second term of the definition of the metric $m(\cdot, \cdot)$ ensures that

$$p_{x_m}(t)$$

converges as $m \to \infty$ uniformly for $t \in [0, \tau]$. Denote the continuous limit function by $p_x(\cdot)$. Then

$$\begin{aligned} | x(n\tau + t) - p_x(t) | &\leq | x(n\tau + t) - x_m(n\tau + t) | \\ &\quad + | x_m(n\tau + t) - p_{x_m}(t) | + | p_{x_m}(t) - p_x(t) |. \end{aligned}$$

With n and $t \in [0, \tau]$ fixed, let $m \to \infty$; the first and third terms on the right go to zero; the second is at most equal to h_n, uniformly in m. It follows that

$$| x(n\tau + t) - p_x(t) | \leq h_n.$$

Finally,

$$\begin{aligned} | x(t + \epsilon) - x(t) | &\leq | x(t + \epsilon) - x_m(t + \epsilon) | \\ &\quad + | x(t) - x_m(t) | + | x_m(t + \epsilon) - x_m(t) |. \end{aligned}$$

Letting $m \to \infty$ for fixed t, the first two terms on the right vanish, and the last is at most $\mu(\epsilon)$. Thus $x(\cdot)$ has modulus of continuity $\mu(\cdot)$, and so belongs to S. Hense S is closed. Also, for $y \in S$,

$$\begin{aligned} | p_y(t + \epsilon) - p_y(t) | &\leq | p_y(t + \epsilon) - y(n\tau + t + \epsilon) | \\ &\quad + | p_y(t) - y(n\tau + t) | + | y(n\tau + t + \epsilon) - y(n\tau + \tau) | \end{aligned}$$

With t and ϵ fixed, let $n \to \infty$; the first two terms on the right vanish, the last is at most $\mu(\epsilon)$. Hence for $y \in S$, $p_y(\cdot)$ has modulus of continuity $\mu(\cdot)$.

Now let x_m be an arbitrary sequence of S. From the associated sequence p_{x_m} of periodic functions we can pick a subsequence converging uniformly on $[0, \tau]$ to a function p_x. From the x_m's associated with this subsequence we can pick a further subsequence $x_{m(i)}$, $i = 1, 2, \ldots$, such that for some $x \in S$

$$x_{m(i)} \to x \text{ uniformly on any compact set,}$$

$$p_{x_{m(i)}} \to p_x \text{ uniformly on } [0, \tau].$$

Then $x_{m(i)}$ converges to x in the m-topology, so that S, being closed and sequentially compact, is compact.

4. Principal Result

Theorem 1: If $\{s_n, n \geq 0\}$, $\{k_n, n \geq 0\}$, and $\{h_n, n \geq 0\}$ are absolutely summable and $\{h_n\}$ satisfies

$$s_n + \beta \sum_{j=0}^{n} h_{n-j} k_j \leq h_n, \qquad n \geq 0 \tag{6}$$

then there exists a solution $x(\cdot)$ of (1) which is ultimately periodic with period τ, and in fact

$$| x(n\tau + t) - p_x(t) | \leq h_n$$

for $t \in [0, \tau]$ and $n \geq 0$. The function $p_x(\cdot)$ is a solution of the equation

$$p_x(t) = p_s(t) + \int_0^\infty F(p_x(t-u))\, k(u)\, du, \qquad t \in [0, \tau].$$

Proof: We define the operators H, J on S:

$$Hx(t) = \int_0^\infty F(x(t-u))\, k(u)\, du, \qquad x \in S$$

$$Jx = s - Hx.$$

Suppose that $x_m \to x$ in m-topology; then $x_m \to x$ uniformly on compact sets, and so the inequality

$$| Hx_m(t) - Hx(t) | \leq 2 \sup_u | F(u) | \int_{u>y} | k(u) | \, du$$

$$+ \beta \int_0^\infty | k(u) | \, du \sup_{0 \leq u \leq y} | x_m(t-u) - x(t-u) |$$

shows that $Hx_m \to Hx$ uniformly on compacta. Also, for $n \geq 0$, $t \in [0, \tau]$,

$$| Hx_m(n\tau + t) - Hx(n\tau + t) |$$

$$\leq \int_0^\infty | F(x_m(n\tau + t - u)) - F(x(n\tau + t - u)) | \cdot | k(u) | \, du$$

$$\leq \sum_{j=0}^{\infty} \int_{j\tau}^{(j+1)\tau} | F(x_m(n\tau - z)) - F(x(n\tau - z)) | \cdot | k(t+z) | \, dz.$$

For $j\tau \leq z < (j+1)\tau$ we have

$$
\begin{aligned}
| x_m(n\tau - z) - x(n\tau - z) | &\leq | x_m(n\tau - z) - p_{x_m}(z - j\tau) | \\
&\quad + | x(n\tau - z) - p_x(z - j\tau) | \\
&\quad + | p_{x_m}(z - j\tau) - p_x(z - j\tau) | \\
&\leq 2h_{n-j} + \sup_{0<u<\tau} | p_{x_m}(u) - p_x(u) |.
\end{aligned}
$$

Hence

$$
\begin{aligned}
&| Hx_m(n\tau + t) - Hx(n\tau + t) | \\
&\leq 2\beta \sum_{j=0}^{n} h_{n-j}k_j + \left(\sum_{j=0}^{\infty} k_j\right) \sup_{0<u<\tau} | p_{x_m}(u) - p_x(u) |.
\end{aligned}
$$

Now let *first* $n \to \infty$ and *then* $m \to \infty$; the right-hand side approaches zero, and so H is continuous on S.

We next show that J preserves the defining property (i) of S. Take $x \in S$; then for $n \geq 0$, $t \in [0, \tau]$,

$$
\begin{aligned}
&\left| s(n\tau + t) + Hx(n\tau + t) - p_s(t) - \int_0^\infty F(p_x(t - u))\, k(u)\, du \right| \\
&\leq s_n + \int_0^\infty | F(x(n\tau + t - u)) - F(p_x(t - u)) | \cdot | k(u) |\, du \\
&\leq s_n + \beta \sum_{j=0}^{n} h_{n-j}k_j \\
&\leq h_n.
\end{aligned}
$$

Now

$$
\begin{aligned}
| Jx(t + \epsilon) - Jx(t) | &\leq \sigma(| \epsilon |) + \sup_u | F(u) | \cdot \sup_{|\delta| \leq |\epsilon|} \int_{-\infty}^{\infty} | k(t + \delta) - k(t) |\, dt \\
&\leq \mu(| \epsilon |)
\end{aligned}
$$

so J preserves the property of having $\mu(\cdot)$ as modulus of continuity. Finally, it is obvious that J preserves the bound σ of property (ii) of S. Hence $JS \subset S$.

Since S is compact and convex, we conclude from Tychonov's theorem [2, p. 456] that there exists an $x(\cdot) \in S$ with $x = Jx$. The equation for the limiting periodic function follows by an Abelian argument, as in formula (3).

REFERENCES

[*1*] Beneš, V. E., A Non-linear Integral Equation from the Theory of Servomechanisms, to be published.

[*2*] Dunford, N., and Schwartz, J. T., "Linear Operators." Interscience, New York, 1958.

[*3*] Widder, D. V., "Advanced Calculus," 2nd ed. Prentice-Hall, Englewood Cliffs, New Jersey, 1961.

Determining the Switching Criterion for Time-Optimal Control[1]

C. A. HARVEY
MPG Research Department
Minneapolis-Honeywell Regulator Company
Minneapolis, Minnesota

In this paper the following problem is considered.

(P) Given a system represented by

$$\frac{dx(t)}{dt} = Ax(t) + bu(t) \tag{1}$$

where $x(t)$ is an n-dimensional vector, $u(t)$ is a scalar, A is a constant $n \times n$ matrix, and b is a constant n-dimensional vector, let G be a given, closed, bounded, convex set in Euclidean n-space, R^n, then for $x(0) = x^0 \notin G$ how should $u(t)$ be chosen with $|u(t)| \leq 1$, $0 < t < T$, such that the corresponding solution, $x(t)$, of (1) defines a trajectory in R^n which reaches G at $t = T$ where T is a minimum.

The case when G is a point (often the origin) has been studied and the results are quite well known. The problem with a more general set G has not received as much attention although certain results are available [*1*]. It is shown in this paper that the problem with a more general set G than a point may naturally arise from consideration of the following single-variable problem.

(P') Suppose $c(t)$ satisfies the equation

$$\sum_{i=0}^{n} a_i c^{(i)}(t) = \sum_{j=0}^{m} b_j u^{(j)}(t); \tag{2}$$

$$c^{(i)}(0) = c_0^{(i)}, \qquad u^{(j)}(0) = u_0^{(j)}, \qquad i = 0, 1, ..., n-1, \qquad j = 0, 1, ..., m-1.$$

and it is desired that $u(t)$ be chosen with $|u(t)| \leq 1$ for $t > 0$ so that the corresponding solution, $c(t)$ of (2) satisfies $c(t) = 0$ for $t > T_R$ where T_R is a minimum.

[1] This work was supported in part by NASA under contract NASr-27.

The problem P' may be reformulated so that its solution is obtained by solving the problem P. In the process of reformulation a target set G with the required properties is obtained.

The study of the single-variable problem was motivated by the following result. The time-optimal control theory was applied to a vehicle attitude control problem, where the attitude component of the state vector was of primary concern. The time-optimal response for the problem of bringing the state vector to zero from a disturbance in attitude only displayed over 1000 % overshoot in the attitude which was unsatisfactory.

The maximum principle of Pontryagin and a transversality condition given below are used to obtain extremal solutions for the problem P. Several examples are then considered where it is not difficult to show the extremal solutions are optimal. Points on the optimal solution curves in phase-space where the control function changes sign can then be determined. These points together with the boundary of G form a switching surface. This provides the desired switching criterion for time-optimal control.

The maximum principle states for the problem, P, that if $u(t)$ is an optimal control, then there exists a non-zero vector $p(t)$ satisfying

$$\frac{dp(t)}{dt} = -A^T p(t) \qquad (A^T = \text{Transpose of } A)$$

such that $u(t)$ maximizes $H(x, p, t) \equiv p \cdot (Ax + bu)$ which in this case means $u(t) = \operatorname{sgn}[p(t) \cdot b]$.

The transversality condition stated for the problem, P, is that if $x(t)$ is the solution of (1) corresponding to an optimal $u(t) = \operatorname{sgn}[p(t) \cdot b]$ and $x(t)$ "intersects" G at $x(T)$, $T > 0$, then $x(T)$ is a boundary point of G and $p(T)$ must have the direction of an "internal" normal to G at $x(T)$ where the meaning of "internal" normal is made precise by the following definition. A vector v is an "internal" normal to a set G at a boundary point x of G if $v \cdot (x - y) \leq 0$ for all $y \in G$.

The method presented for determining the switching criterion is similar to the method of "backing out" from the origin which has been used in that special case. When G contains more than one point solutions are "backed out" from boundary points of G restricting solutions so that they satisfy the transversality condition at the boundary of G.

REFERENCE

[1] Rozonoer, L. I., L. S. Pontryagin's maximum principle in the theory of optimum systems, II. *Avtomat. i Telemekh.* **20**, No. 11, 1441-1458 (1959).

On the Pontryagin Maximum Principle[1]

RICHARD E. KOPP
Grumman Aircraft Engineering Corporation
Bethpage, New York

Summary

This paper gives an expository discussion of the Pontryagin Maximum Principle, comparing it to the more classical variational techniques. In the Pontriagin Approach we see that the auxiliary *p* variables are the adjoint system variables. Further, we find that the Maximum Principle can be derived from an extension of the properties of adjoint systems, which is motivated by one of the well-known linear properties of adjoint systems.

The equivalence of the Weierstrass Condition and the Maximum Principle has been realized by many people for the case of unbounded controls. By adjoining the inequality constraints via the Valentine technique, we find the Weierstrass Condition remains valid when properly interpreted.

Introduction

Considerable attention has been drawn to the Pontryagin Maximum Principle due to the recently available translations of several Soviet papers [*2*, *4*, *5*, *7-10*]. The purpose of this paper is to give an expository discussion of the Maximum Principle and to compare it with some of the more classical variational techniques. First a derivation of the Maximum Principle will be demonstrated by extending the properties of adjoint systems. Following this the equivalence between the Weierstrass Condition and the Maximum Principle will be discussed and shown valid even when the control variables are bounded.

[1] Portions of this material were prepared in connection with Contract Nonr 3384(00) under the cognizance of the Information Systems Branch of the Office of Naval Research.

Problem Formulation

We formulate the optimization problem in terms of a Mayer variational problem [*1*]. That is, given the system of differential equations

$$\dot{x}_i = f_i(x_1 \ldots x_j \ldots x_n, u_1 \ldots u_k \ldots u_r, t), \qquad x_i(0) = x_i^0, \qquad i = 1 \ldots n, \tag{1}$$

$$\text{Max. (Min.) } S = \sum_{i=1}^{n} c_i x_i(t_f), \tag{2}$$

with respect to the control variables $u_k(t)$. The x_j are designated as state variables. Constraints on the control variables are expressed as: $\mathbf{u}(t) \in U$, where U is the admissible class of control functions in r space. The quantity $\mathbf{u}(t)$ is the vector control function with components $u_k(t)$.

Additional state variables may be added to Eqs. (1) to include non-linear and integral relationships in the pay-off function S. We will assume in the development which follows, that Eqs. (1) include any additional state variables required to specify the pay-off function.

Adjoint System

Before deriving the Maximum Principle, let us remove the constraints on the control variables and find stationary solutions to our variational problem. From Eqs. (1) we obtain by linearization, a system satisfied by the state variable increments:

$$\delta\dot{x}_i = \sum_{j=1}^{n} \frac{\partial \bar{f}_i}{\partial x_j}\,\delta x_j + \sum_{k=1}^{r} \frac{\partial \bar{f}_i}{\partial u_k}\,\delta u_k, \qquad \delta x_i(0) = 0, \tag{3}$$

where the bar over the partial derivatives denotes that they are evaluated on the stationary path.

The adjoint system of equations by definition is obtained by deletion of control terms and transposition of elements with change of sign:

$$\dot{p}_i = -\sum_{j=1}^{n} p_j \frac{\partial \bar{f}_j}{\partial x_i}. \tag{4}$$

The solutions of the two systems are related by:

$$\frac{d}{dt}\sum_{i=1}^{n} p_i \delta x_i = \sum_{i=1}^{n}\sum_{k=1}^{r} p_i \frac{\partial \bar{f}_i}{\partial u_k}\,\delta u_k. \tag{5}$$

From Eq. (5) it follows that,

$$\sum_{i=1}^{n} p_i \delta x_i \Big|_{t_0}^{t_f} = \int_{t_0}^{t_f} \sum_{i=1}^{n} \sum_{k=1}^{r} p_i \frac{\partial \bar{f}_i}{\partial u_k} \delta u_k dt. \tag{6}$$

For stationary paths, the first variation in the pay-off function S vanishes:

$$\delta S = \sum_{i=1}^{n} c_i \delta x_i(t_f) = 0. \tag{7}$$

The left member of Eq. (6) is seen to be δS if we impose boundary conditions on the adjoint system such that $p_i(t_f) = - c_i$ for free right right end state variables. Eq. (7) can be satisfied for arbitrary variations in the control vector only if,

$$\sum_{i=1}^{n} p_i \frac{\partial \bar{f}_i}{\partial u_k} = \frac{\partial H}{\partial u_k} = 0, \tag{8}$$

where $H = \Sigma p_i f_i$.

Through a similar development we will now derive the Pontryagin Maximum Principle for local maxima and minima with bounded control variables. A total variation about the optimum path is considered and the time derivative of $\Sigma p_i x_i \Delta$ examined as follows:

$$\frac{d}{dt} \sum_{i=1}^{n} p_i \Delta x_i = \sum_{i=1}^{n} \dot{p}_i \Delta x_i + \sum_{i=1}^{n} p_i \Delta \dot{x}_i. \tag{9}$$

The total variation $\Delta \dot{x}_i$ is evaluated from Eq. (1) giving:

$$\Delta \dot{x}_i = f_i(\bar{\mathbf{x}} + \Delta \mathbf{x}, \bar{\mathbf{u}} + \Delta \mathbf{u}, t) - f_i(\bar{\mathbf{x}}, \bar{\mathbf{u}}, t) \tag{10}$$

where,

$$\mathbf{x} \equiv (x_1 \dots x_j \dots x_n) \qquad \text{and} \qquad \mathbf{u} \equiv (u_1 \dots u_k \dots u_r).$$

The variations in $\Delta \dot{x}_i$ due to the state variables and the control variables are separated and the latter evaluated by a Taylor series expansion:

$$f_i(\bar{\mathbf{x}} + \Delta \mathbf{x}, \bar{\mathbf{u}} + \Delta \mathbf{u}, t) - f_i(\bar{\mathbf{x}}, \bar{\mathbf{u}}, t) = f_i(\bar{\mathbf{x}}, \bar{\mathbf{u}} + \Delta \mathbf{u}, t) - f_i(\bar{\mathbf{x}}, \bar{\mathbf{u}}, t)$$

$$+ \sum_{j=1}^{n} \frac{\partial f_i(\bar{\mathbf{x}}, \bar{\mathbf{u}} + \Delta \mathbf{u}, t)}{\partial x_j} \Delta x_j + \frac{1}{2} \sum_{\substack{j=1 \\ s=1}}^{n} \frac{\partial^2 f_i(\bar{\mathbf{x}} + \xi \Delta \mathbf{x}, \bar{\mathbf{u}} + \Delta \mathbf{u}, t)}{\partial x_j \partial x_s} \Delta x_j \Delta x_s, \tag{11}$$

where $0 \leq \xi \leq 1$. It is tacitly assumed that the first and second partial derivatives of f_i with respect to the state variables exist. Integrating

Eq. (9) after substituting Eqs. (4) and (11) gives for the change in pay-off function when boundary terms are evaluated:

$$\Delta S = -\int_{t_0}^{t_f} \sum_{i=1}^{n} p_i[f_i(\bar{\mathbf{x}}, \bar{\mathbf{u}} + \mathbf{\Delta u}, t) - f_i(\bar{\mathbf{x}}, \bar{\mathbf{u}}, t)]dt$$

$$-\int_{t_0}^{t_f} \sum_{\substack{i=1\\j=1}}^{n} p_i \left\{\frac{\partial}{\partial x_j}[f_i(\bar{\mathbf{x}}, \bar{\mathbf{u}} + \mathbf{\Delta u}, t) - f_i(\bar{\mathbf{x}}, \bar{\mathbf{u}}, t)]\right\} \Delta x_j dt$$

$$-\frac{1}{2}\int_{t_0}^{t_f} \sum_{\substack{i=1\\j=1\\s=1}}^{n} p_i \frac{\partial^2 f_i(\bar{\mathbf{x}} + \boldsymbol{\xi}\mathbf{\Delta x}, \bar{\mathbf{u}} + \mathbf{\Delta u}, t)}{\partial x_j \partial x_s} \Delta x_j \Delta x_s dt. \tag{12}$$

First let us consider the special case of a linear system in which the control variables are separable in f_i. In this case

$$f_i = \sum_{j=1}^{n} a_{ij}(t)\, x_j + \varphi(\mathbf{u}, t). \tag{13}$$

The last two members on the right side of Eq. (12) vanish and ΔS becomes:

$$\Delta S = -\int_{t_0}^{t_f} [H(\bar{\mathbf{x}}, \bar{\mathbf{u}} + \mathbf{\Delta u}, t) - H(\bar{\mathbf{x}}, \bar{\mathbf{u}}, t)]\, dt. \tag{14}$$

A necessary condition for a maximum (minimum) of the pay-off function S is that $\Delta S \leq 0$ ($\Delta S \geq 0$). We will assume a variation Δu_k in only one component of the control vector $\bar{\mathbf{u}}$. For this variation, Δu_k is zero everywhere except in the interval t_1, t_2 ($t_2 > t_1$). This interval is assumed completely adjustable in the interval t_0, t_f. Let us assume that

$$H(\bar{\mathbf{x}}, \bar{\mathbf{u}} + \mathbf{\Delta u}, t) - H(\bar{\mathbf{x}}, \bar{\mathbf{u}}, t) < 0, \tag{15}$$

for some interval between t_0 and t_f. If the interval t_1, t_2 is chosen to include the interval over which Ineq. (15) is satisfied, then $\Delta S > 0$. A similar argument may be presented for all the control variables. Therefore a necessary condition for a maximum of S is,

$$H(\bar{\mathbf{x}}, \bar{\mathbf{u}} + \mathbf{\Delta u}, t) - H(\bar{\mathbf{x}}, \bar{\mathbf{u}}, t) \geq 0, \tag{16}$$

for all t, $t_0 \leq t \leq t_f$. An analogous development may be made for a minimum of S.

To consider nonlinear systems we must evaluate the additional terms in Eq. (12). It is assumed that f_i, $\partial f_i/\partial x_j$, $\partial^2 f_i/\partial x_j \partial x_s$ and u_k are bounded functions. From these assumptions it can be shown that, $|\Delta x_i| \leq M\tau$, where M is a positive bounded constant and τ is the increment of time $t_2 - t_1$. The control vector $\bar{\mathbf{u}}(t)$ is again varied in the special manner discussed for the linear case. By making τ small enough it can be argued that Ineq. (16) remains as a necessary condition for a maximum of S. Therefore, the necessary condition for a maximum (minimum) of the pay-off function S is that H be minimized (maximized) with respect to the control vector at all times.

In the preceding development of the Maximum Principle it was assumed that final time t_f was fixed. For final time opened it can be easily shown that the additional relationship needed is that the Hamiltonian H is zero at t_f [6].

Constraints on the final values of the state variables are included using Lagrangian multipliers. This changes the boundary conditions on the auxiliary p variables [10].

Weierstrass Condition

It has been pointed out in a paper by Boltyanskii, Gamkrelidze, and Pontryagin [2] that the Maximum Principle is equivalent to the Weierstrass Condition when no constraints are imposed on the control variables. We will now demonstrate that this equivalence can be extended even when boundedness constraints are imposed on the control variables.

The constraints on the control variables are designated by the following inequalities:

$$u_{1k} \leq u_k \leq u_{2k}, \qquad k = 1 \ldots r. \tag{17}$$

Using the technique of Valentine [11] we introduce r new real variables,

$$\gamma_k^2 = (u_{2k} - u_k)(u_k - u_{1k}), \qquad k = 1 \ldots r. \tag{18}$$

From the variational approach of Bliss [1] for the Mayer formulation we obtain from the Euler equations in addition to the adjoint system equations,

$$\lambda_{\gamma k}\gamma_k = 0, \qquad k = 1 \ldots r, \tag{19}$$

and

$$\sum_{i=1}^{n} \lambda_i \frac{\partial f_i}{\partial u_k} + \lambda_{\gamma k}(u_{2k} + u_{1k} - 2u_k) = 0, \qquad k = 1 \ldots r. \tag{20}$$

For paths neighboring the extremal path,

$$\sum_{i=1}^{n} \lambda_i \delta x_i \Big|_{t_0}^{t_f} = \int_{t_0}^{t_f} \sum_{i=1}^{n} \sum_{k=1}^{r} \lambda_i \frac{\partial f_i}{\partial u_k} \delta u_k dt. \tag{21}$$

In Fig. 1 is shown a segment *ab* of an extremal path and a neighboring path *acb*. We will assume that the extremal path is one which maximizes a payoff function S, which requires that $\Delta S \leq 0$ for path *acb* by assumption:

$$\Delta S = \sum_{i=1}^{n} c_i \Delta x_i \Big|_b = - \sum_{i=1}^{n} \lambda_i \Delta x_i \Big|_b \leq 0. \tag{22}$$

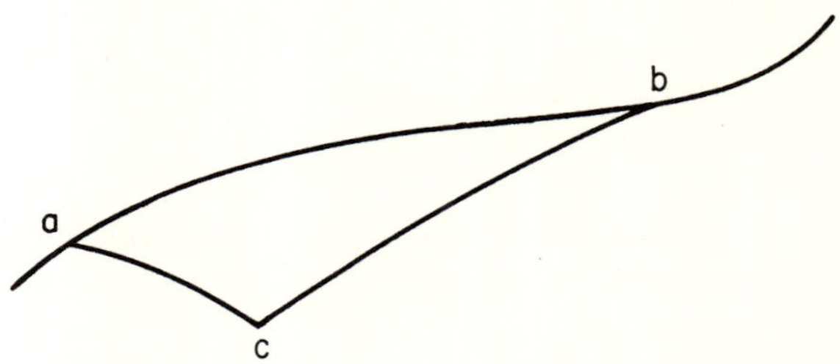

FIG. 1. Development of the Weierstrass excess function

The symbol Δx_i designates the variations in the state variables due to both explicit variations δx_i as well as variations δt in time:

$$\sum_{i=1}^{n} \lambda_i \delta x_i \Big|_a^b = \sum_{i=1}^{n} \lambda_i \Delta x_i \Big|_a^b - \sum_{i=1}^{n} \lambda_i \dot{x}_i \delta t \Big|_a^b. \tag{23}$$

For path *acb* it can be argued that $\sum \lambda_i \dot{x}_i \; \delta t \mid_a^b = 0$, giving,

$$\sum_{i=1}^{n} \lambda_i \delta x_i \Big|_a^b = - \Delta S \geq 0. \tag{24}$$

Since the variation $\delta \mathbf{u}$ in the control vector $\mathbf{u}$ was arbitrary within the constraints imposed by Ineq. (17) we conclude that the integrand in Eq. (21) must be always greater than or equal to zero:

$$\sum_{i=1}^{n} \sum_{k=1}^{r} \lambda_i \frac{\partial f_i}{\partial u_k} \delta u_k \geq 0. \tag{25}$$

Let us assume that at a the control vector $\mathbf{u}$ is varied from its optimum value $\mathbf{u}$ for a period of time δt_0 resulting in arc *ac* and now consider path *cb* to be a neighboring path to the extremal path *ab*. We designate the total time over path *acb* as $t_f - t_0 + \delta t_f$ and the time over path *ac* as δt_0.

From Eqs. (21) and (23) and Ineq (25) we obtain for path cb,

$$\sum_{i=1}^{n} \lambda_i \Delta x_i \Big|_b - \sum_{i=1}^{n} \lambda_i \Delta x_i \Big|_a - \sum_{i=1}^{n} \lambda_i \dot{\bar{x}}_i \Big|_b \delta t_f + \sum_{i=1}^{n} \lambda_i \dot{\bar{x}}_i \Big|_a \delta t_0 \geq 0, \tag{26}$$

where $\dot{\bar{x}}_i$ is evaluated along the extremal path ab. The variations $\Delta x_i \,|_a$ are expressed in terms of $x_i \,|_a$ measured on path ac. From Eq. (24) and Ineq. (26) we obtain,

$$-\Delta S + \sum_{i=1}^{n} \lambda_i (\dot{\bar{x}}_i - \dot{x}_i) \Big|_a \delta t_0 \geq 0, \tag{27}$$

from which we conclude that a necessary condition for $\Delta S \leq 0$ for an arbitrary $\delta \mathbf{u}$ satisfying Ineq. (17) is that

$$\sum_{i=1}^{n} \lambda_i (\dot{\bar{x}}_i - \dot{x}_i) \Big|_a \delta t_0 = E \delta t_0 \leq 0 \tag{28}$$

Weierstrass gave the name Excess Function to E. Since arc ab was an arbitrary arc of an extremal path, a necessary condition for a maximum of S is that $E \leq 0$ in all regions of the optimum path. It is observed that the comparison of $\dot{x}_i$ and $\dot{\bar{x}}_i$ in Eq. (26) is made with respect to the control variables only. With this observation we see that $-E$ is the ΔH in the Pontryagin Maximum Principle:

$$E = -[H(\bar{\mathbf{x}}, \bar{\mathbf{u}} + \Delta u, t) - H(\bar{\mathbf{x}}, \bar{\mathbf{u}}, t)]. \tag{29}$$

Cicala made use of the Weierstrass Condition for bounded controls in what he referred to as "index value".

REFERENCES

[1] Bliss, G. A., "Lectures on the Calculus of Variations." Univ. of Chicago Press, Chicago, Illinois, 1946.

[2] Boltyanskii, V. G., Gamkrelidze, R. V., and Pontryagin, L. S., The theory of optimal processes, I. Maximum principle. *Isvest. Akad. Nauk S.S.S.R., Ser. Mat.* **24**, No. 1 (1960); transl. by L. W. Neustadt, *Trans. AMS* **17**, 341-382 (1961).

[3] Cicala, P., "An Engineering Approach to the Calculus of Variations." Levrotto & Bella, Torino, 1957.

[4] Feldbaum, A. A., Synthesis of optimum systems with the aid of the phase space. *Avtomat. i Telemekh.* **16**, No. 2 (1955).

[5] Gamkrelidze, R. V., On the general theory of optimal processes. *Doklady Akad. Nauk S.S.S.R.* **123**, No. 2 (1958).

[6] Kopp, R. E., *in* "Optimization Techniques" (G. Leitmann, ed.), Chapter 7. Academic Press, in press.
[7] Pontryagin, L. S., Some mathematical problems arising in connexion with the theory of optimal systems of automatic control. *Proc. Acad. Sci. U.S.S.R.* **11** (1957).
[8] Pontryagin, L. S., Optimal control processes. *Uspekhi Mat. Nauk.* **14** (1959).
[9] Pontryagin, L. S., Optimal processes of regulation. *Proc. intern. math. Conf., Edinborough, 1958*, pp. 182-202. Cambridge Univ. Press, 1960.
[10] Rozonoér, L. I., L. S. Pontriagin's principle in the theory of optimum systems. *Avtomat. i Telemekh.* October, November, and December (1959); Transl. *Automation and Remote Control* June, July, and August (1960).
[11] Valentine, F. H., The Problem of Lagrange with Differential Inequalities as Added Side Conditions. Dissertation, Dept. Mathematics, Univ. of Chicago (1937).

The Nonlinear Effect, Arising from White Noise and the Flicker Effect, in Nuclear Resonance Spectrometers of the "Marginal Oscillator" Type[1]

A. BLAQUIÈRE
Maître de Conférences, Université de Paris
P. GRIVET
Professeur, Université de Paris

I. Description of the Self-Oscillator

1. Outline of the Principle of the Self-Oscillator

We shall apply the theory which has been developed in references [*2-4*] to an oscillator due initially to Colpitts which is often employed in spectroscopy as a marginal oscillator (fig. 1). The potential applied to

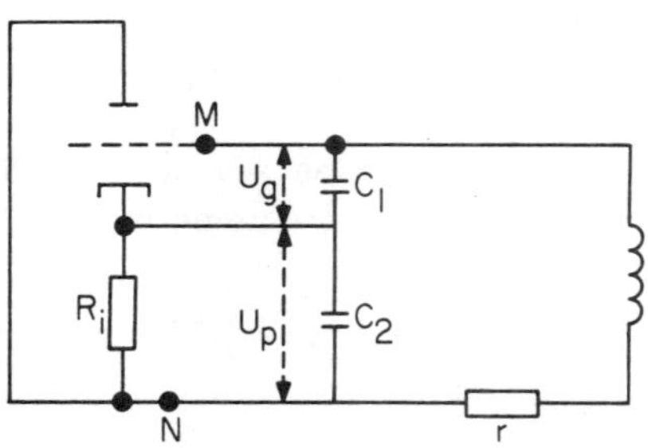

FIG. 1

the valve will be the grid potential U_g at the point M, and the potential at the output will be the potential of the anode U_p, calculated between the point N and the cathode. With the hypothesis that the tube has a very high internal resistance, the intensity of the anode current will be related to the grid potential when the tube is working in the linear region by the relation:

$$i_p = SU_g \qquad (S \text{ is the mutual conductance of the valve}).$$

[1] The full text of this communication will appear in the special number of the journal *Archives des Sciences de Genève* devoted to the 1961 Colloque A.M.P.E.R.E.

We write:

$$Z_0 = r + jL\omega \qquad Z_1 = \frac{1}{j\omega C_1} \qquad Z_2 = \frac{1}{G_1 + j\omega C_2}$$

in which r is the series resistance and L the self-inductance of the tank circuit. G_i denotes the conductance of the load $1/R_i$ which is placed between the anode and the cathode of the valve; the internal resistance is included in this load. Usually, G_i is negligible. It can easily be shown that the circuit equation takes the general form

$$-Z_1 Z_2 S = Z_0 + Z_1 + Z_2 \tag{1}$$

which in this case becomes

$$\frac{LS}{C_1 + C_2} = r + jL\omega + \frac{1}{j\omega}\frac{C_1 + C_2}{C_1 C_2}. \tag{2}$$

Finally, we transform this into the following form which is approximate but very clear:

$$-LC\omega^2 + j\omega[rC - LS(C_1/C_2)] + 1 = 0 \tag{3}$$

with

$$C = \frac{C_1 C_2}{(C_1 + C_2)}.$$

Energy is regenerated, therefore as a result of the introduction of a negative resistance into the tank circuit network

$$\rho = -\frac{1}{C} L \frac{C_1}{C_2} S$$

in series with r, L and C.

The parallel resistance R which is equivalent to this negative resistance is often employed; it is related to the latter by the usual expression:

$$R = Q^2 r = L/rC$$

Q is the quality factor of the tank circuit when the oscillation is not maintained; we write $G = 1/R$ and $C_1/C_2 = m$; with this notation, the circuit equation becomes:

$$LC\omega^2 + j\omega L(G - mS) + 1 = 0 \tag{4}$$

2. The Index of Reaction

The limiting coupling is reached when the negative resistance exactly compensates the natural resistance of the tank circuit. In the following discussion, we shall characterize the degree of coupling by a new parameter, the index of reaction n:[1]

$$n = \frac{m - m_0}{m_0}$$

in which m_0 is the value of the ratio C_1/C_2 *at the coupling point* ($m_0 = G/S$). For $0 < m < m_0$, or $-1 < n < 0$, the circuit is a *quality factor multiplier*, and for $m_0 < m$ or $0 < n$, the circuit is a *self oscillator*. We shall assume that the amplitude of the oscillations settles down to a stable value, U_S, and we shall thus be forced to introduce the non-linearity of the valve characteristic.

II. Background Noise of Thermal Origin

1. Relation Between the Circuit Parameters and the Noise

In an earlier study of the background noise of self oscillators, we have shown how important is the role which is played by the decay time τ of the perturbation produced by a very short disturbance, which is assumed to be unique, and we have developed a method of calculating τ in the general case [*1*, *4*]. Below, the coupling point, this time-lag corresponds to the time taken by the system to return to rest; beyond the coupling point, to return to the steady state. From this all-embracing point of view, therefore, the difference between these two cases is not very great. We shall take the non-linearity of the valve characteristic into account by selecting a function of the following form to describe it:

$$i_p = SU_g + aU_g^2 + bU_g^3$$

(i_p, U_g are the changes in the anode current and the grid potential respectively).

In these conditions the equation describing the behaviour of the circuit takes the form

$$LC\frac{d^2U_g}{dt^2} + L(G - mS - 2maU_g - 3mbU_g^2)\frac{dU_g}{dt} + U_g = E(t) \qquad (5)$$

$E(t)$ represents the thermal noise due to the resistance r; we shall examine the effect of the valve noise later. This is a Van der Pol equation with an additional term on the right hand side, which is a generalized version of equation (4), written in the conventional form.

[1] This convenient notation was introduced by Buyle-Bodin [*7*] and Hasenjäger [*9*].

2. The Effect of Noise on the Circuit, either Passive or Oscillating Continuously

We shall now summarize the conclusions at which we arrive in the various operating conditions of the system.

When the tank circuit is insulated from the whole system which is designed to maintain the oscillations, equation (5) is reduced to:

$$LC\frac{d^2U_g}{dt^2} + LG\frac{dU_g}{dt} + U_g = E(t) \tag{6}$$

The time constant τ of the tank circuit is $\tau = 2(C/G)$, and we obtain the following expression for the power of the noise, integrated over the whole frequency spectrum:

$$\begin{aligned} P = \overline{U_g^2} &= \frac{kTr}{2}\,\omega_0^2\tau \\ &= \frac{kTLG}{2C}\,\omega_0^2\tau \end{aligned} \tag{7}$$

or in this case kT/C (T is the absolute temperature of the noisy resistance, k is Boltzmann's constant; ω_0 is the natural circular frequency of the tank circuit.)

We can also obtain without difficulty the spectral distributions of the envelope of the signal modulated in amplitude by the noise.

$$P(\omega)\,\Delta\omega = \frac{2}{\pi}\,\frac{kT}{C}\,\frac{LG\omega_0^2\Delta\omega}{(1/\tau)^2 + \omega^2} \tag{8}$$

For $-1 < n < 0$, beneath the *coupling limit*, the behaviour is still linear. Formula (7) remains valid, but with the new time constant $\tau = -(1/n)(2C/G)$. The mean square noise potential becomes $P = -(1/n)(kT/C)$ and the power of the noise in the band $\Delta\omega$ is again given by formula (8).

In the self-oscillatory situation, the behaviour is non-linear. From equation (5), the time constant τ [*1*, *4*] and the noise power are deduced by assuming that the return to the equilibrium state follows the same law as a linear pendulum. We have assumed that the noise remains small relative to the amplitude of the oscillation. When this is true, formula (7) is again valid if we replace ω_0 by the steady state frequency ω_S. We obtain:

$$\begin{aligned} \tau &= \frac{1}{n}\frac{C}{G} \\ P &= \frac{kTr}{2}\,\omega_s^2\tau = \frac{kTLG}{2C}\,\omega_s^2\tau = \frac{1}{2n}\,\frac{kT}{C} \end{aligned} \tag{9}$$

and once again, the noise power in the band $\Delta\omega$ of the equivalent linear oscillator which we are using as model is given by formula (8).

Instead of reasoning in terms of τ, we can introduce a quality factor, taking care of the feedback however, a procedure which is more usual in circuit theory. A natural extension of the standard definition

$$Q = \frac{L\omega_0}{r} = \frac{\tau\omega_0}{2}$$

leads us to adopt the following definitions:

— In the case of quality factor multiplier

$$Q^* = -\frac{1}{n}\frac{C\omega_0}{G} = -\frac{1}{n}Q$$

— In the case of self-oscillators

$$Q^* = \frac{1}{n}\frac{C\omega_s}{2G} = \frac{1}{n}\frac{Q}{2}\frac{\omega_s}{\omega_0} \simeq \frac{1}{n}\frac{Q}{2}$$

(Q is the quality factor of the tank circuit when it is insulated from the supply valve).

The expression for the noise power in the band $\Delta\omega$ is then:
For a damped circuit:

$$P(\omega)\,\Delta\omega = \frac{2}{\pi}\frac{kT}{C}\frac{LG}{1/4Q^{*2} + \omega^2/\omega_0^2}\Delta\omega$$

$$= \frac{2}{\pi}\frac{kT}{C}\frac{LG\,\Delta\omega}{n^2/4Q^2 + \omega^2/\omega_0^2} \tag{10}$$

and for a self-oscillatory circuit:

$$P(\omega)\,\Delta\omega = \frac{2}{\pi}\frac{kT}{C}\frac{LG}{1/4Q^{*2} + \omega^2/\omega_0^2}\Delta\omega$$

$$= \frac{2}{\pi}\frac{kT}{C}\frac{LG\,\Delta\omega}{n^2/Q^2 + \omega^2/\omega_0^2} \tag{11}$$

3. Frequency Stability

We have already shown (1) (3) that the Van der Pol equation corresponds to an ideal case for which the frequency of the oscillation is practically independant of changes in the steady state amplitude. Applying this theory, we obtain the "frequency-amplitude" dependance law for a

Colpitts oscillator when the load conductance G_i is no longer neglected. We obtain

$$\nu = \nu_0 \left[1 + \frac{\lambda}{2L} S + \frac{3}{8}\frac{\lambda}{L} b\hat{U}_g^2\right]$$

and the frequency changes are related to the amplitude changes by the expression:

$$\frac{\Delta\nu}{\nu_0} = \frac{3}{4}\frac{\lambda}{L} b\hat{U}_g \Delta\hat{U}_g$$

where

$$\lambda \simeq \frac{C_1 G_i}{C_2^2} L^2 = m \frac{G_i}{C_2} L^2$$

From this, we deduce that the law which defines the frequency instability is of the form:

$$\frac{\Delta\nu}{\nu_0} = \frac{3}{4} mb \frac{LG_i}{C_2} \hat{U}_g \Delta\hat{U}_g$$

III. The Scintillation Effect and Its Influence upon the Self-Oscillatory Mode

We shall now take the valve noise into account by adding a ficticious resistance R_l to the grid (fig. 2); we shall denote the randomly varying electromotive force which it produces by E_l, which corresponds to two types of noise. The Schottky noise of the valve and the scintillation noise.

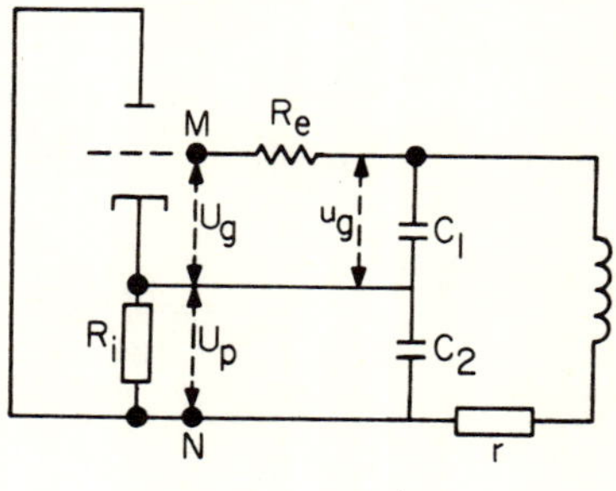

FIG. 2

In theory the effects of the randomly varying excitation E_l can be described by an equation of the form:

$$LC\frac{d^2U_g}{dt^2} + L(G - mS - 2maU_g - 3mbU_g^2)\frac{dU_g}{dt} + U_g = \left(LC\frac{d^2}{dt^2} + LG\frac{d}{dt} + 1\right) E_l \qquad (12)$$

We notice that in practice it is the potential u_g and not the potential U_g which is observed experimentally (fig. 2), which suggests that we should transform this basic equation.

We have shown in earlier articles [2, 3] that the effect of the Schottky noise is negligible in comparison with the thermal effect of the resistance r. This is not true of the scintillation noise, however, the spectrum of which possesses a branch which is nearly infinite at low frequencies [8]; up to about 1000 c/s, the order of magnitude of this abnormal noise is incomparably larger than the thermal noise. Even if we introduce the marked attenuation which we took into account in the earlier theory, and as a result of which we can neglect the Schottky noise, the order of magnitude of the effect of the abnormal noise on the oscillatory mode remains considerable, and in fact, we must combine it with the thermal effect to define the overall noise level of the oscillator.

If we calculate the mean square noise potential as a function of the index of reaction n, we find [5]:

$$\overline{(\delta\hat{u}_g)^2} = -\frac{4a^2}{3bS}\frac{1+n}{n}\overline{E_l^2} \tag{13}$$

A phenomenological theory of the scintillation effect has been given by Buyle-Bodin, in which a random fluctuation of the mutual conductance of the valve is introduced [6]. This theory is made precise in equation (13), where the fluctuations of mutual conductance are ascribed to the randomly varying polarization of the grid. From formula (13), we can deduce the spectral distribution of the perturbation in the amplitude produced by the scintillation effect:

$$P_f(\omega)\,\Delta\omega = -\frac{4a^2}{3bS}\frac{1+n}{n}4kTR_l\Delta f \tag{14}$$

This, we can compare with the perturbation due to thermal noise whose frequency is to be found in formula (11). If, for example, $r = 1\Omega$ (a series resistance in the tank circuit), $Q = 100$, $T = 300°$ K and $k = 1{,}37.10^{-23}$ joules/°K we obtain

$$\frac{2}{\pi}\frac{kT}{C}LG = 4kTr = 1{,}64 \cdot 10^{-20} \text{ joules}$$

For the thermal effect, formula (11) gives

$$P_t(\omega) = \frac{10^4}{n^2} \cdot 1{,}64 \cdot 10^{-20}$$

In formula (14) for the scintillation noise, a, b and S are usually of the of the same order of magnitude. A reasonable value for the coefficient $-4a^2/3bS$, is about 10, for example.

In the neighbourhood of the coupling limit, n is small; if, for example, we take $n = 1/100$, then $(1 + n)/n \simeq 100$, so that the quality factor of the tank circuit in the steady state is $Q^* = 50Q = 5000$.

From these values, we deduce:

$$P_f(\omega) = 10^3 \cdot 1{,}64 \cdot 10^{-20}\, R_l$$

$$P_t(\omega) = 10^8 \cdot 1{,}64 \cdot 10^{-20}$$

For $R_l = 10^5\ \Omega$, the thermal noise and the scintillation noise are of the same order of magnitude; for larger values of R_l, the scintillation noise becomes preponderant. But at a frequency of 10 kc/s, the equivalent resistance of the abnormal noise of a 6AK5 tube is already greater than $10^4\ \Omega$. This example confirms the usefulness of this theory in the low frequency region.

Formula (13) also shows that the influence of the scintillation noise can be suppressed, or at least considerably reduced, by choosing a working point on the characteristic for which the coefficient a vanishes.

On the other hand the triode connection of a tube is the most favourable one, giving a smooth variation of the coefficients a and b with grid bias; in a pentode at the opposite extreme, electron optical effects in the screen grid anode region, give to the b variation, an oscillatory behaviour, with many zeros (a most unfortunate choice as regards scintillation noise).

IV. Robinson's Low Noise Oscillator

A noteworthy method of improving autodyne oscillators has recently been developed by Robinson [*10*], in which the amplitude of the intensity which provides the reaction is limited after a considerable linear amplification. This is a particularly interesting arrangement from the point of view of noise. It is easy to establish that this set of operations is equivalent, to a first approximation,

— to the action of a linear dynamic characteristic of variable slope, where the amplitude obeys the law

$$S(\hat{U}) = K/\hat{U}$$

in which the constant K depends upon the clamping level. This oscillator is of marked originality. Its oscillation equation is

$$LC\frac{d^2U_g}{dt^2} + L\left(G - \frac{mK}{\hat{U}_g}\right)\frac{dU_g}{dt} + U_g = 0 \tag{15}$$

At first sight, therefore it appears to have no coupling condition, as the term $mK/\hat{U}_g$ is very large if the amplitude is very small, and it is this term which forces the oscillations to increase until it is compensated by the positive term G. The stabilized amplitude is therefore:

$$\hat{U}_S = mK/G$$

Despite the radically non-linear reaction, this oscillator possesses almost perfectly linear properties, and in particular, its time constant for the return to the steady state is $\tau = 2C/G$. This coincides with the time constant which the tank circuit would have if it were insulated from the valve which maintains the oscillations. The time constant is thus far smaller than that of the conventional types of oscillator, and the noise is in consequence very slight; the classical thermodynamic noise limit is reached. A more exact examination shows that the static characteristic is even more original; this latter is discontinuous at the working point (fig. 3) but it is legitimate, in theory at least, to regard it as a continuous curve, by virtue of the filtering produced by the selective tank circuit. The working point is clearly a point of inflexion, and in consequence, the oscillator is virtually insensitive, as we have seen, to the scintillation effect.

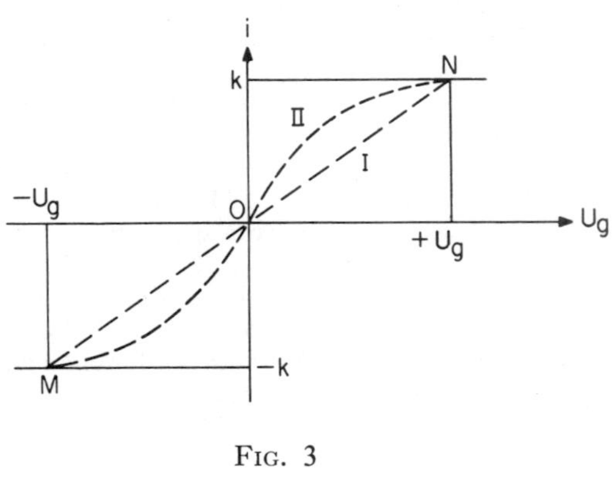

FIG. 3

If we go on to derive the current which is obtained after amplification and clamping, the current which produces the reaction proves to be a succession of alternately positive and negative pulses. Since the height of the original square waves is independant of the potential of the input grid, the same is true of these successive "tops". The oscillatory circuit which "drives" the self oscillator only has the effect of determinating the period of the successive pulses which it receives; this period is selected in such a way that each impulse is transmitted to the circuit both at a point where the elongation is zero, and in the proper sense.

We thus understand why the self-oscillator has the same time constant as the tank circuit; in fact, it is exactly as though the succession of impulses which maintain the oscillation originate in a self-maintained source, while the reaction does no more than transmit *pure information* to the latter and only gives rise to a *phase condition*.

In conclusion we might mention that this mechanism is analogous to that of an anchor escapment in a clock, a Short clock for example; we see therefore that in the course of a long historical evolution, clock-makers have tended towards an arrangement which is, so far as noise is concerned, exceptional.

REFERENCES

[*1*] Blaquière, A., Les oscillateurs non linéaires et le diagramme de Nyquist. *J. phys. radium* **13**, 527 (1952); Adaptation générale de la méthode du diagramme de Nyquist dans le domaine non linéaire. *Ibid.* p. 636.

[*2*] Blaquière, A., Effet du bruit de fond sur la fréquence des auto-oscillateurs à lampes (Thèse de Doctorat de Physique). *Ann. radioélec.* **8**, 36 and 153 (1953).

[*3*] Blaquière, A., Limite imposée par le bruit de fond à la précision des horloges radioélectriques. *Ann. franç. chronométrie* [2] **10**, 15-43 (1956).

[*4*] Blaquière, A., Mécanique non linéaire, les oscillateurs à régimes quasi-sinusoïdaux. *Mém. sci. mat.* **141** (1960).

[*5*] Blaquière, A., Le bruit de scintillation dans les auto-oscillateurs à lampes. *Compt. rend. acad. sci.* **252**, 2390 (1961).

[*6*] Buyle-Bodin, M., Effet des fluctuations de pente sur la sensibilité des auto-oscillateurs. *Compt. rend. acad. sci.* **243**, 1618 (1956).

[*7*] Buyle-Bodin, M., Sensibilité et fidélité des oscillateurs autodynes en spectroscopie hertzienne. *J. phys. radium.* **20**, 159 (1959).

[*8*] Grivet, P., and Blaquière, A., Le bruit de fond. *In* "Cours d'Électronique" (P. Grivet, ed.). Masson, Paris, 1958.

[*9*] Hasenjäger, H. J., Contribution à l'étude de l'oscillateur HF en vue de son application en spectroscopie hertzienne. Thèse d'Ingénieur-Docteur, Faculté des Sciences de Grenoble (1958).

[*10*] Robinson, F. N. H., Nuclear resonance absorption circuit. *J. Sci. Instr.* **36**, 481 (1959).

L'Extension de la Méthode de Nyquist aux Boucles Non-Linéaires. Application à L'Étude de la Stabilité des Réacteurs Nucléaires

A. BLAQUIÈRE
Maître de Conférences, Université de Paris

Nous résumerons ici une méthode d'étude des autooscillateurs introduite par l'auteur [*2*], comme généralisation, pour les conditions de fonctionnement non linéaires, de la méthode classique de Nyquist, valable seulement dans les cas linéaires.

Cette méthode apparaît aussi comme une extension des méthodes développées indépendemment par Dutilh [*7*] et Kochenburger [*9*] dans le domaine des servo-mécanismes, et conduit à une synthèse générale des systèmes bouclés non-linéaires usuels. Postérieurement aux travaux cités [*2*, *3*], elle a été adoptée pour sa commodité par de nombreux chercheurs français et étrangers [*1*, *8*, *10*, *11*, *13*], qui la désignent habituellement sous le nom de "Describing function Method".

1. Equation des Boucles

Du schéma fonctionnel d'une boucle (fig. 1): oscillateur, servo-mécanisme, amplificateur à contre-réaction, ... on déduit l'équation fondamentale:

$$X_s = \frac{G}{1 - GT} X_e \tag{1}$$

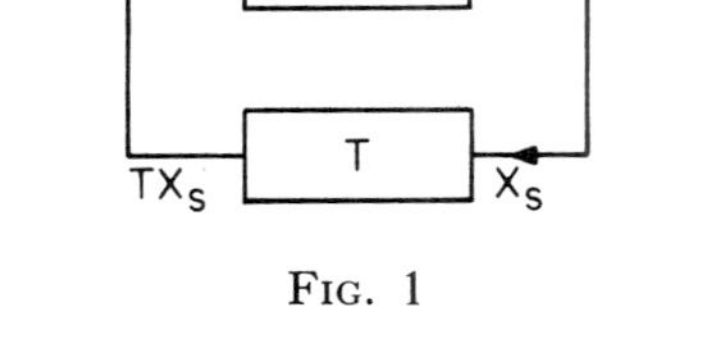

Fig. 1

Dans ce schéma, les éléments passifs et actifs de la boucle sont supposés réduits aux deux quadripôles (T) et (G) ayant respectivement pour *fonction de transfert* T et pour *gain* G. X_e et X_s désignent d'autre-part les grandeurs d'entrée et de sortie, fonctions du temps.

En fait, G et T ne sont généralement pas de simples facteurs algébriques, mais des expressions plus compliquées, faisant intervenir des dérivées par rapport au temps d/dt, d^2/dt^2, etc.... Ce sont des *opérateurs*, aussi appelés *fonctionnelles*. Quant à l'équation (1), mise sous la forme

$$(1 - GT)\, X_s = GX_e$$

c'est généralement une équation différentielle avec second membre.

En premier lieu on se limitera à l'étude de la boucle sans excitation extérieure, $X_e = 0$, et on examinera ensuite l'effet de l'excitation extérieure X_e.

2. Les Boucles Linéaires et les Boucles Non-Linéaires

La boucle est linéaire si les coefficients des termes en X_s, $\dot{X}_s$, $\ddot{X}_s$, ... de l'équation sont constants, fixés par les seuls facteurs géométriques: résistances, selfs, capacités, etc... qui entrent dans la constitution des quadripôles.

Elle est non-linéaire si ces coefficients dépendent du régime qui s'établit. La fonctionnelle fait intervenir alors, non seulement *des facteurs géométriques* que l'on peut maintenir fixes, mais aussi *de paramètres* a_1, a_2, ... a_n, dont la valeur dépend de l'état de régime à l'instant considéré. La principale difficulté de l'étude, dans ce dernier cas, provient de ce que l'évolution du système dépend d'une équation qui évolue avec le temps, et dont la connaissance, à chaque instant, nécessiterait une information préalable sur le régime. Le problème peut être résolu approximativement lorsqu'il est possible de faire certaines hypothèses raisonnables concernant le régime. Ce sera le cas notamment des autooscillateurs dont la loi d'évolution ne s'écarte pas trop, à chaque instant, d'une loi sinusoïdale, et l'on pourra adopter alors pour paramètre de régime a l'amplitude supposée lentement variable au cours du temps.

Si nous désignons par H la fonctionnelle $1 - GT$, elle sera de la forme, dans le cas linéaire

$$H(d/dt,\, d^2/dt^2,\, ...) \tag{2}$$

Les coefficients qui interviennent dans son expression sont des constantes qui ne dépendent que de données géométriques rigides. Et, dans le cas non-linéaire, de la forme

$$H(d/dt,\, d^2/dt^2,\, ...\, a_1,\, a_2,\, ...\, a_n) \tag{3}$$

3. Diagramme à Deux Paramètres; Liaison avec la Méthode de Nyquist

La méthode de Nyquist classique conduit à associer à la fonctionnelle (2) une fonction analytique $H(j\omega)$, (j: symbole imaginaire), où ω est une fréquence angulaire (ou pulsation) réelle ou complexe. On représente alors cette fonction sur le plan complexe en donnant à la variable ω des valeurs réelles $\omega_{\text{réel}}$. On obtient ainsi une courbe: le diagramme de Nyquist du système, d'où on déduit ses conditions de stabilité.

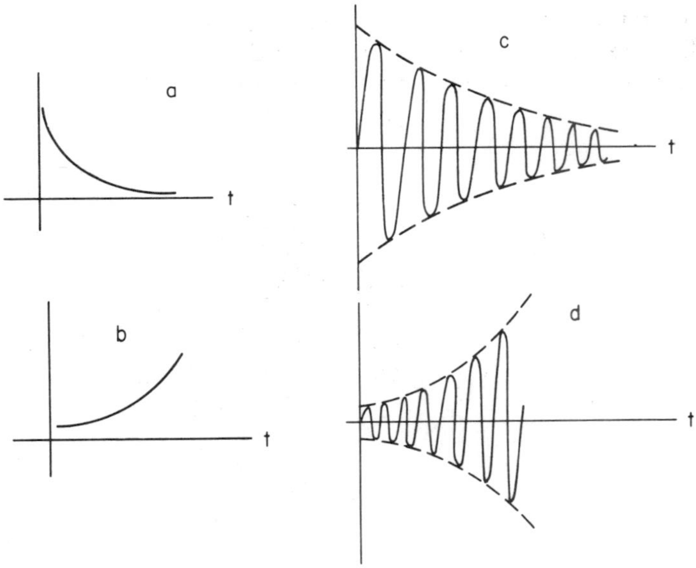

FIG. 2

Ces conditions s'expriment sous une forme particulièrement simple dans le cas des systèmes du second ordre, mais l'énoncé s'étend sans difficulté au cas général.

Soit un observateur qui décrit le diagramme de Nyquist dans le sens des fréquences croissantes, s'il voit l'origine O du plan complexe à sa gauche, le système est amorti. S'il voit le point O à sa droite, il est auto-excité.

Le régime amorti peut revêtir les aspects (a) ou (c) de la figure 2. De même, l'autoexcitation peut être oscillante ou non oscillante (cas (d) ou (b) fig. 2).

Plaçons nous maintenant dans le cas où l'équation du système étudié dépend d'un paramètre de régime a. Pour chaque valeur de a, on peut tracer le diagramme de Nyquist de la fonction complexe $H(j\omega_{\text{réel}}, a)$. La courbe de Nyquist est donc remplacée, en régime non-linéaire, par une famille de courbes dépendant du paramètre a. On voit sur la figure 3 que, suivant les valeurs de a, le système peut être autoexcité ou amorti. Ainsi, dans cet example, la limite d'entretien, atteinte lorsque le diagramme de Nyquist passe par l'origine O, correspond à la valeur a_0 du paramètre de régime. Pour les valeurs de a inférieures à a_0 le système est autoexcité, et il s'ensuit une croissance de a depuis la valeur initiale jusqu'à a_0. Pour les valeurs de a supérieures à a_0, le système est amorti, ce qui entraîne encore un retour au régime limite, mais par décroissance de a. D'autre part, si le régime limite est du type oscillant, la pulsation de l'oscillation est celle qui est attachée au point O.

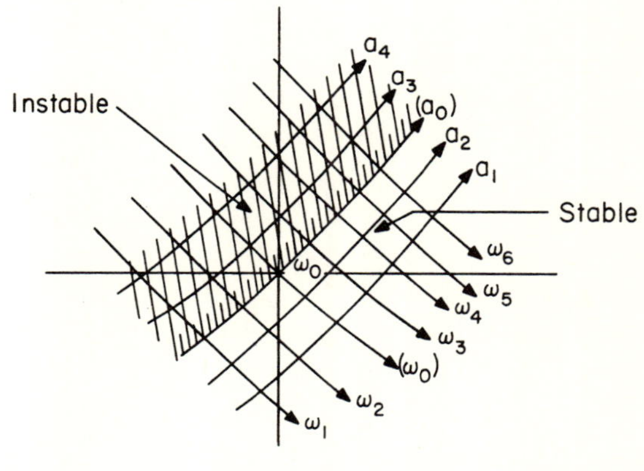

FIG. 3

Les courbes ainsi tracées, pour diverses valeurs du paramètre a, ont été appelées dans nos travaux antérieurs "courbes équiamplitudes". Le diagramme peut être complété par le tracé d'une autre famille de courbes: les "courbes équipulsations". En effet, si on se donne une valeur $\omega_{\text{réel}}$ de la pulsation, il lui correspond un point sur chaque équiamplitude. Le lieu de ce point, lorsque a varie, est la courbe équipulsation (ω). L'équipulsation (ω_0) passant par O définit la pulsation d'accrochage.

Un exemple de discussion montrant comment on peut rechercher en pratique les conditions de régime stationnaire, et étudier la stabilité de cet état, est développé dans la référence [6].

4. Construction Expérimentale du Diagramme

Le diagramme à deux paramètres (a et ω) peut être obtenu expérimentalement de façon simple. Il suffit d'ouvrir la boucle de la figure 1 et d'appliquer à l'entrée du circuit non-linéaire ainsi obtenu (fig. 4) le signal sinusoïdal $X = a \sin \omega t$.

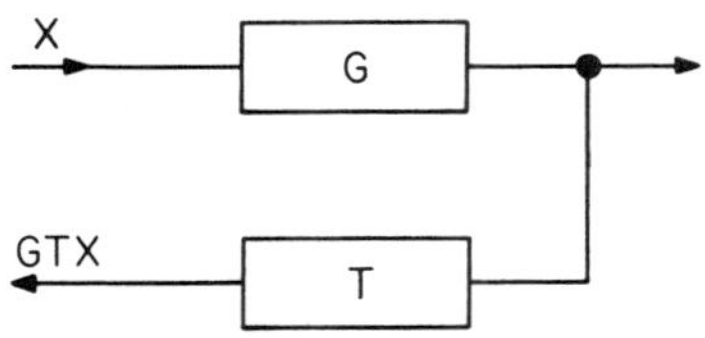

FIG. 4

La mesure de l'amplitude et de la phase du signal de sortie correspondant conduit à la valeur complexe GT ou, à un changement d'origine près, dans le plan complexe, à $H = 1 - GT$.

Ainsi, à chaque signal d'entrée, on fait correspondre un point du plan complexe, image de H, et il suffit de faire varier de façon continue, soit l'amplitude (à pulsation constante), soit la pulsation (à amplitude constante), pour obtenir les familles d'équipulsations et d'équiamplitudes définies plus haut.

La méthode expérimentale permet de prévoir le comportement de la boucle fermée à partir du diagramme à deux paramètres construit en boucle ouverte.

5. Critères Géométriques de Stabilité

Dans la référence [*12*] nous avons donné une démonstration simple du critère géométrique de stabilité suivant:

Soient (a_s) et (ω_s) l'équiamplitude et l'équipulsation passant par l'origine, qui définissent un régime stationnaire. Supposons que ces courbes soient orientées respectivement dans le sens des pulsations et des amplitudes croissantes. La stabilité du régime est assurée lorsque l'angle que fait l'équiamplitude (a_s) avec l'équipulsation (ω_s) est positif. Dans le cas contraire, l'état stationnaire est instable.

Dans la même référence, nous avons étudié à partir du diagramme à deux paramètres la synchronisation produite par une action sinusoïdale imposée, et discuté la stabilité du régime synchronisé, ce qui nous a conduit à un critère géométrique plus général dont le précédent est un cas particulier.

6. Généralisation — Diagramme à *n* Paramètres

La méthode du diagramme à deux paramètres ne peut rendre compte de certains phénomènes tels que, par exemple, la résonance sous-harmonique. Celà provient de ce que, dans la théorie précédente, tous les signaux périodiques sont systématiquement remplacés par leurs fondamentaux, ce qui correspond à une perte d'information.

On enrichira le diagramme si on fait intervenir, en concurrence avec le fondamental, les amplitudes des harmoniques successifs, a_2, a_3, ... a_k. On verra ainsi apparaître les fréquences de battement des harmoniques entr'eux et avec le fondamental, qui étaient précédemment négligées, et le terme final obtenu lorsque l'on reviendra, tous calculs effectués, à l'approximation fondamentale, s'en trouvera généralement modifié.

Du point de vue théorique, on aboutira à une fonctionnelle dépendant de la pulsation ω, de l'amplitude a du fondamental, et d'autres paramètres, plus ou moins nombreux: les amplitudes des harmoniques considérés $H(j\omega, a, a_2, a_3, \dots a_k)$. Expérimentalement, celà revient à améliorer l'analyse du système non-linéaire en admettant à l'entrée de la boucle ouverte un signal formé par la superposition d'un fondamental de pulsation ω et de différents harmoniques (un ou deux en pratique). Ainsi, le signal d'entrée étant par exemple de la forme

$$a \sin \omega t + a_2 \sin (2\omega t - \varphi_2)$$

on étudiera le signal de sortie pour différentes valeurs des paramètres d'entrée. Notamment on pourra maintenir fixe a, a_2, et faire varier ω, comme dans la méthode de Nyquist classique, mais on pourra également fixer les valeurs de ω, a_2, et faire varier a, etc....

Il y a lieu de noter que la condition

$$H(j\omega, a, a_2, a_3, \dots a_k) = 0 \tag{4}$$

qui exprime que le diagramme passe par l'origine du plan complexe ne définit plus un état *absolument stationnaire*. Elle définit un état stationnaire pour le fondamental de l'oscillation, *relativement à un certain lot de paramètres* a_2, a_3, ... a_k. Or si ces paramètres sont évolutifs, le système s'écartera naturellement du régime, à supposer que la condition (4) soit initialement vérifiée.

Il peut néanmoins arriver que les paramètres a_2, a_3, ... a_k soient fixés par différentes conditions physiques simples, c'est notamment ce qui se produit dans l'étude de la résonance sous-harmonique [*12*].

Lorsque les paramètres a_2, a_3, ... a_k sont "flottants", il faut joindre à la condition (4) autant de conditions nouvelles qu'il y a de paramètres, autres que a; ces dernières ayant pour objet de définir les états stationnaires respectifs de a_2, a_3, ... a_k.

7. Domaine des Réacteurs Nucléaires

Enfin, dans un tout autre domaine, celui des piles atomiques, la méthode de Nyquist généralisée s'est montrée particulièrement fructueuse. C'est ainsi que, très postérieurement aux travaux qui viennent d'être cités, de nombreux auteurs ont appliqué cette méthode à l'étude de la stabilité des réacteurs nucléaires. Un exemple particulièrement important et qui montre bien l'efficacité du procédé a été donné par Bethe [*1*].

La fonctionnelle linéaire du réacteur nucléaire $Q(j\omega)$, obtenue à très faible puissance, est remplacée, lorsque l'on tient compte des non-linéarités, par une fonctionnelle plus générale, de la forme

$$Q(j\omega) - PR(j\omega)$$

où Q et R sont deux fonctionnelles linéaires et P la puissance du réacteur qui joue le même rôle que l'amplitude a, dans l'étude précédente.

Nous avons donné dans la référence [*6*] une construction géométrique simple des équipulsations et des équiamplitudes pour ce cas. D'autre part, la méthode conduit à l'énoncé d'une condition nécessaire de stabilité d'un réacteur nucléaire.

Il ne semble pas toutefois que les auteurs du domaine nucléaire aient eu connaissance des publications antérieures dans le domaine de l'électronique et des servo-mécanismes. Ils ont ainsi été amenés à retrouver et à prendre pour base les énoncés précis que nous avions formulés quelques années avant, comme le montre la lecture des articles [*1*, *6*, *13*] qui ne constituent qu'une bibliographie sommaire de ce vaste sujet.

L'importance du danger d'emballement des piles, si actuelle pour les entreprises modernes de propulsion navale, montre bien que l'étude de la méthode du diagramme à deux (ou n) paramètres pourra se développer largement dans un proche avenir.

BIBLIOGRAPHIE

[*1*] Bethe, H. A., Reactor safety and oscillator tests. *APDA*, **117** (1956).

[*2*] Blaquière, A., Extension de la théorie de Nyquist au cas de caractéristiques non linéaires. *Compt. rend. acad. sci.* **233**, 345-349 (1951).

[*3*] Blaquière, A., Les oscillateurs non linéaires et le diagramme de Nyquist. *J. phys. radium* **13**, 527 (1952); Adaptation générale de la méthode du diagramme de Nyquist dans le domaine non linéaire. *Ibid.* p. 636.

[*4*] Blaquière, A., Effet du bruit de fond sur la fréquence des auto-oscillateurs à lampes (Thèse de Doctorat de Physique). *Ann. radioélec.* **8**, 36 and 153 (1953).

[*5*] Blaquière, A., Mécanique non linéaire, les oscillateurs à régimes quasi-sinusoïdaux (Thèse de Doctorat de Mathématiques, 1957). *Mém. sci. mat.* **141** (1960).

[*6*] Blaquière, A., Le diagramme de Nyquist généralisé et son application à l'étude de la stabilité des réacteurs nucléaires. Rapport CEA No. 1700 (1960).

[*7*] Dutilh, J. R., Théorie des servo-mécanismes non linéaires. *Radio (Paris)* No. 5, 1-7 (1950); *Onde élec.* **30**, 438-445 (1950).

[*8*] Klotter, K., How to obtain describing functions for nonlinear feedback systems. *Trans. ASME* (1957).

[*9*] Kochenburger, R. J., Analyse du fonctionnement des servo-mécanismes du type pas à pas, à l'aide d'une méthode fournissant la réponse en fonction de la fréquence. *Elec. Eng.* **69**, 687-692 (1950).

[*10*] Ku, Y. H., "Analysis and Control of Nonlinear Systems." Ronald Press, New York, 1958.

[*11*] Sandmeier, H. A., Nonlinear kinetics and stability studies on analog computers. *Compt. rend. Colloque intern. sur l'électronique nucléaire.* Agence Intern. de l'Énergie Atomique, 1959.

[*12*] Symposium. La méthode du diagramme de Nyquist généralisé dans le domaine non linéaire. *Symposium sur les oscillations non linéaires, Kiev, 1961.*

[*13*] Smets, H. B., Nonlinear stability of nuclear reactors. *Compt. rend. Colloque intern. sur l'électronique nucléaire.* Agence intern. de l'Énergie Atomique, 1959.

A Contribution to Liapunov's Second Method: Nonlinear Autonomous Systems[1]

GEORGE P. SZEGÖ
RIAS, Baltimore, Maryland, and
Control and Information Systems Laboratory
Purdue University, Lafayette, Indiana

1. Introduction

Given the nonlinear autonomous system:

$$\dot{\mathbf{x}} = \mathbf{X}(\mathbf{x}), \tag{1}$$

we want to investigate the stability of its trivial solution $\mathbf{x} = 0$ which is assumed to be the only equilibrium point. This assumption that the system (1) has only one equilibrium point and that this equilibrium point is $\mathbf{x} = 0$, possibly after a linear transformation, will be maintained through the whole work unless otherwise stated.

The following two different possibilities exist.

(a) The stability[2] (instability) of the solution $\mathbf{x} = 0$ is global; that is, the stability property of the solution $\mathbf{x} = 0$ is invariant in the whole n-dimensional euclidean phase-space E_n.

(b) The stability of the solution $\mathbf{x} = 0$ is local; i.e. there are regions of stability and regions of instability with defined boundaries between them.

In the case of higher order systems, very little information about the form of these surfaces, and their physical meaning is available, but in the case of second order system, as the Poincaré-Bendixson Theory states, these surfaces, if closed bounded curves around the origin, are limit cycles.

In this paper the problem of stability of nonlinear autonomous systems will be investigated by means of the second method of Liapunov [*1*, *4*, *7*, *8*].

First of all the following theorems must be given.

[1] This research was supported by the National Science Foundation Grant No. G 16460. The full text of this contribution will appear in *Trans. ASME* (D).

[2] The stability defined is the sense of Liapunov through the whole work. In particular, if the property is global, we mean asymptotic stability.

Theorem 1. The solution $\mathbf{x} = 0$ of the system (1) is asymptotically stable (Uniform-Asymptotically Stable [*10*]) with respect to the coordinates x_i in a region S, defined by $v(\mathbf{x}) \leqq l$, of the n-dimensional Euclidean space E_n, if there exists a positive definite scalar function $v(\mathbf{x})$ with continuous first partial derivatives in S and, if $l \to \infty$, with $v(\mathbf{x}) \to \infty$ as $\| \mathbf{x} \| \to \infty$, such that dv/dt is negative semidefinite in S and not identically zero on any nontrivial solution of the system (1) [*2*]. If dv/dt is positive semidefinite in S and all the other previously stated conditions are satisfied, then the solution $\mathbf{x} = 0$ of the system (1) is completely unstable.[1]

The proof of the second part of this theorem follows immediately by the analogous theorem due to Liapunov [*8*, *9*].

Assume that $v = \varphi(\mathbf{x})$ is a Liapunov function[2] for system (1), in the case (a) dv/dt will be positive semidefinite and will not equal zero on any n.t. solution of the system (1). In the case (b) dv/dt will be positive semidefinite and not equal to zero on a solution of the system (1), in some regions of the phase space around the origin, and will be negative semidefinite and not equal to zero on a solution in some other regions. In this case we have the following theorem.

Theorem 2. A sufficient condition for local stability of the solution $\mathbf{x} = 0$ of system (1) is that there exists a positive definite scalar function $v = \varphi(\mathbf{x})$ with continuous first partial derivatives, such that its total time derivative with respect to the system (1) has the form

$$\frac{dv}{dt} = \theta(\mathbf{x})\, g[\xi(\mathbf{x})] \tag{2}$$

where $\theta(\mathbf{x})$ is a semidefinite function not identically equal to zero on any n.t. solution of the system (1), and $g(\mathbf{x})$ is indefinite on a closed surface.[3] In particular assume that $\xi(\mathbf{x}) = 0$ is a closed bounded surface, or a family of closed surfaces, $g(u)$ is such that $g(0) = 0$ and $g(u)/u > 0$.

Let us indicate by v_c the surface of the family $v = \varphi(\mathbf{x})$ that is circumscribed to the surface $\xi(\mathbf{x}) = 0$ and by v_m the inscribed surface. Then, if we indicate with $v_c' \geqq v_c$ any surface of the family outside v_c, dv_c'/dt will be semidefinite of a particular sign and for every surface with $v \leqq v_m$, dv/dt will be semidefinite of the opposite sign. The closed and

[1] Given any $\epsilon > 0$, $\exists t_1(\epsilon) > t_0$ such that $0 < \| x_0 \| < \epsilon \Rightarrow \| \mathbf{x}(t_1, \mathbf{x}_0) \| = \epsilon$.

[2] A v-function is called a Liapunov's function if it satisfies Theorem 1 for the given system.

[3] We call a function $v = \chi(\mathbf{x})$ indefinite on a closed surface if $\chi(\mathbf{x}) = 0$ is a closed bounded surface and if the sign of $\chi(\mathbf{x})$ inside $\chi(\mathbf{x}) = 0$ is different from the sign of $\chi(\mathbf{x})$ outside $\chi(\mathbf{x}) = 0$.

bounded boundary between the two regions of different stability properties, if such a boundary exists, will lie in the region between the two surfaces v_c and v_m.

In the particular case in which $dv_c'/dt \leqq 0$, if $\varphi(\mathbf{x}) \to \infty$ as $\| x \| \to \infty$, it is possible, according to La Salle [6] to conclude that the solutions of system (1) are bounded. In the case in which $\xi(\mathbf{x}) = 0$ defines a family of concentric closed surfaces the boundedness of solution is defined only if v_c is the v-function circumscribed to the largest closed surface $\xi(\mathbf{x}) = 0$ and $dv_c'/dt \leqq 0$. If $v_c = v_m$, then the area of the region of indefiniteness of the function dv/dt shrinks to zero and the surface $\xi(\mathbf{x}) = 0$ identifies exactly the boundary between stability and instability. This boundary is now a limit set[1] of the system and namely a Γ^+ if $dv_c'/dt \leqq 0$ and Γ^- if $dv_c'/dt \geqq 0$ [6, 11].

If the system is of second order, the Poincaré-Bendixson theory assures the existence of an odd number of limit cycles between v_c and v_m. If the system has only one limit cycle, this will namely be stable and the solutions of the system bounded if $dv_c'/dt \leqq 0$ and unstable if $dv_c'/dt \geqq 0$.

Our approach to the problem of nonlinear systems with local stability is completely different from La Salle's [6] and Zubov's [12], since we impose dv/dt to be indefinite on a closed surface. The most important feature of this theory is the possibility of an approximate identification of the limit surface, a possibility which is not straightforward in the other treatments.

2. The Constructions of Liapunov's Functions

Few attempts [5] have succeeded in constructing Liapunov's functions such that the stability properties of nonlinear systems can be discussed as a function of the nonlinear part of the system.

In the present work a method will be developed for the construction of Liapunov's functions in the case of nonlinear autonomous systems in which the nonlinearity is representable in polynomial form. This method is based upon the use of a v-function with some variable coefficients, that we call a generating v-function. Such a generating v-function is to be a polynomial in $x_1, ..., x_n$ that we represent as a quadratic form with variable coefficients:

$$v = \mathbf{x}' A(\mathbf{x}) \mathbf{x} \tag{3}$$

[1] The set of points $c^+(x_m)$ $(c^-(x_m))$ such that there exist a sequence t_n with $t_n \to +\infty (t_n \to -\infty), x(t_n) \to x_m$ as $n \to +\infty$ $(n \to -\infty)$ is called positive limiting set Γ^+ (negative limiting set Γ^-).

where $A(\mathbf{x}) = \{a_{ij}(x_i, x_j)\}$ and $a_{ij} = a_{ji}$. We shall represent this generating v-function as

$$v = \mathbf{x}' A(x_i, x_j)\mathbf{x} \tag{4}$$

and the system (1) as

$$\dot{\mathbf{x}} = B(\mathbf{x})\mathbf{x} \tag{5}$$

where the matrix $B(\mathbf{x})$ is generally not uniquely defined by the system.

The total time derivative of the v-function (4) with respect to the system (5) has the form:

$$\dot{v} = \mathbf{x}' \left[B'(\mathbf{x})\, A(x_i, x_j) + A(x_i, x_j)\, B(\mathbf{x}) + \frac{dA(x_i, x_j)}{dt}\right] \mathbf{x}. \tag{6}$$

In this particular case the expression (6) can be written as:

$$\dot{v} = \mathbf{x}'[B'(\mathbf{x})\, A^*(x_i, x_j) + A^{*\prime}(x_i, x_j)\, B(\mathbf{x})]\mathbf{x} \tag{7}$$

where the elements of the matrix $A^*(x_i, x_j)$ have the form:

$$a_{ij}^*(x_i, x_j) = a_{ij}(x_i, x_j) + \epsilon_{ij} \frac{\partial a_{ij}(x_i, x_j)}{\partial x_i} x_i \tag{8}$$

where

$$\epsilon_{ij} = \begin{cases} \frac{1}{2} & \text{for } i = j \\ 1 & \text{for } i \neq j \end{cases} \qquad \text{and} \qquad a_{ij}^*(x_i, x_j) \neq a_{ji}^*(x_i, x_j).$$

Since $a_{ij}(x_i, x_j)$ are polynomials functions, the form of $a_{ij}^*(x_i, x_j)$ is the same as the form of $a_{ij}(x_i, x_j)$[1] and $A(x_i, x_j)$ has the same form as $A^*(x_i, x_j)$. It is consequently possible to investigate instead of the exact expression (7) the expression

$$\dot{v}^* = \mathbf{x}'[B'(\mathbf{x})\, A(x_i, x_j) + A(x_i, x_j)\, B(\mathbf{x})]\mathbf{x} \tag{9}$$

which has the same form.

We want, as pointed out in § 1, $\dot{v}$ to have a particular form:

$$\dot{v} = \theta(\mathbf{x})\, [\mathbf{x}' C(\mathbf{x})\mathbf{x} - k] \tag{10}$$

where $\varphi = \mathbf{x}' C(\mathbf{x})\mathbf{x}$ is a definite scalar function for $k - \alpha \leq \varphi \leq k + \beta$; $\alpha, \beta > 0$, and $\theta(\mathbf{x})$ is a semidefinite scalar function not equal to zero on a solution of the system (1), and $k \geqq 0$.

Assuming $\theta(\mathbf{x}) = \mathbf{x}' \Theta(\mathbf{x})\mathbf{x}$ and rearranging the expression (10) one obtains:

$$\dot{v} = \mathbf{x}'[\Theta(\mathbf{x})\, \mathbf{x}\mathbf{x}'\, C(\mathbf{x}) - k\Theta(\mathbf{x})]\mathbf{x}. \tag{11}$$

[1] They are both polynomials of the same terms in x_i.

We want to find the matrix solution $A(x_i, x_j)$ of the equation:

$$\mathbf{x}'[B'(\mathbf{x})\, A^*(x_i, x_j) + A^*(x_i, x_j)\, B(\mathbf{x})]\mathbf{x} = \mathbf{x}'[\Theta(\mathbf{x})\, \mathbf{x}\mathbf{x}'\, C(\mathbf{x}) - k\Theta(\mathbf{x})]\mathbf{x} \tag{12}$$

obtained by equating the derivative (7) and the form (11).

From the previous considerations about the form of $A^*(x_i, x_j)$ we know that the matrix $A(x_i, x_j)$ satisfying eq. (12) has the same form as the solution of the equation:

$$\mathbf{x}'[B'(\mathbf{x})\, A(x_i, x_j) + A(x_i, x_j)\, B(\mathbf{x})]\mathbf{x} = \mathbf{x}'[\Theta(\mathbf{x})\, \mathbf{x}\mathbf{x}'\, C(x) - k\Theta(\mathbf{x})]\mathbf{x} \tag{13}$$

obtained by equating the expression (9) and the form (11).

The elements of the matrix solution $A(x_i, x_j)$ of Eq. (13) will have the form:

$$a'_{ij}(x_i, x_j) = \sum_{k,m=0}^{N} a_{ij\,km} x_i^k x_j^m \tag{14}$$

which will generally not satisfy equation (12), since the coefficients $a_{ijkm}(k, m > 0)$ are not exactly determined.

It will therefore be necessary to replace these coefficients with some arbitrary constant and build the matrix $A''(x_i, x_j)$ whose coefficients have the form:

$$a''_{ij}(x_i, x_j) = a_{ij_\infty} + \sum_{k,m=1}^{N} b_{ij\,km} x_i^k x_j^m \tag{15}$$

where $b_{ijkm} = 0$ for $a_{ijkm} = 0$ and sign $b_{ijkm} =$ sign a_{ijkm} since in expression (8) it is $\epsilon_{ij} > 0$.

By substituting the matrix $A''(x_i, x_j)$ in the equation (12) it is possible to determine the unknown constant coefficients b_{ijkm} in such a way that the matrix $A''(x_i, x_j)$ satisfies Eq. (12).

The problem of the stability of the system (1) is consequently reduced to the investigation of the properties of the matrix $A''(x_i, x_j)$ and the value of the constant k.

In the case of second order systems every polynomial function in $\mathbf{x}$ is representable in the form (3). This is no more true for systems of order $m \geqq 3$. The applicability of the form (3) consequently decreases with the increasing of the order of the system. It is however possible to construct generating v-functions also for higher order systems. These more general v-functions will be used if the simpler form (3) does not yield the wanted answer. Let us consider in the case of the third order system the v-function.

$$v = \mathbf{x}'[D\mathbf{x}\mathbf{x}'E + A]\mathbf{x} \tag{16}$$

where A, D, and E are symmetric matrices and

$$\begin{aligned} A &= \{a_{ij}(x_i, x_j)\} \\ D &= \{d_{ij}(x_i, x_j)\} \\ E &= \{e_{ij}(x_i, x_j)\}. \end{aligned} \tag{17}$$

The total time derivative of the form (16) with respect to the system (5) is:

$$\frac{dv}{dt} = \mathbf{x}' \left[\left(B'D + DB + \frac{dD}{dt}\right) \mathbf{xx}'E + D\mathbf{xx}' \left(B'E + EB + \frac{dE}{dt}\right) + \left(B'A + AB + \frac{dA}{dt}\right)\right] \mathbf{x} \tag{18}$$

By introducing the matrices A^*, D^* and E^* in a similar way as in expression (8) one obtains

$$\frac{dv}{dt} = \mathbf{x}'[(B'D^* + D^{*\prime}B)\,\mathbf{xx}'E + D\mathbf{xx}'(B'E^* + E^{*\prime}B) + (B'A^* + A^{*\prime}B)]\mathbf{x}. \tag{19}$$

The problem of stability of the third order system is therefore reduced to the computation of the 18 elements of the matrices A, D and E which satisfy the following equation:

$$\begin{aligned} \mathbf{x}'[(B'D + DB)\,\mathbf{xx}'E + D\mathbf{xx}'(B'E + EB) + B'A + AB]\mathbf{x} \\ = \mathbf{x}'[\Theta(\mathbf{x})\,\mathbf{xx}'\,C(\mathbf{x}) - k\Theta(\mathbf{x})] \cdot \mathbf{x} \end{aligned} \tag{20}$$

where for the definiteness of (16) it is sufficient that the matrices A, D, E are semidefinite if there does not exist a point $\mathbf{x}_e \neq 0$ of E_n such that $A(\mathbf{x}_e) = D(\mathbf{x}_e) = E(\mathbf{x}_e) = 0$. In the case of the fourth order system the form (16) is still satisfactory. For a fifth order system one can in the general case use the following form:

$$v = \mathbf{x}'[F\mathbf{xx}'G\mathbf{xx}'H + D\mathbf{xx}'E + A]\mathbf{x}. \tag{21}$$

Where A, F, G and H are symmetric matrices and either D or E is symmetrical, and

$$\begin{aligned} A &= \{a_{ij}(x_i, x_j)\} \qquad D = \{d_{ij}(x_i, x_j)\} \qquad E = \{e_{ij}(x_i, x_j)\} \\ H &= \{h_{ij}(x_i, x_j)\} \qquad G = \{g_{ij}(x_i, x_j)\} \qquad F = \{f_{ij}(x_i, x_j)\} \end{aligned} \tag{22}$$

Similar v-functions can be constructed for every system.

The same method for constructing Liapunov's functions can also be used in the "Krasovski approach", by introducing a generating v-function of the type

$$v = \mathbf{X}'(\mathbf{x})\, A(x_i, x_j)\, \mathbf{X}(\mathbf{x}) \tag{23}$$

whose total time derivative with respect to the system (1) has the form:

$$\dot{v} = \mathbf{X}'(\mathbf{x}) \left[J'(\mathbf{x})\, A(x_i, x_j) + A(x_i, x_j)\, J(\mathbf{x}) + \frac{dA(x_i, x_j)}{dt}\right] \mathbf{X}(\mathbf{x}) \tag{24}$$

where $J(\mathbf{x})$ is the Jacobian matrix of the system (1).

The vantage of this approach is that $J(x)$ is uniquely defined by the system, as opposed to $B(\mathbf{x})$ which is not uniquely defined. However the procedure for the construction of Liapunov's functions is formally identical to that previously outlined.

Let us consider for example the van der Pol equation:

$$\begin{aligned} \dot{x}_1 &= x_2 \\ \dot{x}_2 &= \epsilon(1 - x_1^2)\, x_2 - x_1 \end{aligned} \tag{25}$$

where $\epsilon = 1$.

As a generating v-function choose now the particular form

$$v = a_{11}(x_1)\, x_1^2 + 2a_{12}(x_1)\, x_1 x_2 + a_{22} x_2^2. \tag{26}$$

The expression (9) for the v-function (26) and the system (25) becomes

$$\dot{v}^* = x_2^2(a'_{12} + a_{22} - a_{22}x_1^2) + x_1x_2(a'_{11} - a_{22} + a'_{12} - a'_{12}x_1^2) - a'_{12}x_1^2 \tag{27}$$

It is important to notice that there does not exist a non-zero polynomial $a'_{ij}(x_1)$ which makes the expression (27) semidefinite. It is easy to obtain the following expressions for the variables $a'_{11}(x_1)$ and $a'_{21}(x_1)$, which make dv/dt indefinite on a closed surface.

$$\begin{aligned} a'_{11}(x_1) &= x_1^4 a_{22} - 2x_1^2 a_{22} + 2a_{22} \\ a'_{12}(x_1) &= x_1^2 a_{22} - a_{22}. \end{aligned} \tag{28}$$

The dv''/dt with respect to the system (26) of $v'' = \mathbf{x}' A''(x_i, x_j)\, \mathbf{x}$ where $A''(x_i, x_j)$ has the form (15) is:

$$\begin{aligned} \frac{dv''}{dt} = (6b_1 - b_3)\, x_1^5 x_2 + (b_3 - 4b_2 + 2a_{22})\, x_1^3 x_2 + (3b_3 - 2a_{22})\, x_1^2 x_2^2 \\ - b_3 x_1^4 + 2a_{22} x_1^2. \end{aligned} \tag{29}$$

By choosing $b_3 = 6b_1$ and comparing the expressions (29) and (10) it follows

$$\theta(\mathbf{x}) = x_1^2 \tag{30}$$

$$g[\xi(\mathbf{x})] = (6b_1 - 4b_2 + 2a_{22})\, x_1x_2 + (18b_1 - 2a_{22})\, x_2^2 - 6b_1x_1^2 + 2a_{22}. \tag{31}$$

The curve $\xi(\mathbf{x}) = 0$ has consequently the equation

$$3b_1x_1^2 - (3b_1 - 2b_2 + a_{22})\, x_1x_2 + (a_{22} - 9b_1)\, x_2^2 = a_{22}. \tag{32}$$

It is now possible to choose the parameters b_1, b_2 and a_2 in such a way that $\xi(\mathbf{x}) = 0$ and v'' are closed curves.

For example (Fig. 1) choose:

$$b_1 = 1, \quad b_2 = 6, \quad a_{22} = 10 \qquad \text{and} \qquad b_1 = 1, \quad b_2 = 7.5, \quad a_{22} = 12.$$

The curve (32) is for the first set of constants, the ellipse $\xi_1 = 0$ with equation

$$3x_1^2 - x_1x_2 + x_2^2 = 10 \tag{33}$$

and, for the second set, the circle $\xi_2 = 0$ with equation

$$x_1^2 + x_2^2 = 4.$$

It is possible, for example, to verify that the curve of the family circumscribed to (33) is $v'' = 220$ and that the inscribed one is $v'' = 40$.

3. Local Stability in Higher Order Systems

In the case of higher order systems, many aspects of the stability problem are still a matter of research. The identification of limit cycles is in this case, for example, still an unsolved problem. We can point out that in the case of systems with limit cycles the existence of dv/dt indefinite on a closed surface is not only unnecessary, but also extremely unlikely. In some cases however, we are able to prove existence of local stability for systems with only one critical point by applying the outlined procedure.

Example. Let us consider the system:

$$\begin{aligned} \dot{x}_1 &= x_2 \\ \dot{x}_2 &= x_3 \\ \dot{x}_3 &= 3x_3 - 2x_2 + x_1^3 \end{aligned} \tag{34}$$

which has only one critical point at $x = 0$, and let

$$v = a_{11}x_1^2 + a_{22}x_2^2 + x_3^2 + 2a_{12}x_1x_2 + 2a_{23}x_2x_3 + 2a_{13}x_1x_3 \tag{35}$$

where $a_{ij} = a_{ij}(x_i, x_j)$.

By following the outlined procedure, one finally obtains

$$v'' = \frac{3}{2}x_1^4 + 11x_2^2 + x_3^2 + 2x_1^3x_2 + 6x_2x_3 \tag{36}$$

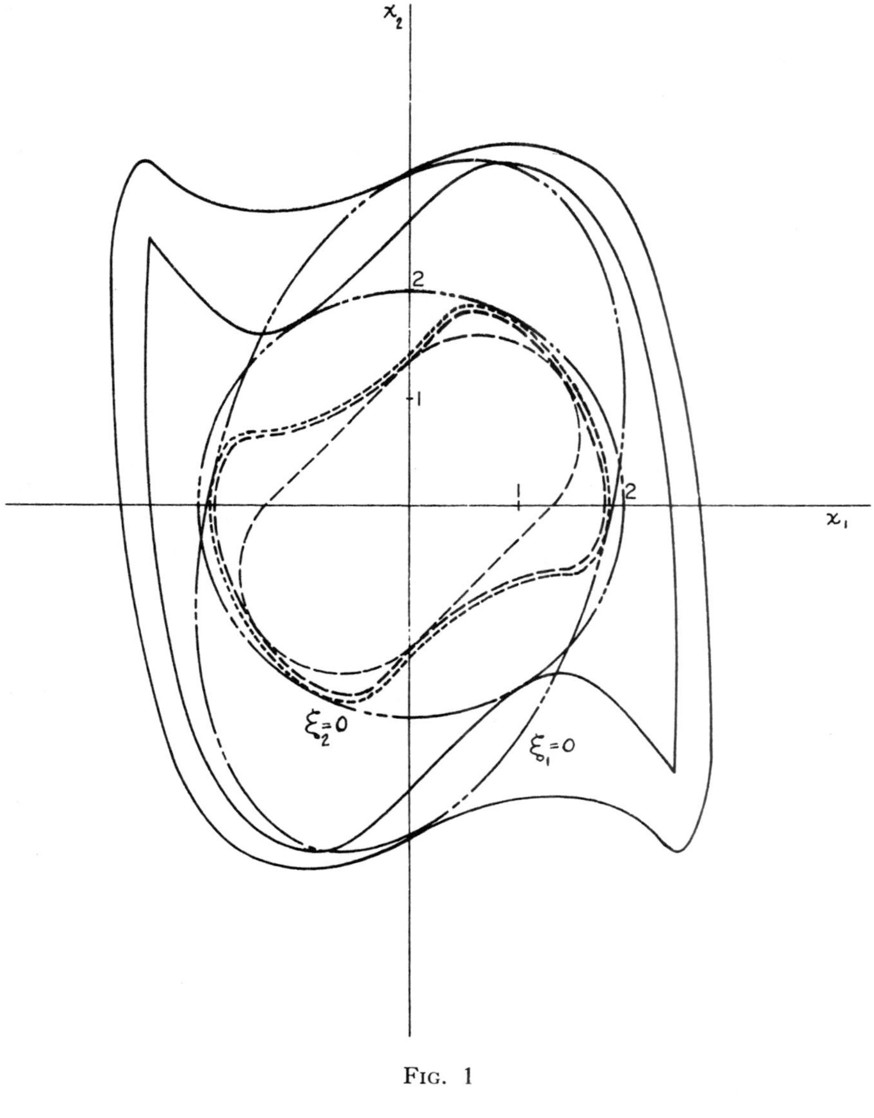

FIG. 1

and

$$\frac{dv''}{dt} = 6x_2^2(x_1^2 - 2). \tag{37}$$

The v-function (36) is positive and closed between the planes $|x_1| = \sqrt{3}$ and its total time derivative (37) is negative semidefinite, not equal to zero on a solution of the system (34) between the two planes $|x_1| = \sqrt{2}$ and positive semidefinite outside them.

The system is consequently stable inside the region defined by the closed surface

$$v'' = \tfrac{3}{2}x_1^4 + 11x_2^2 + x_3^2 + 2x_1^3x_2 + 6x_2x_3 \Big|_{|x_1| \underset{\max}{=} \sqrt{2}} \tag{38}$$

and unstable according to Chetaev's theorem [3] outside the planes $|x_1| = \sqrt{3}$.

REFERENCES

[1] Antosiewicz, H. A., A Survey of Liapunov's second method. *In* "Contributions to the Theory of Nonlinear Oscillations" (S. Lefschetz, ed.), Vol. 4, pp. 141-166. Princeton Univ. Press, Princeton, New Jersey, 1958.

[2] Barbashin, E. A., and Krasovskii, N. N., Ob ustoichivosti dvizhenia b zelom (On the global stability of a motion). *Doklady Akad. Nauk S.S.S.R.* **86**, 453-456 (1952).

[3] Chetaev, N. G., "Ustoichivost' dvizheniia" (Stability of Motion). Gostekhizdat Moskow, 1946; also Un théorème sur l'instabilité. *Compt. rend. acad. sci. U.R.S.S.* **1**, 529-531 (1934).

[4] Hahn, W., "Theorie and Anwendung der direkten Methode von Liapunov," p. 8. Springer, Berlin, 1959.

[5] Kasovskii, N. N., On the stability with large initial perturbations. *Priklad. Mat. i Mekh.* **21**, 309-319 (1957).

[6] La Salle, J. P., Some extensions of Liapunov's second method. *IRE Trans. on Circuit Theory* **CT-7**, 520-527 (1960).

[7] La Salle, J. P., and Lefschetz, S., "Stability by Liapunov's Direct Method with Applications." Academic Press, New York, 1961.

[8] Liapunov, A., "Problème Général de la Stabilité du Mouvement." Princeton Univ. Press, Princeton, New Jersey, 1947.

[9] Malkin, J. G., "Theory of Stability of Motion." Gos. Izdat. tekh.-teoret Lit., Moscow, 1952; English transl., AECtr.-3352 (1958).

[10] Massera, J. L., Contribution to stability theory. *Ann. Math. (Princeton)* **66**, 182-206 (1956).

[11] Zubov, V. I., Problems in the theory of the second method of Liapunov. Construction of the general solution in the region of asymptotical stability. *Priklad. Mat. i Mekh.* **19**, 179-210 (1955).

[12] Zubov, V. I., "Matematicheskie Metody issledovaniia Sistem avtomaticheskogo Regulirovaniia" (Mathematical Methods of Study of Automatic Control Systems). Gos. Izdat. sudoctpoitelnoi promyshlennosti, Leningrad, 1959.

An Analysis of the Bistability of the Parametron

R. A. WILLOUGHBY
International Business Machines Corporation
Thomas J. Watson Research Center, Yorktown Heights, New York

The parametron is a nonlinear and time varying electrical circuit which can oscillate in two distinct steady states. The circuit is that presented by Goto [*1*]. The purpose of the analysis is to establish the bistability on a global basis. The equations are studied in a rotating phase plane and are approximated by use of the time averaging method. While the technique is often applied to only a local study of the periodic solutions, a method is presented here which allows for a global description of the solutions. This is accomplished by the construction of a Liapunov function which has critical points corresponding to the three periodic solutions. La Salle's theorem on the extent of asymptotic stability [*2*] provides the mathematical justification.

The basic circuit is indicated below:

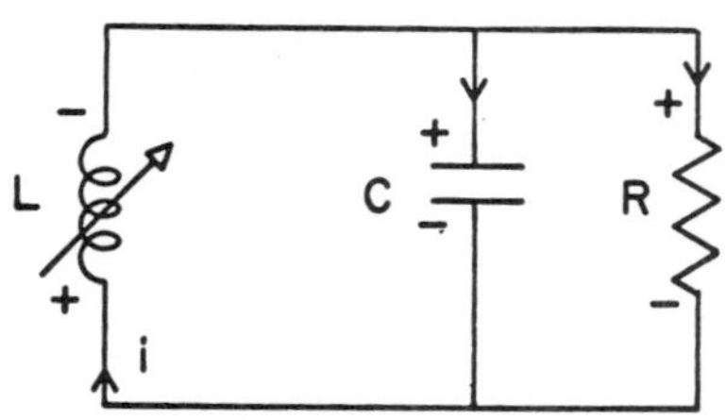

FIG. 1

The differential equation for the circuit is

$$\frac{d^2I}{d\tau^2} + \delta\frac{dI}{d\tau} + (1+\alpha)\,I(1 - 2\Gamma\sin 2\tau + \beta I^2) = 0 \tag{1}$$

where $T = \omega t$, $\omega_0 = (L_0C)^{-1/2}$, $\alpha = (\omega_0/\omega)^2 - 1$, $\delta = 1/\omega CR$, $i = L_0I/L$, $L = L_0(1 + 2\Gamma\sin 2\tau - \beta I^2)$. ω_0 is the (circular) frequency for the linearly operated LC tank circuit and the pumping of L is at the frequency of 2ω where $\omega \approx \omega_0$. 2Γ, δ and $|\beta|\max I^2$ are assumed small along with $|\alpha|$ so (1) can be treated as a perturbation of the equation $d^2I/dT^2 + I = 0$; that is, the equation for simple harmonic motion of period 2π. Under restrictions on the parameters which will be derived there is a resonance phenomenon in which the orgin $dI/dT \equiv I \equiv 0$ is unstable and there is a locking in on a subharmonic oscillation of frequency ω. This frequency is just half the frequency of the pump.

If $I(\tau)$ satisfies (1) then so do $-I(\tau)$ and $I(\tau + \pi)$. Thus one can hardly expect to have a unique stable periodic solution of period 2π. It will be shown that there are exactly two, namely $I(\tau)$ and $-I(\tau) = I(\tau + \pi)$.

Let

$$I = u \cos(\tau + \theta_0) + v \sin(\tau + \theta_0)$$

and

$$\frac{dI}{d\tau} = -u \sin(\tau + \theta_0) + v \cos(\tau + \theta_0)$$

where θ_0 is to be chosen so that $v \to 0$ as $\tau \to \infty$. The time averaged equations for u and v which aproximate (1) are given by[1]

$$\begin{aligned} \frac{du}{dT} &= \frac{v}{2}(\rho + \sqrt{1 - a^2} - b) \operatorname{sgn} \beta \\ \frac{dv}{dT} &= -\frac{u}{2}(\rho - \sqrt{1 - a^2} - b) \operatorname{sgn} \beta - av \end{aligned} \tag{2}$$

where

$$\rho = \frac{3\,|\beta|}{4\Gamma}(u^2 + v^2), \qquad T = \Gamma(1 + \alpha)\tau, \qquad \cos 2\theta_0 = -a = -\frac{\delta}{\Gamma(1 + \alpha)}$$

$$\sin 2\theta_0 = \operatorname{sgn} \beta \sqrt{1 - a^2}, \qquad b = -\frac{\alpha}{\Gamma(1 + \alpha)} \operatorname{sgn} \beta.$$

If $a^2 + b^2 < 1$, that is,

$$\alpha^2 < \Gamma^2(1 + \alpha)^2 - \delta^2, \tag{3}$$

then there are exactly three critical points for (2); namely, $v = 0$, $u = 0$, $\pm u_0$ where

$$u_0^2 = \frac{4\Gamma}{3\,|\beta|}(\sqrt{1 - a^2} + b).$$

Also, the origin is a saddle point.

[1] By definition $\operatorname{sgn} \beta = \begin{cases} +1 \text{ for } \beta > 0 \\ -1 \text{ for } \beta < 0. \end{cases}$

The following Liapunov function[1] is introduced to establish global bistability in the time averaged plane:

$$2S = \frac{2\Gamma}{3\,|\,\beta\,|}\,\rho^2 + (\sqrt{1 - a^2} - b)\,v^2 - (\sqrt{1 - a^2} + b)\,u^2. \tag{4}$$

Since

$$\frac{du}{dT} = \tfrac{1}{2}\frac{\partial S}{\partial v}\,\mathrm{sgn}\,\beta \qquad \text{and} \qquad \frac{dv}{dT} = -\,\tfrac{1}{2}\frac{\partial S}{\partial u}\,\mathrm{sgn}\,\beta - av,$$

then

$$\frac{dS}{dT} = -\,av^2(\rho + \sqrt{1 - a^2} - b) \tag{5}$$

and thus $dS/dT \leq 0$ for all (u, v). S has a saddle point at the origin and minima at $(\pm\, u_0, 0)$. By La Salle's theorem then (2) is asymptotically stable in the entire (u, v) plane. That is, every trajectory must have as a limit for $T \rightarrow +\,\infty$ one of the three points $(0, 0)$, $(u_0, 0)$, $(-\,u_0, 0)$. Since the origin is a saddle point, there are exactly two trajectories having $(0, 0)$ as a limit and these comprise the separatrix. For all other trajectories the limit is determined by which side of the separatrix the initial point lies.

If $b = 0$ (i.e., $\omega = \omega_0$) then the level curves of S are just the ovals of Cassini so that S has a simple geometric interpretation via the product of the distances to the two stable points.

While there is global bistability in the (u, v) plane it is necessary to restrict the trajectories to the region $u^2 + v^2 < 1/|\,\beta\,|$ in order to have $|\,\beta\,|\,\max I^2 < 1$. If it is assumed that at $T = 0$, $S \leq 0$ then it is sufficient to have the level curve $S = 0$ lie inside the circle $u^2 + v^2 = 1/|\,\beta\,|$. Restriction (3) then is replaced by

$$\alpha^2 < \Gamma^2(1 + \alpha)^2 - \delta^2 < \begin{cases} \left(\dfrac{3 - 5\alpha}{8}\right)^2, & \beta < 0 \\[2ex] \left(\dfrac{3 + 11\alpha}{8}\right)^2, & \beta > 0. \end{cases} \tag{6}$$

In determining the separatrix it is convenient to introduce the polar coordinate system (ρ, ϕ) where

$$\rho = \frac{3\,|\,\beta\,|}{4\Gamma}\,(u^2 + v^2) \qquad \text{and} \qquad \phi = 2\left(\arctan\frac{u}{v} + \theta_0\right)\mathrm{sgn}\,\beta.$$

[1] The method of construction is due to Moser [3].

Then from (2) it follows that

$$\frac{d\rho}{dT} = -\rho(\cos\phi + a), \qquad \frac{d\phi}{dT} = \sin\phi + b - \rho. \tag{7}$$

The limiting value for ϕ for the trajectory which approaches the origin as $T \to \infty$ is given by $\sin\phi = -b$, $\cos\phi = \sqrt{1-b^2}$. A nonlocal representation for the separatrix can be obtained by numerical integration of the equation

$$\frac{d\phi}{d\rho} = \frac{\rho - \sin\phi - b}{\rho(\cos\phi + a)} \tag{8}$$

for increasing ρ from $\rho = 0$. Special procedures can be avoided at $\rho = 0$ by noting that there one has

$$\frac{d\phi}{d\rho} = \frac{1}{a + 2\sqrt{1-b^2}}.$$

REFERENCES

[1] Goto, E., The parametron, a digital computing element which utilizes parametric oscillation. *Proc. IRE* **47**, 1304-1316 (1959).

[2] La Salle, J. P., Some extensions of Liapunov's second method. *IRE Trans. on Circuit Theory.* **CT-7**, 520-526 (1960).

[3] Moser, J. K., Bistable Systems of Differential Equations with Applications to Circuits Containing Esaki Diodes. IBM Research Report No. RC-306 (1960).

Liapunov Functions and Comparison Theorems

FRED BRAUER
Mathematics Department
University of Wisconsin, Madison, Wisconsin

1. One of the principal methods used in the study of stability of solutions of differential equations is the second method of Liapunov [*6*]. We will outline an approach, based on this method, which can be used to unify many results on stability and asymptotic behaviour. Since the proofs have already appeared elsewhere [*3*], we will confine ourselves to an outline of the results.

Liapunov's second method involves the construction of a function specially chosen to study the solutions of a system $x' = f(t, x)$. We may assume $f(t, 0) = 0$, so that $x = 0$ is a solution. This solution is said to be stable if, given any $\epsilon > 0$, any solution $x(t)$ of the system with sufficiently small $|x(0)|$ exists and obeys $|x(t)| < \epsilon$ for $0 \leq t < \infty$. The solution $x = 0$ is said to be asymptotically stable if it is stable and if every solution $x(t)$ with sufficiently small $|x(0)|$ tends to zero as $t \to \infty$.

Let $V(t, x)$ be a function defined for $0 \leq t < \infty$ and real n-vectors x, with non-negative real values, satisfying

$$\lim_{|x| \to \infty} V(t, x) = \infty,$$

which is continuous in (t, x) and has one-sided partial derivatives with respect to t and the components of x. These differentiability assumptions can be relaxed without changing the approach in any essential way, as has been done by Antosiewicz [*2*]. We use V_t to denote a partial derivative of V with respect to t, V_x to denote some gradient vector of V with respect to x, and $\cdot$ to denote the usual scalar product of vectors. Any condition which involves V_t or V_x will always be required for all one-sided derivatives. We define

$$V^*(t, x) = V_t(t, x) + V_x(t, x) \cdot f(t, x).$$

The function $V(t, x)$ is said to be positive definite if, given any $\epsilon > 0$, there exists $\delta > 0$, depending only on ϵ, such that $V(t, x) < \delta$ implies $|x| < \epsilon$.

Liapunov proved [6] that if there exists a positive definite function $V(t, x)$ with $V^*(t, x) \leq 0$, then the solution $x = 0$ of $x' = f(t, x)$ is stable. If, in addition, $V(t, x) \to 0$ with $|x|$ uniformly in t, and $V^*(t, x)$ is negative definite, then the solution $x = 0$ is asymptotically stable. Actually, Liapunov assumed that $V(t, x)$ was defined only for x in some bounded set; since we are interested only in solutions starting, close to zero, this assumption is enough.

We will consider the function $V(t, x(t))$ as a measure of the size of the solution $x(t)$ of the system $x' = f(t, x)$. Instead of assuming $V^*(t, x)$ to be negative, we assume a condition of the form

$$V^*(t, x) \leq \omega[t, V(t, x)],$$

where $\omega(t, r)$ is a continuous real-valued function on $t \geq 0$, $r \geq 0$. We will compare the solutions of the system to the solutions of the scalar equation $r' = \omega(t, r)$.

2. Conti [4] proved that if $x(t)$ is a solution of the system, if $r_0 = V(0, x(0))$, and if $r(t)$ is the maximum solution of the scalar equation with $r(0) = r_0$, then $x(t)$ can be continued to the right as far as $r(t)$ can be continued, and $V(t, x(t)) \leq r(t)$ for all such t. If $V(t, x)$ is continuously differentiable, as Conti assumes, then $V^*(t, x(t))$ is the derivative of $V(t, x(t))$. Under our weaker differentiability assumptions, or under the still weaker assumptions of Antosiewicz [2], we can get a similar relation between $V^*(t, x(t))$ and $V(t, x(t))$, which is sufficient to complete the proof in a straightforward manner.

This comparison proves that if all solutions of the scalar equation can be continued for all t, then the same is true for all solutions of the system. Also, boundedness of all solutions, or the tending to zero of all solutions of the scalar equation imply the same properties for all solutions of the system. The continuation and boundedness properties do not even require V to be positive definite, merely non-negative. It is worth noting that if only some of the solutions of the scalar equation have one of these properties, then it still follows that certain solutions of the system have the same property.

If we have a lower bound, of the form

$$V^*(t, x) \geq \theta[t, V(t, x)],$$

we obtain the opposite sort of inequality. If $\rho(t)$ is the minimum solution of $\rho' = \theta(t, \rho)$, $\rho(0) = V(0, x(0))$, then $V(t, x(t)) \geq \rho(t)$ as far to the right as $x(t)$ exists.

3. To study the behaviour of solutions for large t, we begin by assuming a two-sided bound,

$$\theta(t, V(t, x)) \leq V^*(t, x) \leq \omega(t, V(t, x)),$$

where $\theta(t, r)$ and $\omega(t, r)$ are monotone non-decreasing functions of r for each fixed t. If all solutions of both scalar equations $\rho' = \theta(t, \rho)$ and $r' = \omega(t, r)$ tend to limits as $t \to \infty$, then for any solution $x(t)$ of the system, $V(t, x(t))$ tends to a limit as $t \to \infty$. Here we note that the choices $\omega(t, r) = 0$ or $\theta(t, r) = 0$ are permissible. Also, if $\omega(t, r) \leq 0$, no lower bound on $V^*(t, x)$ is required. As before, if some but not all solutions of the scalar equations tend to limits, we can still draw conclusions about certain solutions of the system.

It might be considered desirable to be able to conclude that if all solutions of the scalar equations tend to limits, then so do all solutions of the system. The example $x_1' = x_2$, $x_2' = -x_1$, which has periodic solutions, with $V(t, x) = x \cdot x$, $V^*(t, x) = 0$, shows that such a conclusion would be false. To prove that all solutions of the system tend to limits, we must impose a stronger condition on V, but at the same time we need impose only an upper bound on V^*. Thus we now assume

$$V^*(t, x) \leq \omega(t, V(t, x))$$

for some function $\omega(t, r)$ which is monotone non-decreasing in r for each t, as before. In addition, we assume

$$V\left[t, \int_a^t f(s, x(s))\, ds\right] \leq \int_a^t V^*(s, x(s))\, ds$$

for $t > a \geq 0$ and all continuous vectors $x(t)$. Now, if all solutions of the scalar equation $r' = \omega(t, r)$ tend to limits, then all solutions of the system tend to limits.

If we assume a uniformity condition on all solutions through a given point, we can do still more. Let z be any n-vector. Suppose there is a solution $r(t)$, possibly depending on z, of the scalar equation such that for any solution $y(t)$ of the system with $y(t_0) = z$ for any $t_0 \geq 0$ we have $V(t, y(t)) \leq r(t)$ for $t \geq 0$. This condition is satisfied, for example, if V is non-increasing in t for each x and $\omega(t, r) \geq 0$. It is easy to see that this condition rules out the possibility of asymptotic stability. Then, if the hypotheses made earlier are also satisfied, for any vector x_∞, there is a solution $x(t)$ of the system on $t \geq 0$, such that $\lim_{t\to\infty} x(t) = x_\infty$.

4. Here we indicate some applications of this approach. It is easy to see that Liapunov's results mentioned earlier are included. At the same time, we can treat linear systems and small perturbations of linear systems by using $V(t, x) = x \cdot x$.

First we examine the linear system $x' = A(t)x$. Let $A^T(t)$ be the transpose of $A(t)$ and let $\Lambda(t)$ and $\lambda(t)$ be the largest and smallest characteristic roots respectively of the symmetric matrix $\frac{1}{2}[A(t) + A^T(t)]$. Then

$$V^*(t, x) = x \cdot (A + A^T)x \leq 2\Lambda(t)\, V(t, x).$$

Thus we are led to the scalar equation $r' = 2\Lambda(t)r$ with solutions $r(t) = r_0 \exp [2 \int_0^t \Lambda(s)\, ds]$. If $\int_0^\infty \Lambda(s)\, ds < \infty$, we obtain stability of the zero solution, and if $\int_0^\infty \Lambda(s)\, ds = -\infty$, we obtain asymptotic stability. We also obtain bounds on the Euclidean norm of any solution, namely

$$\| x(t) \|^2 = x(t) \cdot x(t) \leqq \| x(0) \|^2 \exp \left[2 \int_0^t \Lambda(s)\, ds\right],$$

or

$$\| x(t) \| \leq \| x(0) \| \exp \left[\int_0^t \Lambda(s)\, ds\right].$$

In the same way, we obtain a lower bound,

$$\| x(t) \| \geq \| x(0) \| \exp \left[\int_0^t \lambda(s)\, ds\right].$$

For the non-linear system

$$x' = A(t)x + g(t, x),$$

we suppose that given any $\epsilon > 0$ there exists $\delta > 0$ such that $\| g(t, x) \| < \epsilon \| x \|$ for $\| x \| < \delta$. Then the choice $V(t, x) = x \cdot x$ gives

$$\omega(t, r) = 2r[\Lambda(t) + \epsilon], \qquad r(t) = r_0 \exp \left\{2 \left[\int_0^t \Lambda(s)\, ds + \epsilon t\right]\right\}.$$

If

$$\limsup_{t\to\infty} \frac{1}{t} \int_0^t \Lambda(s)\, ds < 0,$$

and ϵ is small enough, $r(t) \to 0$ and we have asymptotic stability of the zero solution.

Another application is to various results of Wintner (see for example ref. [7]). We use

$$V(t, x) = | x | = \sum_{i=1}^{n} | x_i |,$$

and then

$$-|f(t,x)| \leqq V^*(t,x) \leqq |f(t,x)|.$$

The additional condition imposed to prove that solutions tend to limits follows from $|\int_a^t h(s)\,ds| \leq \int_a^t |h(s)|\,ds$ for any continuous h. Now we assume $|f(t,x)| \leq \omega(t,|x|)$, and compare solutions of the system with solutions of the scalar equation. We note that in this situation, if the norm of a solution tends to a limit, then the solution itself tends to a limit vector.

In particular, if $\omega(t,r) = \lambda(t)\,\phi(r)$, where $\phi(r)$ is a monotone non-decreasing function, corresponding to a condition $|f(t,x)| \leq \lambda(t)\,\phi(|x|)$, we can analyse the solutions of the scalar equation in detail. If

$$\int^{\infty} \frac{dr}{\phi(r)} = \infty,$$

every solution $r(t)$ can be continued. If, in addition,

$$\int^{\infty} \lambda(t)\,dt < \infty,$$

every solution $r(t)$ tends to a limit $r(\infty)$, and also

$$r(\infty) - r(t) \leq K \int_t^{\infty} \lambda(s)\,ds$$

for some constant K depending only on $r(0)$. This yields

$$|x(\infty) - x(t)| \leq K \int_t^{\infty} \lambda(s)\,ds$$

for any solution $x(t)$ of the system tending to a limit $x(\infty)$. We also obtain the existence of solutions tending to any prescribed limit.

The problem of finding solutions which tend to a prescribed limit arises in questions of asymptotic equivalence. Consider the two systems

$$x' = A(t)x$$

and

$$y' = A(t)y + g(t,y).$$

If $X(t)$ is a fundamental matrix for the first system, then any solution has the form

$$x(t) = X(t)c,$$

for some constant vector c. If a solution $y(t)$ of the second system is written as

$$y(t) = X(t)\, v(t),$$

then $v(t)$ satisfies

$$v' = X^{-1}(t) f(t, X(t)v).$$

If we can find a solution $v(t)$ of this system which tends to the limit c, then we have a correspondence between solutions $x(t)$ of the first system and solutions $y(t)$ of the second system with

$$y(t) - x(t) = X(t)\, [v(t) - c],$$

or

$$|\, y(t) - x(t) \,| = o(|\, X(t) \,|) \qquad \text{as} \qquad t \to \infty.$$

Finally, suppose we have a function $V(t, x)$ for the system

$$x' = f(t, x),$$

with

$$V^*(t, x) \leq \omega[t, V(t, x)].$$

We may try to use this as a Liapunov function for the perturbed system $y' = f(t, y) + R(t, y)$, as suggested by Antosiewicz [1] and Corduneau [5]. Under the assumptions

$$|\, V_y(t, y) \,| \leq \theta_1[t, V(t, y)], \qquad |\, R(t, y) \,| \leq \theta_2[t, V(t, y)],$$

we obtain

$$V^*(t, y) \leq \omega[t, V(t, y)] + \theta_1[t, V(t, y)]\, \theta_2[t, V(t, y)].$$

Thus we are led to two comparison equations

$$r' = \omega(t, r),$$

and

$$\rho' = \omega(t, \rho) + \theta_1(t, \rho)\, \theta_2(t, \rho).$$

As a very crude example of what can be done by this approach, we can apply it to

$$x' = Ax, \qquad y' = Ay + R(t, y),$$

where A is a constant matrix, using $V(t, x) = x \cdot x$. The result we obtain is that if the characteristic roots of $\frac{1}{2}(A + A^T)$ are no greater than

$-\sigma \leq 0$, if $|R(t, y)| \leq \theta(t, |y|)$, and if $u = 0$ is asymptotically stable for $u' = -\sigma u + \theta(t, u)$, then $y = 0$ is asymptotically stable for $y' = Ay + R(t, y)$. Presumably, better results of this type could be obtained with more detailed analysis.

REFERENCES

[*1*] Antosiewicz, H. A., Stable systems of differential equations with integrable perturbation term. *J. London math. Soc.* **31**, 208-212 (1956).

[*2*] Antosiewicz, H. A., An inequality for approximate solutions of ordinary differential equations. *Math. Z.* **78**, 44-52 (1962).

[*3*] Brauer, F., Global behaviour of solutions of ordinary differential equations, *J. math. Anal. and Appl.* **2**, 145-158 (1961).

[*4*] Conti, R., Sulla prolungabilita delle soluzioni di un sistema di equazioni differenziali ordinarie. *Boll. Unione mat. ital.* **11**, 510-514 (1956).

[*5*] Corduneanu, C., Sur la stabilité asymptotique, II. *Acad. rep. pop. Romîne, Fil. Iasi, Stud. cerc. sti., Ser. I. Mat.* **10**, 209-213 (1959).

[*6*] Liapunov, A. M., "Problème général de la stabilité du mouvement." Princeton Univ. Press, Princeton, New Jersey, 1947.

[*7*] Wintner, A., Ordinary differential equations and Laplace transforms (Appendix). *Am. J. Math.* **79**, 265-294 (1957).

Structures Transverse to a Vector Field

BRUCE L. REINHART[1]
RIAS and University of Maryland
Baltimore, Maryland

In the case in which the lines of flow of a vector field form the fibres of a locally trivial fibre space over a manifold, the algebraic topological and differential geometric invariants of the base space can be considered as properties of the flow. These properties in some sense describe the behaviour of the flow in a direction transverse to the direction of motion. In general, instead of the base space of a fibering we have the quotient space formed by identifying each line of flow to a point, which latter space is usually too degenerate to carry all the transverse structure which the given flow may actually possess. My purpose in this lecture is to point out the importance of developing suitable techniques for the study of such transverse structures. Since I am discussing a nonexistent general theory, I shall proceed by giving examples.

Let us first look at some of the transverse structures which actually arise.

1. The flow defined by a vector field has the tranverse structure of a manifold, except at singular points.

2. There may be a differentiable or analytic structure in the transverse space, possibly even at singular points if suitable definitions are adopted. Analytic transverse structures of a vector field with singularities have in fact been the most important tool in the thesis of Haefliger [*3*].

3. Suppose that on the manifold M of dimension $2n + 1$, there is a contact structure, that is, a one form ω such that $\omega \wedge (d\omega)^n \neq 0$. Then there is a nonsingular flow defined by $(d\omega)^n = 0$, and this flow has a transverse symplectic structure defined by $d\omega$.

4. More generally, if we replace the action of the real line by that of any finite or infinite Lie group, and suppose the action is reducible but not completely reducible, then the orbits of a subgroup which acts

[1] This research was partially supported by the United States Air Force through the Air Force Office of Scientific Research of the Air Research and Development Command, under Contract Number AF 49(638)-382. Reproduction in whole or in part is permitted for any purpose of the United States Government.

irreducibly have the transverse structure of a space acted on by a Lie group. Hermann [4] has studied the related case of a foliation with the transverse structure of a Lie group.

5. The integral invariants of Cartan [1] are differential forms transverse to a flow. In the case of a fibering, they are exactly the differential forms on the base space. Since the ring of integral invariants is closed under exterior differentiation, we may define certain cohomology groups, generalizing the cohomology of the base space.

Let us look more closely at these cohomology groups. One natural question to ask is: do they satisfy the Poincaré duality theorem? They certainly do in the case of a fibration. In the case of a flow with bundle-like metric on a compact manifold, the duality theorem may be proved by use of harmonic integral theory [8], even though the quotient space may have, for example, the indiscrete topology. If the flow comes from an action of the circle group, Koszul [5] has showed that whether or not the space acted on is compact, the cohomology of the integral invariants gives the real cohomology of the quotient space. Hence, in good cases a suitable duality theorem may be expected.

For the rest of the cohomology discussion, I shall limit myself to nonsingular flows. Let T be a nonsingular vector on V^{n+1}, and let ω be a differential form such that $\omega(T) = 1$. Let $i(T)$ be the operator of contraction by T, and $\theta(T)$ be Lie differentiation by T. Then an integral invariant is a differential form ϕ such that $i(T)\phi = \theta(T)\phi = 0$. In terms of local coordinates $(x^1, \ldots, x^n, t)$ such that $T = \partial/\partial t$, ϕ has the form

$$\sum \phi_{i_1 \ldots i_p}(x)\, dx^{i_1} \wedge \ldots \wedge dx^{i_p}.$$

In the case $p = 0$, the closed forms are exactly the constant functions, so the cohomology group has dimension 1. The case $p = n$ is more complicated. There is a one one correspondence between integral invariants of degree n and invariant measures, the mapping being $\phi \to \omega \wedge \phi$ and its inverse being $\psi \to i(T)\psi$. This tells us immediately that we cannot hope in general to find a differentiable ϕ in degree n. However, by the theorem of Kryloff and Bogoliuboff [6, 7], such a ϕ does exist in the sense of measure theory, for any flow on a compact metric space. Hence, if we are to have a duality theorem of any generality, we need to consider something like currents (differential forms with distribution coefficients). Under suitable hypotheses, we would hope to show that the cohomology class of the measure depends only upon the value of its integral over the total space, hence that the cohomology group of dimension n is a vector space of dimension 1. In general, we have a pairing of integral invariants of degree p with those of degree $n - p$ into the reals by

$(\phi^p, \psi^{n-p}) \rightarrow \int \phi \wedge \psi \wedge \omega$. Assuming a reasonable theorem about forms of degree n, we may hope that this pairing gives rise to an isomorphism.

It is important to notice that the cohomology theory of integral invariants is not locally trivial, in the sense that a given line of flow may have no neighborhood which is the union of lines of flow and which has trivial cohomology. Hence, in addition to the cohomology discussed above, there are other types of cohomology groups which may be defined. Definitions and elementary properties may be found in the recent note of Clifton and Smith [*2*], along with an application to finding sufficient conditions for a Lorentz structure to admit a closed time-like path.

In closing, let me remark that I would like to see studied the transverse properties not only of vector fields but of a fairly broad class of foliations with singularities. There are two ways of defining such a foliation: as the level surfaces of a family of locally defined functions, or as the orbits of locally defined Lie algebras of vector fields. The first approach is used by Haefliger in the paper cited above; the transverse structure enters explicitly into his definition. Clifton and Smith base their definition on the local automorphisms of the structure, and so include at least some examples of both kinds of singular foliation.

REFERENCES

[*1*] Cartan, E., "Leçons sur les invariants intégraux." Hermann, Paris, 1922.

[*2*] Clifton, Y. H., and Smith, J. W., The category of topological objects. *Proc. natl. Acad. Sci. U.S.* **47**, 190-195 (1961).

[*3*] Haefliger, A., Structures feuilletées et cohomologie à valeur dans un faisceau de groupoïdes. *Comm. Math. Helv.* **32**, 248-329 (1958).

[*4*] Hermann, R., On the differential geometry of foliations. *Ann. Math. (Princeton)* **72**, 445-457 (1960).

[*5*] Koszul, J. L., Sur certains groupes de transformations de Lie. *In* "Géométrie Différentielle," Colloques intern. C.N.R.S., Strasbourg, pp. 137-142. Éditions du C.N.R.S., Paris, 1953.

[*6*] Kryloff, N., and Bogoliuboff, N., Les mesures invariantes et transitives dans la mécanique nonlinéaire. *Mat. Sbornik* [N.S.] **1**, 707-710 (1936).

[*7*] Kryloff, N., and Bogoliuboff, N., La théorie générale de la mesure dans son application à l'étude des systèmes dynamiques de la mécanique non linéaire. *Ann. Math. (Princeton)* **38**, 65-113 (1937).

[*8*] Reinhart, B. L., Harmonic integrals on foliated manifolds. *Am. J. Math.* **81**, 529-536 (1959).

Systems of Differential Equations without Linear Terms[1,2]

COURTNEY COLEMAN
Mathematics Department
Harvey Mudd College, Claremont, California

1. The qualitative behavior of solutions of a system such as (1) below is often closely connected to the behavior of the solutions of the corresponding system of first approximation, (2) below. If (2) is linear, much is known [*6*]. In this paper we discuss what happens when (2) is not linear. We show that there is a set of real numbers, the radial type numbers of definitions 2 and 3 below, which reduce in the linear case to the real parts of the characteristic numbers of the matrix of coefficients. These numbers are then used to obtain a generalization (theorem 3) of the classical stability theorem of Liapunov (theorem A). To those who view questions such as these from an analytic point of view, our results may seem surprising. From a geometrical point of view, the results are almost self-evident.

2. Consider the systems,

(1) $dx/dt = f(x) + g(x, t)$ and

(2) $dx/dt = f(x)$,

where: (a) x is an n-vector; (b) $f(x)$ is homogeneous of positive degree k; (c) $f(x)$ and $g(x, t)$ are continuously differentable in x and in t for all x in a region D containing the origin and for all t; (d) $g(x, t) = 0(\| x \|^k)$ in D uniformly in t. Conditions (c) and (d) can be weakened (see the first remark at the end of section 5). If $k = 1$, $f(x) = Ax$ for some constant matrix A, and the basic stability theorem holds.

Theorem A (Liapunov) [*6*]. If the characteristic numbers of A have negative real parts, then the trivial solution, $x = 0$, of (1) is asymptotically stable.

First note that theorem A is a logical consequence of the following two theorems.

[1] Presented under the title "Asymptotic Stability" at the OSR-RIAS Symposium, Colorado Springs, August 4, 1961.

[2] This research was supported in part by the National Science Foundation under contract number NSF G 12932.

Theorem B. If the characteristic numbers of A have negative real parts, then the trivial solution of (2) is asymptotically stable.

Theorem C. If the trivial solution of (2) is asymptotically stable, so is the trivial solution of (1).

Under the weakened conditions (c) and (d) on the functions f and g, theorem C has been proved for any positive k by Massera [7]. Thus, the desired generalization of theorem A is obtained once the hypothesis of theorem B is suitably generalized and the resulting statement proved (theorem 2).

3. In order to see how the hypothesis should be generalized, we consider in detail the case that $f(x) = Ax$.

Let the matrix A have distinct characteristic roots $\mu_1, \ldots, \mu_{2r}, \mu_{2r+1}, \ldots, \mu_s$, where:

$$\mu_{2j-1} = \alpha_j + i\beta_j, \qquad \alpha_j \text{ and } \beta_j \text{ real}, \qquad j = 1, \ldots, r;$$

$$\mu_{2j} = \bar{\mu}_{2j-1}, \qquad j = 1, \ldots, r; \qquad \mu_j \text{ is real}, \qquad j = 2r+1, \ldots, s.$$

Let the multiplicity of μ_j be n_j, $j = 1, \ldots, s$, and note that

$$n_{2j-1} = n_{2j}, \qquad j = 1, \ldots, r.$$

Let $\sigma_j = n_0 + \ldots + n_j$, where for indexing consistency $n_0 = 0$. There is a real non-singular matrix P and there are integers $m_1, \ldots, m_s$ such that the linear transformation $x = Px'$ transforms (2) into $dx'/dt = Bx'$, where [2]

$$B = P^{-1}AP = \left\| \begin{array}{ccccccc} B_1 & & & & & & \\ & \cdot & & & & 0 & \\ & & \cdot & & & & \\ & & & B_r & & & \\ 0 & & & & B_{r+1} & & \\ & & & & & \cdot & \\ & & & & & & \cdot \\ & & & & & & B_s \end{array} \right\|$$

For $j = 1, \ldots, r$,

$$B_j = \left\| \begin{array}{ccccccc} B_{j1} & 0_2 & & & & & \\ I_2 & B_{j2} & \cdot & & & 0 & \\ & \cdot & \cdot & \cdot & & & \\ & & \cdot & \cdot & \cdot & & \\ & & I_2 & B_{jm_j} & \cdot & & \\ 0 & & & 0_2 & B_{jm_j+1} & & \\ & & & & \cdot & \cdot & \\ & & & & \cdot & \cdot & 0_2 \\ & & & & & 0_2 & B_{jn_j} \end{array} \right\|$$

$$B_{jk} = \begin{pmatrix} \alpha_j & -\beta_j \\ \beta_j & \alpha_j \end{pmatrix}$$

for $k = 1, ..., n_j$, $I_2 = \begin{pmatrix} 1 & 0 \\ 0 & 1 \end{pmatrix}$, $0_2 = \begin{pmatrix} 0 & 0 \\ 0 & 0 \end{pmatrix}$. For $j = r + 1, ..., n$,

$$B_j = \begin{Vmatrix} u_{j1} & & & & & & \\ 1 & \cdot & & 0 & & & \\ & \cdot & \cdot & & & & \\ & & 1 & u_{jm_j} & & & \\ & 0 & & 0 & \cdot & & \\ & & & & \cdot & \cdot & \\ & & & & & 0 & u_{jn_j} \end{Vmatrix}$$

The characteristic numbers of A coincide with those of B in value and multiplicities. For simplicity, it is assumed that A already has the form of B. In any event, it could be transformed into B before the u, y-coordinates (to be discussed below) are introduced. For such an A, the ith component, $x_i(t)$, of the general solution, $x(t)$, of (2) is as follows:

(a) if for some j_i, $1 \leq j_i \leq r$, it is true that $\sigma_{j_i-1} < i \leq \sigma_{j_i}$ and that $i - \sigma_{j_i-1}$ is odd, then

$$x_i(t) = (P_i(t) \sin \beta_{j_i} t + Q_i(t) \cos \beta_{j_i} t) \cdot \exp(\alpha_{j_i} t),$$

where $P_i(t)$ and $Q_i(t)$ are polynomials of degree $i - 1 - \sigma_{j_i-1}$;

(b) if for some j_i, $1 \leq j_i \leq r$, it is true that $\sigma_{j_i-1} < i \leq \sigma_{j_i}$ and that $i - \sigma_{j_i-1}$ is even, then

$$x_i(t) = (P_i(t) \cos \beta_{j_i} t - Q_i(t) \sin \beta_{j_i} t) \exp(\alpha_{j_i} t),$$

where $P_i(t) = P_{i-1}(t)$, $Q_i(t) = Q_{i-1}(t)$;

(c) if, for some j_i, $r < j_i \leq s$, it is true that $\sigma_{j_i-1} < i \leq \sigma_{j_i}$, then

$$x_i(t) = R_i(t) \exp(\mu_{j_i} t),$$

where $R_i(t)$ is a polynomial of degree $i - 1 - \sigma_{j_i-1}$. The coefficients of the $P_i(t)$, $Q_i(t)$, and $R_i(t)$ are functions of n arbitrary constants.

We now investigate the asymptotic behavior of this general solution of (2) by using the radial and direction coordinates used by Gomory [5]. Actually, these coordinates and the definitions that follow are introduced whether or not (2) is linear. Let $u = \| x \|$, where $\| x \|$ represents the

Euclidean norm of x, and let $y = u^{-1}x$. For $k \geq 1$, $\| x \| > 0$, system (2) becomes

$$dy/d\tau = f(y) - (y, f(y))\, y, \tag{3a}$$

$$du/d\tau = (y, f(y))\, u \tag{3b}$$

where $(y, f(y))$ is the scalar product of y and $f(y)$ and $d\tau = u^{k-1}dt$. System (3) is to be considered for $u = 0$ even though the change of variables is undefined for $u = 0$. Each solution, $y = y(\tau, 0, y^0)$, where $y^0 = y(0, 0, y^0)$, of (3a) determines a one parameter family of solutions of (3b),

$$u = u_0 \exp \int_0^\tau (y(t, 0, y^0), f(y(t, 0, y^0)))\, dt = u(\tau, 0, y^0, u_0),$$

where u_0 is the parameter. It is always assumed that $u_0 \geq 0$ and, hence, that $u(\tau, 0, y^0, u_0) \geq 0$ for all τ. If u_0 and y^0 are given, then

$$y = y(\tau, 0, y^0), \qquad u = u(\tau, 0, y^0, u_0)$$

is the solution of (3) for which

$$y^0 = y(0, 0, y^0), \qquad u_0 = u(0, 0, y^0, u_0).$$

Note that $u = 0$ defines an integral surface I in the u, y-space.

All this suggests that the three definitions that follow would be useful in describing asymptotic behavior in the u, y-space.

Definition 1. If y^0 belongs to the positive limit set of some solution of (3a), then $y = y(\tau, 0, y^0)$, $u = u(\tau, 0, y^0, u_0)$ is called a limit solution of (3).

Definition 2. The radial type number, $\lambda(y^0)$, of a solution, $y = y(\tau, 0, y^0)$, $u = u(\tau, 0, y^0, u_0)$, where $u_0 > 0$, of (3) is the number

$$\overline{\lim_{\tau \to +\infty}}\ \tau^{-1} \ln u(\tau, 0, y^0, u_0).$$

Note that $\lambda(y^0)$ is independent of u_0, $u_0 > 0$, that $\lambda(y^0)$ may be infinite, and that $\lambda(y^0) = \lambda(y^1)$ if $y^1 = y(\tau_1, 0, y^0)$ for some τ^1. An equivalent definition of $\lambda(y^0)$ is as follows.

Definition 3. $\lambda(y^0) = -\sup\,\{\lambda' : u(\tau, 0, u_0, y^0) \exp \lambda'\tau$ is bounded for all τ, $0 \leq \tau < +\infty\}$.

Definitions 2 and 3 are similar to the usual definition of type numbers or their negatives, Liapunov numbers [*1*].

Now we return to the case that $f(x) = Ax$, A having the canonical form given above. Let $x = x(t)$ be a particular solution of (1), and let μ be the largest real part of the characteristic numbers of A which actually appear as exponents in the expression for $x(t)$. Since $k = 1$, $d\tau = dt$ and

$$y(\tau) = y(t) = \| x(t) \|^{-1} x(t)$$

is a solution of the transformed system (3a). Let the components of $x(t)$ be as described previously. First, suppose that for some $j_i \leq r$, $\mu = \alpha_{j_i}$. After factoring the term

$$(P_i(t) \sin \beta_{j_i} t + Q_i(t) \cos \beta_{j_i} t) \exp (\alpha_{j_i} t)$$

out of $x(t)$, it is immediately seen that the solution $y = y(t)$ of (3a) has for its positive limit set the trajectory defined by the periodic solution of (3a), $y = y^*(t)$, where $y_i^*(t) \equiv 0$, $i \neq \sigma_{j_i - 1}$, $i \neq \sigma_{j_i}$, and

$$y^*_{\sigma_{j_i-1}}(t) = (\sin \beta_{j_i} t + K_{\sigma_{j_i-1}} \cos \beta_{j_i} t)\, (1 + K^2_{\sigma_{j_i-1}})^{-1/2},$$

and

$$y^*_{\sigma_{j_i}}(t) = (\cos \beta_{j_i} t - K_{\sigma_{j_i-1}} \sin (\beta_{j_i} t)\, (1 + K^2_{\sigma_{j_i-1}})^{-1/2},$$

where

$$K_{\sigma_{j_i-1}} = \lim_{t \to \infty} Q_i(t)\, P_i^{-1}(t) \neq 0.$$

The radial type number, $\lambda(y^*(0))$, is by definition

$$\overline{\lim_{t \to \infty}}\, t^{-1} \int_0^t (y^*(\tau)\, Ay^*(\tau))\, d\tau.$$

But a simple computation shows that

$$(y^*(\tau), Ay^*(\tau)) = \alpha_{j_i}.$$

Hence

$$\lambda(y^*(0)) = \alpha_{j_i}.$$

Now suppose that for some $j_i > r$, $\mu = \mu_{j_i}$. Then, following a procedure similar to that above, it is seen that $y = y(t)$ has for its positive limit set the point trajectory defined by the constant solution

$y = y^*(t) = (0, ..., 0, 1, 0, ..., 0)$, whose only non-zero component is the σ_{j_i}th component. In this case

$$(y^*(t), Ay^*(t)) = \mu_{j_i},$$

and it follows that

$$\lambda(y^*(0)) = \mu_{j_i}.$$

Thus, it is seen that the limit solutions of (3) correspond to the constant solutions and the periodic solutions of (3a) and that the set of radial type numbers of these is identical with the set of real parts of the characteristic numbers of A.

4. The results of 3, above, suggest that a natural extension of the hypothesis of theorem A to the non-linear case would be the following:

Hypothesis NL. The radial type number of every limit solution of (3) is negative and $k \geq 1$.

We also state

*Hypothesis NL**. The radial type number of every limit solution of (3) is positive and $k \geq 1$.

Theorem 1 (1*). Assuming $NL(NL^*)$, the radial type number of every solution of (3) is negative (positive).

Theorem 2. Assuming NL, the trivial solution of (2) is asymptotically stable.

Theorem 3. Assuming NL, the trivial solution of (1) is asymptotically stable.

Proof of theorem 1. Let $y = y(\tau, 0, y^0)$, $u = u(\tau, 0, y^0, u_0)$ be a solution of (3), where $u_0 > 0$.

If this is a limit solution, there is nothing to prove. Suppose that it is not a limit solution. Then it must be shown that $\lambda(y^0) < 0$. Let p be a point of Ω, the positive limit set of the solution $y = y(\tau, 0, y^0)$ of (3a). Then,

$$y = y(\tau, 0, p), \qquad u = u(\tau, 0, p, u_0)$$

is a limit solution and $\lambda(p) < 0$. From definition 2 there is a $\tau(p)$ such that

$$\int_0^\tau h(t, p)\, dt < 2^{-1}\, \lambda(p)\, \tau < 0$$

for all $\tau > \tau(p)$, where for brevity we let

$$h(t, p) = (y(t, 0, p), f(y(t, 0, p))).$$

Even though $\lambda(p') = \lambda(p)$, $p' = y(\tau', 0, p)$ for some $\tau' > 0$, it may happen that $\tau(p') \neq \tau(p)$. However, $y(\tau, 0, y^0)$ is continuous in τ and in y^0 and hence $h(\tau, y^0)$ is, too. For every $\epsilon > 0$ and for every point p of Ω there is a $\delta = \delta(\epsilon, p) > 0$ such that

$$| h(\tau + \tau', y^0) - h(\tau, p) | < \epsilon$$

for all τ, $0 \leq \tau \leq \tau(p)$ if

$$|| y(\tau', 0, y^0) - p || < \delta(\epsilon, p)$$

for some $\tau' \geq 0$. If $S(\delta)$ denotes the open spheroid, contained in the integral surface I, of radius δ centered at p, then

$$\bigcup_{p \in \Omega} S(\delta) \supset \Omega.$$

Since y is a unit vector, Ω is a bounded, compact set on I, and there is a finite covering of Ω by spheroids $S(\delta_i)$, $i = 1, ..., r$, $\delta_i = \delta(\epsilon, p_i)$. Let

$$\delta^* = \min_i \{\delta_i\}, \qquad \tau^* = \max_i \{\tau(p_i)\}, \qquad \lambda^* = \max_i \{\lambda(p_i)\}.$$

Note that λ^* is negative.

Since (3a) is autonomous and since the point $y(\tau, 0, y^0)$ on I tends to Ω with increasing τ, it can be assumed without loss of generality that for all $\tau \geq 0$ the Euclidean distance between $y(\tau, 0, p)$ and Ω is less than δ^*. Hence, for each $\tau \geq 0$ the point $y(\tau, 0, y^0)$. belongs to one or another of the spheroids $S(\delta_1)$, ..., $s(\delta_r)$. For each non-negative integer j let i_j be an integer, $1 \leq i_j \leq r$, for which $y(j\tau^*, 0, y^0)$ belongs to $S(\delta_{i_j})$. Thus, for $\tau > 0$, $\tau_1 = 0$, $\tau_{j+1} - \tau_j = \tau^*$ and $0 \leq \tau - \tau_{s+1} < \tau^*$,

$$\int_0^\tau h(t, y^0)\, dt = \sum_{j=0}^{s} \int_{\tau_j}^{\tau_{j+1}} h(t, y^0)\, dt + \int_{\tau_{s+1}}^{\tau} h(t, y^0)\, dt$$

$$\leq \sum_{j=0}^{s} \int_0^{\tau^*} h(t, p_{i_j})\, dt + \int_0^{\tau - \tau_{s+1}} h(t, p_{i_{s+1}})\, dt + \epsilon\tau$$

$$\leq s\, 2^{-1} \lambda^* \cdot \tau^* + \int_0^{\tau - \tau_{s+1}} h(t, p_{i_{s+1}})\, dt + \epsilon\tau.$$

Since $h(t, p_{i_{s+1}})$ is bounded for all t, and since $|\tau - \tau_{s+1}| < \tau^*$ and $|\tau - s\tau^*| < \tau^*$ for all τ, it then follows that

$$\lambda(y^0) = \overline{\lim_{\tau\to+\infty}} \tau^{-1} \int_0^\tau h(t, y^0)\, dt$$

$$\leq \overline{\lim_{\tau\to+\infty}} \left\{ \tau^{-1} s\, 2^{-1} \lambda^* \tau^* + \tau^{-1} \int_0^{\tau-\tau_{s+1}} h(t, p_{i_{s+1}})\, dt + \epsilon \right\}$$

$$= 2^{-1}\lambda^* + \epsilon.$$

Since ϵ may be chosen arbitrarily and since $\lambda^* < 0$, it follows that $\lambda(y^0) < 0$, which was to be proved.

The proof of theorem 1* is similar and is omitted.

Proof of theorem 2. First, note that the introduction of the u, y variables defines a differentiable homeomorphism of the x-space with the origin deleted onto that part of the u, y-space with $u > 0$. This homeomorphism maps trajectories of (2) onto trajectories of (3), the parametric sense being preserved since $d\tau = u^{k-1}dt$ and $u > 0$. The inverse of the homeomorphism is also a differentiable, trajectory preserving mapping.

Let $x = x(t, 0, x^0)$ be the solution of (2) for which $x(0, 0, x^0) = x^0$. It is known [8] that the trivial solution of (2) is asymptotically stable if and only if: (a) for all $x^0 \neq 0$ it is not true that $x(t, 0, x^0)$ tends to 0 as t tends to $-\infty$; (b) there is a neighborhood of the origin containing no complete trajectories—those corresponding to a solution $x = x(t, 0, x^0)$, $-\infty < t < +\infty$ — except the trivial one. By virtue of the trajectory homeomorphism defined above conditions (a) and (b) are equivalent to conditions (a′) and (b′) relative to system (3): (a′) for all $u_0 > 0$ and for all y^0 it is not true that $u(\tau, 0, y^0, u_0)$ tends to 0 as τ tends to $-\infty$; (b′) for some positive u_1 the subset of the u, y-space defined by $0 \leq u \leq u_1$ contains no complete trajectories—those corresponding to a solution $y = y(\tau, 0, y^0)$, $u = u(\tau, 0, y^0, u_0)$, $-\infty < \tau < +\infty$-except for those on the integral surface I defined by $u = 0$.

Now all that is necessary is to show that the hypothesis of theorem 2 implies (a′) and (b′). Let (3*) be the system obtained from (3) by replacing τ by $-\tau$. Then the radial type numbers of the limit solutions of (3*) are positive. By theorem 1* applied to system (3*) the radial type numbers of all solutions of (3*) are positive. Referring to definition 3, this implies that every solution of (3) is unbounded in the interval $(-\infty, 0)$. This immediately implies (a′) and (*b′*) and the theorem is proved.

Proof of theorem 3. This is a logical consequence of theorem C for $k \geq 1$ and of theorem 2.

5. Remark 1. Condition (c) on the functions f and g can be replaced by any conditions sufficient to guarantee existence, uniqueness, and continuity in initial values of solutions of (1) and of (2). Condition (d) on g can, in a certain sense, be weakened; for, Massera has shown [7] that if the trivial solution of (2) is asymptotically stable, there is a positive constant M such that for any function $g(x, t)$, for which

$$\| g(x, t) \| \leq M \| x \|^k$$

uniformly in t in some neighborhood of 0, the trivial solution of (1) is asymptotically stable.

Remark 2. Zubov [8] has shown that the trivial solution of (2) is asymptotically stable if and only if for every x^0, $\| x^0 \| = 1$, it is true that $\| x(t, 0, x^0) \| \leq at^{-\alpha}$ for all $t > T$, where a, α, and T are constants and $x = x(t, 0, x^0)$ is the solution of (2) for which $x(0, 0, x^0) = x^0$. His approach is completely different from that given here, and his results are not comparable.

Remark 3. Some comments made by Lefschetz to the author and a previous paper of the author's [3] were the stimuli for this work, which has previously appeared in summary form [4]. In future papers hypotheses NL and NL^* will be used to generalize a number of other classical theorems of stability.

REFERENCES

[1] Cesari, L., "Asymptotic Behavior and Stability Problems in Ordinary Differential Equations." Springer, Berlin, 1959.

[2] Coddington, E., and Levinson, N., "Theory of Ordinary Differential Equations." (McGraw-Hill, New York, 1955).

[3] Coleman, C., Asymptotic stability in 3-space. *In* "Contributions to the Theory of Nonlinear Oscillations" (S. Lefschetz, ed.,) Vol. 5, pp. 257-268. Princeton Univ. Press, Princeton, New Jersey, 1961.

[4] Coleman, C., Systems of differential equations without linear terms. *Proc. nat. Acad. Sci. U.S.* **47**, 1650-1651 (1961).

[5] Gomory, R., Trajectories tending to a critical point in 3-space. *Ann. Math. (Princeton)* **61**, 140-153 (1955).

[6] Lefschetz, S., "Differential Equations: Geometric Theory." Interscience, New York, 1957.

[7] Massera, J., Contributions to stability theory. *Ann. Math. (Princeton)* **64**, 182-205 (1956).

[8] Zubov, V., "Metody A. M. Liapunova i ikh Primenenie" (Methods of A. M. Liapunov and Their Applications). Izdat. Leningrad Univ., 1957.

Prolongations and Generalized Liapunov Functions

JOSEPH AUSLANDER[1] AND PETER SEIBERT[2]
RIAS, Baltimore, Maryland

Introduction

In this paper, we outline a theory of stability in dynamical systems which unites three hitherto unrelated points of view. The first is a generalized version of the second method of Liapunov which was inspired by Zubov [*8*]. A concise exposition of this point of view has been given by Lefschetz [*1*]. The second is the concept of "prolongation", which is a method of continuing orbits beyond their omega limit sets. This notion was first considered by Poincaré, in a special context, in connection with his studies of dynamical systems in the plane. It was developed further and extensively studied by Ura [*5*, *6*]. The third component is the concept of "stability under perturbations" or "total stability" due to Dubošin, Goršin, and Malkin [*2*]. This notion was introduced in order to clarify the effect of permanently acting perturbations upon stability properties.

The concept of prolongation leads naturally to the definition of new concepts of stability, all of which take intermediate positions between ordinary Liapunov stability and asymptotic stability. The most important of these is called absolute stability which can be characterized by the existence of a continuous generalized Liapunov function. In this connection, asymptotic stability can be characterized by the existence of a continuous strictly decreasing Liapunov function.

By using the fundamental properties of prolongations as axioms, we define "abstract prolongations". This enables us to consider total stability from the point of view of prolongations.

Parallel to stability, we consider the related concepts of boundedness. In many cases, there is a kind of duality between stability and boundedness. This analogy was also emphasized by Yoshizawa [*7*]. In the study of boundedness, we make use of generalized Liapunov functions at infinity.

[1] This research was partially supported by the United States Air Force through the Air Force Office of Scientific Research of the Air Research and Development Command, under Contract Number AF 49(638)-382, and [2] ARO (Durham) Contract DA-36-034-ORD-3220. Reproduction in whole or in part is permitted for any purpose of the United States Government.

§ 1. General Concepts and Notations

Consider the system of differential equations

$$\dot{x} = f(x) \tag{1}$$

where x and f are n-vectors and where f is defined in a region X of R^n. Let conditions be placed upon f sufficient to guarantee the existence and uniqueness of solutions through every point x_0 of X, the solutions depending continuously on the initial value and being defined for all real t. Then, as is well known [3], the solution curves define a *dynamical system* or *continuous flow* on X. That is, there is a continuous map π of $X \times R$ onto X satisfying

$$\begin{aligned} &\text{(a)} \quad \pi(x, 0) = x && (x \in X); \\ &\text{(b)} \quad \pi(\pi(x, t_1), t_2) = \pi(x, t_1 + t_2) && (x \in X;\, t_1, t_2 \in R). \end{aligned} \tag{2}$$

The *motions* or *orbits* of the dynamical system (2) are the solution curves of (1).

The notions occurring in the first four sections of this paper may be formulated directly in terms of the dynamical system (2), without explicit reference to the system of differential equations which gives rise to it. Therefore, let us suppose that we are given a locally compact second countable metric space X, with metric d, and a dynamical system (2), on X. We shall suppress notationally the map π; if $x \in X$ and $t \in R$, we shall write xt in place of $\pi(x, t)$.

We adopt the following definitions and notations. If $x \in X$, $xR^+ = [xt \mid t \geqq 0]$, is called the *positive semiorbit* of x. If $A \subset X$, $\bar{A}$ will denote the closure of A in X, and, if $\epsilon > 0$,

$$S_\epsilon(A) = [y \in X \mid d(y, A) < \epsilon].$$

If Q_0 is a map of X into 2^X (the set of all subsets of X) and $A \subset X$, then

$$Q_0(A) = \bigcup_{x \epsilon A} Q_0(x).$$

§ 2. Prolongations

We now define two operators, $\mathscr{D}$ and $\mathscr{S}$, on the maps from X to 2^X. If $Q_0\colon X \to 2^X$, then

$$(\mathscr{D}Q_0)(x) = \bigcap_{W \in \mathscr{N}(x)} \overline{Q_0(W)},$$

where $\mathcal{N}(x)$ denotes the neighborhood filter of x. We define

$$(\mathscr{S}Q_0)(x) = \bigcup_{n=1,2,\ldots} Q_0^n(x),$$

where

$$Q_0^1(x) = Q_0(x), \qquad \text{and} \qquad Q_0^n(x) = Q_0(Q_0^{n-1}(x)).$$

It is readily seen that $y \in \mathscr{D}Q_0(x)$ if and only if there exist sequences $x_n \to x$, and $y_n \in Q_0(x_n)$ such that $y_n \to y$. Also $y \in \mathscr{S}Q_0(x)$ if and only if there exist $x_1, \ldots, x_n$ in X, with $x_1 = x$, $x_n = y$ and $x_{j+1} \in Q_0(x_j)$, $(j = 1, \ldots, n-1)$.

If $\mathscr{S}Q_0 = Q_0$, then the map Q_0 is said to be *transitive*.

We define an *abstract prolongation* (or simply *prolongation*) to be a map Q_0 from X to 2^X with the following properties:

(α) If $x \in X$, then $xR^+ \subset Q_0(x)$.

(β) $\mathscr{D}Q_0 = Q_0$.

(γ) If $A \subset X$, and $x \in A$, then either $Q_0(x) \subset A$ or $Q_0(x)$ meets the boundary of A.

If Q_0 is a prolongation and $A \subset X$, it is clear that $A \subset Q_0(A)$. Moreover, if $x \in X$ and $t \geqq 0$, then $Q_0(x)t \subset Q_0(x)$.

If the map $Q_0: X \to 2^X$ satisfies (α) and (γ) in the definition of prolongation, but not necessarily (β), it will be called a *semi-prolongation*. If Q_0 is a semi-prolongation, then it can be shown that $\mathscr{S}Q_0$ is a semi-prolongation, and that $\mathscr{D}Q_0$ is a prolongation.

Now, let Q_0 be a semi-prolongation. We define $Q_1 = \mathscr{D}\mathscr{S}Q_0$. Then Q_1 is a prolongation. If α is any ordinal number, we define, inductively,

$$Q_\alpha = \mathscr{D}\Big(\bigcup_{\beta<\alpha} \mathscr{S}Q_\beta\Big).$$

It is clear that if $\alpha_1 < \alpha_2$, then

$$Q_{\alpha_1}(x) \subset Q_{\alpha_2}(x).$$

The maps Q_α are prolongations. If γ denotes the first uncountable ordinal, it can be proved that $Q_\alpha = Q_\gamma$, for all $\alpha \geqq \gamma$. This "maximal" prolongation will be denoted by Q.

The following lemmas are crucial for our study of stability and boundedness in dynamical systems.

Lemma 1. Let Q_0 be a semi-prolongation and let M be a compact subset of X. Then the following are equivalent:

(a) $Q(M) = M$

(b) There exists a fundamental system of compact neighborhoods $\{W_\lambda\}$ of M such that $Q(W_\lambda) = W_\lambda$.

(c) There exists a continuous real valued function defined on a neighborhood W of M satisfying

(i) $V(x) \geqq 0$,
(ii) $V(x) = 0$ if and only if $x \in M$, and
(iii) if $y \in Q(x)$, then $V(y) \leqq V(x)$.

Lemma 2. Let Q_0 be a semi-prolongation. Then the following are equivalent:

(a) If A is compact, there exists a compact set $B \supset A$ such that $Q(B) = B$. (Then, in particular, the set $Q(A)$ has this property.)

(b) There exists a continuous real valued function V defined on X such that

(i) if $y \in Q_0(x)$, then $V(y) \leqq V(x)$, and
(ii) given any λ, there is a compact set $W \subset X$ such that $V(x) > \lambda$ for $x \in X - W$.

If Q_0 is a transitive prolongation, then $Q_\alpha = Q_0$, for every ordinal number α, and we obtain the following corollary:

Corollary. A necessary and sufficient condition for invariance of a compact set M under a transitive prolongation Q_0 (i.e., $Q_0(M) = M$), is that every neighborhood of M contains a compact neighborhood which is invariant under Q_0.

In order to apply the foregoing considerations to the study of stability and boundedness in dynamical systems, we define, for $x \in X$, $D_0(x)$ to be the set $\overline{xR^+}$, that is, the positive semi-orbit closure of x. D_0 is a transitive semiprolongation; it is not, in general, a prolongation. It follows from our previous discussion that $D_1 = \mathscr{D}\mathscr{S}D_0$ is a prolongation, as is

$$D_\alpha = \mathscr{D}(\bigcup_{\beta<\alpha} E_\beta),$$

for any ordinal number $\alpha \geqq 1$ where, for convenience, we have written E_β for $\mathscr{S}D_\beta$.

The sets $D_\alpha(x)$ are closely connected with the prolongations of Ura (see refs. [5] and [6]). However, our system of numbering the prolongations differs from the one used by Ura, since he includes among the prolongations the sets $E_\beta(x)$.

§ 3. Stability and Generalized Liapunov Functions

Now, let M be a compact, positively invariant set. (That is, $Mt \subset M$ for all $t \geqq 0$.) We recall that M is said to be *Liapunov stable or stable* if, for any $\epsilon > 0$, there exists $\eta > 0$ such that $S_\eta(M)t \subset S_\epsilon(M)$ for all $t \geqq 0$. It is clear that the set $D_1(M)$ is intimately connected with the stability or instability of the compact set M. Indeed, it is not hard to show that M is Liapunov stable if and only if $D_1(M) = M$. In the general case (M not necessarily stable) $D_1(M)$ may be considered as a measure of the instability of M. In this connection, we remark that $D_1(M)$ is the intersection of all closed positively invariant neighborhoods of M.

The prolongations D_α may be used to generalize the concept of Liapunov stability. If α is any ordinal number, we say that the compact set M is *stable of order* α, or α-*stable*, if $D_\alpha(M) = M$. If M is stable of order α, for every α, then M is said to be *absolutely stable*. It is clear that if M is α-stable and $\beta < \alpha$, then M is β-stable. By the remarks above, stability of order 1 is just Liapunov stability.

Analogously, we say that the dynamical system (2) is *bounded of order* α, or α-*bounded* if, whenever A is compact, $D_\alpha(A)$ is compact. If the dynamical system is α-bounded for every α; it is called *absolutely bounded*.

The study of stability and boundedness in dynamical systems has been facilitated by the introduction of generalized Liapunov functions [1]. A *generalized Liapunov function* for the compact invariant set M is a nonnegative function V, defined in a positively invariant neighborhood W of M, and satisfying the conditions:

(a) If $\epsilon > 0$, then there exists $\lambda > 0$ such that $V(x) > \lambda$, for x not in $S_\epsilon(M)$.

(b) If $\lambda > 0$, there exists $\eta > 0$ such that $V(x) < \lambda$, for $x \in S_\eta(M)$.

(c) If $x \in W$, and $t \geqq 0$, then $V(xt) \leqq V(x)$.

By a *generalized Liapunov function at infinity we mean a nonnegative function* V defined on all of X (or on the complement of some compact set) satisfying:

(a) V is bounded on every compact set.

(b) The set $[x \mid V(x) \leqq \lambda]$ is compact, for all $\lambda \geqq 0$.

(c) If $x \in X$, and $t \geqq 0$, then $V(xt) \leqq V(x)$.

Theorem 1 [*1*]. The compact set M is stable if and only if there exists a generalized Liapunov function for M.

Theorem 2. The dynamical system (2) is 1-bounded if and only if there exists a generalized Liapunov function at infinity.

The next two theorems demonstrate the relation between absolute stability (respectively, absolute boundedness) and generalized Liapunov functions. They are immediate consequences of lemmas 1 and 2.

Theorem 3. Let M be compact. Then the following are equivalent:

(a) M is absolutely stable.

(b) M possesses a fundamental sequence of absolutely stable compact neighborhoods.

(c) There exists a continuous generalized Liapunov function for M.

Theorem 4. The following are equivalent:

(a) The dynamical system (2) is absolutely bounded.

(b) Every compact set is contained in an absolutely stable compact set.

(c) There exists a continuous Liapunov function at infinity.

§ 4. Asymptotic Stability and Ultimate Boundedness

The compact set M is said to be *asymptotically stable* if it is Liapunov stable, and if there exists a neighborhood W of M such that $d(xt, M) \to 0$ as $t \to \infty$, for all $x \in W$. This last condition is equivalent to saying $\Omega(W) \subset M$. If, in addition, $W = X$, M is called *asymptotically stable in the large*.

The dynamical system (2) is said to be *ultimately bounded* if there exists a compact set A such that $\Omega(X) \subset A$.

The relations between asymptotic stability, ultimate boundedness, absolute stability and absolute boundedness are revealed in the following two theorems.

Theorem 5. If the compact set M is asymptotically stable, it is absolutely stable. Indeed, M is asymptotically stable if and only if there exists a continuous Liapunov function V for M such that, if $x \notin M$, and $t > 0$, then $V(xt) < V(x)$.

Theorem 6. If the dynamical system (2) is ultimately bounded, it is absolutely bounded. Moreover, there exists a compact set M which is asymptotically stable in the large.

Asymptotic stability cannot be described in terms of invariance under a prolongation, as the following considerations indicate. Suppose there were an abstract prolongation Q such that $Q(M) \subset M$ if and only if M is asymptotically stable. Now, consider a compact invariant set M, such that

$$M = \bigcap_{n=1,2,\ldots} M_n,$$

where the M_n are asymptotically stable, but M is not asymptotically stable. Such exist (see example 3). Then

$$Q(M) \subset Q(M_n) = M_n,$$

so

$$Q(M) \subset \bigcap M_n = M,$$

and M would have to be asymptotically stable.

However, by using a modification of the prolongation $D_\alpha(x)$, it is possible to obtain a necessary and sufficient condition for asymptotic stability of a compact positively invariant set M. If $x \in X$, we define the set $D_1^{(\infty)}(x)$ as follows: $y \in D_1^{(\infty)}(x)$ if and only if there exist sequences $x_n \to x$, and $t_n \to +\infty$, such that $x_n t_n \to y$. Then for any ordinal number α, define

$$D_\alpha^{(\infty)} = \mathscr{D}\Big(\bigcup_{\beta<\alpha} \mathscr{S} D_\beta^{(\infty)}\Big).$$

As usual, let $D^{(\infty)}$ denote the "maximal" $D_\alpha^{(\infty)}$. In general $D_\alpha^{(\infty)}(x)$ is not a prolongation. However, it can be shown that

$$D_\alpha(x) = D_\alpha^{(\infty)}(x) \cup xR^+.$$

Theorem 7. Let M be a compact positively invariant set. Then M is asymptotically stable if and only if there is a neighborhood W of M with $D^{(\infty)}(W) \subset M$.

§ 5. Stability under Perturbations

Now we consider explicitly the system of differential equations (1). Let $\delta > 0$. By a *δ-perturbed motion* with respect to (1) we mean an absolutely continuous curve $x^*(t)$ satisfying the differential inequality

$$\| \dot{x}^*(t) - f(x^*(t)) \| < \delta$$

for all t for which $\dot{x}^*(t)$ exists. For $x_0 \in X$, let $X^*(x_0; \delta)$ denote the class of δ-perturbed motions x^* with $x^*(0) = x_0$. Let

$$P_\delta(x_0) = \bigcup_{x^* \in X^*(x_0; \delta)} x^*(R^+),$$

and let

$$P(x_0) = \bigcap_{\delta > 0} P_\delta(x_0).$$

It can be shown that P_δ *is a semi-prolongation, and* P *is a transitive prolongation.*

The compact positively invariant set M is said to be *weakly stable under perturbations*, or *totally stable* if, for any $\epsilon > 0$, there exist δ and $\eta > 0$ such that $P_\delta(S_\eta(M)) \subset S_\epsilon(M)$.

Theorem 8. Let M be a compact positively invariant set. Then the following are equivalent:

(a) M is totally stable.

(b) $P(M) = M$.

(c) M contains a fundamental system of compact totally stable neighborhoods.

The system (1) is said to be *bounded under perturbations* or *totally bounded* [*7*], if, for any compact set $A \subset X$, and sufficiently small $\delta > 0$, the set $\overline{P_\delta(A)}$ is compact.

Theorem 9. Total boundedness of system (1) is equivalent to the following: If A is a compact subset of X, then $P(A)$ is also compact.

It can be shown that $D(x)$ is a subset of $P(x)$. This yields the following theorem:

Theorem 10. (a) Total stability implies absolute stability. (b) Boundedness under perturbations implies absolute boundedness.

The converse of Theorem 10 is not true, as can be seen from the example of a center (e.g., the origin in $\ddot{x} + x = 0$). Here $D(0) = 0$, but $P(0) = R^2$.

It is known [*2*] that asymptotic stability implies total stability. For a detailed discussion of the connections between asymptotic stability and stability under perturbations, see Seibert [*4*].

§ 6. Examples.[1]

The following examples are all special cases of the equation

$$\ddot{x} + f(x^2 + \dot{x}^2)\,\dot{x} + x = 0.$$

To every zero of the function $f(r^2) = f(x^2 + \dot{x}^2)$ corresponds a limit cycle $x^2 + \dot{x}^2 = r^2$. The orbits between two neighboring limit cycles are spirals with decreasing or increasing distance from the origin, depending on the sign of f.

(1) Let $f(r^2) = -\sin^2 \pi/r^2$. Then the origin is stable of order 1, but not of order 2.

(2) Let $f(r^2) = 0$, for $r = 1/(n + 1/m)$, $n = 0, 1, 2, \ldots, m = 1, 2, \ldots,$ and let $f(r^2) < 0$ elsewhere. Then the origin is stable of order 2, but not of order 3. If n is any positive integer, it is clear that this example may be suitably modified, so that a system is obtained in which the origin is stable of order n, but not of order $n + 1$.

(3) If $f(r^2) = \sin \pi/r^2$, the origin is totally stable (and therefore absolutely stable) but not asymptotically stable.

(4) When $f(r^2) = -\sin(\pi r^2)$, the system is bounded of order 1, but not of order 2.

(5) If $f(r^2) = \sin(\pi r^2)$, the system is totally bounded (and therefore absolutely bounded) but not ultimately bounded.

REFERENCES

[*1*] Lefschetz, S., Liapunov and stability in dynamical systems. *Bol. soc. mat. mex.* [2] **3**, 25-39 (1958).

[*2*] Malkin, J. G., "Theory of Stability of Motion." Gos. Izdat. tekh.-teoret. Lit., Moscow, 1952; English transl. AECtr-3352 (1958).

[*3*] Nemickii, V. V., and Stepanov, V. V., "Qualitative theory of Differential Equations." Princeton Univ. Press, Princeton, New Jersey, 1960.

[*4*] Seibert, P., Stability, under perturbations in generalized dynamical systems. This volume, p. 00.

[*5*] Ura, T., Sur les courbes définies par les équations différentielles dans l'espace à m dimensions. *Ann. sci. école norm. sup.* [3] **70**, 287-360 (1953).

[*6*] Ura, T., Sur le courant extérieur à une région invariante. *Funk. Ekv.* **2**, 143-200 (1959).

[*7*] Yoshizawa, T., Liapunov's function and boundedness of solutions. *Funk. Ekv.* **2**, 95-142 (1959).

[*8*] Zubov, V. I., "Metody A. M. Liapunova i ikh Primenenie" (Methods of A. M. Liapunov and Their Application). Izdat. Leningrad. Univ., 1957.

[1] Most of these examples are due to Ura [*6*].

Stability under Perturbations in Generalized Dynamical Systems

PETER SEIBERT[1]
Research Institute for Advanced Study (RIAS), Baltimore, Maryland

It is an important problem in the theory of stability, to what extent small but permanently acting disturbances affect stability properties of dynamical systems. While earlier contributions, notably those of Liapunov, were restricted to the case where the perturbation term vanishes at the point of equilibrium itself, the study of the general problem was initiated relatively late by a note of Dubošin [*2*] in 1940, and by the subsequent important work of Malkin [*6*, *7*] and Goršin [*4*].[2] The most concise theorem on stability under perturbations, found independently by the two last mentioned authors, states that uniform asymptotic stability implies stability under persistent perturbations or, as it is now frequently called, "total stability." Recently, Yoshizawa [*9*, *10*] extended the investigation of the influence of perturbations to the concepts of boundedness and ultimate boundedness.

In this note we prove one theorem on ultimate boundedness under perturbations, and a second one which contains the theorem of Malkin-Goršin. We introduce a concept of "strong stability under perturbations" which is equivalent to uniform asymptotic stability and, therefore, implies total stability. We prove all results for a class of systems similar to the "general systems" of Zubov [*11*], allowing more than one solution with given initial conditions. We formulate the restrictions on the perturbations in terms of the maximal deviation within time intervals of given length of a perturbed motion from an unperturbed one with the same initial values. This method allows us to consider also cases where the derivatives of both motions differ strongly from each other while the orbits themselves remain close together, a situation which is encountered, e.g., in the "chattering" phenomenon of relay type control systems (cf. André and Seibert [*1*]). Furthermore, it gives rise to a new

[1] This research was partially supported by the United States Air Force through the Air Force Office of Scientific Research of the Air Research and Development Command, under Contract Number AF 49(638)-382. Reproduction in whole or in part is permitted for any purpose of the United States Government.

[2] For a summary of all previous contributions to the subject, see Hahn [*5*, § 28].

"absolute" concept of total stability, one which is free of the element of arbitrariness inherent in other methods of restricting the perturbations, such as bounds on the norm of the perturbation term.

For studying stability properties of a perturbed system when those of the unperturbed system are already known, the second method of Liapunov is not the most direct one. Indeed, while the stability qualities of the unperturbed system are reflected in the properties of its Liapunov functions, it may sometimes be more convenient to utilize the stability qualities directly. This is the approach we use here. While, numerically, Liapunov's second method yields more precise results, the method used here is probably the shortest way of deriving the basic qualitative theorems, and it produces them under a minimum of hypotheses.

The author wishes to thank Professor J. Auslander for many helpful comments.

1. We denote by R the real line and use the following notations for subsets of R: $R_\alpha^\beta = [\alpha, \beta]$, $R_\alpha^\infty = [\alpha, \infty)$ $(\alpha > 0)$, $R^+ = [0, \infty)$. The systems under consideration will be defined in a metric space X (distance: d). Let $A \subset X$; then we write $S_\alpha(A)$ for the set $\{x \in X: d(x, A) < \alpha\}$. From Zubov's notations we adopt the following, in general unsymmetric functional defined on $2^X \times 2^X$ (2^X being the collection of all subsets of X):

$$\rho(A, B) = \sup_{x \in A} d(x, B).$$

If φ is any function and A is a set, we write $\varphi(A)$ for $\bigcup_{y \in A} \varphi(y)$, and analogously for functions of more than one variable. A continuous, strictly increasing function defined on R^+ and vanishing at 0 will be called "of *class K*".

Throughout the paper, constants denoted by small Greek letters will be understood to be > 0.

2. By a *general flow* (or simply *flow*) we mean a map F of the product space $R \times R \times X$ into 2^X satisfying the following conditions:

I. The set $F(t, t_0, x_0)$ is non-empty for all $x_0 \in X$, $t_0 \in R^+$, $t \geqq t_0$; in particular, $F(t_0, t_0, x_0) = x_0$.

II. Given $t_0 \in R^+$, $x_0 \in X$, $t \in R_{t_0}^\infty$, and an arbitrarily small constant ϵ, there exists $\delta = \delta(t_0, x_0, t, \epsilon) > 0$ so that

$$F(t', t_0, x_0) \subset S_\epsilon(F(t, t_0, x_0))$$

holds for all $t' \in [t, t + \delta]$, and

$$F(t, t_0, x_0) \subset S_\epsilon(F(t'', t_0, x_0))$$

holds for all $t'' \in [t - \delta, t]$.

III. The identity $F(t, t_0, x_0) = F(t, t_1, F(t_1, t_0, x_0))$ holds for all $x_0 \in X$, $t_0 \in R^+$, $t \geqq t_1 \geqq t_0$.

This definition differs slightly from that of Zubov's "general systems" [*11*]; the degree of generality is almost the same.

The following property of general flows is a corollary of condition II:

IIa. Suppose that $x_0 \in A \subset X$ and $F(t', t_0, x_0) \not\subset S_\epsilon(A)$ for some constants t', t_0, and ϵ. Then there exists a $t'' \in (t_0, t')$ such that $F(t'', t_0, x_0) \subset S_\epsilon(A)$, but not $\subset A$.

A collection of sets $F(t, t_0, x_0)$, with t_0 and x_0 fixed and t varying through $R_{t_0}^\infty$, is called a *motion* of the flow F.

We will frequently use the notation

$$F(t, x_0) = \bigcup_{t_0 \in R^+} F(t + t_0, t_0, x_0). \tag{1}$$

A set $M \subset X$ is called *positively invariant* (with respect to F) if it has the property $F(R^+, M) \subset M$.

3. Along with a given flow F we consider "perturbed" flows F^* which are restricted by a condition of the following kind:

Let $D \subset X$, and let φ be a function of class K (§ 1). We say that F^* satisfies a *condition* $[D, \varphi]$ (with respect to F) if, given any $t_0 \in R^+$, $\tau \in R^+$, and $x_0 \in D$ such that

$$F(R_{t_0}^{t_0+\tau}, t_0, x_0) \cup F^*(R_{t_0}^{t_0+\tau}, t_0, x_0) \subset D,$$

the inequality

$$\rho(F^*(t, t_0, x_0), F(t, t_0, x_0)) < \varphi(\tau) \tag{2}$$

holds for all $t \in [t_0, t_0 + \tau]$.

The existence of a function $\varphi(t)$ is well known in the case where the flows F and F^* are given, respectively, by the two differential equations

$$\frac{dx}{dt} = f(x, t), \tag{3}$$

$$\frac{dx}{dt} = f(x, t) + g(x, t) \tag{4}$$

$(x, f, g \in R^n)$, with f satisfying a Lipschitz condition in D, and g satisfying an inequality $\| g(x, t) \| \leqq g(t)$ such that for all t_0

$$\int_{t_0}^{t_0+\tau} g(t)\, dt < G(\tau) < \infty.^1$$

[1] Recently, much more general theorems of this nature have been established, including such, where f may be discontinuous (see for example Filippov [*3*]).

Then:

$$\varphi(\tau) = G(\tau) \cdot e^{L\tau} \qquad (L\text{: Lipschitz constant}).$$

4. The following lemma, concerning the preservation of an ultimate boundedness type property under perturbations, is crucial for our subsequent considerations.

Lemma 1. Consider three sets A, B, C in X which, for a certain constant δ, satisfy the inclusion relations

$$S_\delta(C) \subset B \subset A. \tag{5}$$

For $S_\delta(A)$ and $S_\delta(C)$ we write A^* and C^*. Let F and F^* be two general flows satisfying the following conditions for a suitable constant τ:

(a) $F(R^+, B) \subset A$ [the left-hand side being defined by (1)].

(b) $F(R_\tau^\infty, A^*) \subset C$.

(c) $\rho(F^*(t, t_0, x_0), F(t, t_0, x_0)) < \delta$ for all $x_0 \in A^*$, $t_0 \in R^+$, $t \in [t_0, t_0 + \tau]$. We will refer to this property as "*condition* $\{A^*, \delta, \tau\}$".

Then F^* has the property $F^*(R_\tau^\infty, B) \subset C^*$.

Proof. Consider a motion $F^*(t, t_0, x_0)$ with its initial point x_0 in B. Without loss of generality, we can assume that $t_0 = 0$. Let t' be any number in $[0, \tau]$. Hypothesis (a) implies

$$F(t', 0, x_0) \subset A,$$

and therefore, in view of (5) and (c),

$$F_{t'}^* = \text{def } F^*(t', 0, x_0) \subset A^*.$$

Now (b) implies

$$F(t' + \tau, t', F_{t'}^*) \subset C,$$

which, together with (c), yields

$$F^*(t' + \tau, t', F_{t'}^*) \subset C^*.$$

Due to property III of general flows, we can write:

$$\begin{aligned} F^*(t' + \tau, 0, x_0) &= F^*(t' + \tau, t', F^*(t', 0, x_0)) \\ &= F^*(t' + \tau, t', F_{t'}^*) \subset C^*. \end{aligned}$$

By taking the union over all $t' \in [0, \tau]$, we obtain

$$F^*(R_\tau^{2\tau}, 0, x_0) \subset C^*.$$

We complete the proof by induction. Suppose that

$$F^*(R_\tau^{n\tau}, 0, x_0) \subset C^*$$

is correct. Then select a point $x_{n-1} \in F^*((n-1)\tau, 0, x_0)$. Since $C^* \subset B$, we have $x_{n-1} \in B$. By proceeding as above, only replacing the initial values $t_0 = 0$ and x_0, respectively, by $(n-1)\tau$ and x_{n-1}, we obtain

$$F^*(R_{n\tau}^{(n+1)\tau}, (n-1)\tau, x_{n-1}) \subset C^*.$$

By taking the union over all points of $F^*((n-1)\tau, 0, x_0)$ and applying the identity III, it follows that

$$F^*(R_{n\tau}^{(n+1)\tau}, 0, x_0) \subset C^*.$$

Hence

$$F^*(R_\tau^{\infty}, 0, x_0) \subset C^*,$$

q.e.d.

5. We say that the system F is *uniformly bounded*, if [with the notation (1)] the set $F(R^+, V)$ is bounded, whenever V is bounded.[1]

If there exists a bounded set U such that to every bounded set W we can associate a constant τ such that

$$F(R_\tau^{\infty}, W) \subset U,$$

we say that F is *uniformly ultimately bounded.*

We call F *ultimately bounded under perturbations* if there exists a bounded set U with the following property: Given any bounded set W, there exist constants δ and τ such that the inclusion

$$F^*(R_\tau^{\infty}, W) \subset U \tag{6}$$

holds for all perturbed systems F^* satisfying a condition $[X, \varphi]$ for some function φ with $\varphi(\tau) = \delta$.

Theorem 1. A general flow F is ultimately bounded under perturbations if it is uniformly bounded and uniformly ultimately bounded.

[1] By a bounded set we mean a set of finite diameter.

Proof. Suppose that the hypotheses of the theorem are satisfied. Let C be a set with the property of U in the definition of uniform ultimate boundedness. Denote by δ any positive constant and by B any bounded set containing $C^* = S_\delta(C)$. The set $A = F(R^+, B)$, due to uniform boundedness, is necessarily bounded; let A^* be its neighborhood $S_\delta(A)$. Finally, choose τ sufficiently large so that

$$F(R_\tau^\infty, A^*) \subset C.$$

If F^* satisfies a condition $[X, \varphi]$ with $\varphi(\tau) = \delta$, it will, a fortiori, satisfy condition (c) of lemma 1, which then yields

$$F^*(R_\tau^\infty, B) \subset C^*.$$

Since B can be chosen arbitrarily large without changing C^*, ultimate boundedness under perturbations follow.

Remark. If F and F^* are given by differential equations of the types (3) and (4), satisfying the conditions mentioned at the end of § 3, with $D = R^n$, the concept of uniform ultimate boundedness formulated above implies the following more special one:

(*) There exists a bounded set U with the property: Given any bounded set W, there exist constants χ and τ such that (6) holds whenever

$$\int_{t_0}^{t_0+\tau} \| g(x, t) \| \, dt < \chi$$

for all $t_0 \in R^+$ and $x \in X$.

Indeed, this implication follows from the fact that the condition on g, with χ sufficiently small, implies a condition $[X, \varphi]$ with $\varphi(\tau) = \delta$. As a consequence, theorem 1 remains valid for the definition (*).[1]

6. Let M be a positively invariant set of the flow F. We will denote its neighborhoods $S_\alpha(M)$ simply by S_α.

M is called *uniformly stable* if, given any ϵ, there exists an η such that $F(R^+, S_\eta) \subset S_\epsilon$. If, in addition to this, there exists a constant ρ such that, no matter how small ϵ, the inclusion $F(R_\tau^\infty, S_\rho) \subset S_\epsilon$ holds for sufficiently large $\tau = \tau(\epsilon, \rho)$, M is called *uniformly asymptotically stable*. We say that ρ is within the *radius of attraction* of M.

By *weak stability under perturbations* (or total stability) of M we mean the following: Given any ϵ and any function φ_0 of class K (§ 1), there exist constants η, γ such that $F^*(R^+, S_\eta) \subset S_\epsilon$ holds whenever F^* satisfies a condition $[S_\epsilon, \gamma \cdot \varphi_0(t)]$.

[1] A similar notion of ultimate boundedness under perturbations was introduced by Yoshizawa in his paper [*10*] and studied by means of Liapunov's second method.

Finally, we call M *strongly stable under perturbations* if

(α) it is weakly stable under perturbations and

(β) there exists a constant ρ with the following property: Given any ϵ, there exist τ and δ such that $F^*(R_\tau^\infty, S_\rho) \subset S_\epsilon$ whenever F^* satisfies a condition $[X, \varphi]$ with $\varphi(\tau) = \delta$.

Theorem 2. Uniform asymptotic stability and strong stability under perturbations are equivalent.

That strong stability under perturbations implies uniform asymptotic stability is obvious. The proof of the converse implication will be based on two lemmas.

Lemma 2. Suppose that F has the property

$$F(R_0^\tau, V) \subset U. \tag{A}$$

Then every perturbed system satisfying the condition

$$[S_\beta(U), \varphi], \qquad \varphi(\tau) \leqq \delta < \beta \tag{B}$$

also satisfies the condition $\{V, \delta, \tau\}$ (cf. lemma 1).

Proof. If we can prove $F^*(R_0^\tau, V) \subset S_\beta(U)$, the assertion follows in view of the definitions. Suppose that, on the contrary, there exist $t_0 \geqq 0$ and $t' \in [t_0, t_0 + \tau]$ such that

$$F^*(t', t_0, V) \subset S_\beta(U).$$

Then, due to IIa, there would exist a $t'' \in [t_0, t']$ such that

$$F^*(t'', t_0, V) \not\subset S_\beta(U), \qquad \text{but} \qquad \not\subset S_\delta(U).$$

Moreover, t'' can be so chosen that

$$F^*(R_{t_0}^{t''}, t_0, V) \subset S_\beta(U).$$

[Indeed, let $t_1 = \inf t'$. If $F^*(t_1, t_0, V) \subset S_\beta(U)$, take $t'' = t_1$. Otherwise, apply IIa.] Then, in view of (B), we have $\rho(F^*(t'', t_0, V), F(t'', t_0, V)) < \delta$ which, together with (A), implies $F^*(t'', t_0, V) \subset S_\delta(U)$ in contradiction with the way in which t'' was chosen.

Lemma 3. Let M be a positively invariant set which is uniformly asymptotically stable and let ρ be any positive number. Then there exists a ρ^* with the following property: Given any ϵ, there exist δ and τ such that the condition $\{S_\rho, \delta, \tau\}$ implies $F^*(R_\tau^\infty, S_{\rho^*}) \subset S_\epsilon$.

Proof. Choose $\sigma < \rho$. Due to uniform stability there exists a ρ^* such that

$$F(R^+, S_{\rho^*}) \subset S_\sigma.$$

Let S_ϵ be the given neighborhood. Without loss of generality we can assume $\epsilon < \rho^*$. Furthermore, choose $\delta < \min(\epsilon, \rho - \sigma)$ and $\eta < \epsilon - \delta$. Then there exists a τ such that

$$F(R_\tau^\infty, S_\rho) \subset S_\eta.$$

By putting $A = S_\sigma$, $B = S_{\rho^*}$, $C = S_\eta$, lemma 1 yields the conclusion of the present lemma.

We now return to the proof of theorem 2. The restraint on F^* in condition (β) obviously implies the condition $\{S_\rho, \delta, \tau\}$. Thus property (β) follows from lemma 3.

Proof of (α). Let the neighborhood $S_{\epsilon'}$ and the function $\varphi_0 \in K$ be given, and choose $\beta < \epsilon'$. Due to uniform stability, there exists an ϵ'' such that

$$F(R^+, S_{\epsilon''}) \subset S_{\epsilon'-\beta}. \tag{7}$$

If F^* satisfies a condition $[S_{\epsilon'}, \gamma \cdot \varphi_0]$ with $\gamma = \delta/\varphi_0(\tau)$, $\delta < \beta$, it also satisfies a condition $\{S_{\epsilon''}, \delta, \tau\}$ due to lemma 2. Therefore, due to lemma 3, the constants δ, τ, ρ^* can be chosen in such a way (without violating the condition $\delta < \beta$), that

$$F^*(R_\tau^\infty, S_{\rho^*}) \subset S_{\epsilon'}. \tag{8}$$

On the other hand, (7) implies

$$F^*(R_0^\tau, S_{\epsilon''}) \subset S_{\epsilon'}. \tag{9}$$

By choosing $\eta = \min(\epsilon'', \rho^*)$, the relations (8) and (9) together imply $F^*(R^+, S_\eta) \subset S_{\epsilon'}$; hence (α).

Remarks. (a) From a point of view of applications it is of considerable interest to know how the size of the region of attraction and the time constants are affected by perturbations. Consider the important case where the region of attraction contains a positively invariant set S_σ such that, for arbitrarily small ϵ, sufficiently large τ, and small δ, the set $F(R_\tau^\infty, S_{\sigma+\delta})$ is contained in S_ϵ. Then it follows immediately from the proof of Theorem 2 that for a perturbed system satisfying condition $\{S_{\sigma+\delta}, \delta, t\}$ the relation

$$F^*(R_\tau^\infty, S_\sigma) \subset S_{\epsilon+\delta}$$

holds. Thus the region of attraction and the time constants are not strongly affected if the perturbation is small. [Actually, it can be shown by using a lemma similar to lemma 1, that

$$F^*(R_{2\tau}^\infty, S_\sigma) \subset S_{\epsilon+\delta}$$

holds without the requirements of invariance of S_σ].

(b) The theorem of Malkin and Goršin, mentioned in the introduction, easily follows as a consequence of Theorem 2. In fact, the classical concept of total stability is implied by the one defined at the beginning of this section.

(c) Massera ([*8*], correction) gave an example of an autonomous system in which the origin is totally stable (in the classical sense) but not asymptotically stable. We are going to show that the origin in this example is weakly stable under perturbations, according to our general definition as we have formulated it for general flows. Since the origin cannot be strongly stable under perturbations (because it is not asymptotically stable), it will follow that the two types of stability under perturbations are not identical.

Consider the dynamical system

$$\pi(t, x_0)\ [= F(t + t_0, t_0, x_0)]$$

defined by the differential equation

$$\frac{dx}{dt} = f(x) \qquad (x \in R^+),$$

where $f = 0$ at 0 and at a sequence of points x_n tending monotonically to 0, and $f < 0$ elsewhere.[1] Let a neighborhood $S_\epsilon = [0, \epsilon)$ of the origin and a function φ_0 of class K be given. Then choose $\delta_0 < \epsilon$ and η in such a way that one of the intervals $[x_{n+1}, x_n]$ is contained in $[\eta, \epsilon - \delta_0]$, and select any pair of points x', x'' such that

$$x_{n+1} < x' < x'' < x_n.$$

[1] We restrict ourselves to R^+ as a matter of convenience. Massera's example is obtained by reflecting our system at the origin.

Obviously, every neighborhood S_α is positively invariant. Assume that[1] F^* satisfies the condition $[S_\sigma, \gamma \cdot \varphi_0]$ with $\sigma > \epsilon - \delta_0 + \delta$. If we restrict γ by

$$\gamma \cdot \varphi_0(1) = \text{def } \delta < \min(\delta_0, x'' - x'),$$

this implies the condition $\{S_{\epsilon-\delta_0}, \delta, 1\}$, due to lemma 2. Thus we have

$$F^*(R_0^1, S_{x'}) \subset S_{x''}. \tag{10}$$

Denoting $\pi(1, x'') = x'' - \mu$ $(\mu > 0)$, it is clear that

$$\pi(1, S_{x''}) \subset S_{x''-\mu}.$$

Now we further restrict γ by $\gamma \cdot \varphi_0(1) < \mu$. Then:

$$F^*(1, S_{x''}) \subset S_\mu(\pi(1, S_{x''})) \subset S_\mu(S_{x''-\mu}) = S_{x''}.$$

We assert that

$$F^*(R^+, S_{x'}) \subset S_{x''}. \tag{11}$$

Indeed, take any number $\tau \in R^+$ and split it into $\tau = n + \theta$ (n: integer, $0 \leqq \theta < 1$). Denote: $t_0 + \theta = \tau_0$. Then [in view of *III*, (10), and of $F^*(1, S_{x''}) \subset S_{x''}$] :

$$F^*(t_0 + \tau, t_0, S_{x'})$$

$$= F^*(\tau_0 + n, \tau_0, F^*(t_0 + \theta, t_0, S_{x'})) \subset F^*(\tau_0 + n, \tau_0, S_{x''}) \subset S_{x''},$$

which proves (11). Since

$$S_\eta \subset S_{x'} \subset S_{x''} \subset S_\epsilon,$$

we have

$$F^*(R^+, S_\eta) \subset S_\epsilon,$$

which proves weak stability under perturbations.

REFERENCES

[*1*] André, J., and Seibert, P., On after-endpoint motions of general discontinuous control systems and their stability properties. *Proc. 1st Intern. IFAC* [Internat. Federation of Automat. Control] *Congr. Moscow 1960*, vol. II, 919-922.

[1] Although F in the present case is a dynamical system, F^* may be a general flow.

[*2*] Dubošin, G. N., On the problem of stability of a motion with respect to constantly acting disturbances. *Trudy gos. astron. Inst. Sternberg* **14**, No. 1 (1940).

[*3*] Filippov, A. F., Differential equations with discontinuous right-hand side. *Mat. Sbornik* [N.S.] **51**, 99-128 (1960).

[*4*] Goršin, S., On the stability of motion under permanently acting disturbances. *Izvest. Akad. Nauk Kazakh. S.S.R., Ser. Mat. Mekh.* **2** (56), 46-73 (1948).

[*5*] Hahn, W., "Theorie und Anwendung der direkten Methode von Liapunov." Springer, Berlin, 1959.

[*6*] Malkin, I. G., Stability in the case of permanently acting disturbances. *Priklad. Mat. i Mekh.* **8**, 241-245 (1944).

[*7*] Malkin, I. G., "Theory of Stability of Motion." Gos. Izdat. tekh.-teoret. Lit., Moscow, 1952; English transl., AECtr-3352 (1958).

[*8*] Massera, J. L., Contribution to stability theory. *Ann. Math. (Princeton)* **64**, 182-206 (1956); Correction, *Ibid.* **68**, 202 (1958).

[*9*] Yoshizawa, T., Appendix to the paper "Note on the boundedness and the ultimate boundedness." *Mem. Coll. Sci., Univ. Kyoto,* **A30**, 91-103 (1957).

[*10*] Yoshizawa, T., Liapunov's function and boundedness of solutions. *Funk. Ekv.* **2**, 95-142 (1959).

[*11*] Zubov, V. I., "Metody A. M. Liapunova i ikh Primenenie" (Methods of A. M. Liapunov and Their Applications). Izdat. Leningrad. Univ., 1957.

A Functional-Differential System of Neutral Type Arising in a Two-Body Problem of Classical Electrodynamics

RODNEY D. DRIVER[1]
RIAS, Baltimore, Maryland, and
Mathematics Research Center, U.S. Army,
University of Wisconsin, Madison, Wisconsin

Introduction

Functional-differential equations of the form

$$y'(t) = f(t, y(t), y(g(t)), y'(g(t))) \qquad \text{for} \qquad t > t_0,$$

where f and g are given functions, $g(t) \leqslant t$, and $y'(g(t))$ stands for dy/dt evaluated at $g(t)$, are called equations of neutral type. The term apparently arose because in certain simple cases the equation could equally well be considered as having a retarded argument or an advanced argument. Such a case would occur, for example, if $g(t) = t - \tau$, where τ is a positive constant, and the equation could be solved for $y'(t - \tau)$.

The existence and uniqueness of solutions for equations or systems of equations of this type has been discussed by Èl'sgol'c [3], Kamenskiĭ [5], and others.

The present paper is concerned with a functional-differential system which arises in one approach to the two-body problem of classical electrodynamics. I refer to this as a system of neutral type because of its resemblence to the equation above. However, in this system, the retarded arguments will depend upon the dependent variables as well as the independent variable. We will consider the system

$$y_i'(t) = f_i(t, y(t), y(g_1(t, y(t))), \ldots, y(g_m(t, y(t))), y'(g_1(t, y(t))), \ldots, y'(g_m(t, y(t))))$$
$$\text{for} \quad t > t_0 \qquad (i = 1, \ldots, n),$$

[1] The research reported here was carried out at RIAS, Baltimore, Maryland, and was partially supported by the U. S. Army Research Office (Durham) under contract number DA-36-034-ORD-3220. The present paper was written under sponsorship by the U. S. Army under contract number DA-11-022-ORD-2059 at the Mathematics Research Center, United States Army, University of Wisconsin.

[2] Present adress: Sandia Corporation, Albuquerque, New Mexico.

where $y(t) = (y_1(t), ..., y_n(t))$, the functions $f_1, ..., f_n$ and $g_1, ..., g_m$ are given, each $g_j(t, y) < t$—a strict inequality—for the values of t and y under consideration, and $y'(g_j(t, y(t)))$ stands for dy/dt evaluated at $g_j(t, y(t))$. More briefly we will write

$$y'(t) = f(t, y(t), y(g(t, y(t))), y'(g(t, y(t)))),$$

where $y(g(t, y(t)))$ stands for $(y(g_1(t, y(t))), ..., y(g_m(t, y(t))))$ and similarly for $y'(g(t, y(t)))$.

A *local* existence and uniqueness theorem for such a system is simple. Let $\alpha \leqslant g_j(t, y) < t$ for all values of t and y under consideration with $t \geqslant t_0$. One then sets $y(t) = \varphi(t)$, a given continuously differentiable function, on $[\alpha, t_0]$. Then $y(t)$ can be continuously extended for a short time to the right of t_0 so as to satisfy the functional-differential system, provided that the ordinary differential system

$$y'(t) = f(t, y(t), \varphi(g(t, y(t))), \varphi'(g(t, y(t))))$$

has a solution to the right of t_0 with $y(t_0) = \varphi(t_0)$. If the solution of the ordinary differential system is unique then so is the solution of the functional-differential system. Appropriate continuity and Lipschitz conditions on the given functions, f, g, φ, and φ', will assure the local existence and uniqueness.

Due to the arbitrariness of the initial function, $\varphi(t)$, there is no reason to expect that $y'(t_0 - 0) = y'(t_0 + 0)$. Thus the extension, $y(t)$, of $\varphi(t)$ will not have a continuous derivative at t_0. Thus one can expect some difficulty extending the solution past any instant, $t_1 > t_0$, at which $g_j(t_1, y(t_1)) = t_0$ for some j.

In the case of difference-differential equations of neutral type—i.e. if each $g_j(t, y) = t - \tau_j$, where each τ_j is a positive constant—this difficulty is easily overcome. More generally, the difficulty is easily overcome whenever each $g_j(t, y) = g_j(t)$, a given function of t alone, and, on some finite interval $[t_0, T]$, each $g_j(t) =$ any constant only a finite number of times. One can then define a solution to be a continuous extension of $\varphi(t)$ which satisfies the functional-differential system on (t_0, β) for some $\beta \in (t_0, T]$ except at the sequence of points, $t_0 < t_1 < t_2 < ... < \beta$, defined inductively by the condition $g_j(t_k) = t_0, t_1, ...,$ or t_{k-1} for some $j = 1, ...,$ or m $(k = 1, 2, ...)$. At these exceptional points, $y(t)$ will not, in general, be differentiable.

For systems in which $g_j(t, y)$ really depends upon t *and* y the appropriate hypothesis—in addition to continuity requirements—is not immediately clear. The following two examples illustrate the difficulty.

Consider the scalar equation

$$y'(t) = -y'(t - y^2(t)/4) \qquad \text{for} \qquad t > 0,$$

with

$$\varphi(t) = 1 - t \qquad \text{for} \qquad t \leqslant 0.$$

As long as $y > 0$, $g(t, y) = t - y^2/4 < t$. The functions f, g, and φ are analytic for all values of their arguments, so there is no question about the continuity or Lipschitz conditions. The obvious method of defining a solution gives $y(t) = 1 + t$ for $0 \leqslant t \leqslant 1$. However, for $1 < t < 3$ both $y(t) = 1 + t$ and $y(t) = 3 - t$ are solutions. Moreover, for each of these solutions we find that every equation $g(t, y(t)) =$ a constant has only a finite number of solutions.

Another innocent-looking example is the scalar equation

$$y'(t) = -y'(y(t) - 2) \qquad \text{for} \qquad t > 0,$$

with

$$\varphi(t) = 1 - t \qquad \text{for} \qquad t \leqslant 0.$$

To assure that $g(t, y) < t$ we consider this equation on a domain in which $y < t + 2$. One readily finds $y(t) = 1 + t$ for $0 \leqslant t \leqslant 1$. However, no solution exists for $t > 1$, since $y(t) > 2$ would imply $y'(t) < 0$ and $y(t) < 2$ would imply $y'(t) > 0$.

The additional hypothesis which we will impose to obtain unique solutions for the equations considered in this paper is that each $g_j(t, \eta(t))$ be a strictly increasing function of t whenever $\eta(t)$ belongs to a certain class of functions which includes all functions that could possibly be solutions. This hypothesis, as it will be stated, is satisfied in the electrodynamics problem to be discussed.

1. Extended Existence and Uniqueness

The equations of motion for the two-body problem of classical electrodynamics have certain singularities, e.g. in case of a collision or in case of a charge traveling with the speed of light. Thus it will be important to assume restricted domains of definition for the functions f and g. The work of Èl'sgol'c and Kamenskiĭ, cited in the introduction, essentially assumes that f is defined for all values of its arguments.

Since the solution will not have any more than a piecewise continuous derivative, nothing is gained by imposing more than piecewise continuity conditions on $\varphi'(t)$.

Let n and m be positive integers. Let

$$(t, Y) = (t, y_1, ..., y_n, y_{11}, ..., y_{1n}, ..., y_{m1}, ..., y_{mn}, z_{11}, ..., z_{1n}, ..., z_{m1}, ..., z_{mn})$$

denote a typical vector in $E^{1+n+2nm}$, Euclidean space of $1 + n + 2nm$ dimensions. Let

$$(t, y) = (t, y_1, ..., y_n)$$

be the projection of (t, Y) into E^{1+n}, the space of the first $1 + n$ coordinates of $E^{1+n+2nm}$.

Let $D = D^{1+n+nm}$ be a domain (a connected open set) in E^{1+n+nm}, the space of the first $1 + n + nm$ coordinates of $E^{1+n+2nm}$, and let

$$f(t, Y) = (f_1(t, Y), ..., f_n(t, Y))$$

be a real, n-vector-valued function over $D \times E^{nm}$. Thus $f(t, Y)$ is defined over a domain in $E^{1+n+2nm}$ whose boundary puts no restrictions on the values of the z_{ji} $(i = 1, ..., n,\ j = 1, ..., m)$.

Let D^{1+n} be the domain obtained by projecting D into E^{1+n}, and let $g(t, y) = (g_1(t, y), ..., g_m(t, y))$ be a real, m-vector-valued function over D^{1+n}.

In terms of the above notation we can now state

Problem 1 (A Functional-Differential System of Neutral Type). Let $\varphi(t) = (\varphi_1(t), ..., \varphi_n(t))$ be a continuous, n-vector-valued function of t on an interval $[\alpha, t_0]$—interpreted as $(\alpha, t_0]$ in case $\alpha = -\infty$. Let $\varphi(t)$ be differentiable on (α, t_0) except at a finite number of points, $\alpha = t_{-p} < t_{-p+1} < ... < t_{-1} < t_0$. We seek a continuous extension of $\varphi(t)$ to $y(t)$ on $[\alpha, \beta)$, where $\beta > t_0$, such that

(a) $(t, y(t)) \in D^{1+n}$ for $t_0 \leqslant t < \beta$,

(b) $\alpha \leqslant g_j(t, y(t)) < t$ for $t_0 \leqslant t < \beta, j = 1, ..., m$,

(c) $(t, y(t), y(g(t, y(t)))) \in D$ for $t_0 \leqslant t < \beta$, and

(d) $y'(t) = f(t, y(t), y(g(t, y(t))), y'(g(t, y(t))))$ for $t_0 < t < \beta$ except at points $t_0 < t_1 < t_2 < ... < \beta$, each t_k of which satisfies the condition $g_{j_k}(t_k, y(t_k)) = t_{-p}, t_{-p+1}, ..., t_{-1}, t_0, t_1, ...,$ or t_{k-1} for some $j_k = 1, 2, ...,$ or m.

Definition. A solution, $y(t)$, of problem 1 will be called *unique* if every other solution coincides with $y(t)$ in their common region of definition.

Notation. Throughout this section, it is convenient to take the norm of a vector $z = (z_1, ..., z_p) \in E^p$ to be

$$\| z \| = \max_{i=1,...,p} | z_i |.$$

For a function $h(t, z)$ defined in some region, R, the statement

$$h(t, z) \in \text{Lip in } R$$

will mean that $h(t, z)$ is Lipschitz continuous with respect to variations in all its arguments throughout R, i.e. there exists a (Lipschitz) constant, A, such that $\| h(t, z) - h(\hat{t}, \hat{z}) \| \leqslant A \max (| t - \hat{t} |, \| z - \hat{z} \|)$ whenever $(t, z), (\hat{t}, \hat{z}) \in R$.

Theorem 1. (Extended Existence and Uniqueness Theorem). Let $f(t, Y)$ and $g(t, y) \in$ Lip in each compact subset of $D \times E^{nm}$ and D^{1+n} respectively. Let $g_j(t, y) < t$ for all $(t, y) \in D^{1+n}$ $(j = 1, ..., m)$. Let each $g_j(t, \eta(t))$ be a strictly increasing function of t whenever, for some continuous functions $\bar{y}(t)$ from $(-\infty, \infty)$ into E^n and $z(t, y)$ from D^{1+n} into E^{nm}, $(t, \eta(t), \bar{y}(g(t, \eta(t)))) \in D$ and $\eta'(t) = f(t, \eta(t), \bar{y}(g(t, \eta(t))), z(t, \eta(t)))$. Let $\varphi(t)$ satisfy the conditions imposed on it in problem 1, and let conditions (a), (b), and (c) of that problem be satisfied at the instant t_0. Let $\varphi(t)$ be Lipschitz on $[\alpha, t_0]$ and let $\varphi'(t)$ be Lipschitz on each interval (t_k, t_{k+1}) $(k = -p, ..., -1)$.

Then problem 1 has a unique solution, $y(t)$, on $[\alpha, \beta)$, where $\beta > t_0$ and for any compact set $F \subset D$ there is a sequence of numbers, $t_0 < \xi_1 < \xi_2 < ... \rightarrow \beta$ such that

$$(\xi_i, y(\xi_i), y(g(\xi_i, y(\xi_i)))) \in D - F \text{ for } i = 1, 2, ...$$

Remark. The hypotheses of this theorem could be weakened in various ways. For example, the region of definition of each $\varphi_i(t)$ could be reduced (cf. ref. [2] where the analogous problem for a delay-differential system is discussed). In particular, if some component of $y(t)$, say $y_i(t)$, and its derivative do not appear with retarded arguments on the right hand side of the functional-differential system, then, for that component, the initial value, $\varphi_i(t_0)$, is sufficient initial data. Also, in this case, f_i need not be Lipschitz continuous with respect to t.

Outline of Proof of Theorem 1. (1) Prove that there exists a unique solution on $[\alpha, t_1)$ where either t_1 satisfies the conditions on β asserted by the theorem, or else

$$\lim_{t \to t_1} g_j(t, y(t)) = t_{-p}, t_{-p+1}, ..., \text{ or } t_0 \qquad \text{for some} \qquad j = 1, 2, ..., \quad \text{or} \quad m.$$

In this proof one must allow for the possibility that, for some $j = 1, \ldots,$ or m, it may be the case that

$$g_j(t_0, \varphi(t_0)) = t_{-p}, t_{-p+1}, \ldots, \text{ or } t_{-1}.$$

(2) Prove that this basic existence and uniqueness step can be iterated to any finite number of points, $t_0 < t_1 < t_2 < \ldots < t_q$ as defined in (d) of problem 1, as long as none of these points satisfies the conditions on β asserted in the theorem.

(3) Suppose (for contradiction) that $y(t)$ is a solution on $[\alpha, \beta)$, where $t_0 < \beta < \infty$, β cannot be increased, and yet for some compact set $F \subset D$

$$(t, y(t), y(g(t, y(t)))) \in F \qquad \text{for all} \qquad t \in [t_0, \beta).$$

Then show that this implies that there are only a finite number of points in the sequence defined in (d) of problem 1 and that the solution can be extended beyond β.

(4) Prove that the solution is unique by a method quite analogous to the extended uniqueness argument for ordinary differential equations.

2. Application to a Two-Body Problem of Classical Electrodynamics

Modern physics references make extensive use of a 4-dimensional space-time notation in which time is the intrinsic (or proper) time of the particle under consideration. Here, however, we will use position vectors (indicated by bold-face type) in E^3, a Euclidean reference frame in some given inertial system, and the time, t, of an observer in that system.

Let $\mathbf{r}_i(t)$ be the position of the i'th point charge ($i = 1, 2$) at time t, and let $\mathbf{v}_i(t) = \mathbf{r}_i'(t)$ be its velocity. Let q_i be the (constant) magnitude and m_i the (constant positive) rest mass of the i'th charge. Let c be the speed of light and ϵ_0, the dielectric constant of free space (two universal positive constants). Then, using the M.K.S. system of units, the Lorentz-Abraham equation (or Dirac equation) for the motion of the i'th charge can be written in the form

$$\mathbf{v}_i(t) - \frac{q_i^2}{6\pi\epsilon_0 c^3 m_i}\left\{\frac{\mathbf{v}_i''(t)}{[1 - v_i^2(t)/c^2]^{1/2}} + \frac{3[\mathbf{v}_i(t)\cdot\mathbf{v}_i'(t)]\,\mathbf{v}_i'(t)}{[1 - v_i^2(t)/c^2]^{3/2}\,c^2}\right\}$$
$$= m_i^{-1}[1 - v_i^2(t)/c^2]^{1/2}\,[\mathbf{F}_i - (\mathbf{F}_i\cdot\mathbf{v}_i(t))\,\mathbf{v}_i(t)/c^2], \tag{2.1}$$

where $\mathbf{F}_i = q_i[\mathbf{E}_i + \mathbf{v}_i(t) \times \mathbf{B}_i]$, $\mathbf{E}_i$ being the electric field and $\mathbf{B}_i$ the magnetic field at $(t, \mathbf{r}_i(t))$ produced by all sources other than the i'th charge itself. In the two-body problem, these fields depend upon t, $\mathbf{r}_i(t)$, and the trajectory of the other charge, call it charge j. The above equation does not look the same as the equations found in the literature (cf. Schott [*6*], p. 246). This is because of the different notation, and because in (2.1) the relativistic expression for $\mathbf{F}_i$ has been essentially solved for the highest order derivative $\mathbf{v}_i''(t)$.

The second term on the left hand side of (2.1), { }, is the radiation reaction, representing the action of the i'th charge on itself. Its exact form is not universally accepted by physicists and it is sometimes omitted. Since the analysis of the problem turns out to be easier and the qualitative results appear to be more in agreement with our physical intuition when this term is not present, it will be omitted for the remainder of this paper. Then (2.1) expresses $\mathbf{v}_i'(t)$ in terms of $\mathbf{F}_i$, with all relativistic mass corrections incorporated in the right hand side.

Since electromagnetic effects propagate with speed c, the fields at $\mathbf{r}_i(t)$ at time t depend on the state of charge j at an earlier instant, $t - \tau_{ji}(t)$. The propagation time, $\tau_{ji}(t)$, must satisfy the functional equation

$$\tau_{ji}(t) = | \mathbf{r}_i(t) - \mathbf{r}_j(t - \tau_{ji}(t)) | /c \qquad \text{for} \qquad (j, i) = (2,1), (1,2), \tag{2.2}$$

where | | implies the Euclidean norm or length of a vector. This equation is not given explicitly in most physics references, but it, or an equivalent, is implied.

If (2.2) has a solution, and assuming that the electric and magnetic fields of a moving charge are those calculated from the Liénard-Wiechert potentials (cf. Schott [*6*], pp. 22, 23), $\mathbf{F}_i$ can be expressed as a function of $q_i q_j$, $\mathbf{r}_i(t) - \mathbf{r}_j(t - \tau_{ji}(t))$, $\mathbf{v}_i(t)$, $\mathbf{v}_j(t - \tau_{ji}(t))$, and $\mathbf{v}_j'(t - \tau_{ji}(t))$. The acceleration of charge j at the retarded time, $t - \tau_{ji}(t)$, enters in an important way because it is one of the principle quantities determining the radiation field of a moving charge. If (2.2) has no solution then charge j does not influence charge i, thus $\mathbf{F}_i = 0$.

Without specifying $\mathbf{F}_i$ explicitly, (2.1) can now be written in the form

$$\mathbf{v}_i'(t) = \begin{cases} q_i q_j m_i^{-1} \mathbf{f}(\mathbf{r}_i(t) - \mathbf{r}_j(t - \tau_{ji}(t)), \mathbf{v}_i(t), \mathbf{v}_j(t - \tau_{ji}(t)), \mathbf{v}_j'(t - \tau_{ji}(t))) \\ \qquad \text{whenever (2.2) has a solution,} \\ 0 \qquad \text{whenever (2.2) has no solution,} \end{cases} \tag{2.3}$$

where $(j, i) = (2, 1)$ or $(1, 2)$ and $\mathbf{f}$ is analytic as long as

$$\mathbf{r}_i(t) - \mathbf{r}_j(t - \tau_{ji}(t)) \neq 0, \qquad | \mathbf{v}_i(t) | < c \qquad \text{and} \qquad | \mathbf{v}_j(t - \tau_{ji}(t)) | < c.$$

This system now resembles the system of neutral type considered in section 1 except for the fact that the retarded arguments depend upon the dependent variables according to an implicit functional condition, (2.2), rather than explicitly.

The problem we will pose for the system based upon (2.2) and (2.3) is

Problem 2 (Two-Body Problem Without Radiation Reaction). Let initial trajectories $\mathbf{r}_i(t)$ $(i = 1, 2)$, be given which are continuously differentiable for $\alpha \leqslant t \leqslant t_0$, where $\alpha < t_0$. Let each $\mathbf{r}_i(t)$ also be twice differentiable with a bounded second derivative on (α, t_0) except at a finite number of points, $\alpha = t_{-p} < t_{-p+1} < \dots < t_{-1} < t_0$. We seek an extension, $\mathbf{r}_i(t)$ $(i = 1, 2)$, of the initial trajectories to $\alpha \leqslant t < \beta$, where $\beta > t_0$, such that each $\mathbf{r}_i(t)$ and $\mathbf{v}_i(t) = \mathbf{r}_i'(t)$ is continuous for $t_0 \leqslant t < \beta$, and

(a) $\mathbf{r}_1(t) \neq \mathbf{r}_2(t)$ for $t_0 \leqslant t < \beta$,

(b) $|\mathbf{v}_i(t)| < c$ for $t_0 \leqslant t < \beta$, $i = 1, 2$, and

(c) $\mathbf{r}_i(t)$ and $\mathbf{v}_i(t)$ $(i = 1, 2)$ satisfy the equations of motion, (2.3) for $t_0 < t < \beta$ except at points $t_0 < t_1 < t_2 < \dots < \beta$, each t_k of which satisfies the condition $t_k - \tau_{ji}(t_k) = t_{-p}, \dots, t_{-1}, t_0, t_1, \dots,$ or t_{k-1} for $(j, i) = (2, 1)$ or $(1, 2)$, where $\tau_{ji}(t_k)$ is a solution of (2.2) at $t = t_k$.

We will first state (without proof) conditions under which the functional equations (2.2) have a solution and can be transformed into delay-differential equations. This will lead to a differential system of the type considered in section 1.

Theorem 2.1. (Existence, Uniqueness, and Properties of Solutions of (2.2)). (a) Let $\mathbf{r}_j(t)$ be continuous on $[\alpha, \beta]$, where $\alpha < \beta < \infty$, and continuously differentiable with $|\mathbf{v}_j(t)| < c$ on (α, β). Then, for any $t \in [\alpha, \beta]$, a solution, $\tau_{ji}(t)$, of (2.2) exists if and only if, for some finite $\bar{\alpha} \geqslant \alpha$,

$$|\mathbf{r}_i(t) - \mathbf{r}_j(\bar{\alpha})| \leqslant c(t - \bar{\alpha}).$$

The solution of (2.2), if it exists, is unique and satisfies the inequalities

$$|\mathbf{r}_i(t) - \mathbf{r}_j(t)|/2c \leqslant \tau_{ji}(t) \leqslant t - \bar{\alpha}.$$

(b) Now, in addition to the continuity and differentiability of $\mathbf{r}_j(t)$, let $\mathbf{r}_i(t)$ be continuous on $[t_0, \beta]$ and continuously differentiable with $|\mathbf{v}_i(t)| < c$ on (t_0, β), for some $t_0 \in (\alpha, \beta)$. Let (2.2) have a solution, $\tau_{ji}(t_0)$, at t_0 and let $\mathbf{r}_i(t) \neq \mathbf{r}_j(t)$ on $[t_0, \beta]$. Then for all $t \in [t_0, \beta]$ there is a unique solution, $\tau_{ji}(t)$, of (2.2), $\tau_{ji}(t)$ is continuous on $[t_0, \beta]$, it is continuously differentiable on (t_0, β) and satisfies the delay-differential equation

$$\tau'_{ji}(t) = \frac{[\mathbf{r}_i(t) - \mathbf{r}_j(t - \tau_{ji}(t))] \cdot [\mathbf{v}_i(t) - \mathbf{v}_j(t - \tau_{ji}(t))]}{| \mathbf{r}_i(t) - \mathbf{r}_j(t - \tau_{ji}(t)) | \, c - [\mathbf{r}_i(t) - \mathbf{r}_j(t - \tau_{ji}(t))] \cdot \mathbf{v}_j(t - \tau_{ji}(t))} \quad (2.4)$$

and $\tau_{ji}(\beta) = 0$ if and only if $\mathbf{r}_i(\beta) = \mathbf{r}_j(\beta)$—a collision.

Theorem 2.2. (Extended Existence and Uniqueness Theorem for the Two-Body Problem Without Radiation Reaction). Let each $\mathbf{r}_i(t)$ $(i = 1, 2)$ satisfy the conditions imposed on it in problem 2 for $\alpha \leqslant t \leqslant t_0$ and let each $\mathbf{v}'_i(t)$ be Lipschitz continuous on each of the open intervals (t_k, t_{k+1}) $(k = -p, \ldots, -1)$. Also let

(1) $\mathbf{r}_1(t_0) \neq \mathbf{r}_2(t_0)$,

(2) $| \mathbf{v}_i(t) | < c$ for $\alpha \leqslant t \leqslant t_0$, $i = 1, 2$, and

(3) equations (2.2) have a solution $\tau_{ji}(t_0)$, at t_0 for $(j, i) = (2, 1)$ and $(1, 2)$.

Then problem 2 has a unique solution for $\alpha \leqslant t < \beta$, where either $\beta = + \infty$ or else

$$\lim_{t \to \beta - 0} \mathbf{r}_1(t) = \lim_{t \to \beta - 0} \mathbf{r}_2(t)$$

—a collision.

Method of Proof. One defines an appropriate problem 1, based on the system composed of (2.3) and (2.4), and shows that this problem 1 has a unique solution. One then shows an equivalence between the problem 1 and problem 2 under consideration, which implies the existence of a unique solution for problem 2. To show that the solution does indeed exist until a collision occurs, one must prove, among other things, that no delays, $\tau_{ji}(t)$, approach zero and that no speeds approach the speed of light in a finite time unless there is a collision. This latter calculation depends upon the specific form of (2.3) and hence requires a more detailed statement of the equations than is presented in this paper.

3. Further Remarks and Unsolved Problems

The N-body problem without radiation reaction can be treated quite similarly, in fact one can even allow the presence of a suitably restricted external field not due to the charges under consideration. This paper was restricted to the two-body problem partly for simplicity and partly to emphasize the fact that the existence theorems for even this most basic problem had not been given before.

When the radiation reaction term of (2.1) is included, a higher order derivative, $\mathbf{v}_i''(t)$, is introduced and the equation no longer appears to be of neutral type. However, due to the apparent necessity, for physical reasons, of allowing the accelerations to be discontinuous at t_0, the system still exhibits behavior something like that of a system of neutral type. It is possible to modify section 1 so that it applies to such system. However, the existence theorem for the two-body problem in this case is not yet in such a satisfactory state as Theorem 2.2.

Another unsettled question of existence is the following. If one omitted condition (3) of Theorem 2.2 would a unique solution exist? This would amount to assuming that the past trajectories had been such that one or both charges had not yet come under the influence of the other. See the example discussed by Havas [*4*].

For the simpler special case of the two-body problem without radiation reaction in which the initial trajectories of both charges are restricted to the x-axis the functional-differential system is no longer of neutral type, since the acceleration terms disappear from the right hand side of (2.3). In this case one can obtain rather specific qualitative information about the future trajectores [*1*]. One would expect that similar information could be obtained in the case of three-dimensional motion as considered here.

A priori, there is at least one other formulation of the two-body problem using the equations of motion discussed in this paper. Instead of specifying rather arbitrary initial trajectories as in problem 2, one might hope to find trajectories which satisfy the equations of motion for all $t < \beta$. Then a unique solution might be determined by specifying appropriate "initial" data at one point, t_0, or at a different point for each charge. Synge [*7*] discussed this approach heuristically, but apparently no existence theorem has been proved. If this could be treated successfully, one might also be able to deal with equations having both retarded and advanced arguments.

Returning to section 1, it is dissatisfying not to have some theorem on the dependence of the solution on the initial data and on the right hand sides of the equations. Continuous dependence can be asserted for delay-differential equations [*2*] and, with an appropriate norm, for equations of neutral type when the retarded arguments depend only on t.

REFERENCES

[*1*] Driver, R. D., Delay-Differential Equations and an Application to a Two-Body Problem of Classical Electrodynamics. Technical Report, Univ. of Minnesota, Minneapolis, 1960; Abstr. in *AMS Notices* **7**, 522 (1960).

[2] Driver, R. D., Existence theory for a delay-differential system. *Contribs. Differential Equations* **1**, to appear.

[3] Èl'sgol'c, L. È., "Qualitative Methods in Mathematical Analysis" (in Russian), Chapter 5. Gos. Izdat. Tekh.-Teoret. Lit., Moscow, 1955.

[4] Havas, P., Bemerkungen zum Zweikörperproblem der Elektrodynamik. *Acta Phys. Austr.* **3**, 342-351 (1949).

[5] Kamenskiĭ, G. A., Towards a general theory of equations with a deviating argument. *Doklady Akad. Nauk S.S.S.R.* **120**, 697-700 (1958) (in Russian). There are a number of mistakes in this paper.

[6] Schott, G. A., "Electromagnetic Radiation." Cambridge Univ. Press, London and New York, 1912.

[7] Synge, J. L., On the electromagnetic two-body problem. *Proc. roy. Soc. (London)* **A177**, 118-139 (1940).

Asymptotic Solutions of a Linear Ordinary Differential Equation of nth Order about a Simple Turning Point

YASUTAKA SIBUYA[1]
RIAS, Baltimore, Maryland, and
Ochanomizu University, Tokyo, Japan

Let us consider an equation:

$$\epsilon^{n-m}L_n(y) + L_m(y) = 0, \tag{1}$$

where

$$L_n(y) = y^{(n)} + \sum_{h=m+1}^{n-1} R_h y^{(h)}, \tag{2}$$

and

$$L_m(y) = (z + \epsilon R_m)\, y^{(m)} + \sum_{h=0}^{m-1} (P_h + \epsilon R_h)\, y^{(h)}. \tag{3}$$

Here ϵ is a complex parameter, z is a complex variable, $y^{(k)}$ denotes the kth derivative of y with respect to z, R_h are holomorphic functions of z and ϵ for

$$|z| \leqq \delta_0, \qquad 0 < |\epsilon| \leqq \rho_0, \qquad |\arg \epsilon| \leqq \alpha_0, \tag{4}$$

where δ_0, ρ_0, α_0 are positive constants, and P_h are holomorphic functions of z for

$$|z| \leqq \delta_0. \tag{5}$$

P_h are supposed to be independent of ϵ, and R_h are supposed to admit for (5) uniformly asymptotic expansions

$$R_h(z, \epsilon) \simeq \sum_{v=0}^{\infty} R_{hv}(z)\, \epsilon^v, \tag{6}$$

as ϵ tends to zero through the region

$$0 < |\epsilon| \leqq \rho_0, \qquad |\arg \epsilon| \leqq \alpha_0, \tag{7}$$

the coefficients R_{hv} being assumed to be holomorphic in z for (5).

[1] The author was a Fulbright grantee. This work was done during his stay in the Institute of Mathematical Sciences as a Temporary Member.

We shall assume that

$$n - 2 \geqq m \geqq 2. \tag{8}$$

Because of (8) and (3), $z = 0$ is a turning point. Among the equations which satisfy the conditions given above, the following equation is the simplest one:

$$\epsilon^{n-m} u^{(n)} + \tilde{L}_m(u) = 0, \tag{9}$$

where

$$\tilde{L}_m(u) = z u^{(m)} + \sum_{h=0}^{m-1} B_k u^{(k)}. \tag{10}$$

Here B_k are holomorphic functions of ϵ for (7), and they are independent of z. B_k are also assumed to admit asymptotic expansions

$$B_k(\epsilon) \simeq \sum_{v=0}^{\infty} B_{kv} \epsilon^v, \tag{11}$$

where B_{kv} are constant. The equation (9) can be integrated by the use of Laplace transform. Therefore, if we can have (1) reduced to an equation of the form (9), then we can construct asymptotic solutions of (1). Concerning this problem, we obtained the following two results:

(I) Let us assume that the equation (1) is formally reduced to (9) by a formal transformation

$$u = \sum_{h=0}^{m-1} A_h y^{(h)} + \sum_{h=m}^{n-1} \epsilon^{h+1-m} A_h y^{(h)}, \tag{12}$$

where A_h are given by formal power series in ϵ with coefficients holomorphic in z:

$$A_h(z, \epsilon) \approx \sum_{v=0}^{\infty} A_{hv}(z)\, \epsilon^v. \tag{13}$$

When u is given by (12), derivatives of u are also given in the following form:

$$\begin{aligned} u^{(k)} &= \sum_{h=0}^{m-1} Q_{kh} y^{(h)} + \sum_{h=m}^{n-1} \epsilon^{h+1-m} Q_{kh} y^{(h)}, \qquad (k = 1, 2, \ldots, m-1), \\ \epsilon^{k+1-m} u^{(k)} &= \sum_{h=0}^{m-1} Q_{kh} y^{(h)} + \sum_{h=m}^{n-1} \epsilon^{h+1-m} Q_{kh} y^{(h)}, \qquad (k = m, m+1, \ldots, n-1). \end{aligned} \tag{14}$$

(14) are derived from (12) by the differentiation with respect to z and the use of the equation (1). We shall assume that Q_{kh} are also formal power series in ϵ with coefficients holomorphic in z:

$$Q_{kh}(z, \epsilon) \approx \sum_{v=0}^{\infty} Q_{khv}(z)\, \epsilon^v.$$

Put

$$Q_{0h} = A_h \qquad (h = 0, 1, ..., n-1), \tag{15}$$

$$Q = \begin{pmatrix} Q_{00} & Q_{01}, & ..., & Q_{0n-1} \\ Q_{10}, & Q_{11}, & ..., & Q_{1n-1} \\ ..., & ..., & ..., & ... \\ Q_{n-10}, & Q_{n-11}, & ..., & Q_{n-1n-1} \end{pmatrix}, \tag{16}$$

and

$$Q(z, \epsilon) \approx \sum_{v=0}^{\infty} Q_v(z)\, \epsilon^v. \tag{17}$$

Then we shall assume that the matrix $Q_0(0)$ is nonsingular.

Let us put

$$\lambda_k = \begin{cases} \exp\,(i(2k+1)\,\pi/p) & (k = 0, 1, ..., p-1), \\ 0 & (k = p), \end{cases} \tag{18}$$

where

$$p = n - m. \tag{19}$$

Then put

$$\omega_{jk} = \arg\,(\lambda_j - \lambda_k) \qquad (j \neq k),$$
$$\omega = \arg z. \tag{20}$$

We shall denote by $\mathscr{D}_{jk}(\gamma, q)$ the region defined by

$$(2q - \tfrac{3}{2})\,\pi + \gamma \leqq \omega_{jk} + \frac{p+1}{p}\,\omega \leqq (2q + \tfrac{3}{2})\,\pi - \gamma \qquad (j \neq k),$$
$$|\, z \,| \leqq \delta_0, \tag{21}$$

where q is an integer and γ is a positive number.

For each ordered pair (j, k), there exists at least one integer q_{jk} such that the line-segment in the z-plane, which is given by

$$0 < z < \delta_0, \tag{22}$$

is contained in the interior of $\mathscr{D}_{jk}(\gamma, q_{jk})$. Put

$$\mathscr{D}(\gamma) = \bigcap_{(j,k)} \mathscr{D}_{jk}(\gamma, q_{jk}). \tag{23}$$

Under the conditions given above, we can show the existence of an analytic transformation in $\mathscr{D}(\gamma)$ by which (1) is reduced to (9) and which admits the formal transformation (12) as its asymptotic form.

(II) There exist an infinite number of polynomials $\mathscr{P}_v$ of a variable σ with integral coefficients such that, if

$$\sigma = P_{m-1}(0) \tag{24}$$

is not a root of any of the equations

$$\mathscr{P}_v(\sigma) = 0 \qquad (v = 1, 2, ...), \tag{25}$$

then we can construct a formal transformation (12) which satisfies the conditions given in (I).

The case when $n = 4$ and $m = 2$ was treated by Lin and Rabenstein [*1*]. Our method is quite similar to theirs.

REFERENCE

[*1*] Lin, C. C., and Rabenstein, A. L., On the asymptotic solutions of a class of ordinary differential equations of the fourth order. *Trans. AMS* **94**, 24-57 (1960).

A Boundary Value Problem for a Singularly Perturbed System of Nonlinear Differential Equations

W. A. HARRIS, JR.[1]
Mathematics Department
University of Minnesota, Minneapolis, Minnesota

This paper is concerned with the existence and asymptotic character of solutions of the nonlinear boundary value problem

$$\begin{aligned} \frac{dx}{dt} &= f(t, x, y, \epsilon) \\ \epsilon \frac{dy}{dt} &= g(t, x, y, \epsilon) \end{aligned} \tag{1}$$

$$x(\alpha, \epsilon) = l_1, \qquad x(\beta, \epsilon) = l_2 \tag{2}$$

as $\epsilon \to 0+$ when the *degenerate* differential system

$$\begin{aligned} \frac{dx}{dt} &= f(t, x, y, 0) \\ 0 &= g(t, x, y, 0) \end{aligned} \tag{3}$$

has a solution $x = \varphi(t)$, $y = \psi(t)$ which satisfies one of the boundary conditions (2)

The main objective of this paper is to construct a solution of the differential system (1) satisfying the boundary conditions (2) so that

$$\lim_{\substack{\epsilon \to 0+ \\ 0<t\leq 1}} \{| x(t, \epsilon) - \varphi(t) | + | y(t, \epsilon) - \psi(t) |\} = 0.$$

The boundary value problem cannot be expected to have a solution of this type unless f and g are severely restricted. As pointed out by Coddington and Levinson [*1*], even $f = y, g = - y - y^3$ is not restricted enough.

[1] Sponsored by the Mathematics Research Center, U.S. Army, Madison, Wisconsin, under Contract No.: DA-11-022-ORD-2059.

We make the following assumptions.

H 1: The functions f and g are linear in y;

$$f(t, x, y, \epsilon) = f_1(t, x, \epsilon) + f_2(t, x, \epsilon)y,$$
$$g(t, x, y, \epsilon) = g_1(t, x, \epsilon) + g_2(t, x, \epsilon)y.$$

H 2: The functions f_1, f_2, g_1, g_2, are regular analytic with respect to x and ϵ and of class $C^{(\infty)}$ with respect to t in a region R of x, t, ϵ—space that contains in its interior all points, $x = \varphi(t)$, $\alpha \leq t \leq \beta$, $\epsilon = 0$.

H 3: The degenerate differential system (3) possesses a solution, $x = \varphi(t)$, $y = \psi(t)$ of class $C^{(\infty)}$ for which

$$\varphi(\beta) = l_2 \quad \text{and} \quad g_2(t, \varphi(t), 0) < 0, \quad \alpha \leq t \leq \beta. \tag{4}$$

Without loss of generality, we may assume that

$$\begin{gathered} \alpha = 0, \quad \beta = 1, \quad \varphi(t) \equiv 0, \quad \psi(t) \equiv 0, \\ g_2(t, 0, 0) = -1, \quad f_1(t, 0, 0) = g_1(t, 0, 0) = \frac{\partial}{\partial x} g_1(t, 0, 0) = 0, \end{gathered} \tag{5}$$

since the transformation of variables,

$$t^* = \int_\alpha^t g_2(\tau, \varphi(\tau), 0)\, d\tau \Big/ \int_\alpha^\beta g_2(\tau, \varphi(\tau), 0)\, d\tau \qquad \epsilon^* = -\epsilon \Big/ \int_\alpha^\beta g_2(\tau, \varphi(\tau), 0)\, d\tau$$

$$x^* = x - \varphi(t) \qquad y^* = y - \psi(t) + G(t)\, x^*$$

$$G(t) = \frac{\{(\partial/\partial x)\, g_1(t, \varphi(t), 0) + (\partial/\partial x)\, g_2(t, \varphi(t), 0)\, \psi(t)\}}{g_2(t, \varphi(t), 0)} \tag{6}$$

will give an equivalent problem for which (5) holds.

With these assumptions, the differential system (1) may be written in the form

$$\begin{aligned} \frac{dx}{dt} &= a_{11}(t, \epsilon)\, x + a_{12}(t, \epsilon)\, y + \epsilon a_1(t, \epsilon) + f_{11}(t, x, \epsilon)\, x^2 + f_{12}(t, x, \epsilon)\, xy \\ \epsilon \frac{dy}{dt} &= \epsilon a_{21}(t, \epsilon)\, x + \{-1 + \epsilon a_{22}(t, \epsilon)\}\, y \\ &\quad + \epsilon a_2(t, \epsilon) + f_{21}(t, x, \epsilon)\, x^2 + f_{22}(t, x, \epsilon)\, xy, \end{aligned} \tag{7}$$

where all functions are analytic with respect to x and ϵ and of class $C^{(\infty)}$ with respect to t in R.

Our construction of the solution of the boundary value problem is in two parts. In the first part we construct a particular solution $x = p(t, \epsilon)$, $y = q(t, \epsilon)$ which satisfies the boundary condition $p(1, \epsilon) = 0$, and has for $0 \leq t \leq 1$ a uniformly asymptotic expansion.

$$p(t, \epsilon) \simeq \sum_{k=1}^{\infty} p_k(t)\, \epsilon^k \qquad q(t, \epsilon) \simeq \sum_{k=1}^{\infty} q_k(t)\, \epsilon^k \tag{8}$$

as $\epsilon \to 0+$, $0 < \epsilon \leq \epsilon_0$.

We can determine a formal solution

$$\hat{p}(t, \epsilon) = \sum_{k=1}^{\infty} p_k(t)\, \epsilon^k, \qquad \hat{q}(t, \epsilon) = \sum_{k=1}^{\infty} q_k(t)\, \epsilon^k, \qquad \hat{p}(1, \epsilon) = 0,$$

of (7) by solving the sequence of equations

$$\left\{\begin{array}{c} \dfrac{d}{dt} p_1(t) = a_{11}(t, 0)\, p_1(t) + a_{12}(t, 0)\, q_1 + a_1(t, 0) \\ q_1(t) = a_2(t, 0) \qquad p_1(1) = 0 \end{array}\right\}$$

$$\left\{\begin{array}{l} \dfrac{d}{dt} p_k(t) = a_{11}(t, 0)\, p_k(t) + a_{12}(t, 0)\, q_k + b_k(t) \\ q_k(t) = -\dfrac{d}{dt} q_{k-1}(t) + c_k(t) \qquad p_k(1) = 0 \end{array}\right\} \quad k \geq 2$$

successively, where

$$\begin{aligned} \sum_{k=2}^{\infty} b_k(t)\, \epsilon^k = {} & \epsilon a_1(t, \epsilon) + f_{11}(t, \hat{p}(t, \epsilon), \epsilon)\, \hat{p}(t, \epsilon)^2 + f_{12}(t, \hat{p}(t, \epsilon)\, \epsilon)\, \hat{p}(t, \epsilon)\, \hat{q}(t, \epsilon) \\ & + a_{11}(t, \epsilon)\, \hat{p}(t, \epsilon) + a_{12}(t, \epsilon)\, \hat{q}(t, \epsilon) \\ & - \epsilon a_1(t, 0) - a_{11}(t, 0)\, \hat{p}(t, \epsilon) - a_{12}(t, 0)\, \hat{q}(t, \epsilon), \end{aligned}$$

$$\begin{aligned} \sum_{k=2}^{\infty} c_k(t)\, \epsilon^k = {} & a_2(t, \epsilon) + f_{21}(t, \hat{p}(t, \epsilon), \epsilon)\, \hat{p}(t, \epsilon)^2 + f_{22}(t, \hat{p}(t, \epsilon), \epsilon)\, \hat{p}(t, \epsilon)\, \hat{q}(t, \epsilon) \\ & + \epsilon a_{21}(t, \epsilon)\, \hat{p}(t, \epsilon) + \epsilon a_{22}(t, \epsilon)\, \hat{q}(t, \epsilon) - \epsilon a_2(t, 0). \end{aligned}$$

Thus we obtain the following result.

Theorem 1. Let hypotheses $H\,1$, $H\,2$ and $H\,3$ hold. Then, there exists a solution of the differential system (7)

$$x = p(t, \epsilon), \qquad y = q(t, \epsilon)$$

such that $p(1, \epsilon) = 0$ and such that $p(t, \epsilon)$, $q(t, \epsilon)$ admit for $0 \leq t \leq 1$ uniformly asymptotic expansions

$$p(t, \epsilon) \cong \sum_{k=1}^{\infty} p_k(t)\, \epsilon^k, \qquad q(t, \epsilon) \cong \sum_{k=1}^{\infty} q_k(t)\, \epsilon^k$$

as $\epsilon \to 0+$, $0 \leq \epsilon \leq \epsilon_1 \leq \epsilon_0$.

These asymptotic expansions may be termwise differentiated with respect to t.

By a transformation of the form

$$x = p(t, \epsilon) + z \qquad y = q(t, \epsilon) + w \tag{9}$$

the differential system (7) is reduced to the system

$$\begin{aligned} \frac{dz}{dt} &= \hat{a}_{11}(t, \epsilon)\, z + \hat{a}_{12}(t, \epsilon)\, w + \hat{f}_1(t, z, w, \epsilon) \\ \epsilon \frac{dw}{dt} &= \epsilon \hat{a}_{21}(t, \epsilon)\, z + \{-1 + \epsilon \hat{a}_{22}(t, \epsilon)\}\, w + \hat{f}_2(t, z, w, \epsilon) \end{aligned} \tag{10}$$

where

$$\hat{f}_i(t, z, w, \epsilon) = \hat{f}_{i1}(t, z, \epsilon)\, z^2 + \hat{f}_{i2}(t, z, \epsilon)\, zw \qquad (i = 1, 2),$$

and

$$\hat{a}_{11}(t, 0) = a_{11}(t, 0), \qquad \hat{a}_{12}(t, 0) = a_{12}(t, 0).$$

A solution of (7) satisfying the boundary conditions corresponds to a solution of (10) with

$$z(0, \epsilon) = l_1 - \varphi(\alpha) - p(0, \epsilon), \qquad z(1, \epsilon) = 0. \tag{11}$$

For convenience, set

$$\mu = l_1 - \varphi(\alpha) - p(0, \epsilon).$$

We can determine a formal solution

$$z(t, \epsilon) = \sum_{k=1}^{\infty} z_k(t, \epsilon)\, \mu^k \qquad w(t, \epsilon) = \sum_{k=1}^{\infty} w_k(t, \epsilon)\, \mu^k \tag{12}$$

of (10) by solving the sequence of equations

$$\begin{aligned} \frac{dz_k}{dt} &= \hat{a}_{11}(t, \epsilon)\, z_k + \hat{a}_{12}(t, \epsilon)\, w_k + h_{1k}(t, \epsilon) \\ \epsilon \frac{dw_k}{dt} &= \epsilon \hat{a}_{21}(t, \epsilon)\, z_k + \{-1 + \epsilon \hat{a}_{22}(t, \epsilon)\}\, w_k + h_{2k}(t, \epsilon) \end{aligned}$$

successively, where

$$h_{11}(t, \epsilon) \equiv h_{21}(t, \epsilon) \equiv 0,$$

and

$$\sum_{k=2}^{\infty} h_{ik}(t, \epsilon) \mu^k = \hat{f}_i \left(t, \sum_{\nu=1}^{\infty} z_k(t, \epsilon) \mu^\nu, \sum_{\nu=1}^{\infty} w_k(t, \epsilon) \mu^\nu, \epsilon\right).$$

The formal solution so determined will satisfy the boundary conditions (11) if we require that

$$z_1(0, \epsilon) = 1, \qquad z_1(1, \epsilon) = 0$$

$$z_k(0, \epsilon) = 0, \qquad z_k(1, \epsilon) = 0, \qquad k > 1.$$

We use the method of majorant series to show the convergence of these series under the additional hypothesis H 4.

H 4: The function $f_2(t, x, \epsilon)$ as given in H 1 has the property

$$f_2(a, \varphi(a), 0) \neq 0,$$

or equivalently,

$$\hat{a}_{12}(0, 0) = a_{12}(0, 0) \neq 0.$$

Thus we have established the following theorem which we state in terms of the original differential system (1) with minor changes of notation.

Theorem 2. Let hypotheses H 1, H 2, H 3 and H 4 hold. Then, there exist constants $\epsilon_2 > 0$, $\mu_1 > 0$ such that the differential system (1) possesses, for $0 < \epsilon \leq \epsilon_2$, $| l_1 - \varphi(\alpha) | \leq \mu_1$, in $\alpha \leq t \leq \beta$ a solution $x(t, \epsilon)$, $y(t, \epsilon)$ satisfying the boundary conditions

$$x(\alpha, \epsilon) = l_1, \qquad x(\beta, \epsilon) = l_2.$$

This solution is representable by series of the form

$$x(t, \epsilon) = \sum_{r=0}^{\infty} x_r(t, \epsilon) \mu^r \qquad y(t, \epsilon) = \sum_{r=0}^{\infty} y_r(t, \epsilon) \mu^r, \tag{13}$$

where

$$\sum_{r=0}^{\infty} x_r(t, \epsilon) \mu^r, \qquad \sum_{r=0}^{\infty} \epsilon y_r(t, \epsilon) \mu^r$$

converge uniformly and absolutely with respect to ϵ, μ, and t for $0 \leq \epsilon \leq \epsilon_2$, $| l_1 - \varphi(\alpha) | \leq \mu_1$, $\alpha \leq t \leq \beta$.

The terms of this series have the following properties:

(i) $x_0(t, 0) = \varphi(t)$, $y_0(t, 0) = \psi(t)$, i.e. the solution of the degenerate differential system with $\varphi(\beta) = l_2$.

(ii) $x_0(t, \epsilon)$, $y_0(t, \epsilon)$ have asymptotic expansions in powers of ϵ. These expansions are a formal solution of the differential system and satisfy the boundary condition $x_0(\beta, \epsilon) = l_2$.

(iii) $\mu = l_1 - x_0(\alpha, \epsilon)$.

(iv) The functions $x_r(t, \epsilon)$, $y_r(t, \epsilon)$, $r \geq 1$, are of the form

$$x_r(t, \epsilon) = \omega_{1r}(t, \epsilon) \exp \left\{ \frac{1}{\epsilon} \int_\alpha^t g_2(\tau, \varphi(\tau), 0)\, d\tau \right\}$$

$$y_r(t, \epsilon) = \frac{1}{\epsilon}\, \omega_{2r}(t, \epsilon) \exp \left\{ \frac{1}{\epsilon} \int_\alpha^t g_2(\tau, \varphi(\tau), 0)\, d\tau \right\}$$

with $\omega_{ir}(t, \epsilon)$ bounded, $0 \leq \epsilon \leq \epsilon_2$, $\alpha \leq t \leq \beta$.

The series (13) may be termwise differentiated with respect to t, $\alpha \leq t \leq \beta$, $0 < \epsilon \leq \epsilon_2$. The series

$$\sum_{r=0}^{\infty} \epsilon \frac{d}{dt} x_r(t, \epsilon)\, \mu^r, \qquad \sum_{r=0}^{\infty} \epsilon^2 \frac{dy_r(t, \epsilon)}{dt}\, \mu^r$$

converge uniformly for $0 \leq \epsilon \leq \epsilon_2$, $| l_1 - \varphi(\alpha) | \leq \mu_1$, $\alpha \leq t \leq \beta$.

If $\epsilon < 0$, or if the degenerate differential system (3) has a solution $x = \varphi(t)$, $y = \psi(t)$ such that

$$\varphi(\alpha) = l_1, \qquad g_2(t, \varphi(t), 0) > 0, \qquad \alpha \leq t \leq \beta,$$

then similar results will hold with the roles of the left and right boundaries, $t = \alpha$ and $t = \beta$, interchanged.

Boundary problems for second order differential equations which can be written as special cases of the form (1) have been treated by Coddington and Levinson [*1*], von Mises [*6*] and Wasow [*7*]. Our treatment of the present problem is in the spirit of Wasow [*7*] as our aim is a constructional procedure for the solution of the boundary value problem by convergent and asymptotic series.

There are points of resemblance between our results and those of Wasow [*7, 8*], Sibuya [*4*], and Volk [*5*].

REFERENCES

[*1*] Coddington, E. A., and Levinson, N., A boundary value problem for a nonlinear differential equation with a small parameter. *Proc. AMS* **3**, 73-81 (1952).

[*2*] Harris, W. A., Jr., Singular perturbations of two-point boundary problems for systems of ordinary differential equations. *Arch. rational Mech. Anal.* **5**, 212-225 (1960).

[*3*] Harris, W. A., Jr., Singular perturbations of a boundary value problem for a nonlinear system of differential equations. *Duke math. J.* to be published.

[*4*] Sibuya, Y., On nonlinear ordinary differential equations containing a parameter. *J. Math. and Mech.* **9**, 369-398 (1960).

[*5*] Volk, I. M., On the periodic solutions of non-autonomous systems depending on a small parameter. *Priklad. Mat. i Mekh.* **10**, 559-574 (1946).

[*6*] von Mises, R., Die Grenzschichte in der Theorie der gewöhnlichen Differentialgleichungen. *Acta Univ. Szeged., Sect. Sci. Math.* **12**, 29-34 (1950).

[*7*] Wasow, W., Singular perturbations of boundary value problems for nonlinear differential equations of the second order. *Communs. pure appl. Math.* **9**, 93-113 (1956).

[*8*] Wasow, W., Solutions of nonlinear differential equations with a parameter by asymptotic series. *Ann. Math. (Princeton)* **69**, 486-509 (1959).

Author Index

Numbers in parentheses are reference numbers and indicate that an author's work is referred to, although his name is not cited in the text. Numbers in italic show the page on which the complete reference is listed.

Subject Index